CONTEMPORARY POLYMER CHEMISTRY

HARRY R. ALLCOCK

FREDERICK W. LAMPE

Department of Chemistry
The Pennsylvania State University

PRENTICE-HALL, INC., Englewood Cliffs, New Jersey 07632

Library of Congress Cataloging in Publication Data

ALLCOCK, H R
  Contemporary polymer chemistry.

  Bibliography: p.
  Includes index.
  1. Polymers and polymerization.   I. Lampe, Frederick
Walter, joint author.   II. Title.
QD381.A44   547.8'4   80-11706
ISBN 0-13-170258-0

© 1981 by PRENTICE-HALL, INC.,
Englewood Cliffs, New Jersey 07632

Printed in the United States of America

10   9   8   7   6   5   4   3   2   1

Editorial/production supervision by Linda Mihatov and Kathleen Lafferty
Interior design by Linda Mihatov
Cover design by Edsal Enterprises, Inc.
Manufacturing buyers: Edmund W. Leone
    and John Hall

PRENTICE-HALL INTERNATIONAL, INC., *London*
PRENTICE-HALL OF AUSTRALIA PTY. LIMITED, *Sydney*
PRENTICE-HALL OF CANADA, LTD., *Toronto*
PRENTICE-HALL OF INDIA PRIVATE LIMITED, *New Delhi*
PRENTICE-HALL OF JAPAN, INC., *Tokyo*
PRENTICE-HALL OF SOUTHEAST ASIA PTE. LTD., *Singapore*
WHITEHALL BOOKS LIMITED, *Wellington, New Zealand*

Contents

part III

PHYSICAL CHARACTERIZATION OF POLYMERS

part IV

FABRICATION, TESTING, AND USES OF POLYMERS

APPENDICES

Preface

This book is designed as an introduction to polymer chemistry for students of chemistry, physics, chemical engineering, engineering, materials science, biophysics, and biochemistry. It assumes a basic knowledge compatible with courses taught in university undergraduate programs in the above disciplines. Specifically, the book aims to broaden the perspective of specialists in different technical areas to the point where they can appreciate the scope and importance of polymer chemistry and polymer technology. Thus, in writing this book we have kept in mind the student who has a sound knowledge of basic science but who knows relatively little about polymer chemistry or polymer science. For this reason, many topics that are well known to practicing polymer scientists are handled here from first principles.

More rigorous and more comprehensive treatments exist for nearly all the topics discussed in this book. However, few attempts have been made to bring together synthetic, structural, kinetic, and use-oriented material in one volume. Our aim has been to provide a broad, coherent introduction to modern polymer chemistry and to direct the reader to more detailed sources for advanced study.

We have prepared this volume with the knowledge that large numbers of chemists and other scientists graduate from universities each year woefully unprepared for careers that will, in most cases, involve work with polymers. It is our hope that this book will stimulate an interest in polymer chemistry and will provide the groundwork for more comprehensive reading. Throughout the text, we have attempted to relate fundamental matters to their practical ramifications, for we believe that an intelligent understanding of this field is not possible without a knowledge of the uses to which polymers can be put.

The book is divided into four parts. *Part I* (Chapters 1–9) provides an introduction to the different classes of polymers and to the way they are synthesized and modified. Individual chapters deal with condensation-, free radical-, and ionic- or coordination polymerization, with photolytic-, radiation-, and electrolytic polymerization, polymerization of cyclic compounds, inorganic polymers, biological macromolecules, and with the ways that synthetic polymers can be modified chemically. The emphasis in these chapters is on descriptive chemistry, general principles, and experimental problems. The material in this section should be understandable to students who have taken elementary undergraduate courses in chemistry. These chapters form the groundwork for the sections that follow.

Part II (Chapters 10–13) deals with thermodynamics, equilibria, and polymerization kinetics. Chapter 10 provides an elementary overview of the underlying principles that determine whether a monomer or a cyclic compound will polymerize or a polymer will depolymerize. Chapters 11, 12, and 13 deal respectively with the kinetics of condensation-, free radical-, and ionic polymerization. A unique feature of these chapters is the full derivation of the kinetic expressions, with every attempt made to explain the underlying principles for each step. This should enable the treatments to be fully understood by anyone with a basic understanding of third year undergraduate physical chemistry.

Part III (Chapters 14–19) covers the physical methods that are employed for the characterization of polymers. Individual chapters cover "absolute" molecular weight measurements by osmometry, light scattering, and ultracentrifugation, secondary molecular weight methods such as solution viscosity and gel permeation chromatography, thermodynamics of high polymer solutions, morphology, glass transitions and crystallinity, conformational analysis, and X-ray diffraction techniques. In all these chapters, the reader is introduced to the underlying theory and, where appropriate, to the practical approaches used. Each chapter provides the basic groundwork for elementary experimental work in these areas or for further, more detailed study.

Finally, in *Part IV* (Chapters 20–22) we discuss the practical, applied aspects of polymer science, including the fabrication of polymers and testing techniques, general uses of polymers, and their biomedical applications. These chapters will be understandable to all who have an interest in science and technology. These topics will provide a useful background for anyone who intends to seek employment in industry or who wishes to interact with industrial polymer chemists.

Appendix I is a brief review of polymer nomenclature, and Appendix II is a compilation of physical property data and uses for a number of important polymers. This latter appendix provides perspective and serves as a reference source as the reader encounters new polymers at different points in the book.

We recommend that the book be read in the sequence outlined above, although specialized topics, such as those discussed in Chapters 10–13, might be absorbed best during a second reading. Because the book was written as a textbook intended for use in introductory polymer courses, as well as for those studying alone, a number of study questions and suggested sources of reading have been included at the end of each chapter. Appendix III consists of a comprehensive list of references for further reading on topics that are not included in this book.

We wish to express our appreciation to the many graduate and undergraduate students who have helped us to refine this material during the introductory polymer chemistry courses from which this book was developed.

Much of the material in Chapters 7 and 10 was written by one of us for articles that appeared in *Scientific American,* the *Journal of Macromolecular Science Reviews in Macromolecular Chemistry,* and *Chemistry in Britain.* We are indebted to the publishers of these journals for their permission to reproduce sections of those articles and several of the diagrams.

We are particularly grateful to Cosette Rodefeld for her assistance with the preparation of the manuscript.

HARRY R. ALLCOCK

FREDERICK W. LAMPE

Harry R. Allcock is a Professor of Chemistry at The Pennsylvania State University. Professor Allcock received his B.Sc. and Ph.D. degrees from The University of London. He has held postdoctoral positions at Purdue University and the National Research Council of Canada and spent five years as a research scientist in American industry before joining The Pennsylvania State University in 1966. Trained initially as a mechanistic organometallic chemist, his research interests have included the synthesis of new organic and inorganic polymers, the use of inorganic and organometallic compounds as polymerization initiators, radiation-induced polymerization, organosilicon compounds, and the structural examination of polymers by X-ray diffraction and conformational techniques. His current research activities are in the fields of inorganic high polymers and the synthesis of new macromolecules for biological studies.

Frederick W. Lampe is a Professor of Chemistry at The Pennsylvania State University. Professor Lampe received a B.S. degree from Michigan State University and A.M. and Ph.D. degrees from Columbia University. He spent seven years as a research scientist with the Humble Oil and Refining Company before taking up his present position in 1960. Professor Lampe is a physical chemist whose polymer interests are in the areas of radiation-induced polymerizations, kinetics of polymerication processes, applications of mass spectrometry to polymer degradation processes, statistical mechanics, and molecular weight methods. His current research activities include mass spectrometry, gaseous ion reactions, photochemistry, and the effects of ionizing radiation on materials.

part I

SYNTHESIS
AND REACTIONS
OF POLYMERS

1

The Scope
of Polymer Chemistry

INTRODUCTION

During the past 30 years, polymer chemistry has had a marked and very direct practical impact on the way of life of people in nearly every region of the earth. Before the beginning of World War II, relatively few materials were available for the manufacture of the articles needed for a civilized life. Steel, glass, wood, stone, brick, and concrete accounted for most of the construction and manufacturing needs of the population, while cotton, wool, jute, and a few other agricultural products provided the raw materials for clothing or fabric manufacture.

The rapid increase in the range of manufactured products following World War II resulted directly from the development of a broad range of new fibers, plastics, elastomers, adhesives, and resins. These new materials are polymers, and their impact on our present way of life is almost incalculable. Products made from polymers are all around us—clothing made from synthetic fibers, polystyrene cups, Fiberglas boats, nylon bearings, plastic bags, polymer-based paints, epoxy glue, polyurethane foam cushions, silicone heart valves, Teflon-coated cookware—the list is almost endless.

It is not surprising, therefore, that more than 50% of all chemists and chemical engineers, large numbers of physicists and mechanical engineers, and nearly all materials scientists and textile technologists are involved with research or development work with polymers. Add to this the fact that biochemistry, biophysics, and molecular biology are fields in which polymer chemistry plays a paramount role, and it is clear why the study of macromolecules is one of the most important and rapidly growing branches of science.

Polymer chemistry is not a specialized side branch of traditional chemistry. Instead, it is a uniquely broad discipline that *encompasses* the whole of chemistry and several other fields as well. Areas of science have always prospered when research workers trained in one specialized area turn their attention to a related area. This has been and still is especially true in polymer research. The challenge in polymer chemistry is the application of fundamental chemical and physical techniques and ideas to large and complex molecules. This is a demanding task, and it requires the very best approaches that traditional chemistry can provide.

It will be clear that polymer chemistry, perhaps more than any other research area, cuts across the traditional lines of organic, inorganic, physical, and analytical chemistry; physics; engineering; biology; and even medicine. A newcomer to polymer science needs to be able to blend together knowledge from all these fields. It is to assist in that process that this book has been written.

DEFINITIONS

Many of the terms and definitions used in polymer chemistry are not encountered in conventional chemical textbooks, and for this reason the following summary of terminology is given. Some of these definitions will seem fairly obvious, but others will need explanation.

Monomers

A monomer is any substance that can be converted into a polymer. For example, ethylene is a monomer that can be polymerized to polyethylene (reaction 1). An

$$CH_2{=}CH_2 \longrightarrow -CH_2{-}CH_2{-}CH_2{-}CH_2{-}CH_2{-} \tag{1}$$

amino acid is a monomer which, by loss of water, can polymerize to give poly-peptides (reaction 2). The term *monomer* is used very loosely—sometimes it applies to dimers or trimers if they, themselves, can undergo further polymerization.

$$n\,H_2N{-}\underset{\underset{H}{\mid}}{\overset{\overset{R}{\mid}}{C}}{-}\overset{\overset{O}{\parallel}}{C}{-}OH \xrightarrow{-H_2O} \left[{+}\underset{\underset{H}{\mid}}{\overset{\overset{H}{\mid}}{N}}{-}\underset{\underset{H}{\mid}}{\overset{\overset{R}{\mid}}{C}}{-}\overset{\overset{O}{\parallel}}{C}{+}\right]_n \tag{2}$$

Dimers, Trimers, and Oligomers

The polymerization of a monomer often occurs in a sequential manner. In other words, two monomer molecules first react together to form a *dimer*. The dimer may then react with a third monomer to yield a *trimer*, and so on. Dimers are usually linear molecules, but trimers, tetramers, pentamers, and so on, can be linear or cyclic.

The reactions outlined in schemes (3) to (5) illustrate the relationship between monomers, dimers, and trimers for three systems.

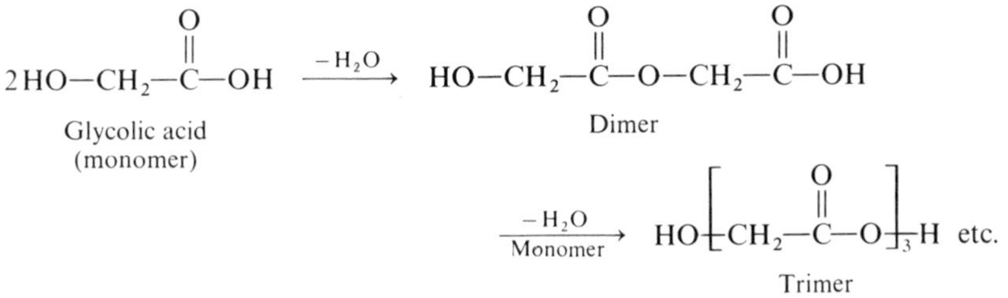

$$2\,HO-CH_2-\overset{\overset{\textstyle O}{\|}}{C}-OH \xrightarrow{\;-H_2O\;} HO-CH_2-\overset{\overset{\textstyle O}{\|}}{C}-O-CH_2-\overset{\overset{\textstyle O}{\|}}{C}-OH \qquad (3)$$

Glycolic acid
(monomer)

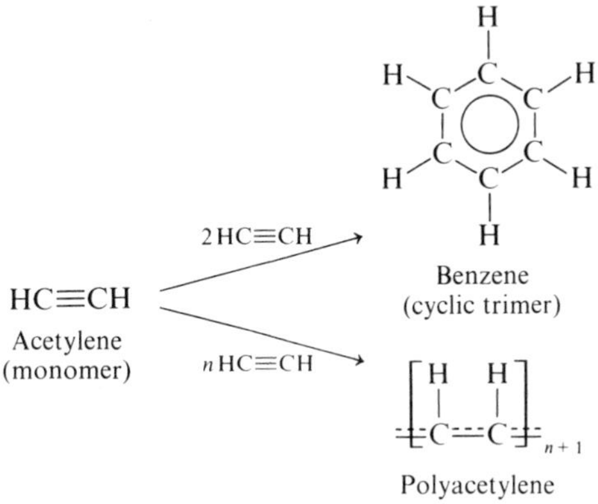

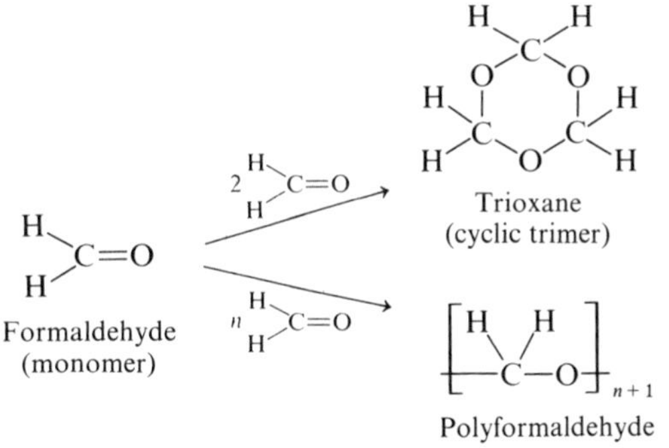

Dimer

$$\xrightarrow[\text{Monomer}]{-H_2O} \quad HO\left[CH_2-\overset{\overset{\textstyle O}{\|}}{C}-O\right]_3 H \quad \text{etc.}$$

Trimer

$$HC\equiv CH \qquad \xrightarrow{2\,HC\equiv CH} \qquad \text{Benzene (cyclic trimer)} \qquad (4)$$

Acetylene
(monomer)

$$\xrightarrow{n\,HC\equiv CH} \quad \left[\overset{\overset{\textstyle H}{|}}{C}\!=\!=\!\overset{\overset{\textstyle H}{|}}{C}\right]_{n+1}$$

Polyacetylene

$$\overset{H}{\underset{H}{>}}C=O \qquad \xrightarrow{2\,\overset{H}{\underset{H}{>}}C=O} \qquad \text{Trioxane (cyclic trimer)} \qquad (5)$$

Formaldehyde
(monomer)

$$\xrightarrow{n\,\overset{H}{\underset{H}{>}}C=O} \quad \left[\overset{H}{\underset{H}{>}}C-O\right]_{n+1}$$

Polyformaldehyde

Low-molecular-weight polymerization products, for example, dimers, trimers, tetramers, pentamers, etc.—cyclic or linear—are known as *oligomers*. Some care should be taken to avoid the use of the term "polymer" to describe materials that are really oligomers, because these two types of products have very different properties.

Polymers

The term *polymer* is used to describe *high-molecular-weight substances*. However, this is a very broad definition and in practice it is convenient to divide polymers into subcategories according to their molecular weight and structure. Although there is no general agreement on this point, in this book we will consider *low polymers* to have molecular weights below about 10,000 to 20,000 and *high polymers* to have molecular weights between 20,000 and several million. Obviously, this is a rather arbitrary dividing line, and a better definition might be based on the number of repeating units in the structure. For example, since polymer properties become almost independent of molecular weight when more than 1000 to 2000 repeating units are present, this point could also constitute a satisfactory dividing line between low and high polymers.

Linear Polymers

A *linear polymer* consists of a long chain of skeletal atoms to which are attached the substituent groups. Polyethylene **(1)** is one of the simplest examples. Linear

$$
\begin{array}{ccccccc}
H & H & H & H & H & H & \quad\quad H \\
| & | & | & | & | & | & \quad\quad | \\
-C-&C-&C-&C-&C-&C-\cdots\cdots\cdots-&C- \\
| & | & | & | & | & | & \quad\quad | \\
H & H & H & H & H & H & \quad\quad H
\end{array}
$$

1

polymers are usually *soluble* in some solvent, and in the solid state at normal temperatures they exist as elastomers, flexible materials, or glasslike thermo-plastics. In addition to polyethylene, typical linear-type polymers include poly(vinyl chloride) or PVC **(2)**, poly(methyl methacrylate) (also known as PMMA, Lucite, Plexiglas, or Perspex) **(3)**, polyacrylonitrile[1] (Orlon or Creslan) **(4)**, and Nylon 66 **(5)**.

$$
\left[\begin{array}{cc}H & Cl \\ | & | \\ C - C \\ | & | \\ H & H \end{array}\right]_n
\qquad
\left[\begin{array}{cc} & O{-}CH_3 \\ & | \\ H & C{=}O \\ | & | \\ C - C \\ | & | \\ H & CH_3 \end{array}\right]_n
\qquad
\left[\begin{array}{cc}H & C{\equiv}N \\ | & | \\ C - C \\ | & | \\ H & H \end{array}\right]_n
$$

$$\quad\quad\quad\quad\text{2}\quad\quad\quad\quad\quad\quad\quad\quad\text{3}\quad\quad\quad\quad\quad\quad\quad\quad\text{4}$$

$$
\left[\begin{array}{c}H \quad\quad\quad\quad H \;\; O \quad\quad\quad O \\ | \quad\quad\quad\quad | \;\; || \quad\quad\quad || \\ N{-}(CH_2)_6{-}N{-}C{-}(CH_2)_4{-}C \end{array}\right]_n
$$

5

Branched Polymers

A *branched polymer* can be visualized as a linear polymer with branches of the same basic structure as the main chain. A branched polymer structure is illustrated in Figure 1.1. Branched polymers are often soluble in the same solvents as the cor-

---

[1] Note that parentheses are not normally used in the name of a polymer if the monomer has a one-word name. The parentheses simply avoid ambiguity.

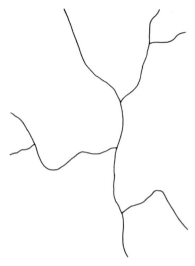

Fig. 1.1   Branched polymer.

responding linear polymer. In fact, they resemble linear polymers in many of their properties. However, they can sometimes be distinguished from linear polymers by their lower tendency to crystallize or by their different solution viscosity or light-scattering behavior. Heavily branched polymers may swell in certain liquids without dissolving completely.

Crosslinked Polymers

A *crosslinked* or *network polymer* is one in which chemical linkages exist between the chains, as illustrated in Figure 1.2. Such materials are usually swelled by

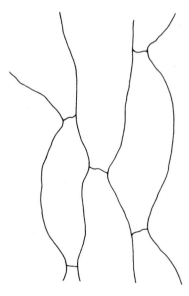

Fig. 1.2   Crosslinked macro-molecule.

"solvents," but they *do not dissolve*. In fact, this insolubility can be used as a cautious criterion of a crosslinked structure. Actually, the amount by which the polymer is swelled by a liquid depends on the density of crosslinking: the more crosslinks present, the smaller is the amount of swelling. If the degree of crosslinking is high enough, the polymer may be a rigid, high-melting, unswellable solid, such as diamond. Light crosslinking of chains favors the formation of rubbery elastomeric properties.

Cyclolinear Polymers

Cyclolinear polymers are a special type of linear polymer formed by the linking together of ring systems. Benzene rings are often incorporated into polymers of this type (**6**), but heterocyclic and inorganic rings can also be utilized in the same way.

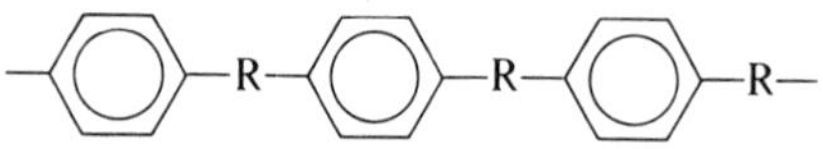

6

The properties of cyclolinear polymers resemble those of conventional linear polymers, except that the solubility of the cyclolinear species is often low. The tendency for crystallization may be very high.

Ladder Polymers

As the name suggests, a *ladder polymer* consists of linear molecules in which two skeletal strands are linked together in a *regular* sequence by crosslinking units, as illustrated diagramatically in **7**. In practice, aromatic rings may constitute the linking units (**8**) or, in ladder-type silicone polymers, silicon–oxygen units serve

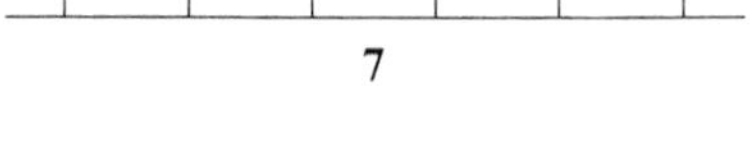

7

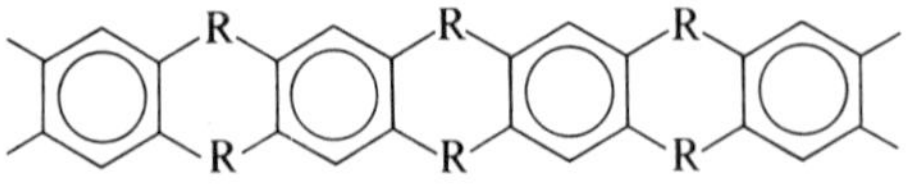

8

the same function (see Chapter 7). As might be expected, ladder polymers have a more rigid molecular structure than do conventional linear polymers, and they are often much less soluble. However, they frequently display very good thermal stability, because molecular-weight decreases must be preceded by the cleavage of

Fig. 1.3   Schematic representation of a spiropolymer structure.

two bonds at each cleavage site. Spiropolymers of the type shown in Figure 1.3 are sometimes included in the ladder polymer classification.

Cyclomatrix Polymers

Many polymer systems are known in which ring systems are linked together to form a three-dimensional matrix of connecting units. These materials are known as *cyclomatrix polymers*. Organic and inorganic rings can be incorporated into such systems, and typical examples are provided by the structures found in some silicate minerals and silicone resins (**9**). Since a three-dimensional network of bonds is

9

formed in these systems, the polymers are highly insoluble, rigid, very high melting, and usually stable at elevated temperatures. Structures of this type are often found in thermosetting resins (see page 13) and in heat-resistant wire coatings.

   Graphite is a special example of a cyclomatrix polymer. It has a structure made up of sheets of fused aromatic rings. Individual sheets or layers are sandwiched between the neighboring layers and are held in place by weak van der Waals forces.

Copolymers

A *copolymer* is a polymer made from two or more different monomers. For example, if styrene and acrylonitrile are allowed to polymerize in the same reaction vessel, a copolymer will be formed which contains both styrene and acrylonitrile residues (reaction 6). Many commercial synthetic polymers are copolymers. It should be

$$ \quad (6) $$

Copolymer

noted that the sequence of monomer units along a copolymer chain can vary according to the method and mechanism of synthesis. Three different types of sequencing arrangements are commonly found.

1. *Random copolymers.* In random copolymers, no definite sequence of monomer units exists. A copolymer of monomers A and B might be depicted by the arrangement shown in **10**. Random copolymers are often formed when olefin-type mono-

$$-A-B-B-B-A-A-B-A-A-A-A-B-A-B-B-B-$$

10

mers copolymerize by free-radical-type processes (see Chapter 3). The properties of random copolymers are usually quite different from those of the related homopolymers.

2. *Regular copolymers.* As the name implies, regular copolymers contain a regular alternating sequence of two monomer units (**11**). Olefin polymerizations

$$-A-B-A-B-A-B-A-B-$$

11

that take place through ionic-type mechanisms (see Chapter 4) can yield copolymers of this type. Again, the properties of the copolymer usually differ markedly from those of the two related homopolymers.

3. *Block copolymers.* Block copolymers contain a block of one monomer connected to a block of another, as illustrated in sequence **12**. Block copolymers are

$$-A-A-A-A-A-A-A-A-B-B-B-B-B-B-B-$$

12

usually formed by ionic polymerization processes. Unlike other copolymers, they retain many of the physical characteristics of the two homopolymers.

Terpolymers

A *terpolymer* contains three different monomer units. These can be sequenced randomly or in blocks.

Graft Copolymers

A *graft copolymer* is usually prepared by linking together two different polymers. For example, a homopolymer derived from monomer A may be induced to react with a homopolymer derived from monomer B to yield the graft copolymer shown in **13**. Graft polymers of this type can often be prepared by the gamma or X-irradiation of a mixture of the two homopolymers, or even by mechanical blending of the two homopolymers. Alternatively, a graft copolymer may be prepared by the

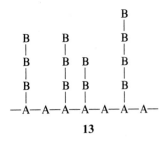

13

polymerization of monomer B from initiation sites along the chain of polymer A. Graft copolymers often display properties that are related to those of the two homopolymers.

Thermoplastics

Basically, a *thermoplastic* is any material that softens when it is heated. However, the term is commonly used to describe a substance that passes through a definite sequence of property changes as its temperature is raised. In Figure 1.4 the thermoplastic characteristics of an amorphous and a crystalline polymer are compared.

Both amorphous and crystalline thermoplastics are glasses at low temperatures, and both change from a glass to a rubbery elastomer or flexible plastic as the temperature is raised. This change from glass to elastomer usually takes place over

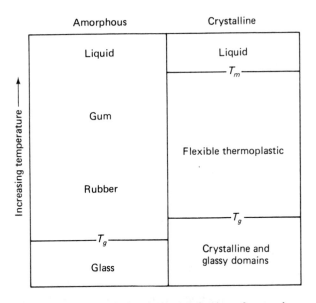

Fig. 1.4 Comparison of the thermal behavior of amorphous and crystalline polymers. The main property differences exist in the middle temperature range, where an amorphous polymer is an elastomer or a gum, and the crystalline polymer is a tough, flexible material.

a fairly narrow temperature range (2 to 5°C), and this transition point is known as the *glass transition temperature* (T_g). For many polymers, the glass transition temperature is the most important characterization feature. It can be compared to the characteristic melting point of a low-molecular-weight compound, although care should be taken to remember that T_g is definitely *not* a melting temperature in the accepted sense of the word. It is more a measure of the ease of torsion of the backbone bonds rather than of the ease of separation of the molecules.

At temperatures above T_g, amorphous polymers behave in a different manner from crystalline polymers. As the temperature of an amorphous polymer is raised, the rubbery elastomeric phase *gradually* gives way to a soft, extensible elastomeric phase, then to a gum, and finally to a liquid. No sharp transition occurs from one phase to the other, and only a gradual change in properties is perceptible.

Crystalline polymers, on the other hand, retain their rubbery elastomeric or flexible properties above the glass transition, until the temperature reaches the melting temperature (T_m). At this point the material liquefies. At the same time, melting is accompanied by a loss of the optical birefringence and crystalline X-ray diffraction effects that are characteristic of the crystalline state.

The amorphous and crystalline behavior described above is characteristic of linear and branched polymers, copolymers, or cyclolinear polymers. In general, these characteristics are not shown by heavily crosslinked polymers or cyclomatrix materials. These latter substances retain their rigidity when heated. Melting phenomena occur only when the crosslink units or backbone bonds become thermally broken. Lightly crosslinked polymers show many of the conventional thermoplastic properties, with the exception that the true liquid phase may not be formed.

Elastomers

In view of the information just given, it will be clear that an *elastomer* is a polymer that is in the temperature range between its glass transition temperature and its liquefaction temperature. In practice, elastomeric properties become more obvious if the polymer chains are lightly crosslinked. In particular, the liquefaction temperature may be raised by crosslinking, and the polymer may exhibit elastomeric properties over a wider temperature range.

Elastomeric properties appear when the backbone bonds can readily undergo torsional motions to permit uncoiling of the chains when the material is stretched (Figure 1.5). Crosslinks between the chains prevent the macromolecules from slipping past each other and thus prevent the material from becoming permanently elongated when held under tension. An important question connected with elasticity is this: Why do the chains revert to the highly coiled state when the tension on the elastomer is released? The answer lies in the fact that a highly coiled polymer system has a higher degree of disorder and, therefore, a higher entropy than a stretched, oriented sample. Thus, the elastic behavior is a direct consequence of the tendency of the system to assume spontaneously a state of maximum entropy. Since free energy, enthalpy, and entropy are related by the usual expression, $\Delta G =$

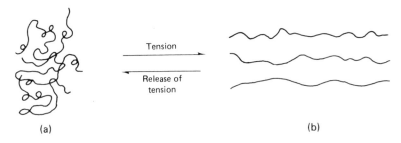

Fig. 1.5 Rubbery elastomeric properties result from the stress-induced uncoiling and recoiling of polymer chains. (a) Relaxed: high entropy. (b) Stretched: ordered—low entropy.

$\Delta H - T\Delta S$, a stretched rubber band immediately held to the lips is warm, and the same material appears cold immediately after contraction.

Plasticizers

Many commercial polymers are too rigid for use as flexible films. Poly(vinyl chloride) in the pure state is a rather rigid material. Only when this polymer is softened by the addition of liquids, such as phthalate esters, can it be used as a flexible-film or Tygon tubing. Such liquid additives are known as *plasticizers*.

Thermosetting Resin

The term *thermosetting polymer* refers to a range of systems which exist initially as liquids but which, on heating, undergo a reaction to form a solid, highly crosslinked matrix. A typical example is provided by the condensation of methylol melamine to give the hard, tough, crosslinked melamine resin (see Chapter 2). Partly polymerized systems which are still capable of liquid flow are called *prepolymers*. Prepolymers are often preferred as starting materials in technology. In practical terms, an uncrosslinked thermoplastic material can be reformed into a different shape by heating; a *thermosetting* polymer cannot.

Polymer Blends

When two or more polymers are mixed together mechanically, the product is known as a *polymer blend*. Many polymer blends display properties that are different from those of the individual polymers. Polymer blends can be of two types: (1) simple cellular mixtures of the polymers, and (2) genuine block or graft copolymers formed by the physical breaking of bonds, followed by bonding between the different polymeric fragments. The latter type of process can occur when two or more polymers are milled or masticated together. The mechanical shearing can result in the homolytic cleavage of bonds, followed by cross-recombination.

Tacticity

Olefin molecules that contain one unique side group, such as propylene, $CH_2=CH(CH_3)$, or styrene, $CH_2=CH(C_6H_5)$, yield polymers that possess an asymmetric center at each monomer residue. The tacticity of a polymer describes the sequencing of these asymmetric centers along the chain. Three primary possibilities exist, called isotactic, syndiotactic, or atactic (heterotactic) sequencing. This subject is discussed in more detail in Chapters 4 and 17. Here it is sufficient to note that the tacticity of a polymer markedly affects the bulk physical properties.

DIFFERENT TYPES OF POLYMERS

It is convenient to classify polymers according to the types of reactions involved in their synthesis. The three main polymerization reaction types are (1) condensation reactions, (2) addition reactions, and (3) ring opening polymerizations.

Condensation Polymers (Step Reactions)

Condensation processes take place when two or more molecules react with each other with the concurrent loss of water or ammonia. This type of reaction is used as a basis for the synthesis of many important polymers, such as nylon, polyesters, phenol–formaldehyde, and urea–formaldehyde resins. It is also the basis for the laboratory formation of silicates and polyphosphates.

Nylon is a condensation polymer formed by the reaction of a diamine with a dicarboxylic acid. In particular, Nylon 66 is made by the condensation of hexamethylenediamine with adipic acid as shown in equation (7).

$$H_2N(CH_2)_6NH_2 + HOC-(CH_2)_4-COH \longrightarrow \text{1:1 Salt}$$

$$\xrightarrow{-H_2O} H[NH(CH_2)_6NHC(CH_2)_4C]_nOH \qquad (7)$$

Polyesters, such as Dacron, Terylene, or Mylar are made by the condensation of a dicarboxylic acid with a diol (reaction 8) (typically $R = C_6H_4$ and $R' = CH_2CH_2$).

$$H-O-C-R-C+O-H + H+O-R'-OH$$

$$\xrightarrow[-H_2O]{\text{Catalyst}} H[O-C-R-C-O-R']_nOH \qquad (8)$$

The water formed in this reaction is removed by distillation. A variant of this synthesis is the use of the methyl ester of the diacid and the removal of methanol. Typical products are made from terephthalic acid ($R=C_6H_4$) and ethylene glycol ($R'=CH_2CH_2$). Alkyd paint resins are manufactured from phthalic anhydride and glycerol, and the trifunctionality of the latter reagent ensures the formation of a crosslinked structure (see Chapters 2 and 11).

Phenol–formaldehyde (Bakelite) resins are hard, rigid polymers made by alkylation–condensation reactions between phenols and formaldehyde, as discussed in Chapter 2.

Urea–formaldehyde resins are formed by the "methylolation" reaction between urea and formaldehyde, and condensation of the methylol units yields the polymer shown in reaction (9).

$$H_2N-\overset{\overset{\displaystyle O}{\|}}{C}-NH_2 + CH_2O \longrightarrow HOCH_2NH-\overset{\overset{\displaystyle O}{\|}}{C}-NHCH_2OH$$

$$\xrightarrow{-H_2O} \left[OCH_2NH-\overset{\overset{\displaystyle O}{\|}}{C}-NHCH_2 \right]_n \quad (9)$$

Inorganic condensation polymers, such as polysilicate glasses and polyphosphates, are formed by removal of water from the appropriate di-, tri-, or tetrahydroxy monomer, as illustrated by reactions (10) to (13).

$$HO\!\!-\!\!\overset{\overset{\displaystyle O}{\|}}{\underset{}{P}}\!\!-\!\!O\!\!\mid\!\!H + HO\!\!-\!\!\overset{\overset{\displaystyle O}{\|}}{\underset{}{P}}\!\!-\!\!OH \xrightarrow[-H_2O]{\text{Heat}} -H\!\!\left[O-\overset{\overset{\displaystyle O}{\|}}{\underset{}{P}}\right]_n\!\!-OH \quad (10)$$

$$(M = \text{An alkali metal}) \qquad\qquad\qquad \text{Polyphosphate}$$

$$HO\!\!-\!\!\overset{MO\ \ \ OM}{\underset{}{Si}}\!\!-\!\!O\!\!\mid\!\!H + HO\!\!-\!\!\overset{MO\ \ \ OM}{\underset{}{Si}}\!\!-\!\!O\!\!\mid\!\!H \xrightarrow[-H_2O]{\text{Heat}} H\!\!\left[O-\overset{MO\ \ \ OM}{\underset{}{Si}}\right]_n\!\!-OH \quad (11)$$

$$\text{Polysilicate (glass)}$$

$$\overset{HO}{\underset{HO}{\diagdown}}\!\!\overset{OM}{\underset{OH}{Si}} \xrightarrow[-H_2O]{\text{Heat}} \begin{array}{l}\text{Crosslinked silicate (glass)}\\ \text{(e.g., structure 9, page 9)}\end{array} \quad (12)$$

$$\overset{HO}{\underset{HO}{\diagdown}}\!\!\overset{OH}{\underset{OH}{Si}} \xrightarrow[H_2O]{\text{Heat}} \text{Silica} \quad (13)$$

Biological condensation polymers. Enzyme-catalyzed condensation reactions are responsible for the polymerization of some amino acids to proteins; for the condensation of sugars to polysaccharides such as starch, cellulose, and glycogen, and for the synthesis of nucleic acids such as DNA and RNA. A separate chapter

(Chapter 8) is devoted to a consideration of biological polymers. From the view-point of fundamental polymer chemistry it should be remembered that the physi-ological properties of biological macromolecules are often determined as much by the *molecular conformation* as by the chemical composition. This aspect is discussed in Chapters 8 and 18. It should also be noted that many naturally occurring con-densation polymers, particularly proteins and cellulose, are chemically modified before they are used in technology. This subject is covered in Chapter 8.

Addition Polymers

Addition polymers are macromolecules formed by the addition reactions of olefins, acetylenes, aldehydes, or other compounds with "unsaturated" bonds. These reactions are summarized by the scheme shown in (14). Many well-known thermo-

$$
m \quad {}^{H}_{H}\!\!>\!\!C{=}C\!\!<^{R}_{H} + n \quad {}^{H}_{H}\!\!>\!\!C{=}C\!\!<^{R}_{H} \longrightarrow \left[\begin{matrix} H & R & H & R \\ | & | & | & | \\ C{-}C{-}C{-}C \\ | & | & | & | \\ H & H & H & H \end{matrix} \right]_{n+m} \tag{14}
$$

plastics are addition-type polymers, the differences between the various materials being mainly connected with the presence of different substituent groups attached to the main chain. For example, high polymers are known (and manufactured on a large scale) in which Cl, CN, C_6H_5, $CH_3OC(O)-$, CH_3, and a variety of other units are present as side groups to the main chain. The following equations and names shown in (15) to (24) illustrate some addition polymerization processes.

(a) $n CH_2{=}CH_2 \longrightarrow +CH_2{-}CH_2)_n$ Polyethylene   (15)

(b) $n CH_2{=}\overset{\overset{\text{Cl}}{|}}{CH} \longrightarrow \left(CH_2{-}\overset{\overset{\text{Cl}}{|}}{CH} \right)_n$ Poly(vinyl chloride)   (16)
(PVC)

(c) $n CH_2{=}\overset{\overset{\text{C}\equiv\text{N}}{|}}{CH} \longrightarrow \left(CH_2{-}\overset{\overset{\text{C}\equiv\text{N}}{|}}{CH} \right)_n$ Polyacrylonitrile   (17)
(acrylic fiber, Creslan)

(d) $n CH_2{=}CH \longrightarrow \left(CH_2{-}CH \right)_n$ Polystyrene   (18)

(e) $n CH_2{=}\overset{\overset{\text{CH}_3}{|}}{\underset{\underset{\text{CH}_3}{|}}{C}} \longrightarrow \left(CH_2{-}\overset{\overset{\text{CH}_3}{|}}{\underset{\underset{\text{CH}_3}{|}}{C}} \right)_n$ Polyisobutylene   (19)
(butyl rubber)

$$(f) \quad nCH_2=\underset{\underset{CH_2=CH}{|}}{\overset{\overset{CH_3}{|}}{C}} \longrightarrow \left(\underset{\overset{\overset{|}{CH_2-CH}}{}}{\overset{\overset{CH_3}{|}}{C}-CH_2} \right)_n \qquad \textit{Trans}\text{-1,4-polyisoprene} \quad (20)$$

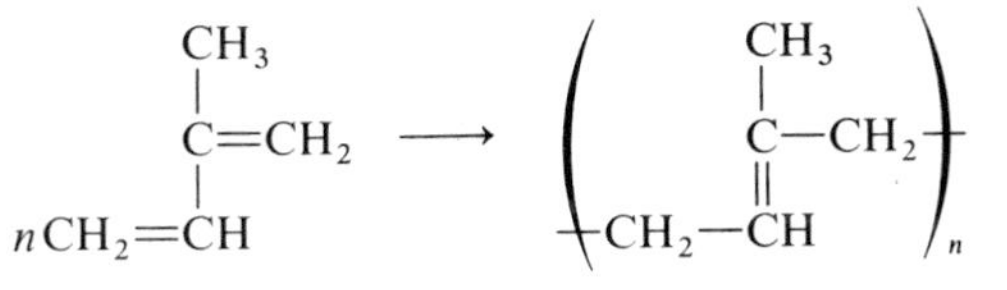

$$(g) \quad nCH_2=\underset{\underset{CH=CH_2}{|}}{CH} \longrightarrow \left(\underset{CH_2-CH}{CH-CH_2} \right)_n \quad \textit{Trans}\text{-1,4-polybutadiene} \quad (21)$$

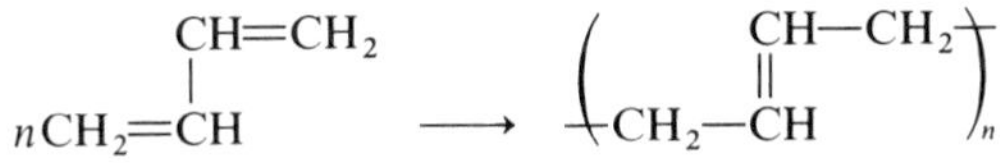

$$(h) \quad nCH_2=\underset{\underset{Cl}{|}}{\overset{\overset{Cl}{|}}{C}=CH_2} \longrightarrow \left(\underset{CH_2-CH}{\overset{\overset{Cl}{|}}{C}-CH_2} \right)_n \quad \begin{array}{l} \textit{Trans}\text{-1,4-polychloroprene} \quad (22) \\ \text{(neoprene rubber)} \end{array}$$

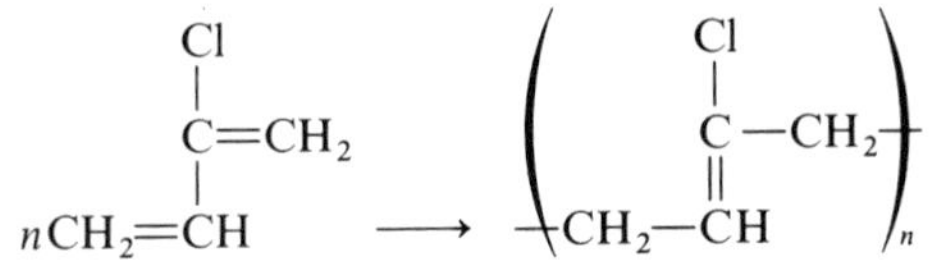

$$(i) \quad nCF_2=CF_2 \longrightarrow +CF_2-CF_2 \rightarrow_n \qquad \begin{array}{l} \text{Poly(tetrafluoroethylene)} \quad (23) \\ \text{(Teflon)} \end{array}$$

$$(j) \quad nCH_2=O \longrightarrow +CH_2-O \rightarrow_n \qquad \begin{array}{l} \text{Polyformaldehyde} \qquad (24) \\ \text{(polyoxymethylene, Delrin)} \end{array}$$

It should be recognized that polymers such as polyacrylonitrile, polybutadiene, and polychloroprene contain additional unsaturation that can be utilized in subsequent high-temperature or crosslinking reactions. Addition polymerizations are considered in more detail in Chapters 3, 4, 5, 12, and 13.

Ring-Opening Polymerizations

The treatment of some cyclic compounds with catalysts brings about cleavage of the ring followed by polymerization to yield high-molecular-weight polymers.

For example, as shown in reaction (25), trioxane polymerizes to yield polyformaldehyde (polyoxymethylene). Caprolactam polymerizes to Nylon 6 (reaction 26), and epoxides undergo ring-opening reactions to yield polyethers (27).

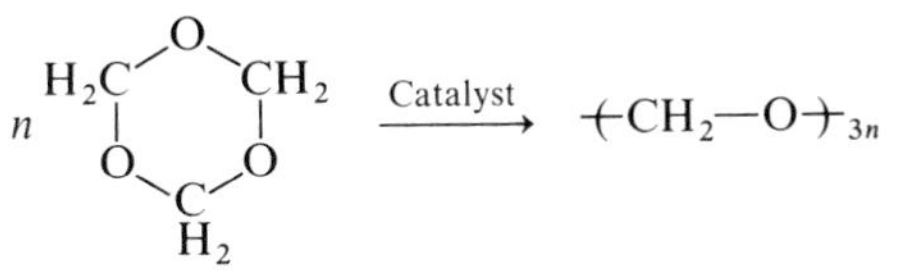

$$n \; \underset{\underset{H_2}{C}}{\overset{O}{\underset{O}{\overset{}{\bigcirc}}}} \xrightarrow{\text{Catalyst}} +CH_2-O\rightarrow_{3n} \qquad (25)$$

$$\underset{(CH_2)_5-C=O}{\overset{\overline{\quad\quad}NH}{|}} \longrightarrow \left[NH-(CH_2)_5-\overset{\overset{O}{\|}}{C} \right]_n \qquad (26)$$

$$\underset{CH_2-CHR}{\overset{O}{\triangle}} \xrightarrow{\text{Tertiary amines}} \left[O-CH_2-\overset{\overset{R}{|}}{CH} \right]_n \qquad (27)$$

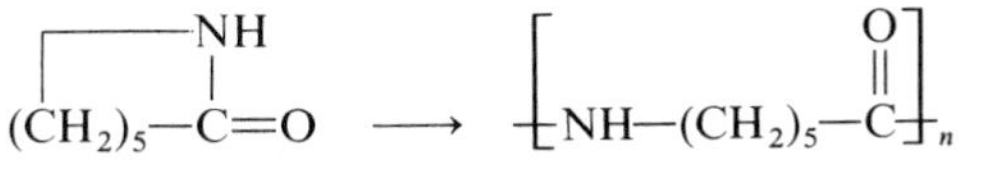

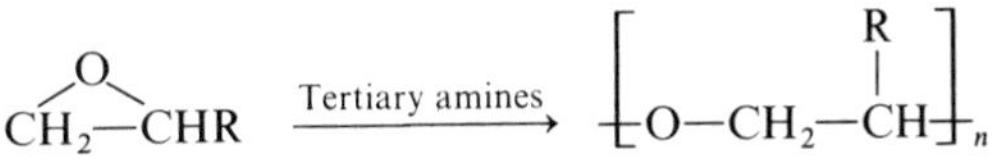

A number of inorganic ring systems also polymerize by ring-opening reactions. Rhombic sulfur, cyclic siloxanes, and cyclic chlorophosphazenes behave in this way (reactions 28 to 30).

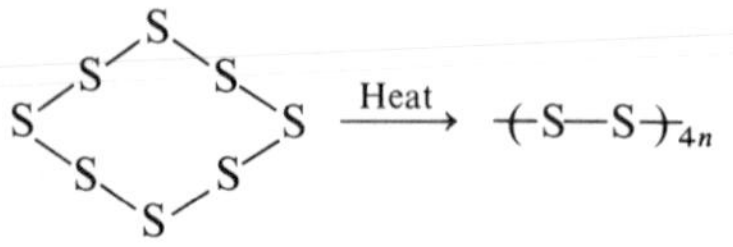

$$\text{Rhombic sulfur} \quad \xrightarrow{\text{Heat}} \quad +S-S\!\!+_{4n} \tag{28}$$

Rhombic sulfur   Plastic sulfur

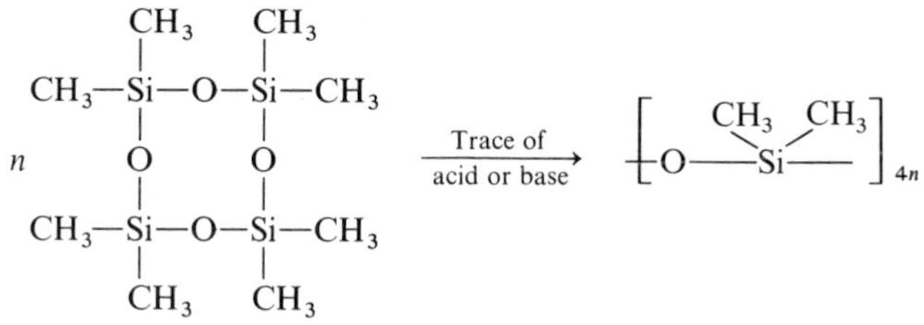

$$n\ \begin{array}{c}\text{CH}_3\quad\text{CH}_3\\ |\qquad\ |\\ \text{CH}_3-\text{Si}-\text{O}-\text{Si}-\text{CH}_3\\ |\qquad\ |\\ \text{O}\qquad\text{O}\\ |\qquad\ |\\ \text{CH}_3-\text{Si}-\text{O}-\text{Si}-\text{CH}_3\\ |\qquad\ |\\ \text{CH}_3\quad\text{CH}_3\end{array} \quad \xrightarrow[\text{acid or base}]{\text{Trace of}} \quad \left[\text{O}-\!\!\!\!\begin{array}{c}\text{CH}_3\ \text{CH}_3\\ \text{Si}\end{array}\!\!\!-\right]_{4n} \tag{29}$$

Octamethylcyclotetrasiloxane   Poly(dimethylsiloxane)

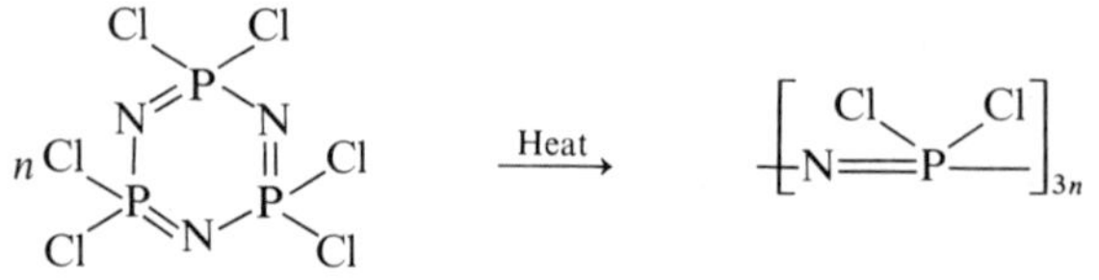

$$n\ \text{Hexachlorocyclotriphosphazene} \quad \xrightarrow{\text{Heat}} \quad \left[\begin{array}{c}\text{Cl}\quad\text{Cl}\\ \text{N}\!\!=\!\!\text{P}\end{array}\right]_{3n} \tag{30}$$

Hexachlorocyclotriphosphazene   Poly(dichlorophosphazene)
                                            (inorganic rubber)

Ring-opening polymerizations will be considered further in Chapters 6, 7, and 10.

HISTORICAL OVERVIEW

The Macromolecular Hypothesis

Although natural polymers have been used by human beings since antiquity, their structure was not understood, even in elementary terms, until the late 1800s. The problem encountered by the earliest investigators was a general unwillingness on the part of the scientific community to believe that giant covalently bonded molecules could exist. Indeed, the physical tools for the measurement of molecular weights in solution (based on the work of Raoult and van't Hoff) did not exist until the 1800s. In 1888, Brown and Morris used a cryoscopic technique to estimate the molecular weight of a starch hydrosylate at about 30,000. The same technique was used by Gladstone and Hibbert to estimate that the molecular weight of rubber was between 6000 and 12,000, or perhaps even higher.

However, the demonstration that materials such as starch, rubber, or proteins had high molecular weights did not convince the scientific community that these materials had polymeric structures. On the contrary, it was generally assumed that the high values for the molecular weights resulted from defects in the cryoscopic method, or were due to association of smaller molecules. This view persisted even as the rubber industry prospered after 1839 and synthetic polymers came into limited use. Evidence about the structure of polymers began to accumulate between 1890 and 1919 from the work of Emil Fischer on proteins. However, it was not until 1920 that Staudinger put forward the idea of covalently bonded macromolecular structures for polystyrene, rubber, and polyoxymethylene and began the process of providing convincing evidence in favor of this structure. Even in the following 10 years, this hypothesis was subjected to intense criticism.

Structural Work

The application of physics and physical chemistry to macromolecular systems dates back to the early attempts made in the late 1800s to understand the unusual properties of natural polymers, such as rubber, polysaccharides, and proteins. However, it was not until the 1920–1930 period that Meyer and Mark in Germany began to establish the structure of cellulose and rubber with the use of X-ray diffraction techniques. Explanations of rubbery elasticity in terms of polymer conformations were put forward by Kuhn, Guth, and Mark between 1930 and 1934. Kuhn, in particular, was the first to apply statistical methods to the study of macromolecules.

The application of light scattering to macromolecular systems was made by Debye during World War II. It was also during this period that Flory began a series of investigations into the applications of statistical methods, conformational analysis, and other fundamental physicochemical techniques, to polymer science. During the early 1950s, Watson, Crick, Wilkins, Franklin, Kendrew, and Hodgkin successfully applied X-ray diffraction analysis to the structure determination of biological polymers, such as DNA, hemoglobin, and insulin.

Single crystals of polyethylene were first reported by Keller and Till in 1957. Increasingly, during the 1960s and 1970s, NMR analysis of polymers has proved to be a valuable physical tool.

Synthetic Polymers

The beginnings of polymer chemistry can be traced to the years 1838 and 1839. Vinyl chloride was first polymerized (photochemically) in 1838. Polystyrene was prepared in 1839. In the same year, MacIntosh and Hancock in Britain and Goodyear in the United States discovered the vulcanization (crosslinking) of rubber by sulfur. This process enabled the soft, gumlike natural material to be modified for use in tires and rainwear. These isolated discoveries constitute the start of an astonishing series of synthetic and manufacturing developments that are continuing to the present day.

In the 90 or so years between 1840 and 1930, a number of key discoveries were made. The plastics industry really began in 1868 with Hyatt's discovery that cellulose nitrate could be mixed with camphor to yield a hard, plasticlike material, known as *Celluloid*. By 1870, the process had been commercialized and celluloid was soon used in a wide range of products, ranging from shirt collars to toys. The manufacture of rayon fibers began between 1893 and 1898 in England, and the first rayon plant was introduced into the United States in 1910. Styrene–diene copolymers were first synthesized in the early 1900s, and phenolic resins (Bakelite) were developed in Germany around 1907.

Cellulose acetate solutions were used as aircraft "dope" for the plywood-fabric aircraft in World War I. The same polymer was later used for fiber manufacture (1924) and for cellulose acetate plastics (1927). Cellulose nitrate lacquers were used on automobiles in 1920, and alkyd resins were introduced in 1926. Poly(vinyl chloride) was produced commercially in 1927, and urea–formaldehyde polymers were introduced in 1929. It is perhaps astonishing to realize that most of these developments occurred during the time when the polymeric nature of the products was not recognized or believed.

The 10 years between 1930 and 1940 represent the springboard for the development of modern synthetic polymer chemistry. By 1930, the macromolecular hypothesis was beginning to be accepted. Moreover, in 1929, Carothers at the DuPont Company, began his classical series of attempts to prepare high polymers from well-characterized, low-molecular-weight compounds. The success of these studies provided a final verification of the macromolecular theory. The work also led to the synthesis of polyamides and polyesters. Nylon 66 was, in fact, produced commercially in 1938. Much of the spadework in polymer synthesis that was eventually responsible for the polymer-based way of life we have today took place during this period. Poly(methyl methacrylate) was prepared in 1931. Poly(vinyl acetate) and poly(vinyl butyrate) were used in laminated safety glass in 1936. Polystyrene manufacture was initiated in Germany in 1937. Both poly(vinyl chloride) and melamine–formaldehyde resins were introduced in 1939. Polysulfide rubber (Thiokol) and neoprene rubber were also introduced in the United States.

The outbreak of World War II hostilities in 1939 precipitated an incredible burst of technological development in polymer chemistry. Polyethylene, first produced in Britain in 1941, proved to be an indispensible insulator for radar and other electronic developments. Moreover, the production of new plastics, such as polyethylene, PVC, and polystyrene, was accelerated as the monomers became available from the petrochemicals industry.

However, the critical chemical drama of this period was the drive to develop synthetic substitutes for rubber. It provides a captivating illustration of the way in which scientific developments are fueled by political and military necessities. Before the outbreak of the war, Germany had foreseen the need to become independent of natural rubber. The main rubber-producing areas—Malaya and Ceylon—were controlled by Britain. By 1939, Germany had completed the development of polybutadiene as a synthetic rubber. Acrylonitrile–butadiene (buna N) had also been developed. Polyurethanes were also first synthesized in Germany during the war. The Japanese conquest of Malaya in 1941 cut off the main supply of natural rubber to Britain and the United States. An incipient military disaster was avoided only by the crash development in the United States of styrene–butadiene (SBR, GR-S) copolymers, and butyl rubber (polyisobutylene) as synthetic elastomers. Fluorocarbon polymers were also first produced during this period.

From 1945 through the 1960s, synthetic polymer chemistry and polymer technology expanded at an accelerating pace. Epoxy resins and ABS (acrylonitrile–butadiene–styrene) plastics appeared in 1947 and 1948. Polyester fibers and polyacrylonitrile fibers were introduced in 1950. Polysiloxanes (silicones), which had been discovered during the war, were commercialized in the United States during the late 1940s and early 1950s. The 1950s also saw the pivotal discovery of linear polyethylene and stereospecific polymerization from the catalytic studies of Ziegler in Germany, and the structural work of Natta in Italy.

Also during the 1950s a variety of inorganic and organometallic compounds were developed as ionic polymerization catalysts. The recognition of nonterminated ("living") polymers by Szwarc occurred during this period. Polyoxymethylene (acetal polymers), polypropylene, polycarbonates, and polyurethane foams appeared during the mid-1950s, and these were followed in the 1960s by synthetic *cis*-polyisoprene rubber, *cis*-polybutadiene rubber, ethylene–propylene rubber, polyimides, poly(phenylene oxide), polysulfones, and styrene–butadiene block copolymers.

The most recent phase of synthetic polymer chemistry has involved a subtle shift in emphasis. Interest is now focused on the synthesis and development of entirely new polymers that have specialized, high-performance properties, such as flame resistance, high-temperature stability, oil and fuel resistance, semi- or superconductivity, biomedical compatibility, or low-temperature flexibility. Aromatic ladder polymers are receiving considerable attention. Polymers that contain inorganic elements are now being studied closely: polyphosphazenes, in particular, are undergoing a rapid technological development. Other systems, such as carborane–siloxanes, carbon fibers, pyrolyzed polyacrylonitrile, and poly(sulfur nitride) are being examined in detail. Polymers are now being used as stationary substrates for the binding of transition metal catalyst systems or enzymes. Presumably, this trend toward the synthesis of new polymers for critical applications will continue in the foreseeable future.

Engineering and Materials Science

Polymer chemistry is a mature science in the sense that all phases of its development now exist, from synthesis and molecular structural work, through theoretical physics and physical chemistry, to engineering. Polymers have always been used in technology because of their engineering advantages (strength, elasticity, light weight, corrosion resistance, low cost). As will be evident from some of the later chapters, few polymers are used technologically in their chemically pure form. Thus, much of the current research and development work connected with polymers involves studies of the ways in which pure polymers can be modified to make them more suitable for specific applications. Such modifications can involve attempts to physically modify the crystallinity of a polymer, or result from the addition of plasticizers, reinforcement agents, or even the addition of other polymers. This interface between polymer chemistry, engineering, rheology, and solid-state science is often described under the general umbrella term of *materials science*. The period since the mid-1960s has seen a striking growth of interest in the materials science aspects of polymer science as investigators seek to improve the properties of well-known synthetic polymers. This type of work is likely to become even more important as technology continues to demand new materials with improved properties at a time when most of the readily available organic polymers have already been commercialized. It seems clear that the technological future of polymer science will require a close interaction between synthetic chemists on the one hand and physical chemists and materials scientists on the other.

STUDY QUESTIONS

1. Refer to an organic chemistry textbook and suggest synthetic routes that could be used to produce monomers such as vinyl chloride, tetrafluoroethylene, styrene, ethylene glycol, and terephthalic acid from petroleum.

2. What types of "crash programs" can you foresee being implemented if the supply of monomers from oil were to be interrupted for several years as a result of international upheavals?

3. Excluding the specific reactions mentioned in this chapter, suggest examples, based on your own prior experience and study, of an addition polymerization, a condensation polymerization, a ring-opening polymerization, and a biological condensation polymerization.

4. How would you distinguish experimentally between an amorphous, a crystalline, and a crosslinked polymer?

5. Devise several reactions that might lead to the synthesis of cyclolinear, ladder, and spiro polymers. After you have spent 10 minutes on this problem, refer to Chapter 2 and reassess your ideas.

6. Based on your intuitive knowledge of chemistry or physics, suggest why rubber is an elastomer at room temperature but glass is not.

7. Without referring to later chapters, suggest ways in which a graft copolymer might be prepared.

8. Before starting to read the next chapter, spend 30 minutes scanning the rest of the book to obtain some perspective of its contents. Glance at the section headings, structural formulas, figures, and the questions at the end of each chapter. Then study individual chapters in detail.

2

Condensation and Other Step-Type Polymerizations

GENERAL FEATURES

Definitions

As discussed in Chapter 1, polymerization processes can be classified very roughly into three categories: condensation, addition, and ring-opening polymerizations. In condensation reactions, monomer molecules react to release a small molecule such as water. Addition polymers are formed by the polyaddition reactions of olefins or carbonyl compounds. Ring-opening polymerizations take place by cleavage of a ring with concurrent or subsequent addition of the linear product to the end of a growing chain. Chapters 3 to 5 deal with addition polymerizations and Chapter 6 covers ring-opening processes. Here we are concerned with polymers of the type shown in Table 2.1, formed by condensations and other step reactions.

The polymerization categories just described reflect different monomer structures. However, a related but distinct classification is based on the *general mechanistic pathways* that are involved. This classification divides polymerization processes into step reactions and chain reactions. *Step reactions* are those in which the chain growth occurs in a slow, stepwise manner. Two monomer molecules react to form a dimer. The dimer can then react with another monomer to form a trimer, or with another dimer to form a tetramer. Thus, the average molecular weight of the system increases slowly over a period of time. Condensation polymerizations and

TABLE 2.1 Summary of Step-Type Polymers Described in This Chapter

NAME	GENERAL REPEATING STRUCTURE

Polymers Formed by Condensation Reactions

Polyesters

$$-O-R-O-\overset{\overset{O}{\|}}{C}-R'-\overset{\overset{O}{\|}}{C}-$$

Polycarbonates

$$-O-R-O-\overset{\overset{O}{\|}}{C}-$$

Polyanhydrides

$$-O-\overset{\overset{O}{\|}}{C}-R-\overset{\overset{O}{\|}}{C}-$$

Polyamides

$$-\overset{\overset{O}{\|}}{C}-R-\overset{\overset{O}{\|}}{C}-\overset{\overset{H}{|}}{N}-R'-\overset{\overset{H}{|}}{N}-$$

Polyimides

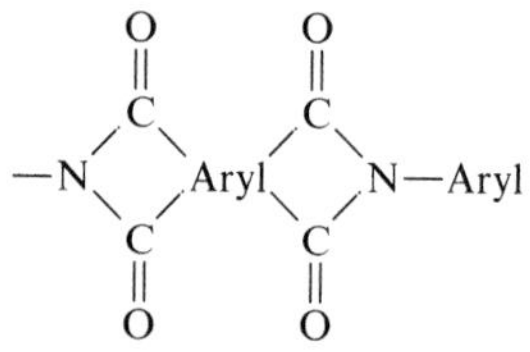

$$-N\underset{\underset{O}{\overset{\|}{C}}}{\overset{\overset{O}{\overset{\|}{C}}}{}}Aryl\underset{\underset{O}{\overset{\|}{C}}}{\overset{\overset{O}{\overset{\|}{C}}}{}}N-Aryl$$

Polybenzimidazoles

$$-R-C\underset{\overset{}{N}}{\overset{\overset{H}{N}}{}}Aryl C\underset{\overset{}{N}}{\overset{\overset{H}{N}}{}}-$$

Polyquinoxalines

$$-Aryl \text{[quinoxaline structures]} Aryl-$$

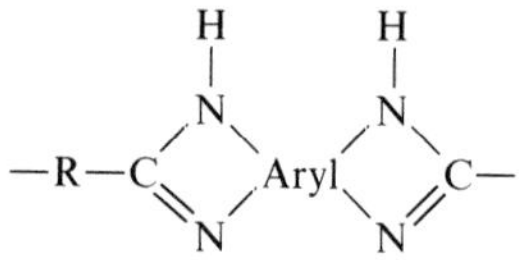

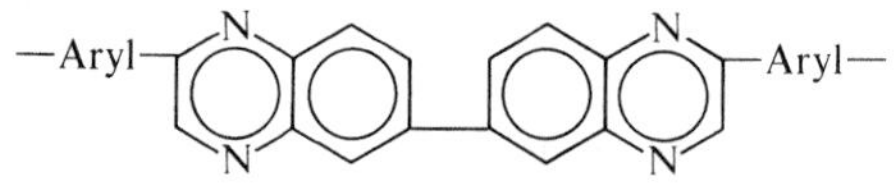

Aromatic ladder polymers

Various structures, often based on

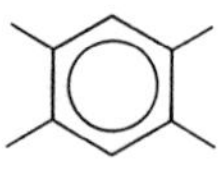

[benzene ring] or [pyrazine ring] units

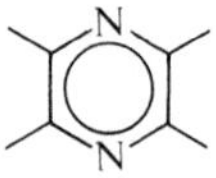

Phenol–formaldehyde polymers

$$-CH_2-Aryl-CH_2-O-$$

Urea–formaldehyde polymers

$$-N(H)-C(O)-N(H)-CH_2-O-$$

Melamine–formaldehyde polymers

$$-CH_2-N-CH_2-O-$$
$$-CH_2-N-\overset{}{C}\underset{N}{\overset{N=C-N}{}}C-N-CH_2-O-$$
$$-CH_2 \quad CH_2-O-$$

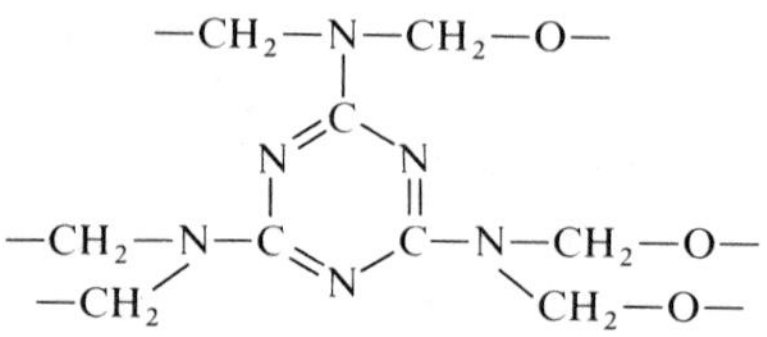

Polyacetals

$$-CH_2-O-R-O-$$

TABLE 2.1 (Continued)

Name	General Repeating Structure
	Polymers Formed by Noncondensation, Step-Type Reactions

Polysulfones (polyethers)

$$-O-R-O-\!\!\bigcirc\!\!-\overset{\overset{O}{\|}}{\underset{\underset{O}{\|}}{S}}-\!\!\bigcirc\!\!-$$

Poly(phenylene oxide) (polyethers) $-Aryl-O-$

Diels–Alder formed polymers

Polyurethanes $-C(O)-O-R-O-C(O)-N(H)-R'-N(H)-$

Polyarylenes $-Aryl-$

some noncondensation reactions, such as Diels–Alder additions, fall into this category. On the other hand, *chain polymerizations* take place by a *rapid* addition of olefin molecules to a growing chain end. Because chain growth occurs rapidly, the system usually contains only unreacted monomer and high polymer. Intermediate polymers usually cannot be isolated. A rather clear-cut distinction exists between these two mechanistic types.

Comparison of Step and Chain Polymerizations

The distinction between step and chain polymerizations is an important concept in polymer chemistry. Hence, it is worthwhile at this point to summarize the observable differences between the two types of systems.

First, in step polymerization, any two molecules in the system can react with each other. Initially, the monomers react to form dimers, dimers can react with dimers, and so on. On the other hand, in chain polymerization, chain growth takes place only at the ends of a few "initiated" chains.

Second, as a result of this difference, step reactions are characterized by a disappearance of the monomer at an early stage in the polymerization, and by the existence of a broad molecular-weight distribution in the later stages of the reaction. With chain polymerizations, the monomer concentration decreases steadily throughout the reaction and, ideally, at any stage the reaction mixture should contain only monomer and high polymer.

Third, the variation in polymer molecular weight at different stages in the reaction provides another distinguishing feature. In step reactions the polymer molecular weight rises steadily during the reaction. In a chain reaction, high polymer is formed rapidly from each "initiated" monomer. Hence, the molecular weight of each polymer molecule does not increase appreciably after the initial rapid propagation. At longer reaction times there may be an increase in the number of polymer molecules, but not in the molecular weight of those already formed.

Types of Condensation Reactions

Typical condensation polymerizations are those which involve the elimination of a water molecule at each condensation step. The formation of polyesters and polyamides are two examples, as illustrated by the general reactions (1) and (2).

$$
HO-R-OH + HO-\overset{O}{\overset{\|}{C}}-R'-\overset{O}{\overset{\|}{C}}-OH \xrightarrow{-H_2O} H\!\!\left[O-R-O-\overset{O}{\overset{\|}{C}}-R'-\overset{O}{\overset{\|}{C}}\right]_n\!\!O-H
$$

(1)

$$
H_2N-R-NH_2 + HO-\overset{O}{\overset{\|}{C}}-R'-\overset{O}{\overset{\|}{C}}-OH \xrightarrow{-H_2O} H\!\!\left[\overset{H}{\overset{|}{N}}-R-\overset{H}{\overset{|}{N}}-\overset{O}{\overset{\|}{C}}-R'-\overset{O}{\overset{\|}{C}}\right]_n\!\!OH
$$

(2)

Variations include condensation by species which possess both hydroxyl and carboxylic acid or amino and carboxylic acid units on the same molecule (reactions 3 and 4). Also included in the condensation or step-type category are the trans-

$$
HO-R-\overset{O}{\overset{\|}{C}}-OH \xrightarrow{-H_2O} H\!\!\left[O-R-\overset{O}{\overset{\|}{C}}\right]_n\!\!O-H
$$

(3)

$$
H_2N-R-\overset{O}{\overset{\|}{C}}-OH \xrightarrow{-H_2O} H\!\!\left[N-R-\overset{O}{\overset{\|}{C}}\right]_n\!\!O-H
$$

(4)

esterification-type processes that occur when carboxylic acid esters react with alcohols (5). Acid chlorides can also be allowed to react with diamines in the synthesis of polyamides (6). A cyclization-type condensation may be used to yield

$$
HO-R-OH + R'O-\overset{O}{\overset{\|}{C}}-R''-\overset{O}{\overset{\|}{C}}-OR'
$$

$$
\xrightarrow{-R'OH} H\!\!\left[O-R-O-\overset{O}{\overset{\|}{C}}-R''-\overset{O}{\overset{\|}{C}}\right]_n\!\!O-R'
$$

(5)

$$
H_2N-R-NH_2 + Cl-\overset{O}{\overset{\|}{C}}-R'-\overset{O}{\overset{\|}{C}}-Cl \xrightarrow{-HCl} H\!\!\left[\overset{H}{\overset{|}{N}}-R-\overset{H}{\overset{|}{N}}-\overset{O}{\overset{\|}{C}}-R'-\overset{O}{\overset{\|}{C}}\right]_n\!\!Cl
$$

(6)

polybenzimidazoles, as illustrated by the reaction shown in (7). These and other step-type polymerizations are discussed in later sections.

$$HO-\overset{\overset{\displaystyle O}{\|}}{C}-R-\overset{\overset{\displaystyle O}{\|}}{C}-OH \;+\; \begin{matrix} H_2N \\ \\ H_2N \end{matrix}\!\!\diagdown\!\!\underset{\diagup}{\overset{\diagdown}{Aryl}}\!\!\diagup\!\!\begin{matrix} NH_2 \\ \\ NH_2 \end{matrix} \quad\xrightarrow{-H_2O}\quad \left[\!R-C\underset{N}{\overset{NH}{\diagup\diagdown}}Aryl\underset{N}{\overset{NH}{\diagdown\diagup}}C\!\right]_n \quad (7)$$

Mechanism of Condensation Polymerization

The interactions illustrated by reactions (1) to (6) are almost certainly more complex than is implied by these equations. In fact, such reactions fall into a category known as *carbonyl addition–elimination reactions*, illustrated by the following sequence:

$$R-\overset{\overset{\displaystyle O}{\|}}{C}-X + Y^{\ominus} \;\rightleftharpoons\; \left[R-\overset{\overset{\displaystyle O^{\ominus}}{|}}{\underset{X}{C}}-Y\right] \;\longrightarrow\; R-\overset{\overset{\displaystyle O}{\|}}{C}-Y + X^{\ominus}$$

The presence of the carbonyl group is believed to stabilize the tetracoordinate transition state, perhaps to such a degree that it has a finite existence as a transient intermediate. However, the speed of the overall reaction may depend on the ease of elimination of $X^{\ominus}$ rather than $Y^{\ominus}$ from the intermediate as well as on the initial formation of the intermediate. The equilibrium formation of the intermediate is usually favored by the presence of metal cations or protonic or Lewis acids that can coordinate to the $C-O^{\ominus}$ unit. Hence, catalysts such as metal oxides or acetates or sulfonic acids are added, especially to polyesterification reactions, to speed up the process.

Requirements for High Molecular Weight

High-molecular-weight condensation polymers can be obtained if the starting materials are pure and the functional groups (NH_2 and COOH, OH and COOH) are present in exactly equal amounts. If one or the other group exists in excess, that group will remain unreacted at the chain ends of low- or medium-molecular-weight polymers. Thus, the establishment of an exact 1:1 ratio of the two functional groups constitutes a critical practical requirement. Three options exist. First, and most obviously, the two reagents (for example, the diol and dicarboxylic acid) can be added together in as close to a 1:1 ratio as is experimentally feasible. Second, a monomer can be chosen which contains both of the functional groups. Glycolic acid, $HOCH_2COOH$, or an amino acid, $H_2NRCOOH$, are examples. Third, use can be made of the fact that some dicarboxylic acids form 1:1 salts with diamines. Hexamethylenediamine, $H_2N(CH_2)_6NH_2$, forms a 1:1 salt with adipic acid, $HOOC(CH_2)_4COOH$. This salt can be purified and used directly for the preparation of Nylon 66.

Sometimes a need exists for the preparation of low- or medium-molecular-weight polymers. This can be accomplished by the addition to a condensation polymerization system of a small amount of a mono-carboxylic acid, alcohol, amine, and so on. For example, if 1 mol % of acetic acid is added to a polyester condensation system, the degree of polymerization will be restricted to values of about 200 or less. Of course, the presence of a slight excess of a difunctional reagent will serve a similar purpose. However, the use of a monofunctional reagent has the advantage that the polymer chain ends are no longer active. Hence scrambling and molecular-weight changes are less likely to occur during subsequent heating of the polymer.

Scrambling Reactions

Condensation polymers that contain active groups may undergo molecular weight changes at elevated temperatures. For example, the carboxylic acid end groups of one polymer could attack the amide linkage of another (8). The total number of molecules remains the same, but the average molecular weight may change.

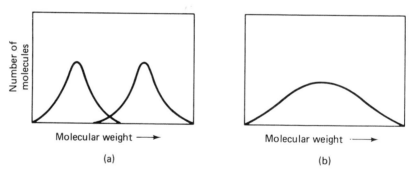

$$(8)$$

Similarly, if two different polymer samples which have different molecular-weight distributions are mixed and heated, the system will scramble to achieve the most probable molecular-weight distribution. This is illustrated in Figure 2.1.

Fig 2.1   Mixture of two polymers with different molecular-weight distributions: (a) will undergo scrambling reactions at elevated temperatures to yield an equilibrium mixture that represents the most probable molecular-weight distribution (b).

Ring Formation Versus Polymerization

At every step in a polymerization reaction the prospect exists that *cyclization* may occur at the expense of linear chain growth. This is especially true with step-type polymerizations, where the chain growth is slow. Consider, for example, the initial steps in the polymerization of an amino acid such as glycine (1). Following the initial condensation to give **2**, two courses of action are available to the system. Linear propagation may continue to yield **4**. Alternatively, **2** may cyclize to yield the diketopiperazine (**3**).

$$2\,H_2N-CH_2COOH \xrightarrow{-H_2O} \underset{\textbf{2}}{H_2N-CH_2-\overset{\displaystyle O}{\overset{\|}{C}}-\overset{\displaystyle H}{\underset{\displaystyle}{N}}-CH_2COOH}$$

1

$$\underset{\textbf{3}}{\begin{array}{c} H_2C-C \diagdown{}^{O} \\ HN \qquad NH \\ C-CH_2 \\ O \end{array}}$$

$\xleftarrow{-H_2O}$

$\downarrow \begin{array}{c} H_2NCH_2COOH \\ -H_2O \end{array}$

$$\underset{\textbf{4}}{H_2N-CH_2-\overset{\displaystyle O}{\overset{\|}{C}}-\overset{\displaystyle H}{N}-CH_2-\overset{\displaystyle O}{\overset{\|}{C}}-\overset{\displaystyle H}{N}-CH_2COOH}$$

Cyclization can yield dimers, trimers, or higher cyclic oligomers. It can occur as a side reaction in the synthesis of polyamides or polyesters. As discussed later in Chapter 10, the ease of cyclization depends on the skeletal bond angles, bond lengths, side-group steric effects, the torsional mobility of the bonds, and ring strain. In general, once the degree of polymerization has exceeded that needed to generate a 12- or 15-membered ring, the probability of cyclization declines. Linear propagation can be facilitated relative to cyclization by the use of high concentrations of the reactants. This effect is a consequence of the fact that linear propagation is a bimolecular process, whereas cyclization is unimolecular.

SPECIFIC CONDENSATION POLYMERIZATIONS

Linear Polyesters

A representative *linear polyester*, poly(ethylene terephthalate), will be discussed here. Poly(ethylene terephthalate) (**7**) can be prepared by a two-step ester interchange reaction between the dimethyl ester of terephthalic acid (**5**) and ethylene

$$CH_3O-\overset{\displaystyle O}{\overset{\|}{C}}-\underset{\textbf{5}}{\bigcirc}-\overset{\displaystyle O}{\overset{\|}{C}}-OCH_3 + \underset{\textbf{6}}{2\,HOCH_2CH_2OH}$$

$$\xrightarrow[①]{-CH_3OH} HOCH_2CH_2O-\overset{\displaystyle O}{\overset{\|}{C}}-\bigcirc-\overset{\displaystyle O}{\overset{\|}{C}}-OCH_2CH_2OH$$

$$\xrightarrow[②]{-HOCH_2CH_2OH} \underset{\textbf{7}}{\left[OCH_2CH_2O-\overset{\displaystyle O}{\overset{\|}{C}}-\bigcirc-\overset{\displaystyle O}{\overset{\|}{C}} \right]_n}$$

glycol (**6**). The polymer can be prepared in the laboratory from a $1:2.4$ molar ratio of dimethyl terephthalate to ethylene glycol.[1] The mixture is heated at temperatures up to 197°C for about 3 h in the presence of condensation catalysts such as calcium acetate and antimony trioxide. During this stage, methanol is evolved. The second step takes place at about 283°C with evolution of ethylene glycol, facilitated by a reduction of the pressure or by a stream of inert gas. The stoichiometric balance between the two monomers is achieved during this step. Failure to remove the last trace of excess ethylene glycol would yield a low-molecular-weight product. This phase (the true polymerization) requires about 3 h of reaction. The product is a tough, flexible polymer that can be melt-pressed into films or drawn from the melt into strong, orientable fibers. The polymer melts at 260 to 270°C.

Poly(ethylene terephthalate) is only one representative of a substantial class of known polyesters. Some are crystalline, others are amorphous. For the crystalline polymers, the melting temperature is often lowered by the asymmetric introduction of substituent groups into the aromatic ring. Methyl groups attached to the diol residue may prevent crystallization.

Branched and Crosslinked Polyesters

The formation of a linear polyester from *di*functional reagents has just been described. However, if one of the reagents is a tri- or multifunctional species, polymerization will generate a branched polymer. For example, if glycerol (**8**) is allowed to react with a diacid or its anhydride, each glycerol residue will generate one branch point (structure **9**, where G = glycerol residues and A = acid residues).

$$
\begin{array}{c}
CH_2OH \\
|\\
CHOH \\
|\\
CH_2OH
\end{array}
$$

8

9

Such molecules can grow to very high molecular weights and an infinitely large polymer network will be formed. If internal coupling occurs (reaction of a hydroxyl group and an acid function from branches of the same or different molecule), the polymer will become crosslinked. In practice, extensive branching and crosslinking cause "gelation" of the polymer. In this state the polymer is swellable by solvents, but it does not dissolve. Highly crosslinked polymers are totally unaffected by solvents.

Clearly, the degree of branching or crosslinking in a polyester system can be controlled by the amount of triol added relative to diol.

[1] W. R. Sorenson and T. W. Campbell, *Preparative Methods of Polymer Chemistry*, 2nd ed. (New York: Wiley–Interscience, 1968), p. 131.

Polycarbonates

Polycarbonates are polymers with the general structure shown in **10**. They are polyesters derived from carbonic acid, $(HO)_2C=O$. Of particular importance are the polycarbonates derived from 2,2-bis(4-hydroxyphenyl)propane (**11**), also

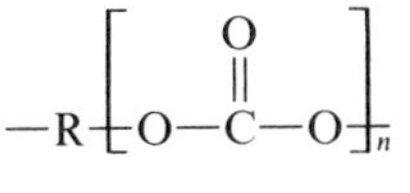

10

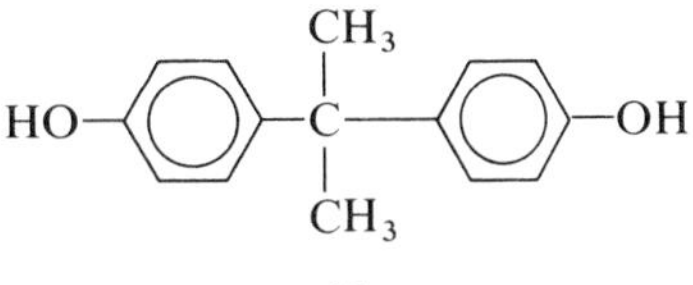

11

known as Bisphenol A. This compound is synthesized on a large scale by the condensation of phenol with acetone. Two convenient reaction routes are available for the preparation of polycarbonates from this monomer. In the first, an ester exchange reaction is performed between molten Bisphenol A and an organic carbonate, such as diphenyl carbonate (reaction 9).

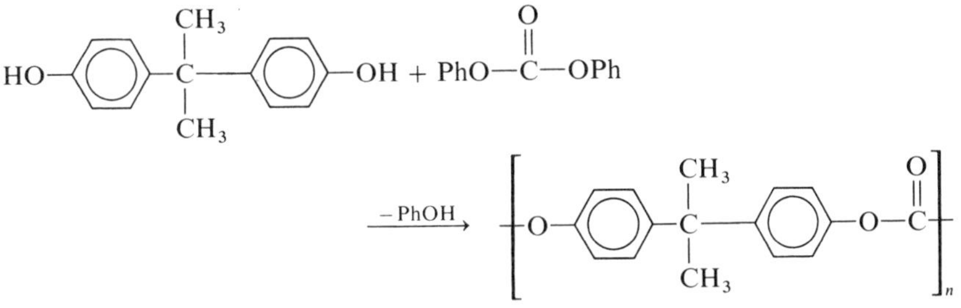

(9)

Initially, a prepolymer is formed by heating of the mixture at 180 to 220°C in vacuum for 1 to 3 h. The temperature is then raised slowly to 280 to 300°C and the pressure is lowered to remove the last traces of phenol. The residual molten polymer is highly viscous. It sets to a transparent solid when cooled. This can be melt-pressed or solution-cast (see Chapter 20) to fabricate tough films.

A second synthetic route to polycarbonates involves a reaction of Bisphenol A with phosgene (10). This can be accomplished[1] by bubbling phosgene into a

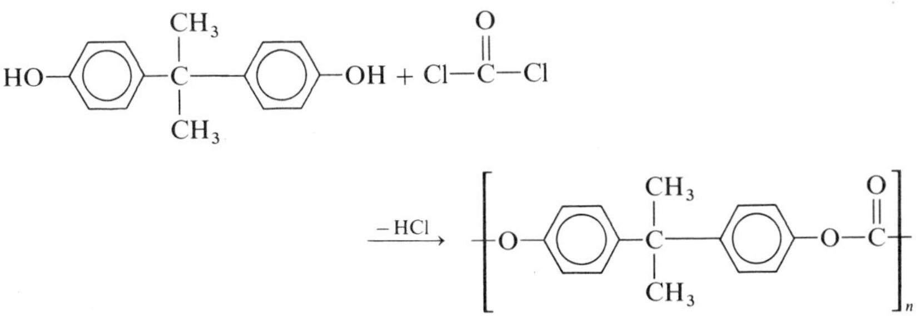

(10)

[1] With suitable precautions. Phosgene is a highly toxic gas.

solution of Bisphenol A in pyridine at 25 to 30°C. The pyridine functions as a hydrogen chloride acceptor. The polymer can be isolated by precipitation in water or methanol. An alternative procedure makes use of a heterophase emulsion system. Phosgene is passed into a rapidly stirred emulsion of Bisphenol A in methylene chloride, aqueous sodium hydroxide, and quaternary ammonium halide catalyst. Recovery of the polymer is effected by separation of the organic phase and evaporation of the methylene chloride.

Polyanhydrides

Polyanhydrides can be prepared by the reaction of a dicarboxylic acid with an excess of acetic anhydride or acetyl chloride. With acetic anhydride, the overall reaction is shown in (11). For example, sebacic acid, $HOOC-(CH_2)_8-COOH$, can be

$$
\underset{\substack{\| \\ O}}{HO-C}-R-\underset{\substack{\| \\ O}}{C}-OH + CH_3-\underset{\substack{\| \\ O}}{C}-O-\underset{\substack{\| \\ O}}{C}-CH_3 \tag{11}
$$

$$
\xrightarrow{-CH_3\overset{O}{\underset{\|}{C}}-OH} \left(\underset{\substack{\| \\ O}}{C}-R-\underset{\substack{\| \\ O}}{C}-O \right)_n
$$

polymerized by heating in boiling acetic anhydride for 6 h. The acetic acid liberated is distilled continuously from the reaction mixture. The product is of only moderate molecular weight, but heating at temperatures up to 200°C in vacuum brings about a removal of the crystalline cyclic dimer. The residue is the high-molecular-weight, so-called ω-anhydride. This polymer can be melt-drawn to yield strong fibers. Unfortunately, like most other aliphatic polyanhydrides, the polymer hydrolyzes rapidly in contact with atmospheric moisture.

Aromatic polyanhydrides are much more stable to moisture and they have higher crystalline melting points. They are synthesized by heating mixed aliphatic–aromatic anhydrides, as shown in the general reaction (12). It is speculated that

$$
CH_3\overset{O}{\underset{\|}{C}}-O-\overset{O}{\underset{\|}{C}}-\langle\bigcirc\rangle-R-\langle\bigcirc\rangle-\overset{O}{\underset{\|}{C}}-O-\overset{O}{\underset{\|}{C}}CH_3 \tag{12}
$$

$$
\xrightarrow[\displaystyle \left(CH_3\overset{O}{\underset{\|}{C}} \right)_2 O]{\text{Vacuum}} \left[\overset{O}{\underset{\|}{C}}-\langle\bigcirc\rangle-R-\langle\bigcirc\rangle-\overset{O}{\underset{\|}{C}}-O \right]_n
$$

the hydrolytic stability of such polymers is partly a consequence of the high degree of crystallinity, since the amorphous modifications hydrolyze more rapidly.

Polyamides

General features

Polyamides or "nylons" were among the first synthetic high polymers to be made and used on a large scale. Their discovery by Carothers and coworkers in 1935 served to usher in the new era of synthetic polymers, which has altered technology and everyday life almost beyond recognition.

Four principal methods are available for the synthesis of high-molecular-weight polyamides: (1) the reaction between a dicarboxylic acid and a diamine, (2) the dehydration–condensation of an amino acid, (3) the reaction between a diacid chloride and a diamine, and (4) the ring-opening polymerization of cyclic amides. This fourth route is discussed in Chapter 6. The first three routes will be considered here.

Melt polymerization

The direct interaction between a dicarboxylic acid and a diamine is the classical method for the synthesis of polyamides. In practice, it is preferable to ensure the existence of a 1:1 ratio of the two reactants by the prior isolation of a 1:1 salt of the two. The overall procedure is summarized by the reaction scheme shown in (13).

$$
\begin{array}{c}
\underset{\substack{\text{HO-C-R-C-OH}}}{\overset{\substack{O \quad\quad O \\ \| \quad\quad \|}}{}} + H_2N-R'-NH_2 \longrightarrow
\end{array}
$$

$$
\begin{array}{c}
\overset{\displaystyle O \quad\quad O}{\underset{\displaystyle \ominus O-C-R-C-O^\ominus}{\| \quad\quad \|}} \\[4pt]
\underset{\oplus \quad\quad\quad \oplus}{H_3N-R'-NH_3} \\[4pt]
\text{Salt}
\end{array}
\tag{13}
$$

$$
\xrightarrow[\text{Heat}]{-H_2O} \quad
\left[\begin{matrix} O & & O & H & & H \\ \| & & \| & | & & | \\ C-R & - & C-N & - & R'-N \end{matrix} \right]_n
$$

The actual polymerization process is known as a *melt polymerization* because the reaction takes place above the melting points of both the reactants and the polymer. This type of reaction should be distinguished from "interfacial polymerizations," to be discussed later.

Probably the best known melt polymerization involves the reaction between hexamethylenediamine, $H_2N(CH_2)_6NH_2$, and adipic acid, $HOOC(CH_2)_4-COOH$, to yield Nylon 66, $\{NH-(CH_2)_6-NH-CO-(CH_2)_4-CO\}_n$. The numerals in the trivial name refer to the numbers of carbon atoms in the two monomers. The first number gives the number of carbon atoms in the diamine. The following description outlines the procedure that can be followed on a laboratory scale.[1]

A 1:1 salt is first prepared by the addition of a warm solution of hexamethylenediamine in dry ethanol to an ethanolic solution of an equimolar amount of

---

[1] W. R. Sorenson and T. W. Campbell, *Preparative Methods of Polymer Chemistry*, 2nd ed. (New York: Wiley–Interscience, 1968), p. 74.

adipic acid, followed by cooling to room temperature. The white, crystalline salt melts at about 196 to 197°C. The salt is then introduced into a polymerization tube. Air is displaced from the tube by nitrogen and the tube is sealed. The next step requires that the tube be heated at 215°C for 1.5 to 2 h. However, during this step, the tube is a *potential bomb*. The high internal pressures could cause it to explode violently. Hence, this reaction should never be carried out unless the tube is shielded by an open steel tube container and a steel explosion shield. Even when cooled after this stage of the reaction, it should be handled only behind shielding and with a leather or asbestos glove. Failure to observe such precautions could have disastrous consequences, and, for this reason, only those who already have substantial synthetic experience in the laboratory should attempt this reaction. After the initial step, the tube is opened (again with adequate shielding), and the contents are heated to 270°C first in a nitrogen atmosphere and later in a vacuum for about 1h. Even while the tube is cooling, it should be shielded to prevent shattered glass from being hurled out as the tube cracks. On a large scale, the reaction is carried out in an autoclave. The tough, white polymer can be melt-cast into films or melt-drawn into fibers. It has a melting point of 267°C, and it is soluble in formic acid, phenols, and cresols.

Polymerization of amino acids

The polymerization of amino acids, such as 11-aminoundecanoic acid, $H_2N-(CH_2)_{10}-COOH$, can be accomplished by melt polymerization techniques at 220°C. However, an alternative procedure often used to prepare "model" polypeptides involves the use of dicyclohexylcarbodiimide (**12**) as a dehydrating agent. In this case, the polymerizations are carried out at moderate temperatures.

Poly(glutamic acid) (**13**) is prepared by the sequence shown in reaction (14).

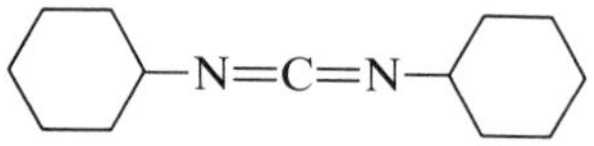

12

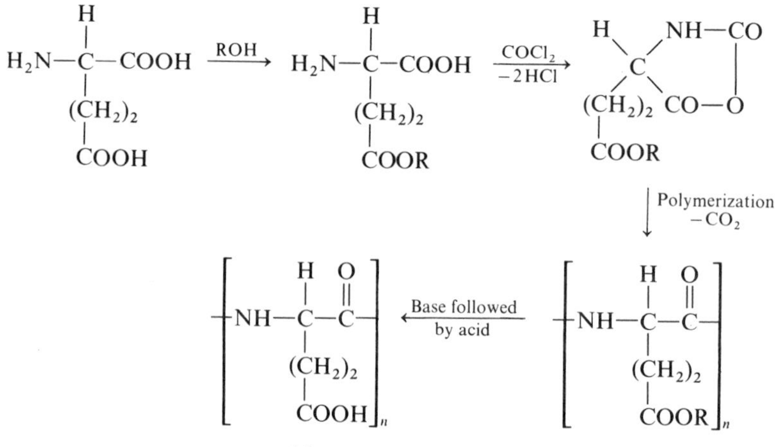

(14)

13

This polymer can be solution-spun to yield silk-like fibers. These are susceptible to decomposition by microorganisms.

Interfacial polymerization

The process of interfacial polymerization can best be illustrated by the reaction between a diamine and a diacid chloride. An interfacial polymerization takes place when the two monomers are present in two immiscible solvents. Reaction then occurs at the interface between the two liquids. In practice, the procedure can be carried out effectively only if the polymerization reaction is rapid at moderate temperatures. However, it provides a valuable method for the synthesis of polymers at temperatures far below those required for melt polymerization. The polymer molecular weights obtained by the interfacial procedure are generally higher than those obtained by the melt method. Monomer molecules tend to react more readily with growing polymer molecules than with other monomer molecules because the reaction is too rapid to allow the monomer to diffuse through the layer of polymer. Hence, rigorous step-type polymerization is not maintained. Furthermore, an exact 1:1 balance of the two monomers is not required.

Two examples will serve to illustrate the technique. First, ethylenediamine (14) can be allowed to react with terephthaloyl chloride (15) to yield poly(ethylene terephthalamide) (16) (reaction 15). The diamine is dissolved in a solution of

$$H_2N(CH_2)_2NH_2 + Cl-\overset{\overset{O}{\|}}{C}-\langle\bigcirc\rangle-\overset{\overset{O}{\|}}{C}-Cl$$

$$\begin{matrix} \textbf{14} & \textbf{15} \end{matrix} \qquad\qquad (15)$$

$$\xrightarrow[\text{Base}]{-\,HCl} \quad \left[\overset{\overset{H}{|}}{N}-(CH_2)_2-\overset{\overset{H}{|}}{N}-\overset{\overset{O}{\|}}{C}-\langle\bigcirc\rangle-\overset{\overset{O}{\|}}{C}\right]_n$$

$$\textbf{16}$$

potassium hydroxide (the hydrochloride acceptor) in water, and the diacid chloride is dissolved in methylene chloride. The two layers are emulsified in a kitchen blender or by means of a high-speed laboratory stirrer at room temperature for 10 min. The polymer is then simply removed by filtration, washed, and dried.

A second method, often used in lecture demonstrations, makes use of a combined polymerization and fiber drawing technique. The preparation of poly(hexamethyl-sebacamide) (Nylon 610) by this method is a good example. A solution of sebacoyl chloride, $Cl-C(O)-(CH_2)_8-C(O)-Cl$, in tetrachloroethylene is placed in the bottom of a beaker. Over it is carefully poured a solution of hexamethylenediamine in water. Polymer forms at the interface. The polymer may then be grasped with tweezers or tongs and raised from the beaker. This causes a continuous filament of polymer to be drawn from the mixture. Mechanical devices are sometimes used (Figure 2.2) to pull a substantial length of monofilament from the reaction.

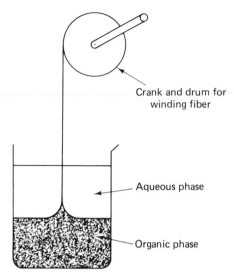

Crank and drum for
winding fiber

Aqueous phase

Organic phase

Fig. 2.2 Monofilament fiber being drawn directly from an interfacial polymerization experiment.

Polyimides

Polyimides are polymers formed by the condensation reactions of dianhydrides with diamines. A typical structure is shown in **18** (reaction 16). Polyimides have been synthesized from aromatic anhydrides and aliphatic diamines, and from

$$-H_2O \rightarrow$$

(16)

17

$$-H_2O \rightarrow$$

18

aromatic anhydrides and aromatic diamines. The latter products are called *aromatic polyimides*. Aromatic polyimides are often infusible and insoluble. Hence, precipitation occurs during the intermediate stages of polymerization. To circumvent this problem, the condensation is performed in two stages. First, a noncyclized, soluble, linear high polymer (**17**) is generated by a fast reaction in a suitable polar solvent at temperatures below 70°C. Polymer **17** can be fabricated into a suitable shape and then cyclized to **18** by heating at temperatures up to 300°C. Such polymers are rigid, high melting, thermally stable materials. Some flexibility can be introduced into the polymer structure by the use of diamines which contain a flexible linkage unit. An example is 4,4′-diaminodiphenyl ether, $H_2N—C_6H_4—O—C_6H_4—NH_2$.

Polyimides are used for the fabrication of machined devices and in laminates for high-temperature applications.

Polybenzimidazoles[1]

Polybenzimidazoles are polymers formed by the condensation of dicarboxylic acids with aromatic tetramines (17). An example is the polymer obtained by the condensation of a dicarboxylic acid with 3,3′-diaminobenzidine (**19**). Again a two-stage reaction is preferred, although the reaction between sebacic acid,

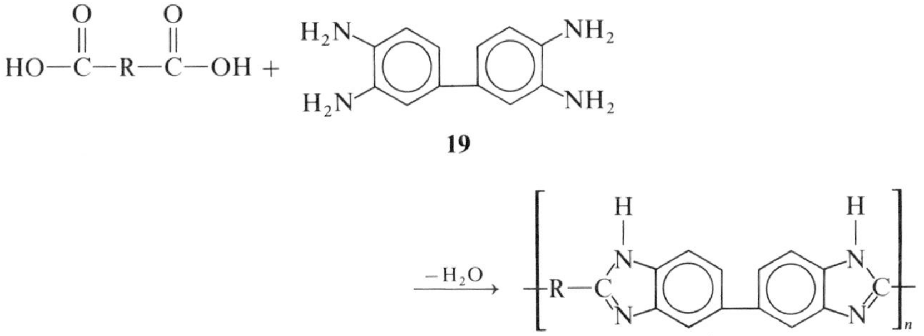

19

20

(17)

$HOOC(CH_2)_8COOH$, and 3,3′-diaminobenzidine (**19**) can apparently be carried out at 265°C for 3.5 h in a one-step process. Normally, however, the condensation is not complete until the polymer is heated to 350 to 400°C. Higher-molecular-weight, thermally stable polymers can often be obtained by the use of the diphenyl esters of aromatic diacids.

Polybenzimidazoles are used for the manufacture of heat-resistant components and in wire enamels and coatings for high-temperature uses.

Polyquinoxalines

Polyquinoxalines are aromatic polymers that are prepared by the condensation of aromatic diglyoxals with aromatic tetramines. A typical example is the reaction

[1] C. S. Marvel in *High Temperature Polymers*, C. L. Segal, ed. (New York: Dekker, 1967), p. 1.

shown in scheme (18). This type of polymer has very high thermal stability. Again it is preferable to carry out the reaction in two stages. The first phase comprises a solution polymerization in a solvent such as hexamethylphosphoramide. The second stage is a high-temperature pyrolysis at about 375°C in vacuum.

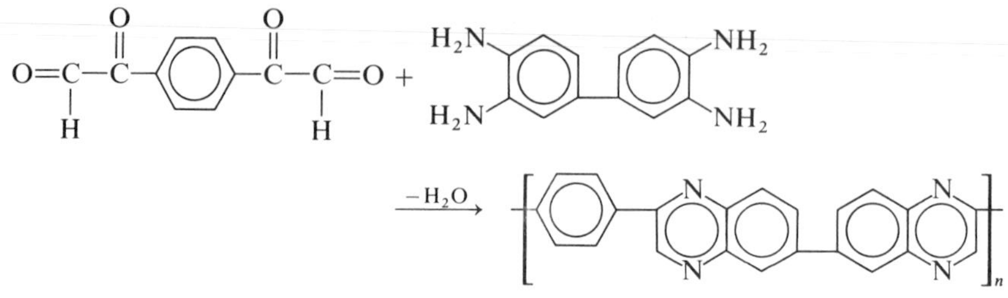

(18)

Aromatic Ladder Polymers

The continuing drive to synthesize organic polymers with very high thermal stabilities has led to the development of aromatic ladder polymers. Every backbone cleavage reaction in a single-strand polymer causes a decrease in chain length and an eventual deterioration of the polymer properties. In a ladder polymer *two* bonds must cleave in the same connecting residue before the chain can break. The likelihood that this will happen is low. Moreover, if one bond in a ladder polymer breaks, the two cleaved ends will be held in juxtaposition by the remaining intact strand and "healing" of the broken bond is likely to occur. Hence, aromatic ladder polymers are capable of withstanding very high temperatures. Typical aromatic ladder polymers are polybenzimidazoles of the type discussed above, and poly-imidazopyrrolones or polyquinoxalines with structures such as those shown in **21** and **22** (schemes 19 and 20). The condensation of quinones with aminophenols

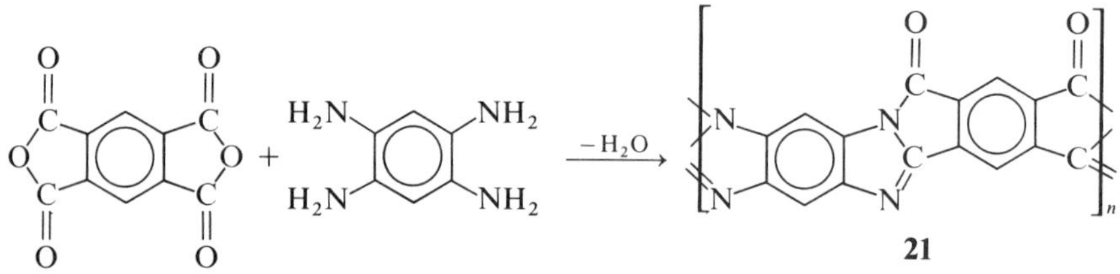

21

(19)

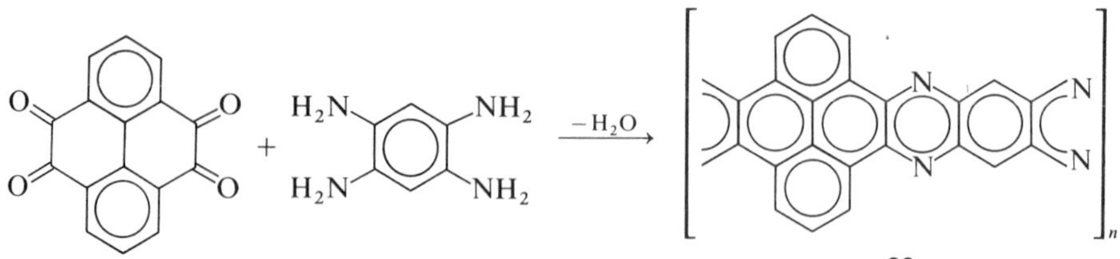

22

(20)

leads to the formation of polymers such as **23**, as shown in reaction (21). Similarly, the condensation of tetraketones or their precursors with tetramines leads to the formation of ladder polymers, such as **24** (reaction 22). A number of aromatic ladder polymers are attracting attention as "rigid rod" macromolecules because of their unusual rheological properties.

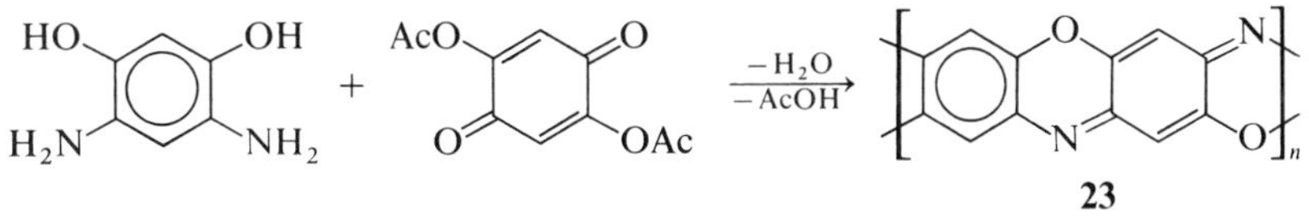

$$\text{(21)}$$

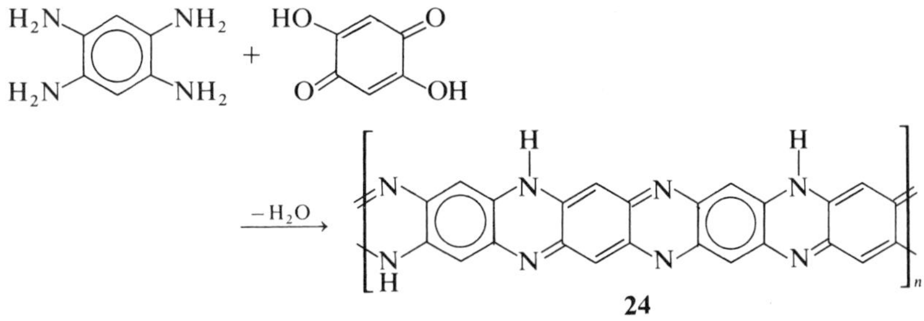

$$\text{(22)}$$

It should be noted that other ladder polymer systems have been synthesized from siloxane residues and by the pyrolysis of polyacrylonitrile. These reactions are discussed in Chapters 4 and 9, respectively.

Phenol–, Urea–, and Melamine–Formaldehyde Resins

The reactions involved in the formation of resins from the reaction of formaldehyde with phenol, urea, or melamine are complex. The initial steps in these processes are not condensations, but the later "curing" stages involve elimination of water. Hence, these reactions are considered here under the classification of condensation polymerizations.

Phenol–formaldehyde polymers

Resins based on the reactions of phenol and formaldehyde were known as early as 1872. During the first half of the twentieth century, in the form of "Bakelite resins," they provided one of the few synthetic polymeric materials available for fabrication into moldable, thermosetting objects. Three stages can be identified in the preparation of a network polymer from this system.

First, a phenol is allowed to react with formaldehyde to yield a mixture of methylolphenols (reaction 23). This reaction can be carried out under acidic or

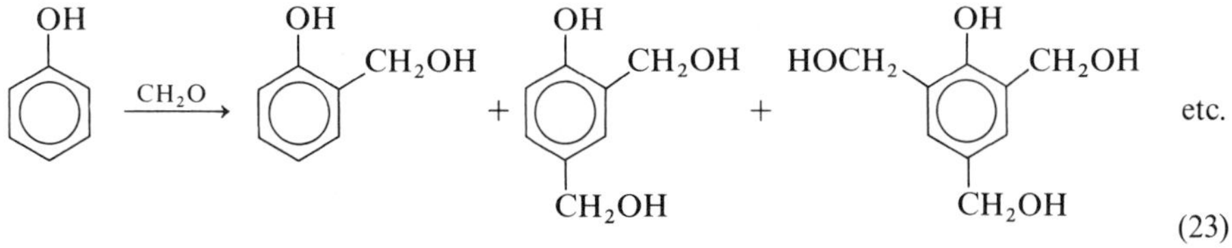

$$\text{(23)}$$

basic conditions. If acidic conditions are used, spontaneous condensation occurs to form low polymers. The products from the base-catalyzed reactions are soluble prepolymers called "resoles."

Heating of the prepolymer at temperatures up to about 105°C under neutral or acidic conditions brings about condensation to generate moldable cyclolinear or branched-type polymers by reactions such as (24). At higher temperatures, methylene bridges are established by reactions of the type shown in (25), which

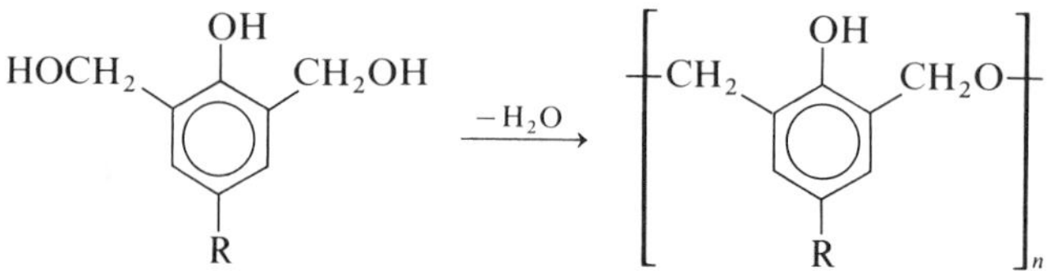

(24)

(25)

involve the elimination of water and formaldehyde. These structures are believed to predominate in the final crosslinked resin. However, the chemistry of these steps is much more complex than is implied by these reactions.

Novolac resins are phenol–formaldehyde polymers made from the acid-catalyzed low polymers discussed earlier. They may be crosslinked by the addition of hexamethylene–tetramine, $(CH_2)_6N_4$, with the concurrent elimination of ammonia.

Urea–formaldehyde resins

Formaldehyde reacts with amino-compounds to generate methylol derivatives. Condensation of the methylol units can then occur, often at moderate temperatures. Urea and melamine are amino-compounds that participate in these types of reactions.

Urea reacts with aqueous formaldehyde in alkaline media to yield mono-methylol– and dimethylol–urea derivatives (**25** and **26**) (reaction 26). Such products

$$H_2N-\overset{\overset{\displaystyle O}{\|}}{C}-NH_2$$

$$\xrightarrow{-CH_2O} \underset{\mathbf{25}}{H_2N-\overset{\overset{\displaystyle O}{\|}}{C}-NH-CH_2OH} + \underset{\mathbf{26}}{HOCH_2-NH-\overset{\overset{\displaystyle O}{\|}}{C}-NH-CH_2OH} \qquad (26)$$

can either be isolated as discrete compounds or the mixture can be used directly for resin formation. Water is removed from the mixture to form a syrup, which may then be acidified and heated near 100°C to bring about condensation and gelation.

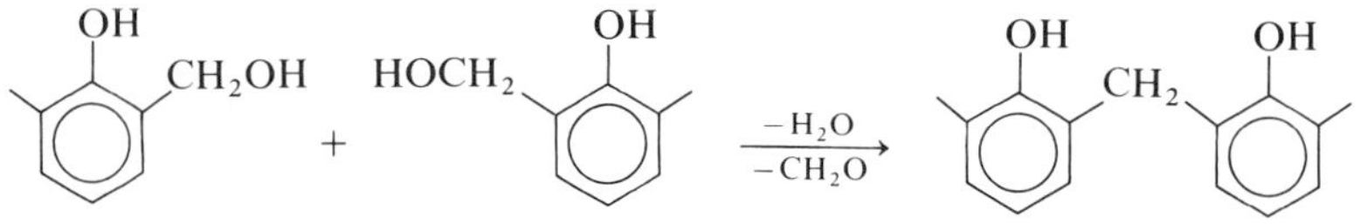

The exact crosslinking mechanism is uncertain, but condensation could occur between —CH_2OH and NH_2 groups, or between two CH_2OH groups, with liberation of water. Methylene bridges could also be formed by loss of both water and formaldehyde, and numerous other reactions have been postulated. Urea–formaldehyde resins are clear and colorless. They are harder than phenol–formaldehyde resins and they can be reinforced with fillers or can be colored for different uses.

Melamine–formaldehyde polymers

Melamine (triamino-*s*-triazine) (**27**) can also be methylolated (27) to yield a range of products which contain up to six methylol groups (**28**). These materials

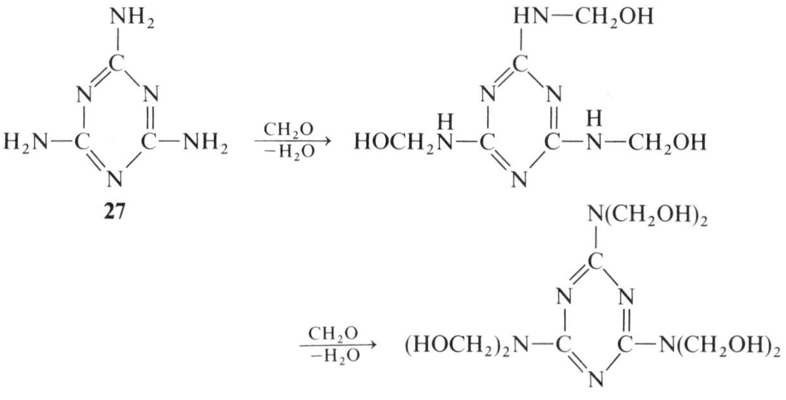

(27)

are soluble in water and can conveniently be applied to textile fabrics as aqueous solutions. Crosslinked resins are obtained at elevated temperatures via the formation of methylene or methylene–ether linkages. Fabrics containing cured methylol-melamine have "permanent press" characteristics. Bulk polymers made from molding powders constitute the well-known "melamine resins."

Polyacetals

Formaldehyde reacts with aliphatic diols to eliminate water and form polyethers known as *polyacetals*. The overall process is shown in reaction (28). This condensation process is, in reality, a carbonyl addition–substitution reaction which

$$CH_2{=}O + HO{-}R{-}OH \xrightarrow{-H_2O} {+}CH_2{-}O{-}R{-}O{+}_n \qquad (28)$$

takes place in the presence of acid catalysts. Long-chain diols are normally required to reduce the tendency for cyclization. High-melting crystalline poly-formals are produced when formaldehyde reacts with cyclic diols. It should be noted that the term "polyacetal" is also sometimes applied to aldehyde *addition* polymers (see Chapter 4).

OTHER STEP-TYPE POLYMERIZATIONS

Condensation reactions form only one class of step-type polymerizations. A number of other organic reaction processes have also been employed for the step synthesis of polymers, and a few examples are mentioned briefly in the following sections.

Polyethers by Aromatic Substitution

Although condensation reactions between aromatic diols can yield low-molecular-weight polyethers, high polymers may be prepared by nucleophilic substitution or oxidative coupling routes. The nucleophilic substitution route is used for the syntheses of the so-called *polysulfone* (**31**) from the disodium salt of Bisphenol A (**29**) and 4,4′-dichlorodiphenyl sulfone (**30**) (reaction 29). The sulfone group in **30**

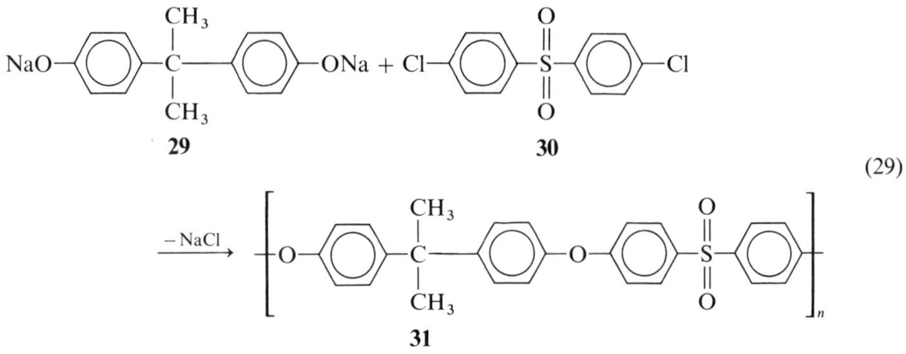

(29)

serves to activate the chlorine atoms to nucleophilic replacement in a way that would not be possible in a conventional chloroaromatic compound. Dimethyl sulfoxide is the preferred solvent for such reactions, and the interaction is carried out at 160°C. The polymer softens near 200°C. It can be cast or pressed into films at 280°C and fibers can be obtained by melt techniques. Polysulfones are used for electrical insulating applications and for the fabrication of heat-resistant articles.

Polyethers by Oxidative Coupling Reactions

Aromatic polyethers are also produced when oxygen is bubbled through solutions of 2,6-disubstituted phenols in an organic solvent. A catalyst complex, formed from a cuprous salt and a tertiary amine, is required for the reaction. The overall process is illustrated by the scheme shown in (30). Polymer **32** is known commercially as "poly(phenylene oxide)." It is used for the manufacture of machined parts, especially when heat stability is needed.

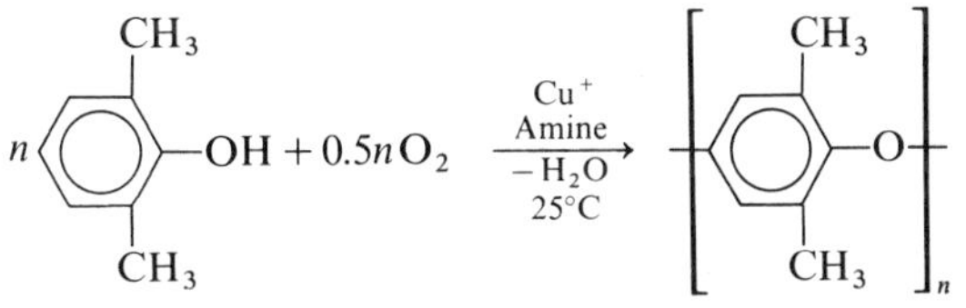

(30)

Diels–Alder Addition Polymers

The importance of ladder polymers has been discussed earlier in this chapter. Another attractive route to polymers of this type involves Diels–Alder-type additions between, for example, quinones and vinyl compounds, as shown in the formation of **33** in reaction (31). Such reactions occur rapidly at only moderate

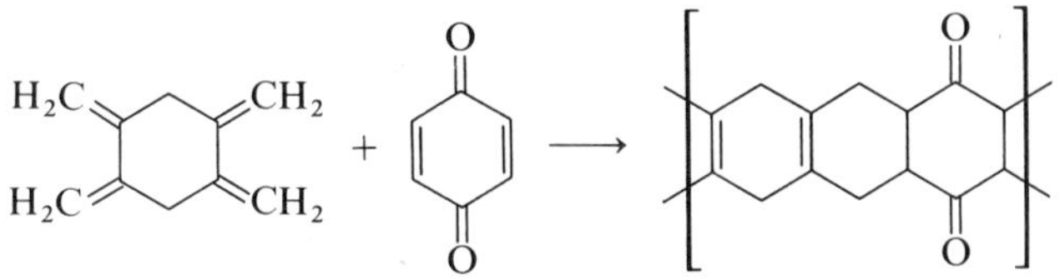

(31)

temperatures. However, side reactions often lead to the formation of structural irregularities.

Polyurethanes

Polyurethanes have the general structure shown in **34**. Two main methods of synthesis are available. First, polyurethanes can be formed by the reaction of bis-chloroformates (**35**) with diamines (reaction 32). Second, they can be synthesized

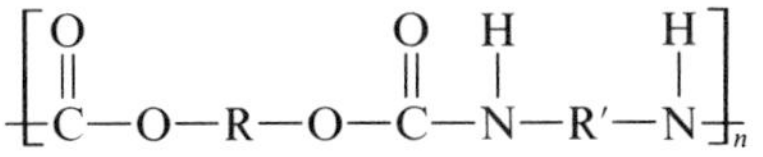

34

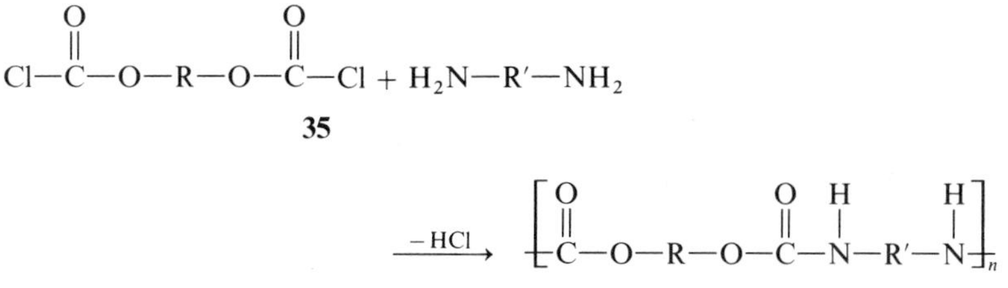

(32)

by the addition of a diol to a diisocyanate (reaction 33). The latter reactions are carried out in the melt or in solution.

$$\begin{matrix} O & & O \\ \| & & \| \\ C{=}N{-}R{-}N{=}C \end{matrix} + HO{-}R'{-}OH \longrightarrow \begin{bmatrix} O & H & & H & O \\ \| & | & & | & \| \\ C{-}N{-}R{-}N{-}C{-}O{-}R'{-}O \end{bmatrix}_n \tag{33}$$

Polyurethanes are widely used as elastic fibers, rubber foams, and as coatings. A typical elastic fiber material is made from 1,4-butanediol and hexamethylene diisocyanate. Elastomers are also synthesized from aromatic diisocyanates, such as 2,4-toluenediisocyanate and long-chain aliphatic diols. Crosslinking (vulcanization) of the elastomers can be effected by reaction with diols, diamines, dicarboxylic acids, and so on.

Polymers from Electrophilic Aromatic Substitutions

Experimental aromatic cyclolinear polymers have been prepared by Friedel–Crafts-type substitution reactions and related processes. For example, poly(arylene-alkylenes) with molecular weights as high as 12,000 can be obtained from the reaction of benzene with 1,2-dichloroethane in the presence of aluminum chloride. Insoluble poly(p-phenylenes) (36) can be isolated from the reaction of benzene with combined Lewis acid and oxidizing agent systems.

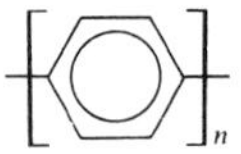

36

CONCLUSIONS

Several step-type systems have been excluded from this chapter because they involve compounds of the inorganic elements. These are more conveniently treated in Chapter 4. However, it will be clear from the multiplicity of reactions mentioned in this chapter that the synthesis of new polymers by condensation and other step-type reactions has been and probably will continue to be a very fertile field for pioneering synthetic research. The current trend is obvious—that future work in this field will almost certainly emphasize the synthesis of new polymers that are stable at high temperatures, that resist burning, or which have unusual electrical properties.

Because of the great variety of (mostly organic) step reactions that have been discovered, this subject is still dominated by descriptive reaction chemistry. Only the best known systems have been analyzed mechanistically and kinetically by detailed physical chemical techniques. This mechanistic aspect is considered in

Chapter 11. By contrast, vinyl-addition-type polymerizations involve the synthesis of many different polymers by the use of only one type of reaction—the linear addition of olefins. Hence, greater opportunities have existed in the olefin polymerization field for the analysis of detailed reaction mechanisms and for the application of physical techniques. This will be clear from the treatment of olefin polymerization in the following chapters.

STUDY QUESTIONS

1. Draw diagrams to compare the changes in molecular-weight distribution that occur with time for (a) a step-type polymerization, and (b) a chain-reaction polymerization.

2. For what reasons might you wish to deliberately limit the molecular weight of a condensation polymerization product by the addition of an excess of, say, one of the reactant species?

3. Draw up a list of different dicarboxylic acids and diols that might be used in polyester formation and attempt to predict advantages or potential problems for each pairwise combination. What property differences do you predict for the final polymers?

4. Give possible reasons why Bisphenol A is a widely used monomer in step-type polymerization syntheses.

5. If branching and crosslinking in a polyester can be induced by the addition of glycerol, what specific reagents might be employed to achieve a similar result in polyamide or polyether synthesis?

6. Suggest a mechanism for the oxidative coupling of phenols to yield polyaromatic ethers. Would phenols that contain more than two methyl groups attached to the phenyl ring be expected to react faster or slower than those with two methyl groups or less?

7. Draw up a list of molecular structural requirements that would be needed for an organic polymer that could withstand temperatures above 350°C.

8. Compile a list of amino compounds other than melamine or urea that might be used for methylolation reactions and subsequent conversion to resins. Speculate on the advantages and disadvantages of each compound for this particular use.

9. Glance through an organic chemistry textbook [e.g., J. D. Roberts and M. C. Caserio, *Basic Principles of Organic Chemistry* (New York: Benjamin, 1964)] and identify any general reactions that are accessible for small molecule compounds but are not mentioned here as step-type polymerization routes. Suggest possible advantages or problems that might be found if these reactions could be used as polymerization processes.

SUGGESTIONS FOR FURTHER READING

BRAUNSTEINER, E. E., AND H. F. MARK, "Aromatic Polymers," *J. Polymer Sci. (D) (Macromol. Rev.)*, **9**, 83 (1974).

CRITCHLEY, J. P., "A Review of the Poly(azoles)," *Progr. Polymer Sci.* (A. D. Jenkins, ed.), **2**, 1 (1970).

D'ALELIO, G. F., and R. K. SCHOENIG, "The Synthesis of Thermally Stable Polymeric Azomethines by Polycondensation Reactions," *J. Macromol. Sci—Rev. Macromol. Chem.* **C3**, 105 (1969).

DE WINTER, W., "Double Strand Polymers," *J. Macromol. Sci—Rev. Macromol. Chem.*, **1**, 329 (1966).

GOODMAN, I., AND J. A. RHYS. *Polyesters*, New York: Elsevier, 1975.

HAY, A. S., "Aromatic Polyethers," *Adv. Polymer Sci.*, **4**, 496 (1967).

HERGENROTHER, P. M., "Linear Polyquinoxalines," *J. Macromol. Sci—Rev. Macromol. Chem.*, **C6**, 1 (1971).

KORSHAK, V. V., AND M. M. TEPLYAKOV, "Synthesis Methods and Properties of Polyazoles," *J. Macromol. Chem—Rev. Macromol. Chem.*, **C5**, 409 (1971).

LENZ, R. W., *Organic Chemistry of Synthetic High Polymers.* New York: Wiley–Interscience, 1967.

LIVINGSTON, H. K., M. S. SIOSHANS, AND M. D. GLICK, "Nylons—Known and Unknown: A Comprehensive Index of Linear Aliphatic Polyamides of Regular Structure," *J. Macromol. Sci.—Rev. Macromol. Chem.*, **C6**, 29 (1971).

LYMAN, D. L., "Polyurethanes," *Rev. Macromol. Chem.*, **1**, 191 (1966).

MARVEL, C. S., "Thermally Stable Polymers," *Pure Appl. Chem.*, **16**, 351 (1968).

MARVEL, C. S., "Thermally Stable Polymers with Aromatic Recurring Units," *Soc. Plastics Eng. J.*, **20**(3), 220 (1964).

MARVEL, C. S., "Trends in High Temperature Polymer Synthesis," *Macromol. Chem.*, **C13**(2), 219 (1975).

MORGAN, P. W., *Condensation Polymers by Interfacial and Solution Methods*, (Polymer Reviews, Vol. 10). New York: Wiley–Interscience, 1965.

NARTISISSOV, B., "Surveys on Heat Resistant Polymers: Pyrrones," *J. Macromol. Sci—Rev. Macromol. Chem.*, **C11**, 143 (1974).

NOREN, G. K., AND J. K. STILLE, "Polyphenylenes," *J. Polymer Sci. (D) (Macromol. Rev.)*, **5**, 385 (1971).

OVERBERGER, C. G., AND J. A. MOORE, "Ladder Polymers," *Adv. Polymer Science*, **7**, 113 (1970).

SCHNELL, H., *Chemistry and Physics of Polycarbonates* (Polymer Reviews, Vol. 9). New York: Wiley–Interscience, 1964.

SORENSON, W. R., AND T. W. CAMPBELL, *Preparative Methods in Polymer Chemistry* (2nd ed.). New York: Wiley–Interscience, 1968.

SPEIGHT, J. G., P. KOVACIC AND F. W. KOCH, "Synthesis and Properties of Polyphenyls and Polyphenylenes," *J. Macromol. Sci.—Rev. Macromol. Chem.*, **C5**, 295 (1971).

SROOG, C. E., "Polyamides," *J. Polymer Sci. (D) (Macromol. Rev.)*, **11**, 161 (1976).

STILLE, J. K., "Diels-Alder Polymerization," *Fortschr. Hochpolym.-Forsch.*, **3**, 48 (1961).

STILLE, J. K., "Cycloaddition Polymerization," *Makromol. Chem.*, **154**, 49 (1972).

STILLE, J. K., "Polycondensation Synthesis of High Temperature Macromolecules," *Macromol. Chem.*, **8**, 373 (1973).

STILLE, J. K., "The Synthesis of Rigid Chain Polymers," *Proc. Intern. Symp. Macromolecules, Rio de Janeiro* (E. B. Mano, ed.), p. 95. New York: Elsevier, 1975.

3

Free-Radical Polymerization

ADDITION REACTIONS

A great many synthetic polymers are prepared by the polyaddition reactions of unsaturated organic compounds. In general terms, such polymerizations can be described by the scheme shown in reaction (1). Such reactions can be induced either

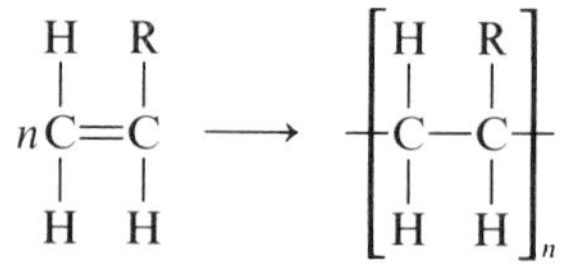

$$
n \begin{array}{c} \text{H} \quad \text{R} \\ | \quad | \\ \text{C}=\text{C} \\ | \quad | \\ \text{H} \quad \text{H} \end{array} \longrightarrow \begin{bmatrix} \text{H} \quad \text{R} \\ | \quad | \\ \text{C}-\text{C} \\ | \quad | \\ \text{H} \quad \text{H} \end{bmatrix}_n \tag{1}
$$

by the addition of free-radical-forming reagents or by ionic initiators such as acids or organometallic species. Ionic polymerizations are discussed in Chapter 4. Here we will concentrate on the use of free-radical polymerization processes.

FREE-RADICAL ADDITION REACTIONS

In a free-radical addition polymerization, the growing chain end bears an unpaired electron (1). Addition of each monomer molecule to the chain end involves an

1

attack by the radical site on the unsaturated monomer. Thus, the unpaired electron is transferred to the new chain end at each addition step.

Such polymerizations are very common. In fact, many olefins will undergo "spontaneous" polymerization reactions during storage. In many cases, a major problem is to *prevent* the polymerization of an olefin. This is usually accomplished by the addition of an "inhibitor" to the system. The inhibitor stabilizes the olefin until such time as the laboratory experiment or the manufacturing process can be effected. Free-radical polymerizations can be carried out in the bulk liquid phase (without a solvent), or in solution.

Free-radical polymerization reactions are of enormous importance in technology. The monomers for these reactions are available in large quantities from the petrochemical industry (e.g., from reaction sequences that start from ethylene, acetylene, or acetone), and the polymers obtained from these monomers form the foundation of much of the polymer industry. Low-density polyethylene, poly(methyl methacrylate), polystyrene, polyacrylonitrile, poly(vinyl chloride), and many other commercially important polymers are manufactured by free-radical processes.

Free-radical polymerizations are, on the whole, much better understood in a fundamental, mechanistic sense than are most of the step-type polymerizations discussed in Chapter 2. This is, in large measure, a consequence of the fact that different free-radical polymerization reactions constitute minor variants of one specific reaction type. Hence, subtle comparisons can be made of the effects of side-group changes, initiator variations, solvent effects, and so on. For this reason, a more detailed mechanistic approach will be taken in this chapter and in Chapter 12.

INITIATORS FOR FREE-RADICAL POLYMERIZATION

Table 3.1 summarizes the types of reagents that can induce the free-radical polymerization of vinyl compounds. The first four categories in Table 3.1 include

TABLE 3.1 Free-Radical Initiators

Initiator Type	Formula or Example
1. Organic peroxides or hydroperoxides	Benzoyl peroxide, $Ph\overset{O}{\overset{\|}{C}}OO\overset{O}{\overset{\|}{C}}Ph$
2. Redox agents	Persulfates + reducing agents, hydroperoxides + ferrous ion
3. Azo compounds	Azobisisobutyronitrile, $Me_2C(CN)N{=}NC(CN)Me_2$
4. Organometallic reagents	Silver alkyls
5. Heat, light, ultraviolet-, or high-energy radiation	
6. Electrolytic electron transfer	

compounds that dissociate into free radicals when heated or subjected to radiation. Categories 5 and 6 include those *physical* influences that generate free radicals from the monomer itself or the solvent. In general, these initiator systems can be handled without the rigorous removal of atmospheric moisture. (This contrasts with most ionic initiators—see Chapter 4.) However, atmospheric oxygen must be excluded. Some care must be exercised in the use of organic peroxides because a number of these reagents detonate when subjected to shock or high temperatures. Benzoyl peroxide is commonly used because it is among the least shock sensitive of these reagents. The role of the initiator is discussed in more detail in a later section.

MONOMERS FOR FREE-RADICAL POLYMERIZATION

A wide variety of unsaturated organic compounds can be induced to undergo free-radical polymerization. Some of these are listed in Table 3.2. In general, the monomer structure can be represented by the formula CH_2=CHR, where the group R is

TABLE 3.2 Monomers for Free-Radical Polymerization

Styrene	CH_2=CHPh	Ethylene	CH_2=CH_2
α-Methylstyrene	CH_2=C(Me)Ph	Vinyl chloride	CH_2=CHCl
1:3-Butadiene	CH_2=CH—CH=CH_2	Vinylidene chloride	CH_2=CCl_2
Methyl methacrylate	CH_2=C(Me)COOMe	Tetrafluoroethylene	CF_2=CF_2
Vinyl esters	CH_2=CHOOCR	Acrylonitrile	CH_2=CHC≡N
		Acrylamide	CH_2=CHCOONH$_2$
N-Vinyl pyrrolidone			

an organic unit, a halogen or pseudo-halogen ligand (C≡N), or even an inorganic residue. A few suitable monomers have the formula CH_2=C(R)(R'), in which two substituent groups are attached to the α-carbon atom. Many of these monomers contain electron-withdrawing substituent groups, although the electron-directing effect of the substituent has a less critical influence on a free-radical polymerization than it does on ionic-type polymerizations.

SOLVENTS AND SYSTEMS

Free-radical polymerizations are often carried out in the bulk phase. In such cases, the monomer itself functions as a solvent in the initial stages of polymerization. This method is useful for the direct polymerization to polymer castings. However, if the polymerization is very exothermic, the reaction may become violent, or bubbles or char may result from the local release of heat.

Polymerizations in solution serve to minimize these problems. However, many organic solvents function as chain transfer agents[1] for free-radical reactions and the polymer molecular weights may be lowered accordingly. Water is a satisfactory solvent for the polymerization of monomers such as acrylonitrile or acrylamide that are soluble or partly soluble in this medium, especially when water-soluble persulfate initiators are used.

Water may also be used as a reaction medium if the monomer is insoluble, since the monomer can be suspended as fine droplets in a stirred aqueous-organic heterophase medium. Alternatively, a detergent may be added to the system to generate an emulsion of the monomer in water. A water-soluble initiator is used and this penetrates the emulsion particles to initiate polymerization.

TYPICAL EXPERIMENTAL PROCEDURES

Thermal Polymerization of Styrene (Reaction 2)

Commercial styrene (2) must first be freed from trace impurities and inhibitors. This can be accomplished by vacuum distillation or by passage of the liquid monomer through an alumina chromatography column. Once purified, the

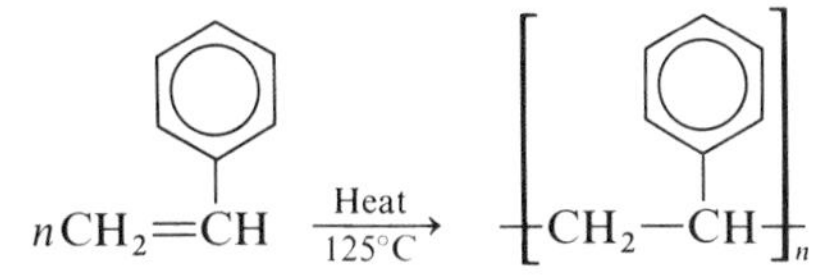

$$n\,CH_2{=}CH \xrightarrow[125°C]{Heat} \left[CH_2{-}CH\right]_n \qquad (2)$$

$$\text{2} \qquad\qquad\qquad \text{3}$$

monomer is sealed under nitrogen in a glass polymerization tube and heated at 125°C for 1 to 7 days. The tube contents become progressively more viscous as polymerization continues. The product (3) can be purified from monomer and oligomers by dissolving it in benzene, followed by precipitation of the polymer in methanol. The product made in this way often has a molecular weight near 150,000.

Bulk Polymerization of Methyl Methacrylate Initiated by AIBN (Reaction 3)

Freshly distilled methyl methacrylate monomer (4) is treated with a small amount of azobisisobutyronitrile (in an approximately 200:1 monomer/initiator weight

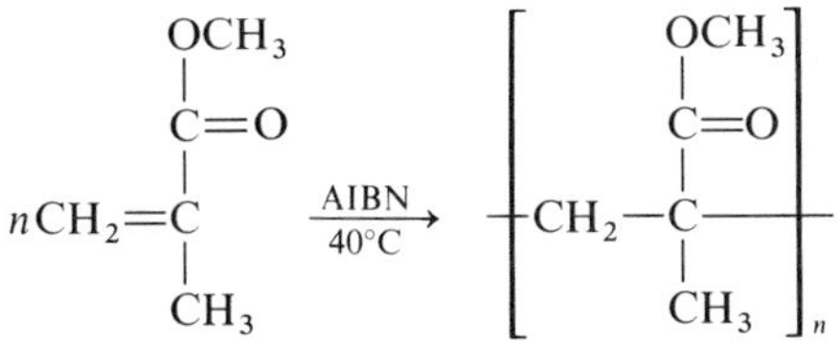

$$n\,CH_2{=}\underset{\underset{CH_3}{|}}{\overset{\overset{OCH_3}{|}\;\;\;\overset{|}{C}{=}O}{C}} \xrightarrow[40°C]{AIBN} \left[CH_2{-}\underset{\underset{CH_3}{|}}{\overset{\overset{OCH_3}{|}\;\;\;\overset{|}{C}{=}O}{C}}\right]_n \qquad (3)$$

$$\text{4} \qquad\qquad\qquad\qquad \text{5}$$

[1] See page 62.

ratio) and a trace of methacrylic acid. The mixture is then allowed to polymerize at about 40°C during one day to give a transparent, glassy polymer (**5**). The polymerization can be effected in a mold (or the bottom of a glass beaker), in which case poly(methyl methacrylate) may be added to the initial reaction mixture to raise the viscosity. Sheets of poly(methyl methacrylate) glass can be prepared by the use of a mold of two glass plates clamped together and separated by a rubber gasket (see Chapter 20).

Emulsion Polymerization of Acrylonitrile (*Reaction 4*)

Acrylonitrile (**6**) must first be freed of inhibitor by passage through a column of silica gel. Polymerization takes place in a stirred mixture of acrylonitrile and water to which has been added a detergent, potassium persulfate, and a trace of sodium bisulfite. Oxygen is excluded by a nitrogen atmosphere. A white emulsion of polyacrylonitrile (**7**) in water is formed during 2 to 3 h at about 35°C, and the polymer can be isolated by coagulation and filtration.

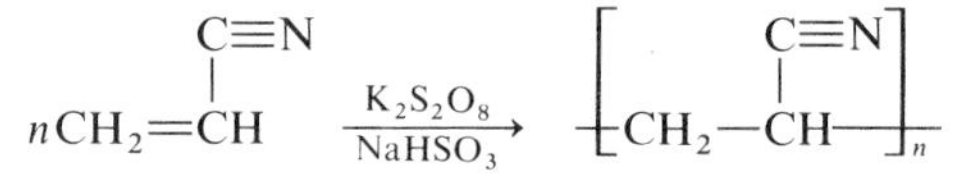

$$ n\,CH_2{=}CH\overset{\displaystyle C{\equiv}N}{\underset{\displaystyle |}{}} \quad \xrightarrow[\text{NaHSO}_3]{\text{K}_2\text{S}_2\text{O}_8} \quad \left[CH_2{-}CH\overset{\displaystyle C{\equiv}N}{\underset{\displaystyle |}{}} \right]_n \tag{4} $$

6 **7**

Poly(*p*-xylylene) by Oxidative Pyrolysis (*Reaction 5*)

Finally, as an example of an unusual type of addition reaction, which probably proceeds by a free-radical mechanism, we can consider the oxidative pyrolysis of *p*-xylene. *p*-Xylene can be pyrolyzed at temperatures up to 950°C to give a cyclic dimer known as di-*p*-xylylene (**8**). At 550 to 600°C this compound dissociates in vacuum to yield *p*-xylylene (**9**), and this monomer polymerizes spontaneously on a surface in vacuum at temperatures near 30°C to yield poly(*p*-xylylene) (**10**). Thus, the key step is an addition polymerization, presumably operating by a radical mechanism. The rate of polymerization is exceedingly fast, and the process can be used

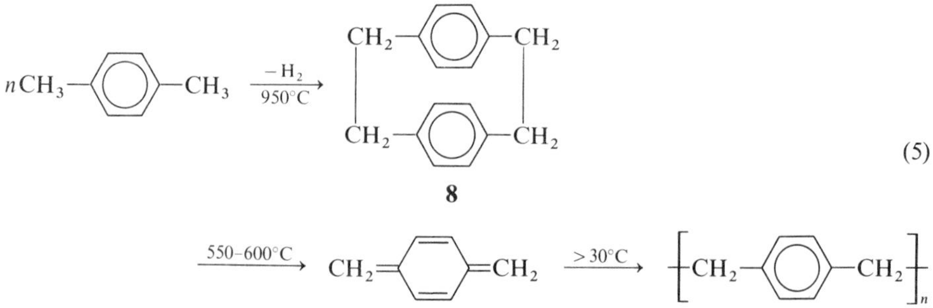

$$ n\,CH_3{-}\!\!\bigcirc\!\!{-}CH_3 \xrightarrow[950°C]{-H_2} \tag{5} $$

8

$$ \xrightarrow{550\text{–}600°C} \quad CH_2{=}\!\!\bigcirc\!\!{=}CH_2 \xrightarrow{>30°C} \left[CH_2{-}\!\!\bigcirc\!\!{-}CH_2 \right]_n $$

9 **10**

to coat objects (especially biomedical devices) essentially by vapor deposition (see Chapter 22).

CHAIN REACTIONS

Free-radical polymerizations are chain reactions. The addition of a monomer molecule to an active chain end regenerates the active site at the chain end. Hence, a large number of monomer molecules are "consumed" for each active site introduced into the system. In chain-reaction polymerizations we may recognize four distinct types of processes. They are:

1. *Chain initiation*—a process in which highly reactive transient molecules or active centers are formed.
2. *Chain propagation*—the addition of monomer molecules to the active chain end, accompanied by regeneration of the terminal active site.
3. *Chain transfer*—involving the transfer of the active site to another molecule (e.g., monomer). The molecule that has lost the active site is now "dead" from a chain-propagation point of view. The molecule that has accepted the active site can start a new chain.
4. *Chain termination*—a reaction in which the active chain centers are destroyed. Chain reactions are found in free-radical, anionic, and cationic vinyl-type polymerizations. In free-radical processes all of the four steps listed above can usually be identified.

Schematically, a free-radical polymerization sequence can be represented by the reactions shown in (6) to (12). In these equations M represents a molecule of

$$\text{Initiator} \longrightarrow 2R'\cdot \qquad \left.\right\} \text{Initiation} \qquad (6)$$
$$R'\cdot + M \longrightarrow R_1^{\cdot} \qquad \qquad (7)$$
$$R_1^{\cdot} + M \longrightarrow R_2^{\cdot} \qquad \qquad (8)$$
$$\cdots\cdots\cdots\cdots\cdots\cdots\cdots\cdots\cdots \left.\right\} \text{Propagation}$$
$$R_n^{\cdot} + M \longrightarrow R_{n+1}^{\cdot} \qquad \qquad (9)$$
$$R_n^{\cdot} + YZ \longrightarrow R_n Y + Z\cdot \qquad \text{Chain transfer} \qquad (10)$$
$$R_n^{\cdot} + R_m^{\cdot} \longrightarrow P_{n+m} \qquad \left.\right\} \text{Termination} \qquad (11)$$
$$R_n^{\cdot} + R_m^{\cdot} \longrightarrow P_n + P_m \qquad \qquad (12)$$

monomer; $R'\cdot$ is an initiating free radical from the initiator; $R\cdot_n$ is the propagating free radical with a degree of polymerization, n; YZ is a chain transfer agent which may be solvent, monomer, initiator, or polymer molecules; and P_n is the final inactive polymer. It should be noted that in some cases the initiator molecule may be the monomer itself.

In practical terms, this general chain-reaction sequence can be identified in the polymerization of vinyl chloride. This reaction can be initiated by di-*t*-butyl peroxide. The sequence for this system is shown in reactions (13) to (18). The chain-transfer steps are not included in this sequence.

$$Me_3COOCMe_3 \xrightarrow{\text{heat}} 2\,Me_3CO\cdot \qquad\qquad \text{Initiator dissociation} \quad (13)$$

$$Me_3CO\cdot + CH_2{=}CHCl \longrightarrow Me_3COCH_2{-}\overset{\displaystyle H}{\underset{\displaystyle Cl}{\overset{|}{\underset{|}{C}}}}\cdot \qquad\qquad \text{Initiation} \quad (14)$$

$$Me_3COCH_2{-}\overset{\displaystyle H}{\underset{\displaystyle Cl}{\overset{|}{\underset{|}{C}}}}\cdot + CH_2{=}CHCl \longrightarrow Me_3CO{-}CH_2{-}\overset{\displaystyle H}{\underset{\displaystyle Cl}{\overset{|}{\underset{|}{C}}}}{-}CH_2{-}\overset{\displaystyle H}{\underset{\displaystyle Cl}{\overset{|}{\underset{|}{C}}}}\cdot$$

Initial propagation   (15)

$$Me_3CO{\Big[}CH_2{-}\overset{\displaystyle H}{\underset{\displaystyle Cl}{\overset{|}{\underset{|}{C}}}}{\Big]}_n CH_2{-}\overset{\displaystyle H}{\underset{\displaystyle Cl}{\overset{|}{\underset{|}{C}}}}\cdot + CH_2{=}CHCl \longrightarrow Me_3CO{\Big[}CH_2{-}\overset{\displaystyle H}{\underset{\displaystyle Cl}{\overset{|}{\underset{|}{C}}}}{\Big]}_{n+1} CH_2{-}\overset{\displaystyle H}{\underset{\displaystyle Cl}{\overset{|}{\underset{|}{C}}}}\cdot$$

Propagation   (16)

$$Me_3CO{\Big[}CH_2{-}\overset{\displaystyle H}{\underset{\displaystyle Cl}{\overset{|}{\underset{|}{C}}}}{\Big]}_n CH_2{-}\overset{\displaystyle H}{\underset{\displaystyle Cl}{\overset{|}{\underset{|}{C}}}}\cdot + \cdot\overset{\displaystyle H}{\underset{\displaystyle Cl}{\overset{|}{\underset{|}{C}}}}{-}CH_2{\Big[}\overset{\displaystyle H}{\underset{\displaystyle Cl}{\overset{|}{\underset{|}{C}}}}{-}CH_2{\Big]}_m OCMe_3$$

$$\longrightarrow Me_3CO{\Big[}CH_2{-}\overset{\displaystyle H}{\underset{\displaystyle Cl}{\overset{|}{\underset{|}{C}}}}{\Big]}_n CH_2{-}\overset{\displaystyle H}{\underset{\displaystyle Cl}{\overset{|}{\underset{|}{C}}}}{-}\overset{\displaystyle H}{\underset{\displaystyle Cl}{\overset{|}{\underset{|}{C}}}}{-}CH_2{\Big[}\overset{\displaystyle H}{\underset{\displaystyle Cl}{\overset{|}{\underset{|}{C}}}}{-}CH_2{\Big]}_m OCMe_3$$

Termination   (17)

$$Me_3CO{\Big[}CH_2{-}\overset{\displaystyle H}{\underset{\displaystyle Cl}{\overset{|}{\underset{|}{C}}}}{\Big]}_n CH_2{-}\overset{\displaystyle H}{\underset{\displaystyle Cl}{\overset{|}{\underset{|}{C}}}}\cdot + \cdot\overset{\displaystyle H}{\underset{\displaystyle Cl}{\overset{|}{\underset{|}{C}}}}{-}CH_2{\Big[}\overset{\displaystyle H}{\underset{\displaystyle Cl}{\overset{|}{\underset{|}{C}}}}{-}CH_2{\Big]}_m OCMe_3$$

$$\longrightarrow Me_3CO{\Big[}CH_2{-}\overset{\displaystyle H}{\underset{\displaystyle Cl}{\overset{|}{\underset{|}{C}}}}{\Big]}_n CH_2{-}\overset{\displaystyle H}{\underset{\displaystyle Cl}{\overset{|}{\underset{|}{C}}}}{-}H + \overset{\displaystyle H}{\underset{\displaystyle Cl}{\overset{|}{\underset{|}{C}}}}{=}CH{\Big[}\overset{\displaystyle H}{\underset{\displaystyle Cl}{\overset{|}{\underset{|}{C}}}}{-}CH_2{\Big]}_m OCMe_3$$

Termination   (18)

FREE-RADICAL INITIATORS

Initiation of a free-radical polymerization requires the production of free radicals in the presence of the unsaturated monomer. The initiating free radicals can be produced directly from the monomer (by irradiation with high-energy radiation, for example), but it is more normal for the radicals to be generated from an added initiator. Such an initiator is usually a molecule that can be decomposed thermally or by irradiation to yield a pair of initiating radicals. The thermal decomposition of di-*t*-butyl peroxide mentioned in the preceding section is an example of this type of reaction. Alternatively, some initiators function by means of redox reactions, in which case one initiator molecule may yield only one initiating radical. The following sections outline the alternative mechanisms that are available for initiation.

Thermal Decomposition of Initiators

Numerous substances decompose to free radicals when heated. If the decomposition temperature corresponds to a convenient temperature range for polymerization, the substance may be useful as an initiator. In fact, it is the dependence of the initiator decomposition rate on the temperature which determines the usefulness of the compound as an initiator. Such thermal decompositions usually yield two free radicals from one initiator molecule by a first-order reaction process.

As will be shown in detail in Chapter 12, the rate of polymerization is proportional to the square root of the rate of initiation. Hence, an increase in the initiator concentration or an elevation of the temperature has a dramatic accelerating effect on the polymerization rate. On the other hand, the average chain length is lowered by an increase in the rate of initiation. Thus, the choice of an initiator and the selection of a temperature for initiation requires a compromise to be made between the need to obtain a reasonably fast polymerization and the need to make high-molecular-weight polymer.

An arbitrary, but useful, rule that can be used to estimate the practical temperature range for a thermal initiator is as follows: *the rate of formation of initiating radicals should be in the range of 10^{-7} to 10^{-6} mol-liter^{-1}-s^{-1} at an initiator concentration of 0.1 M.* In other words, its useful operating range is that temperature range in which the first-order decomposition rate constant is in the range of 10^{-6} to 10^{-5} s^{-1}. Because most free-radical polymerizations must be carried out at temperatures below 150°C to prevent side reactions, the list of useful thermal initiators is largely restricted to organic peroxides, organic hydroperoxides, azo compounds, and metal alkyls. Each of these classes of initiator will now be considered in turn.

Dialkyl peroxides, ROOR, decompose thermally by cleavage of the oxygen–oxygen bond to yield two alkoxy radicals, RO·, as shown by (19). An example of such a decomposition was given earlier for the specific case where R was Me₃C. As

$$\text{ROOR} \longrightarrow \text{RO·} + \text{RO·} \qquad (19)$$

shown in the earlier example, the alkoxy radicals may then initiate the polymerization by reaction with the monomer. However, the alkoxy radicals may decompose to alkyl radicals and aldehydes or ketones before initiation can occur. For example, in the case of di-t-butyl peroxide, this additional decomposition yields methyl radicals and acetone, as shown in (20). The methyl radical then completes the initiation

$$Me_3CO\cdot \longrightarrow (Me)_2C{=}O + CH_3\cdot \tag{20}$$

process by reaction with the monomer. The first-order rate constant for the decomposition of di-t-butyl peroxide is[1] $6.3 \times 10^{15}e^{-37,500\,cal/RT}$ s^{-1}. Application of the rule described above indicates that the predicted useful initiator temperature range for di-t-butyl peroxide is 100 to 120°C.

Diacylperoxides, RC(O)OO(O)CR, decompose similarly by an initial cleavage of the oxygen–oxygen bond. In general, these compounds provide a somewhat lower useful temperature range for initiation than do the dialkyl peroxides. For example, benzoyl peroxide undergoes the initial bond scission shown in (21) with a

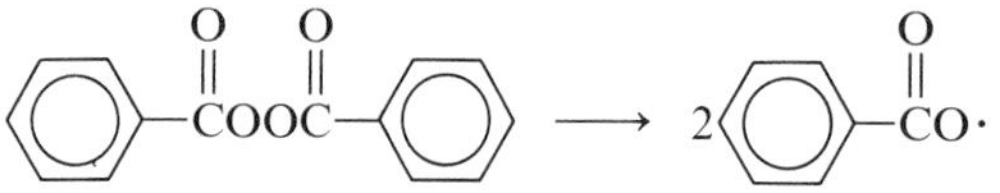

$$\text{(21)}$$

first-order rate constant[2] of $1 \times 10^{14}e^{-29,900/RT}$ s^{-1}. This rate constant indicates that benzoyl peroxide is a useful thermal initiator in the temperature range of 60 to 80°C. Despite the relatively low temperatures at which this initiator is useful, the benzoyl radicals formed by the initial bond rupture (21) may decompose by the reaction shown in (22) before they can react with the monomer. However, the

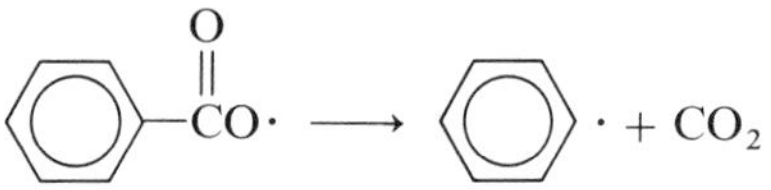

$$\text{(22)}$$

occurrence of reaction (22) has very little effect on the overall rate of initiation, because the phenyl radicals formed in (22) can themselves add to the monomer to bring about initiation.

Organic hydroperoxides, ROOH, have also been used as thermal initiators. Here also the initial decomposition involves rupture of the oxygen–oxygen bond, as shown for cumylhydroperoxide in reaction (23). Both the cumyloxy radical and the hydroxyl radical may add to the monomer to initiate polymerization. The cumyloxy

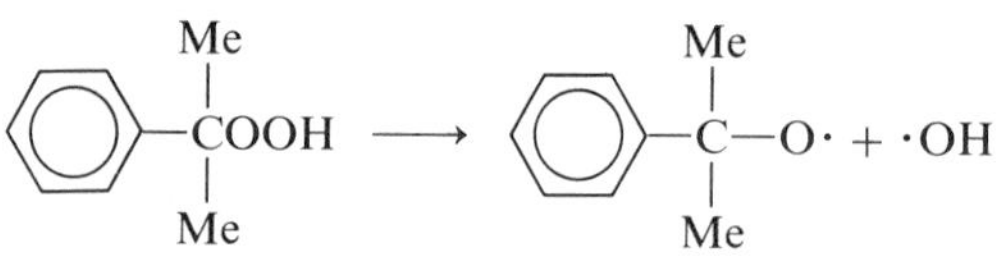

$$\text{(23)}$$

[1] W. H. Richardson and H. E. O'Neal, *Comprehensive Chemical Kinetics*, Vol. 5, C. H. Bamford and C. F. H. Tipper, eds. (Amsterdam: Elsevier, 1972).
[2] Ibid.

radical may itself decompose to acetophenone and a methyl radical. However, a disadvantage exists to the use of hydroperoxides as initiators. Hydroperoxides are susceptible to attack by HO· and RO· (or R·) radicals at the temperatures normally used for initiation. The free-radical products of these radical reactions, as shown in (24) to (26), are peroxy radicals. Often, these are not sufficiently reactive to add to the monomer. Thus, they do not initiate polymerization.

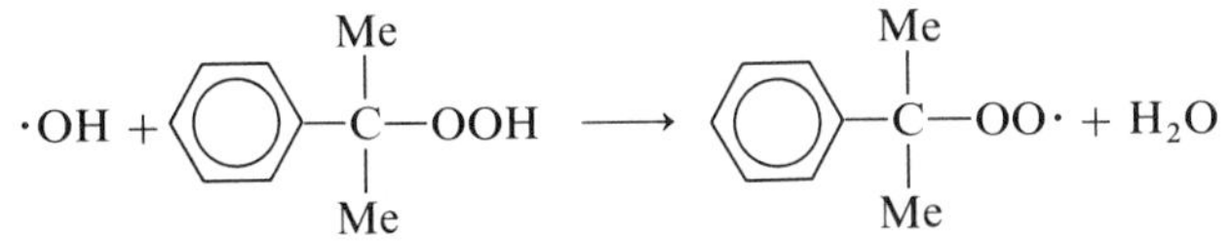

$$\cdot OH + \text{(structure)} \longrightarrow \text{(structure)} + H_2O \quad (24)$$

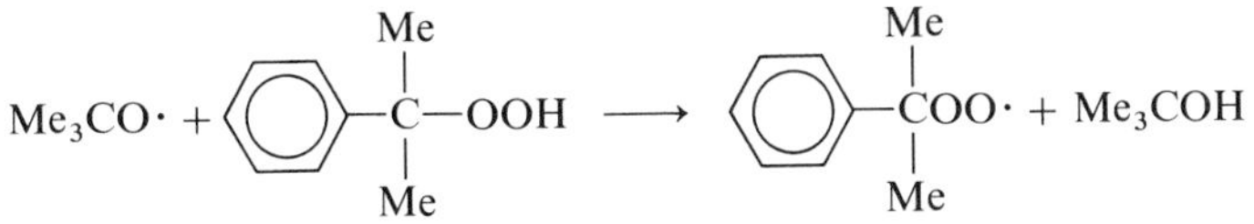

$$Me_3CO\cdot + \text{(structure)} \longrightarrow \text{(structure)} + Me_3COH \quad (25)$$

$$CH_3\cdot + \text{(structure)} \longrightarrow \text{(structure)} + CH_4 \quad (26)$$

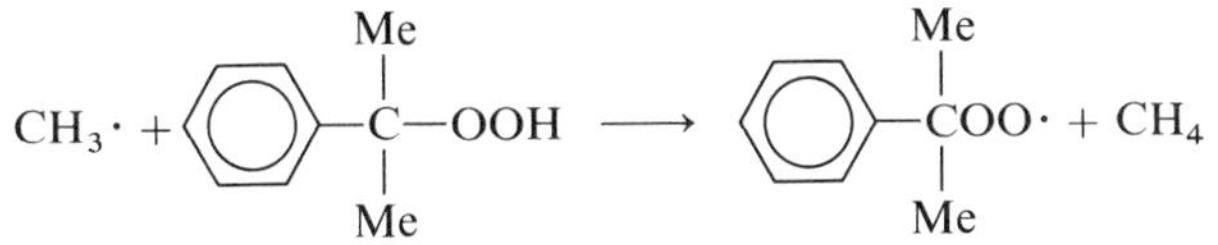

Azo compounds, RN=NR, such as azobis*iso*butyronitrile, decompose thermally to give nitrogen and two alkyl radicals, as shown in reaction (27). Azo compounds are useful and popular thermal initiators because they offer a wide

$$(CH_3)_2\overset{CN}{\underset{}{C}}N{=}N\overset{CN}{\underset{}{C}}(CH_3)_2 \longrightarrow 2(CH_3)_2\overset{CN}{\underset{}{C}}\cdot + N_2 \quad (27)$$

range of useful operating temperatures. Moreover, because nitrogen gas is formed during the decomposition, measurement of the amount of nitrogen produced gives the experimenter a measure of the number of initiating radicals formed. This, in turn, allows an estimate to be made of the efficiency of initiation versus radical recombination (within the solvent cage).

Table 3.3 lists activation energies and useful operating temperature ranges (Δt) (as defined previously) for a number of azo compounds. Azomethane and azo-*iso*propane are much too stable to be important as thermal initiators at normal polymerization temperatures. The effective temperature range shown for azobis-*iso*butyronitrile (AIBN) illustrates why it is one of the most popular thermal initiators.

All the free-radical initiators discussed above are appropriate only for polymerizations carried out at room temperature or above. However, silver alkyls have been used to initiate radical polymerizations at temperatures as low as -20 to $-60°C$. Ethylsilver, for example, is believed to decompose by the reaction shown in

TABLE 3.3 Activation Energies and Useful Temperature Ranges for Initiation by R'N=NR' Compounds

R'	E_{act} (kcal/mol)	USEFUL INITIATION TEMPERATURE RANGE (°C)*
CH_3	50.2	225–250
$(CH_3)_2CH \cdot$	40.8	180–200
$C_6H_5(CH_3CH \cdot)$	36.5	105–125
$(C_6H_5)_2CH$	26.6	20–35
$(CH_3)_2(CN)C \cdot (AIBN)$	30.8	40–60

* Defined by the condition that k_1 is in the region 10^{-6} to 10^{-5} s^{-1}.

Source: W. H. Richardson and H. E. O'Neal, *Comprehensive Chemical Kinetics*, Vol. 5, C. H. Bamford and C. F. H. Tipper, eds. (Amsterdam: Elsevier, 1972).

(28). Because silver alkyls decompose below room temperature they must, of course, be prepared at low temperatures and used as soon as possible.

$$Ag-C_2H_5 \longrightarrow Ag + C_2H_5^{\cdot} \tag{28}$$

Initiation by Redox Reactions

In one of the example polymerizations given earlier, the water-soluble initiator used was a persulfate salt together with bisulfite ion. Initiation by this system falls into a category generally described as *redox reactions*. The bisulfite ion (HSO_3^-) reduces the persulfate ion ($S_2O_8^{2-}$) to yield sulfate (SO_4^{2-}) and the $SO_4 \cdot^-$ radical ion. This latter species reacts with water to generate the bisulfate ion (HSO_4^-) and a hydroxyl radical ($\cdot OH$). Thiosulfate ion ($S_2O_3^{2-}$) can also be used as a reducing agent. Moreover, traces of ferric ion may also participate. The reactions shown in (29) to (31) summarize these processes. The radicals formed in these processes then initiate polymerization.

$$S_2O_8^{2-} + HSO_3^- \longrightarrow SO_4^{2-} + SO_4^- + HSO_3^{\cdot} \tag{29}$$

$$S_2O_8^{2-} + S_2O_3^{2-} \longrightarrow SO_4^{2-} + SO_4^- + S_2O_3^{\cdot -} \tag{30}$$

$$HSO_3^- + Fe^{3+} \longrightarrow HSO_3^{\cdot} + Fe^{2+} \tag{31}$$

Related redox reactions can be carried out in aqueous media with the use of alkyl hydroperoxides and a reducing agent, such as ferrous ion. Thus, cumyl hydroperoxide reacts with ferrous ion to yield cumyloxy radicals, as shown in reaction (32). The reaction is sufficiently fast that cumyloxy radicals are formed at temperatures as low as 15 to 50°C, whereas temperatures of 85 to 105°C are needed

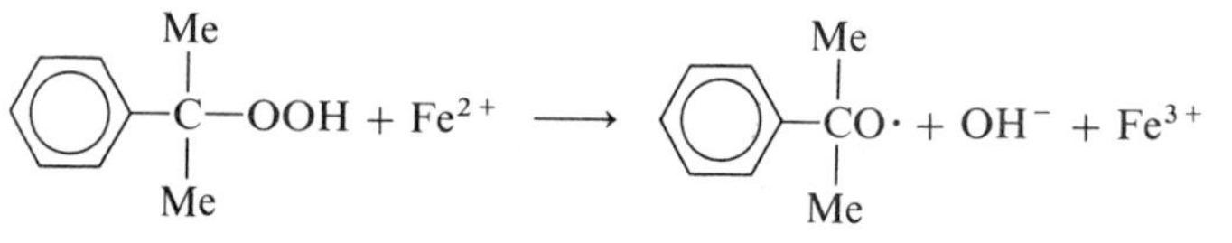

$$\tag{32}$$

for the direct thermal cleavage of this peroxide. Moreover, side reactions and further cleavage of the cumyloxy radical are unlikely at this lower temperature. A further advantage of this system is that the peroxide cleavage process can be monitored by the conversion of ferrous to ferric ion.

Direct Thermal and Photolytic Initiation

Some monomers undergo free-radical polymerization simply when heated or when exposed to light. Styrene is a monomer which is particularly well known for this type of behavior. The mechanism of initiation is still not completely understood.

It appears possible that initiation involves first the formation of a triplet-state diradical by the collision of two monomer molecules. However, strong evidence exists that a monoradical is the chain-propagating species. Hence, the following tentative mechanism has been proposed, (33) to (35). Species **11** and **12** may be

$$PhHC=CH_2 + PhHC=CH_2 \longrightarrow PhHC=CH_2 + \cdot\underset{\underset{Ph}{|}}{\overset{\overset{H}{|}}{C}}-CH_2 \tag{33}$$

Triplet

$$\cdot\underset{\underset{Ph}{|}}{\overset{\overset{H}{|}}{C}}-CH_2 + CH_2=CHPh \longrightarrow \cdot\underset{\underset{Ph}{|}}{\overset{\overset{H}{|}}{C}}-CH_2-CH_2-\underset{\underset{Ph}{|}}{\overset{\overset{H}{|}}{C}}\cdot$$

Triplet

$$\longrightarrow PhCH=CH_2 + CH_2=CHPh \tag{34}$$

$$\cdot\underset{\underset{Ph}{|}}{\overset{\overset{H}{|}}{C}}-CH_2-CH_2-\underset{\underset{Ph}{|}}{\overset{\overset{H}{|}}{C}}\cdot + PhHC=CH_2 \longrightarrow \cdot\underset{\underset{Ph}{|}}{\overset{\overset{H}{|}}{C}}-CH_3 + PhHC=CH-CH_2-\underset{\underset{Ph}{|}}{\overset{\overset{H}{|}}{C}}\cdot$$

$$\underset{\textbf{11}}{} \qquad\qquad \underset{\textbf{12}}{} \tag{35}$$

the real initiators for polymerization. It is interesting that the mechanism of this, one of the oldest known polymerizations,[1] is still somewhat obscure.

The polymerization of many monomers can be induced by irradiation with light if a suitable "photosensitizer" is also present in the system. This method of initiation will be discussed in more detail in Chapter 5.

Initiation by Ionizing Radiation

Free-radical polymerizations can also be induced by irradiation of the monomer with high-energy radiation, such as X-rays, γ-rays, α-particles, high-energy electrons, protons, and so on. The monomer can be in the bulk phase or in solution. The absorption of energy by the monomer is far less selective when ionizing radiation is

[1] E. Simon, *Ann.*, **31**, 265 (1839).

used than when light is employed. All the species present in the system (monomer and solvent) absorb energy and decompose to yield free radicals. The radicals then initiate polymerization.

However, the process is far more complicated than that found for the absorption of light. Ionizing radiation produces positive and negative ions as well as free radicals. The ions may react with the monomer or other ions to yield secondary ions or free radicals, or they may initiate ionic chain polymerization. Thus, the reaction mechanisms can be extremely complicated. This subject is discussed further in Chapter 5.

REACTIONS OF INITIATOR RADICALS WITH THE MONOMER

Clearly, the second important step in a radical chain polymerization is the addition of the initiating radical to a monomer molecule. Only a fraction of the initiating radicals formed behave in this way, because it is generally found that the rate of chain initiation is lower than the rate at which the initiating radicals are formed.

This wastage of initiating radicals occurs because of competition from alternative fast reactions. First, it must be recognized that initiator radicals are usually formed in pairs by the cleavage of a peroxide or azo compound. At the instant of dissociation, the two radicals are imprisoned together in a "cage" of monomer or solvent molecules. The mean free path of the radicals is no more than the average diameter of a small molecule. Hence, the two initiating radicals will have excellent opportunities for collision and recombination before they are separated by diffusion. This effect is important even in a cage of monomer molecules because a reaction between the radical and a monomer generally requires many more collisions than does recombination of two free radicals. *Primary* or *geminate recombination* is the name applied to this process.

Second, those initiating radicals that do escape from their original partners in the solvent cage may combine with radicals from other solvent cages. This further lowers the efficiency of initiation. This process is known as *secondary recombination*. Its occurrence may be reduced by decreasing the rate of radical formation or by increasing the monomer concentration.

Third, initiator radicals may be wasted by reactions with propagating polymer radicals in such a way as to terminate the chains; and fourth, the initiator radicals may react with the original initiator to generate radicals of low reactivity. Add to this the fact that radicals can abstract chlorine or other radicals from solvents or hydrogen from terminated polymer molecules, and it is clear why the efficiency of initiation is generally low (Table 3.4). The maximum initiation efficiency can usually be obtained by the use of low temperatures (to reduce the rate of radical formation) and by the use of low concentrations of initiator relative to the monomer (to improve the probability of a successful initiation).

Quantitative studies[1] indicate that the efficiency of initiation (usually given the

[1] P. E. M. Allen and C. R. Patrick, *Kinetics and Mechanisms of Polymerization Reactions* (New York: Wiley, 1974), p. 114.

TABLE 3.4 Possible Fates of Initiator Radicals

R· →
- Radical recombination within the solvent cage: R· + ·R → R—R
- Secondary recombination outside the cage: R· + ·R → R—R
- Reaction with polymer radicals: R· + ·—W— → R—W—
- Reaction with initiator: R· + R'—R' → R—R' + ·R
- Radical abstraction: R· + H—R' → R—H + ·R'
- Reaction with solvent: R· + CCl₄ → R—Cl + Cl₃C·
- Chain initiation: R· + monomer → R-monomer·

symbol f) lies most often between 0.1 and 0.8. It correlates approximately with the viscosity of the medium and is approximately independent of temperature.

RADICAL CHAIN PROPAGATION

Chain propagation involves the addition of a free radical to the double bond of a monomer molecule. The product must itself be a free radical and the process can be repeated. In fact, it is common for thousands of monomer molecules to add successively to the end of the chain.

The most likely form of monomer addition is called *head-to-tail addition*. Reaction (36) illustrates this behavior. Alternatively, the addition may involve a *head-to-head* (reaction 37) or a *tail-to-tail* reaction (reaction 38). Although it might be supposed that species **14**, **16**, and **18** might be distributed randomly through the molecular chain, it is, in fact, found that head-to-tail linkages (**14**) are in great excess. For example, in poly(vinyl alcohol), $+CH_2CH(OH)\frac{}{}_n$, it

$$R\text{—W—}CH_2\text{—}\underset{X}{\overset{H}{C}}\cdot + CH_2\text{=}\underset{X}{\overset{H}{C}} \longrightarrow R\text{—W—}CH_2\text{—}\underset{X}{\overset{H}{C}}\text{—}CH_2\text{—}\underset{X}{\overset{H}{C}}\cdot \qquad (36)$$

13 **14**

$$R\text{—W—}CH_2\text{—}\underset{X}{\overset{H}{C}}\cdot + \underset{X}{\overset{H}{C}}\text{=}CH_2 \longrightarrow R\text{—W—}CH_2\text{—}\underset{X}{\overset{H}{C}}\text{—}\underset{X}{\overset{H}{C}}\text{—}CH_2 \qquad (37)$$

15 **16**

$$R\text{—W—}\underset{X}{\overset{H}{C}}\text{—}CH_2 + CH_2\text{=}\underset{X}{\overset{H}{C}} \longrightarrow R\text{—W—}\underset{X}{\overset{H}{C}}\text{—}CH_2\text{—}CH_2\text{—}\underset{X}{\overset{H}{C}}\cdot \qquad (38)$$

17 **18**

Chapter 3 / Free-Radical Polymerization

has been found[1] that the percentage of **16** (and, therefore, also of **18**) is only 1.1% in polymer prepared at 25°C, and only 1.8% in polymer prepared at 100°C.

The principal reason for the preference of head-to-tail addition lies in the greater thermodynamic stability of a free radical such as **13** relative to that of one such as **17**, and perhaps also to steric inhibition of steps such as (37). It can, in fact, be shown that the addition of methyl radicals to propylene, CH_2=CHMe, is 3.8 kcal/mol more favorable if the methyl group adds to the CH_2 unit rather than to the CHMe component. Thus, attack on the CH_2 unit should be favored up to a temperature of about 225°C. A similar preference for attack on the CH_2— group exists when butyl radicals are used. However, although these thermodynamic effects would be all-important when equilibrium conditions prevail, it seems more likely that kinetic factors (such as steric effects and activation energies) may exert the strongest influence.

CHAIN TRANSFER REACTIONS

In an "ideal" free-radical polymerization, chains become initiated, they propagate linearly, and they are then terminated. Such ideal circumstances are often found in ionic polymerizations, but only rarely in free-radical processes. The deviation from ideality occurs when a propagating oligomer or polymer radical reacts with another molecule, not by addition, but by abstraction. By "abstraction" we mean a process in which a radical fragment is removed from the second molecule with concurrent generation of a radical residue from that second molecule. This is illustrated by reaction (39), in which YZ represents the second molecule, which could be monomer, solvent, initiator, polymer molecules, or other molecules deliberately or accidentally incorporated into the reaction mixture.

$$R-\!\!\wedge\!\!\!\wedge\!\!-CH_2-\overset{\overset{\displaystyle H}{\displaystyle |}}{\underset{\underset{\displaystyle X}{\displaystyle |}}{C}}\cdot + YZ \longrightarrow R-\!\!\wedge\!\!\!\wedge\!\!-CH_2-\overset{\overset{\displaystyle H}{\displaystyle |}}{\underset{\underset{\displaystyle X}{\displaystyle |}}{C}}-Y + Z\cdot \qquad (39)$$

19

It is important to note that although the polymer chain (**19**) is now effectively terminated, $Z\cdot$ may initiate a new chain if its reactivity is comparable to that of a normal propagating radical. Such a process is called *chain transfer*. The number of radicals growing at any instant is unchanged, but *the average chain length of the polymer produced will be reduced*.

When the second molecule is a polymer molecule, the ultimate result is generally the formation of a *branched* polymer. Thus, if hydrogen radical abstraction from a polymer leads to the formation of a residue such as **20**, the radical site on this residue can initiate growth of a branch, such as the one shown in **21**.

[1] P. J. Flory and F. S. Leutner, *J. Polymer Sci.*, **3**, 880 (1948); **5**, 267 (1950).

$$
\begin{array}{cc}
& \begin{array}{c}
\text{H}-\overset{\displaystyle \cdot}{\text{C}}-\text{X} \\
| \\
\text{CH}_2 \\
| \\
\text{H}-\text{C}-\text{X} \\
| \\
\text{CH}_2 \\
|
\end{array} \\
\text{R}-\!\!\sim\!\!-\text{CH}_2-\overset{\displaystyle \cdot}{\underset{\displaystyle |}{\text{C}}}-\!\!\sim\!\!-\text{R}' \qquad & \text{R}-\!\!\sim\!\!-\text{CH}_2-\underset{\displaystyle |}{\text{C}}-\!\!\sim\!\!-\text{R}' \\
\quad\; \text{X} & \quad\; \text{X} \\
\mathbf{20} & \mathbf{21}
\end{array}
$$

It should also be noted that if the second molecule, YZ, yields a radical fragment, Z·, which is unreactive toward the monomer, the substance YZ is termed an *inhibitor* and the process is termed "inhibition." Such compounds are often added to monomers to inhibit polymerization during storage or transportation. If, on the other hand, Z· is reactive toward the monomer but less reactive than the normal propagating radicals, then YZ is called a *retarder*.

FREE-RADICAL CHAIN TERMINATION

Free-radical chains can be terminated by reaction of a growing polymer radical with some other free radical in the system. First, in the simplest sense, the polymer radical may react with initiator radicals (which will still be produced after the start of the reaction) as shown in (40). However, this is a wastage of initiating radicals that, in practice, should be avoided by keeping the rate of initiation sufficiently low.

$$
\text{R}-\!\!\sim\!\!-\text{CH}_2-\overset{\displaystyle \text{H}}{\underset{\displaystyle \text{X}}{\text{C}}}\cdot \;+\; \cdot\text{R}' \;\longrightarrow\; \text{R}-\!\!\sim\!\!-\text{CH}_2-\overset{\displaystyle \text{H}}{\underset{\displaystyle \text{R}}{\text{C}}}-\text{R}' \tag{40}
$$

Second, and more important, the termination may occur either by combination with another polymer radical (reaction 41), or by transfer of an atom (usually hydrogen) from one polymer radical to another (reaction 42). If reaction (41)

$$
\text{R}-\!\!\sim\!\!-\text{CH}_2-\overset{\displaystyle \text{H}}{\underset{\displaystyle \text{X}}{\text{C}}}\cdot \;+\; \cdot\overset{\displaystyle \text{H}}{\underset{\displaystyle \text{X}}{\text{C}}}-\text{CH}_2-\!\!\sim\!\!-\text{R} \;\longrightarrow\; \text{R}-\!\!\sim\!\!-\text{CH}_2-\overset{\displaystyle \text{H}}{\underset{\displaystyle \text{X}}{\text{C}}}-\overset{\displaystyle \text{H}}{\underset{\displaystyle \text{X}}{\text{C}}}-\text{CH}_2-\!\!\sim\!\!-\text{R} \tag{41}
$$

$$
\text{R}-\!\!\sim\!\!-\text{CH}_2-\overset{\displaystyle \text{H}}{\underset{\displaystyle \text{X}}{\text{C}}}\cdot \;+\; \cdot\overset{\displaystyle \text{H}}{\underset{\displaystyle \text{X}}{\text{C}}}-\text{CH}_2-\!\!\sim\!\!-\text{R} \;\longrightarrow\; \text{R}-\!\!\sim\!\!-\text{CH}_2-\overset{\displaystyle \text{H}}{\underset{\displaystyle \text{X}}{\text{C}}}-\text{H} \;+\; \overset{\displaystyle \text{H}}{\underset{\displaystyle \text{X}}{\text{C}}}\!\!=\!\!\text{C}-\!\!\sim\!\!-\text{R} \tag{42}
$$

TABLE 3.5 Termination of Free-Radical Polymerization at 60°C

Monomer	Formula	Disproportionation	Combination
Acrylonitrile*	$CH_2{=}CH{-}CN$	~0	~100
Methyl methacrylate†	$CH_2{=}\underset{\underset{O}{\overset{\|\|}{C}}}{\overset{\overset{CH_3}{\|}}{C}}{-}OCH_3$	79	21
Styrene‡	⟨benzene ring⟩$-CH{=}CH_2$	23	77
Vinyl acetate§	$CH_2{=}CH{-}O\overset{\overset{O}{\|\|}}{C}CH_3$	~100	~0

* C. H. Bamford, A. D. Jenkins, and R. Johnston, *Trans. Faraday Soc.*, **55**, 179 (1959).
† J. C. Bevington, H. W. Melville, and R. P. Taylor, *J. Polymer Sci.*, **12**, 449; **14**, 463 (1954).
‡ K. C. Berger, *Makromol. Chem.*, **176**, 3575 (1975).
§ At 90°C. C. H. Bamford and A. D. Jenkins, *Nature*, **176**, 78 (1955).

predominates, at least one head-to-head linkage must be found in the final terminated polymer. This amounts to 1% of the total linkages for a polymer, with an average degree of polymerization of 100 units. This in itself is sufficient to account for nearly all the head-to-head units in many polymers. Note also that termination by reaction (41) can bring about a considerable increase in the molecular weight of the final polymer.

When reaction (42) (known as "disproportionation") predominates, two chemically different types of polymer molecules are produced. The predominance of one of the processes shown in (41) or (42) depends on the nature of the monomer and on the temperature. The dependence on temperature results from the fact that the combination reaction (step 41) usually has a lower activation energy than one which requires the breaking of a chemical bond (step 42). Hence, combination will normally be preferred at low temperatures, but disproportionation (42) should become more significant at high temperatures. This has been demonstrated for methyl methacrylate[1] and styrene[2] polymerization. The relative probabilities of occurrence of reactions (41) and (42) will equal the relative reaction rates of these two processes. Hence, the relative probabilities can be calculated from

$$\frac{\text{Probability of combination }(c)}{\text{Probability of disproportionation }(d)} = \frac{k_c}{k_d} = \frac{A_c}{A_d}e^{(E_d - E_c)/RT} \tag{43}$$

where k and A are the rate constants and frequency factors, respectively, and E is the activation energy.

In a practical sense, the relative occurrence of combination and disproportionation can, in principle, be measured from the average number of initiator fragments,

[1] J. C. Bevington, H. W. Melville, and R. P. Taylor, *J. Polymer Sci.*, **12**, 449 (1954).
[2] K. C. Berger, *Makromol. Chem.*, **176**, 3575 (1975).

R', per polymer molecule that are found in the final product. Reaction (41) should yield polymer with two initiator fragments per molecule, whereas reaction (42) should give polymer with only one R' unit per molecule. Although such measurements are difficult to make accurately, a few definitive experiments have been carried out with the use of ^{14}C radioactively labeled initiator molecules. The results for several polymer systems are shown in Table 3.5.

STUDY QUESTIONS

1. Write the set of elementary reactions analogous to (6) to (12) that comprise the free-radical chain polymerizations of vinyl acetate $(CH_2=CHOC(O)CH_3)$ and methyl methacrylate $(CH_2=C(CH_3)C(O)OCH_3)$ when the process is initiated by azobisisobutyronitrile. If carbon tetrachloride is added to the systems, describe the effects of chain transfer with carbon tetrachloride on the polymer composition and average molecular weight.

2. Show by equations analogous to those in Problem 1 how branching may arise in the free-radical polymerization of isoprene $(CH_2=C(CH_3)CH=CH_2)$ initiated by di-t-butyl peroxide even in the absence of chain transfer. How do you think the presence of dissolved oxygen (which may be viewed as a diradical) would affect the polymerization?

3. The first-order rate constant for the decomposition of diethyl peroxide has been reported to be $1.0 \times 10^{14} e^{-35,000\,cal/RT}$ s^{-1} [W. A. Pryor, D. M. Huston, T. R. Fiske, T. L. Pickering, and E. Ciuffarin, $J.\,Am.\,Chem.\,Soc.$, **86**, 4237 (1964)]. Predict the temperature range in which diethyl peroxide would be a useful initiator.

4. J. C. Bevington [$Trans.\,Faraday\,Soc.$, **51**, 1392 (1955)] has shown that the fraction of radicals produced in the thermal decomposition of azobisisobutyronitrile that actually initiate the polymerization of styrene in benzene increases with increasing monomer concentration from a value of 0.17 at the lowest monomer concentration used, to an apparent limiting value of 0.65. Discuss possible reasons for the increase in this fraction as the monomer concentration is increased and explain why an efficiency of greater than 0.65 cannot be achieved.

5. The following results [J. C. Bevington, H. W. Melville, and R. P. Taylor, $J.\,Polymer\,Sci.$, **12**, 449 (1954)] have been obtained for the temperature dependence of the percentage of chain termination by disproportionation in methyl methacrylate:

t (°C)	40	60	80
% Disproportionation	50	59	70

(a) Calculate the difference in activation energy between disproportionation and combination.

(b) At what temperature will termination be 90% by combination?

6. K. C. Berger [$Makromol.\,Chem.$, **176**, 3575 (1975)] has recently reported the following percentages of termination by combination in styrene polymerization:

t (°C)	30	52	62	70	80
% Combination	86	80	77	68	60

(a) Evaluate the difference in activation energy between disproportionation and combination in the termination step.

(b) Calculate the temperature at which the rates of combination and termination are equal.

7. Using the chemical literature for the necessary data, evaluate the "useful temperature range" for ethylsilver as an initiator.

8. Discuss the difference between free-radical chain termination and the termination of a growing free radical. How can the latter be proceeding without the former?

SUGGESTIONS FOR FURTHER READING

ALEXANDER, E. A., AND D. H. NAPPER, "Emulsion Polymerization," *Progr. Polymer Sci.* (A. D. Jenkins, ed.), **3**, 145 (1971).

BEVINGTON, J. C., *Radical Polymerization*. New York: Academic Press, 1961.

BLACKLEY, D. C., *Emulsion Polymerization*. New York, Wiley: 1975.

BOUNDY, R. H., and R. F. BOYER (eds.), *Styrene: Its Polymers, Copolymers, and Derivatives* (Parts I–III), ACS Monograph Series. Darien, Conn.: Hafner, 1970.

EHRLICH, P., AND G. A. MORTIMER, "Fundamentals of Free Radical Polymerization of Ethylene," *Adv. Polymer Sci.*, **7**, 386 (1970).

FLORY, P. J., *Principles of Polymer Chemistry*. Ithaca, N.Y.: Cornell University Press, 1953.

KOLESKE, J. V., AND L. H. WARTMAN, *Poly(vinyl chloride)*, New York: Gordon and Breach, 1969.

O'DRISCOLL, K. F., AND T. YONEZAWA, "Application of Molecular Orbital Theory to Vinyl Polymerization," *Rev. Macromol. Chem.*, **1**, 1 (1966).

TAKEMOTO, K., "Preparation and Polymerization of Vinyl Heterocyclic Compounds," *J. Macromol. Sci.—Rev. Macromol. Chem.*, **C5**, 29 (1970).

THOMAS, W. M., "Mechanism of Acrylonitrile Polymerization," *Fortschr. Hochpolymer.-Forsch.*, **2**, 401 (1961).

VOLLMERT, B., *Grundriss der makromolekularen Chemie*, pp. 37–60. Berlin: Springer-Verlag, 1962.

WALL, L. A. (ed.), *Fluoropolymers*. New York: Wiley–Interscience, 1971, 1972.

WALLING, C., *Free-Radicals in Solution*, Chap. 3. New York: Wiley, 1957.

YOKUM, R. H., AND E. B. NYQUIST, *Functional Monomers*. New York: Dekker, 1973.

4

Ionic and Coordination Polymerization

An ionic polymerization is an addition polymerization in which the growing chain ends bear a negative or a positive charge:

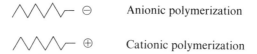

If the growing chain end bears a negative charge, the process is called an *anionic polymerization*. If the chain end bears a positive charge, the reaction is a *cationic polymerization* process. These two alternative mechanisms are associated with different catalyst systems and different reaction conditions. Some polymerization reactions, particularly those initiated by Ziegler–Natta or transition metal complexes, are best described as "coordination polymerizations," since the reaction mechanisms are believed to involve complexes formed between the transition metal and the π-electrons of the monomer. However, most of the known reactions of this type bear many similarities to anionic polymerizations and they are considered here as part of that classification. Thus, the rest of this chapter is divided into a discussion of first anionic, and then cationic processes. It should be emphasized from the beginning that the reaction conditions for ionic polymerizations differ markedly from those discussed in Chapter 3 for free-radical reactions.

ANIONIC POLYMERIZATION

Initiators[1] for Anionic Polymerization

Table 4.1 lists different classes of initiators that are known to initiate anionic- or coordination-type polymerizations. Of these initiators the alkali metal suspensions are prepared by dispersion of a molten alkali metal in an inert organic solvent; organolithium reagents are prepared by the reaction of lithium metal with an organic halide; and Grignard reagents are obtained by the reaction of magnesium

TABLE 4.1 Anionic and Coordination Initiators

Initiator Class	Formula or Example
1. Alkali metal suspensions	Sodium in tetrahydrofuran or in liquid ammonia
2. Alkyl or aryllithium reagents	nC_4H_9Li
3. Grignard reagents	RMgX (R = Alkyl or aryl, X = Halogen)
4. Aluminum alkyls	AlR_3
5. Organic radical anions	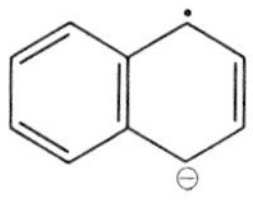 Na$^+$ (Sodium naphthalenide)
6. Some Ziegler–Natta catalysts	$TiCl_4 + AlR_3$
7. Transition metal π-allyl complexes	$(\pi C_4H_7)_2Ni$
8. Transition metal oxides	Oxides of V, Ti, Cr, Co, Ni, or W
9. Ionizing radiation	X-rays, γ-rays, or electrons

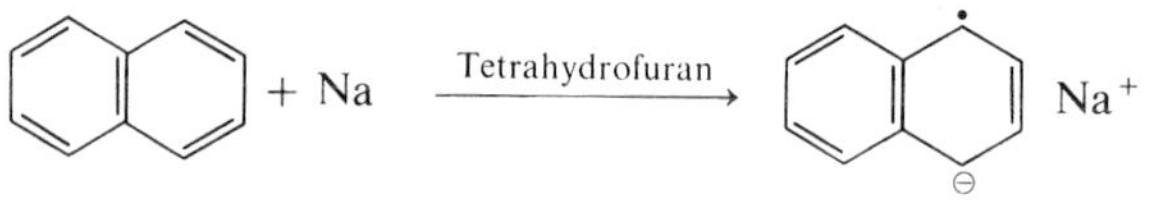

turnings with an organic halide. These catalysts and the aluminum alkyls can be obtained commercially. However, radical anions such as sodium naphthalenide must be prepared from naphthalene and a sodium mirror in an ether-type solvent (reaction 1). The appearance of an intense green color accompanies the formation of this species. Water must be rigorously excluded from the system.

(1)

As a precautionary measure, it should be noted that aluminum alkyls are spontaneously flammable in the atmosphere, and Grignard reagents and organ-

---

[1] The terms "initiator" and "catalyst" are often used interchangeably in polymer chemistry. However, the initiating species is usually not recoverable at the end the reaction, and the word "catalyst" is, therefore, something of a misnomer.

olithium compounds may explode if isolated in the solid state. Hence, these reagents are nearly always handled as solutions in organic solvents under an inert atmosphere.

Monomers for Anionic Polymerization

Typical monomers that can be polymerized by anionic mechanisms include *styrene* (**1**), *methyl methacrylate* (**2**), and *acrylonitrile* (**3**). The important thing to remember

is that monomers which are suitable for anionic polymerization generally contain *electron-withdrawing substituent groups*. Many anionic polymerizations are characterized by *highly colored* reaction mixtures.

A Typical Experimental Procedure

As an example, we will consider the polymerization of styrene using *n*-butyllithium as a catalyst. The reaction takes place in an organic solvent such as ether or tetrahydrofuran. Water and carbon dioxide must be *rigorously* excluded from the solvent, the monomer, the catalyst, and from the inside of the reaction apparatus. The monomer itself should be very pure and free from inhibitors. Furthermore, some means must be utilized for the introduction of very small measured quantities of the catalyst species.

An apparatus that can be used is shown in Figure 4.1. It consists of a glass three-necked flask fitted with a rubber serum cap stopper, a mercury or silicone oil bubbler, and with the flask connected to a solvent distillation system. A Teflon-covered magnetic stirrer bar is contained in the reaction flask. A stream of dry nitrogen or helium can be used to sweep the air from the system.

The solvent (in this case tetrahydrofuran) is dried by distillation from calcium hydride or lithium aluminum hydride[1] into the reaction flask. Care should be taken that some tetrahydrofuran remains behind in the distillation flask to avoid the possibility that explosive peroxides will be concentrated in the residue. Once the solvent has been distilled into the flask, the condenser system can be removed and replaced by a glass stopper.

Commercial monomers, such as styrene, always contain small amounts of "inhibitors"—compounds that retard the free-radical polymerization process. Usually, the inhibitor can be removed by the passage of the liquid monomer through

---

[1] Under no circumstances should these hydrides be allowed to come into contact with water, or a fire or explosion could result. Lithium aluminum hydride in particular should be handled with great care, and distillations of solvent from this drying agent should be carried out behind a shield.

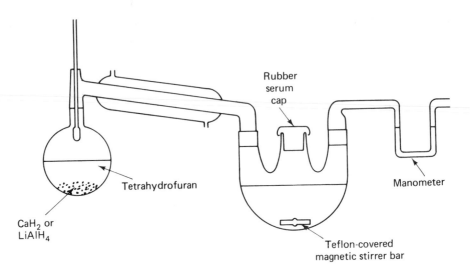

Fig. 4.1 Apparatus for the anionic polymerization of styrene with the use of *n*-butyllithium as an initiator.

a chromatography tube that has been packed with alumina. Vacuum distillation of the monomer can also be used as a purification process. The styrene can be introduced into the reaction vessel from a hypodermic syringe through the serum cap stopper. At this stage, the temperature of the reaction mixture is often lowered.

The catalyst—in this case, *n*-butyllithium—can be obtained commercially as a solution of known concentration in a hydrocarbon such as pentane. Its concentration should be checked by acid–base titration or from the amount of butane liberated on hydrolysis. A known amount of catalyst solution is then added by hypodermic syringe through the serum cap stopper into the stirred solution of the monomer. A color change occurs rapidly from colorless to orange. After completion of the reaction, the active chain ends are destroyed by the addition of water or Dry Ice, the solution becomes colorless, and the polymer can be isolated. If Dry Ice is used for termination, a subsequent treatment with dilute aqueous acid will be needed to convert the lithium carboxylate end groups to free carboxylic acid residues.

It should be noted that, in many anionic polymerizations, the amount of catalyst added determines the chain length of the polymer (see Chapter 13, equation 9). In principle, each catalyst molecule will generate one polymer chain. Thus, the more catalyst that is added, the lower will be the molecular weight. In fact, the polystyrene prepared by this method has a very narrow molecular-weight distribution, and this is a consequence of the small amount of chain transfer in the system.

Similar experimental procedures can be employed when other catalysts are used. Sodium naphthalenide is more sensitive to moisture and carbon dioxide than alkyllithium reagents: consequently, a glass high-vacuum system is frequently employed for the reaction of naphthalene with a sodium mirror. The solvent must be ultra-dry before sodium naphthalenide can be used.

70   *Synthesis and Reactions of Polymers*   /   *Part I*

The Mechanism of Anionic Polymerization
(*General*)

The overall reaction can be divided into initiation, propagation, and termination steps.

1. *Initiation* takes place by addition of the catalyst across the double bond of the monomer (reaction 2). In this sequence, R′ represents an electron-withdrawing group such as phenyl or cyano. Sequence (2) is actually an oversimplification

$$
\begin{array}{ccc}
 & R' & & R' \\
 & | & & | \\
CH_2\!=\!C + R\!-\!M & \longrightarrow & R\!-\!CH_2\!-\!C^{\ominus} \;\; M^{\oplus} \\
 & | & & | \\
 & H & & H
\end{array}
\tag{2}
$$

because some organometallic initiators, such as *n*-butyllithium, exist as aggregates in solution. Thus, breakdown of, say, a tetrameric aggregate may be necessary before initiation can occur. Most anionic catalysts have the capacity to react as an organometallic ion pair. For example, although *n*-butyllithium is generally considered to have a covalent type of structure, heterolytic cleavage of the lithium–carbon bond to give ions can occur in the presence of a suitable monomer (3). Note that, in

$$
nC_4H_9Li \longrightarrow nC_4H_9^{\ominus} \;\; Li^{\oplus}
\tag{3}
$$

undergoing addition to the monomer, the ion pair adds in such a way that the anion from the catalyst becomes attached to the carbon atom farthest from the electron-withdrawing group, R′. This is, of course, the carbon atom that possesses the lower electron density (4). Note also that in initiation and in the subsequent steps, the

$$
\begin{array}{ccc}
 & R' & & R' \\
 & \uparrow & & | \\
^{\delta+}CH_2\!=\!CH^{\delta-} + R^{\ominus} \;\; M^{\oplus} & \longrightarrow & R\!-\!CH_2\!-\!C^{\ominus} \;\; M^{\oplus} \\
 & & & | \\
 & & & H
\end{array}
\tag{4}
$$

active site at the end of the chain is accompanied by a counterion. The presence of the counterion explains many of the differences from free-radical polymerization.

2. *Propagation* then involves the successive insertion of monomer molecules into the terminal "ionic" bond (5). This process of chain growth continues until all the monomer has been consumed, or until the reaction is terminated.

$$
\begin{array}{ccc}
 R' & & R' \quad\quad R' \\
 | & & | \quad\quad\; | \\
R\!-\!CH_2\!-\!C^{\ominus} \; M^{\oplus} & \longrightarrow & R\!-\!CH_2\!-\!C\!-\!CH_2\!-\!C^{\ominus} \; M^{\oplus} \;\; \text{etc.} \\
 | & & | \quad\quad\; | \\
 H & & H \quad\quad H
\end{array}
$$

$$
\begin{array}{c}
 R' \\
 | \\
 CH_2\!=\!C \\
 | \\
 H
\end{array}
\tag{5}
$$

3. *Chain transfer or chain branching* does not occur to any appreciable extent with anionic systems, and this is especially true if the reaction is carried out at low temperatures.

4. *Termination* of the chains occurs either accidentally or deliberately when the active chain end reacts with a molecule of carbon dioxide or with water, alcohols, or other protonic reagents (6 or 7). However, it is important to note that, if termination

$$R-\text{\footnotesize$\mathcal{W}$}\overset{\displaystyle R'}{\underset{\displaystyle H}{C}}\text{—H} + \text{MOH} \qquad (6)$$

$$R-\text{\footnotesize$\mathcal{W}$}\overset{\displaystyle R'}{\underset{\displaystyle H}{C}^{\ominus}}\ M^{\oplus}$$

$\xrightarrow{H_2O}$

$\xrightarrow[\text{HCl}]{\substack{CO_2 \\ \text{followed by}}}$

$$R-\text{\footnotesize$\mathcal{W}$}\overset{\displaystyle R'}{\underset{\displaystyle H}{C}}\text{—COOH} + \text{MCl} \qquad (7)$$

reagents are absent, the chains could remain active indefinitely. In practice, it is impossible to remove all traces of water molecules from the inside of glass equipment and, even if that were possible, the Si—OH groups at the glass surface could presumably function as termination agents. However, in a well-dried system the brilliant colors that indicate the integrity of the active end groups may persist for days.

"Living" Polymers

The term "*living*" *polymer* is applied to ionic polymerizations which are not terminated. These systems have the capability of continuing chain growth if more of the same monomer or, indeed, a second monomer is added. Provided that chain terminators are absent, there is no reason in theory why chain growth should stop if more and more monomer is added. In practice, small amounts of terminators are inevitably introduced. Chain growth also slows down eventually because of the high viscosity of the system or because the chains become insoluble.

Initiators that yield some of the most dramatic "living" polymerizations are those which form radical anions during the initiation process. Metallic sodium or sodium naphthalenide function in this way. Consider first the role of sodium metal. The polymerization process can be divided into discrete steps, as follows:

1. *Initiation* occurs when an electron is transferred from sodium to the monomer to generate a radical anion (8).

$$\text{CH}_2\!\!=\!\!\overset{\displaystyle R}{\underset{\displaystyle H}{C}} + \text{Na} \longrightarrow \cdot\text{CH}_2\!\!-\!\!\overset{\displaystyle R}{\underset{\displaystyle H}{C}^{\ominus}}\ \text{Na}^{\oplus} \qquad (8)$$

2. *Dimerization* of the radical anion then takes place to form a *di*anion (9).

$$\text{Na}^{\oplus} \quad \overset{\overset{\displaystyle R}{\displaystyle |}}{\underset{\overset{\displaystyle |}{\displaystyle H}}{\overset{\ominus}{C}}}{-}CH_2 + \cdot CH_2{-}\overset{\overset{\displaystyle R}{\displaystyle |}}{\underset{\overset{\displaystyle |}{\displaystyle H}}{\overset{\ominus}{C}}} \quad \text{Na}^{\oplus} \quad \longrightarrow \quad \text{Na}^{\oplus} \quad \overset{\overset{\displaystyle R}{\displaystyle |}}{\underset{\overset{\displaystyle |}{\displaystyle H}}{\overset{\ominus}{C}}}{-}CH_2{-}CH_2{-}\overset{\overset{\displaystyle R}{\displaystyle |}}{\underset{\overset{\displaystyle |}{\displaystyle H}}{\overset{\ominus}{C}}} \quad \text{Na}^{\oplus} \quad (9)$$

3. *Propagation* can then occur *at both ends* of the dimer (9) by the insertion of monomer molecules into the ionic bonds. Thus, the growing polymer chain has the structure shown in (10).

$$\text{Na}^{\oplus} \quad \overset{\overset{\displaystyle R}{\displaystyle |}}{\underset{\overset{\displaystyle |}{\displaystyle H}}{\overset{\ominus}{C}}}{-}W{-}CH_2{-}CH_2{-}W{-}\overset{\overset{\displaystyle R}{\displaystyle |}}{\underset{\overset{\displaystyle |}{\displaystyle H}}{\overset{\ominus}{C}}} \quad \text{Na}^{\oplus} \quad (10)$$

Sodium naphthalenide functions as a catalyst in a very similar way. The naphthalene radical anion (**4**) either transfers an electron to the monomer to form a mono-

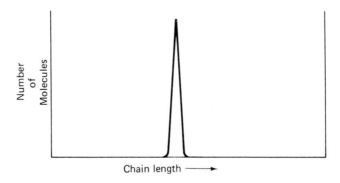

4

mer radical anion (as described above), or the naphthalene radical anion itself may initiate monomer polymerization directly.

One of the characteristics of "living" polymer systems is that they yield polymers with very narrow molecular-weight distributions (Figure 4.2). This phenomenon is discussed in more detail in Chapter 13. However, from a qualitative point of view, it is sufficient to note that the narrow molecular-weight distribution is largely a consequence of the rapidity of the initiation step. All the chains are initiated essentially at the same instant (at the point of catalyst injection), and all of them grow

Fig. 4.2 Exremely narrow molecular-weight distribution that is characteristic of anionic and radical–anionic polymerization products.

at the same rate until the monomer is consumed. The fact that chain transfer (e.g., proton abstraction) does not occur at low temperatures also prevents a broadening of the molecular-weight distribution. Thus, anionic reactions are much "cleaner" than free-radical polymerizations.

Copolymerization

Because many anionic chain ends remain "alive," particularly at low temperatures, even though the monomer has been consumed, chain growth can be restarted by the addition of more of the same monomer—or addition of a different monomer. If a different monomer is added, a copolymer is formed. If A represents the first monomer and B the second, a growing copolymer might be represented by the structure shown in **5**. Subsequently, either more of the monomer A, or a third monomer, C, might be added. In this way it is possible to synthesize *block copolymers*.

$$^{\ominus}BBBBBBBBAAAAAAAACH_2-CH_2AAAAAAAABBBBBBBB^{\ominus}$$

5

In practice some restrictions exist with respect to which monomers can be used in a block-copolymerization reaction. The two monomers should have similar electron affinities if mutual reinitiation is to take place. However, if monomer B has a much higher electron affinity (i.e., it contains a more powerful electron-withdrawing substituent) than monomer A, then, although a terminal unit of an A-block will initiate the propagation of a B-block, the terminal unit of the B-block will not initiate the addition of more molecules of A. In other words, terminal residues of type B cannot bond to monomer A. However, they will initiate propagation of an even more polar monomer, C.

This is illustrated schematically in (11). For example, an active chain made up of

$$-\text{w}-BBBBB^{\ominus} \quad \begin{matrix} \overset{A}{\nearrow} & \text{No initiation} \\ \underset{C}{\searrow} & \\ & -\text{w}-BBBBBCCCCCC^{\ominus} \end{matrix} \qquad (11)$$

methyl methacrylate units (B) will undergo continued growth in the presence of acrylonitrile (C), but not in the presence of styrene (A). The polarographic reduction potential of a monomer provides a good measure of its electron affinity. Thus, a block copolymerization series can be worked out from the relative order of the reduction potentials.

Catalysis by Grignard Reagents

Grignard reagents, of formula, RMgX, have the ability to initiate the polymerization of some vinyl compounds. Many of these reactions are believed to proceed by an anionic mechanism.

One of the structures that exists in an etheric solution of a Grignard reagent is shown in **6**. The structure consists basically of a coordination complex of a

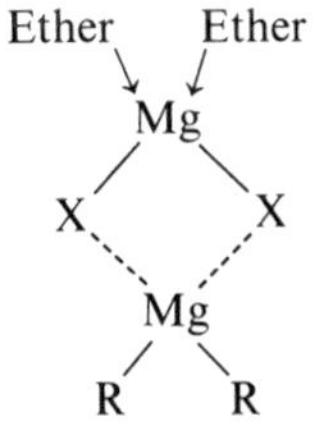

6

magnesium dialkyl or diaryl (MgR_2) and a magnesium halide (MgX_2). The complex is stabilized by coordination with the oxygen atoms of two ether molecules.

A monomer, such as methyl methacrylate (**2**) can itself coordinate with the complex by displacement of one of the ether molecules. Thus, the *initiation* process is shown in reaction (12). In the monomer–catalyst complex (**7**) *the catalyst is held*

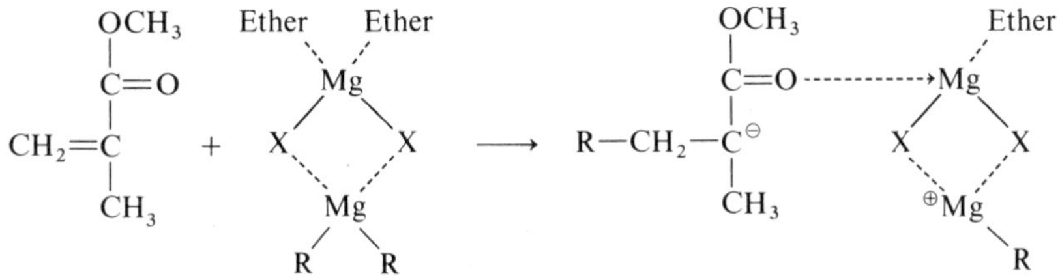

     2   **7**   (12)

in a special orientation relative to the monomer. This occurs because the carbonyl group coordinates to a magnesium atom. The overall result is that the complex is able to direct an incoming monomer molecule into a specific orientation prior to addition of the new monomer to the chain end (**8**). Such species are called *stereospecific initiators.*

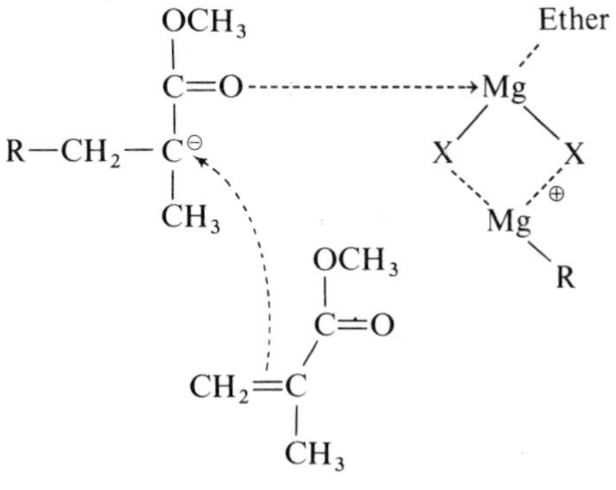

8

Ziegler–Natta Catalysts

In the early 1950s, Karl Ziegler and his coworkers discovered that ethylene reacts with aluminum alkyls in a high-pressure system to yield organometallic oligomers or polyethylene (reaction 13). It was subsequently found that the addition of

$$AlEt_3 + 3n\,CH_2{=}CH_2 \longrightarrow Al[(CH_2{-}CH_2)_n{-}Et]_3 \qquad (13)$$

transition metal "*cocatalysts*", such as $TiCl_4$ or VCl_4, to the aluminum alkyl generated a system that would polymerize ethylene at *atmospheric pressure* and *room temperature*. High-molecular-weight polyethylene can, in fact, be made by bubbling ethylene gas into a suspension of the catalyst in a liquid-hydrocarbon medium. Natta and coworkers in Italy found that catalysts of this type induce the formation of crystalline, stereoregular polymers. Since that time an enormous amount of subsequent research has been carried out on these polymerizations, and their industrial utility has been exploited widely. Ziegler–Natta catalysis has such wide ramifications that a separate volume would be needed to do it justice. However, in the following sections, we will consider briefly some of the general characteristics of these reactions, the stereoregular nature of the polymerization, the composition of the catalyst, the reaction mechanisms, and an extension of this catalytic pathway to the polymerization of dienes and cycloolefins.

General features of the reaction

First, it must be emphasized that Ziegler–Natta catalysis has the capability to yield unbranched and stereospecific polymers. The polyethylene produced by this process is linear and has a higher density than that prepared by free-radical techniques. Ziegler–Natta polymerization of propylene can yield an isotactic stereoregular form of polypropylene that has extremely useful technological properties. Many nonpolar unsaturated organic monomers can be polymerized in a similar way. The following example illustrates how such polymerizations are carried out on a laboratory scale.[1]

A catalyst system can be prepared by the successive addition of solutions of titanium tetrachloride and triisobutylaluminum[2] in decahydronaphthalene to decahydronaphthalene diluent under an atmosphere of dry nitrogen. A brown-black suspended precipitate is formed which changes to a deep violet color when the mixture is heated to 185°C for 40 min. This is called the "*aging*" step. The mixture is then cooled, cyclohexane is added as a reaction solvent, and additional triisobutylaluminum is introduced to yield a purplish-black suspension. The polymerization of ethylene can be accomplished simply by allowing ethylene gas to bubble through this mixture. The flask must be cooled by an external water–ice bath because heat is evolved during the polymerization. The polyethylene precipitates

---

[1] For specific details see, for example, W. R. Sorenson and T. W. Campbell, *Preparative Methods of Polymer Chemistry*, 2nd ed. (New York: Wiley–Interscience, 1967), pp. 289–312.

[2] Aluminum alkyls are dangerously pyrophoric when exposed to atmospheric oxygen or water. They should be handled only in an inert atmosphere and with appropriate safety precautions.

for as long as ethylene is introduced into the system. The polymer can be isolated by pouring the slurry into stirred isopropanol, followed by filtration. Isotactic polypropylene is prepared in the same way by allowing propylene gas to bubble into the catalyst suspension.

Stereoregularity

As discussed in a later section of this chapter, monomers that are asymmetric with respect to the disposition of side groups about the double bond can yield polymers with a specific stereochemical arrangement of the side groups. Ziegler–Natta catalysts generate a high degree of stereoregularity. Vinyl monomers usually yield isotactic polymers with these initiator systems, although syndiotactic polypropylene can be formed under certain conditions. The degree of stereoregularity generated depends on a number of factors, including the homogeneous or heterogeneous nature of the catalyst, the detailed composition and history of the catalyst, and the nature of the side group in the monomer. The influence of the catalyst will be discussed in the next section.

Composition of the catalyst

A Ziegler–Natta catalyst is made from two components; (1) a transition metal compound from groups IVB to VIIIB of the periodic table, and (2) an organometallic compound, usually derived from a group IA to IIIA metal. The transition metal component employed is usually a halide or oxyhalide of titanium, vanadium, chromium, molybdenum, or zirconium, and the second component often consists of an alkyl, aryl, or hydride of aluminum, lithium, magnesium, or zinc. Perhaps the best known systems are those derived from $TiCl_4$ or $TiCl_3$ and an aluminum trialkyl. The catalyst systems may be heterogeneous (some titanium-based systems) or soluble (most vanadium-containing species). The stereoregularity of the catalytic process can also be altered by the addition of Lewis bases, such as amines.

Changes in the catalyst system affect the yield of polymer, the chain length, and the degree of stereoregularity. The effects of gross catalyst changes on the stereoregularity of polypropylene are shown in Table 4.2. For one particular aluminum alkyl, changes in the transition metal halide affect the percentage stereoregularity, as shown in Table 4.3. The stereoregularity of the polypropylene *decreases* with increasing size of the organic group attached to aluminum.

The nature of Ziegler–Natta catalyst systems is still a subject for debate. The soluble catalysts appear to have well-defined structures. For example, the catalyst system generated from bis(cyclopentadienyl)titanium dichloride and triethylaluminum has a halogen-bridged structure (**9**).

$$\begin{array}{ccc} \mathrm{Cp} & \mathrm{Cl} & \mathrm{C_2H_5} \\ & \mathrm{Ti} \quad \mathrm{Al} & \\ \mathrm{Cp} & \mathrm{Cl} & \mathrm{C_2H_5} \end{array}$$

9

TABLE 4.2 Stereoregularity of Polypropylene with Different Catalyst Systems

Catalyst	Stereoregularity (%)
R_3Al^* + $TiCl_4$	35.2
R_3Al + α-$TiCl_3$	84.7
R_3Al + β-$TiCl_3$	45
R_3Al + $TiCl_4$ + NaF	97
R_3Al + $TiCl_4$ + compounds of P, As, or Sb	98
R_3Al + $TiCl_3$ + amine	81
R_3Al + $Ti(O$-iso-$Bu)_4^\dagger$	20
R_3Al + $V(acac)_3^\ddagger$	0
R_3Al + $Ti(C_5H_5)Cl_2$	70–90
R_3Al + $Ti(C_5H_5)_2Cl_2$	85
$R_2AlX\S$ + $TiCl_3$	90–99
$RAlX_2$ + γ-$TiCl_3$ + amine	>99
$RAlX_2$ + $TiCl_3$ + HPT$^\parallel$	97
RNa + $TiCl_3$	90
RNa + $TiCl_4$	90
RLi + $TiCl_4$	90
R_2Zn + $TiCl_3$	65
R_2Zn + $TiCl_3$ + amine	93

* R = alkyl.
† Bu = butyl.
‡ (acac) = acetylacetonate.
§ X = halogen.
∥ HPT = hexamethyl phosphoric triamide.

Source: Reprinted by permission. D. O. Jordan, *The Stereochemistry of Macromolecules*, Vol. I, A. D. Ketley, ed. (New York: Marcel Dekker, Inc., 1967).

TABLE 4.3 Influence of the Transition Metal on the Stereoregularity of Polypropylene*

Transition Metal Compound	Stereoregularity (%)
$TiCl_4$	48
$TiBr_4$	42
$TiCl_3$, α, γ, or δ	80–92
$TiCl_3$, β	40–50
$ZrCl_4$	55
VCl_3	73
$CrCl_3$	36
VCl_4	48
$VOCl_3$	32

* The organometallic compound is $Al(C_2H_5)_3$ in each case.

Source: Reprinted by permission. D. O. Jordan, *The Stereochemistry of Macromolecules*, Vol. I, A. D. Ketley, ed. (New York: Marcel Dekker, Inc., 1967).

However, $TiCl_4$ or $TiCl_3$ give rise to much more complex initiator systems, the structure of which is still not clear. One fact does appear to be certain, the true catalysts are *not* simple coordination adducts formed from the original metal halide and aluminum alkyl. A critical "aging" period for the catalyst is often needed before it achieves its highest activity, and complex reactions occur during this period. These reactions probably included an initial exchange of substituent groups between the two metals to form transition metal–carbon bonds by interactions such as those shown in (14) and (15). These organotitanium halides are unstable

$$AlR_3 + TiCl_4 \rightleftharpoons R_2AlCl + RTiCl_3 \qquad (14)$$

$$R_2AlCl + TiCl_4 \rightleftharpoons RAlCl_2 + RTiCl_3 \qquad (15)$$

and can undergo reductive decomposition processes, such as those shown in (16) and (17). Note that $TiCl_3$ can be used as an initial catalyst component in place of

$$RTiCl_3 \longrightarrow R\cdot + TiCl_3 \qquad (16)$$

$$R_2TiCl_2 \longrightarrow R\cdot + RTiCl_2 \qquad (17)$$

$TiCl_4$. Further reduction may yield $TiCl_2$, which can itself react with the aluminum trialkyl ligand-exchanged species. For the heterogeneous catalysts, the reactions are more complicated than is implied by these equations. However, the analogous vanadium-containing systems are soluble and may well be represented fairly accurately by reactions such as those shown in (14) to (17).

In all these systems one of the most important steps is the *reduction* of the transition metal to a low-valency state in which the metal possesses unfilled ligand sites. These low-valency transition metal species are believed to be the real catalysts or precursors of the real catalysts.

The polymerization mechanism

Convincing evidence now exists that Ziegler–Natta catalysts function by the formation of transient π-complexes between the olefin and the low-valence transition metal species. Stable π-complexes between olefins and transition metals are well known. The complexes are held together by an overlap of the d-orbitals of the transition metal with the π-orbitals of the olefin (Figure 4.3). As shown in Figure 4.3a, the d_{xy} orbital of the metal can overlap the π-antibonding orbitals of the olefin. Overlap is also possible between one lobe of a $d_{x^2-y^2}$ orbital and an olefin π-orbital, as shown in Figure 4.3b. In Ziegler–Natta catalysis, π-complex formation could not occur if the transition metal occupied its highest valence state (in other words, if all the coordination sites were occupied by strongly bound ligands). Hence, the reduction process outlined above is critically important for catalyst activity. However, the exact details of the coordination behavior and the olefin addition mechanism are still obscure.

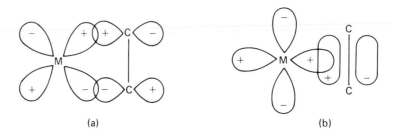

(a)   (b)

Fig. 4.3   Orbital overlap scheme for the formation of a π-type interaction between an olefin and a transition metal. (a) Use of the π-antibonding orbitals of the olefin. (b) Overlap of one lobe of a $d_{x^2-y^2}$ orbital from the metal and a π-bonding orbital of the olefin.

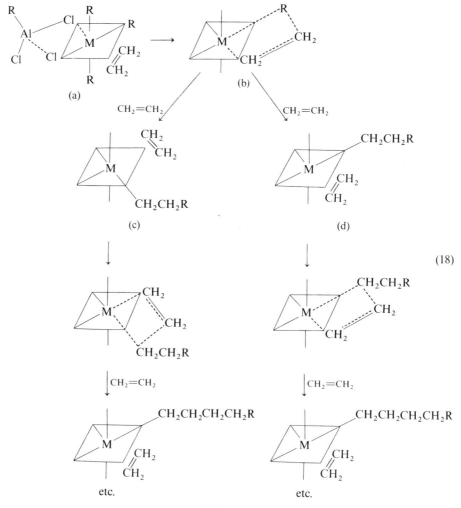

Scheme 1

Two general mechanisms have been proposed, one involving coordination of the olefin to a vacant site on the transition metal (the "monometallic" mechanism), and the other supposing a participation by both the transition metal and the aluminum atom (the "bimetallic" mechanism).

The monometallic mechanism is summarized in Scheme 1. The olefin (in this case, ethylene) becomes coordinated to a vacant site on the transition metal coordination sphere (a). A transfer of an organic group, R, then takes place from the metal to the olefin (b). In other words, the olefin becomes inserted into the weak metal–carbon bond. An incoming olefin molecule then coordinates to the new vacant site just formed, ready for insertion into the metal–alkane bond. The mode of insertion will depend on the geometry of the switch to a metal–carbon σ-bond and on the location of the vacant coordination site (c or d). It will be clear that this mechanism allows a partial understanding of the reasons for stereospecific addition. The side groups on the olefin sterically determine the coordination and insertion geometries. The presence of the π-bond would prevent orientational scrambling once the monomer is attached to the catalyst.

The bimetallic mechanism is summarized in Scheme 2. The alkyl groups in the catalyst (a) are presumed to function as bridging units. Such behavior is well documented in other organometallic species. The initial π-coordination of the olefin [as shown in (b)] is followed by an insertion of the olefin into a titanium–carbon bond to yield a new bridged species (d). The coordination and insertion of successive olefin molecules can then take place by the same pathway.

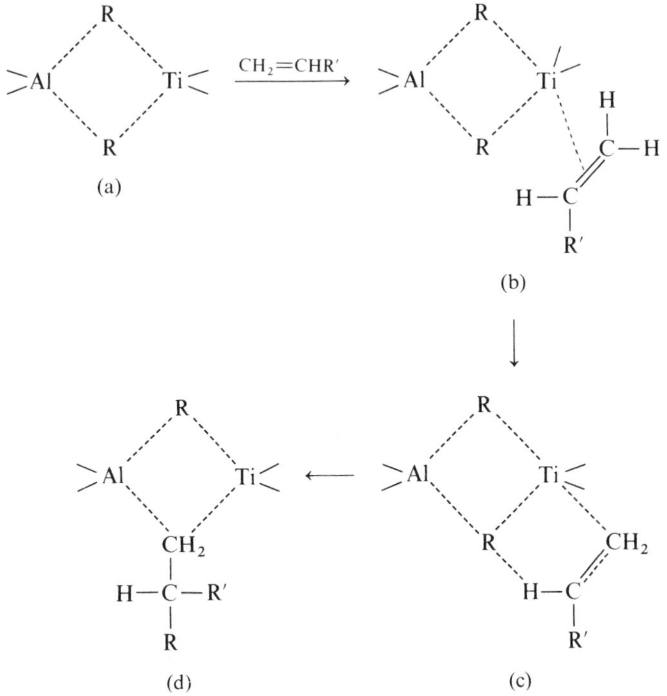

Scheme 2

Both mechanisms seem plausible for catalysis in solution. When a hetero-geneous-type system is present, defects in the crystal surface may also be responsible for the stereoregular character of a polymerization.

Diene and cycloolefin polymerization

Ziegler–Natta initiators are effective catalysts for the polymerization of dienes as well as monoolefins. 1,3-Butadiene or isoprene undergo stereospecific poly-merization with Ziegler–Natta systems to yield *trans*-1,4-, *cis*-1,4-, syndiotactic, or isotactic-1,2-polybutadiene, or *trans*-1,4-, *cis*-1,4-, or 3,4-polyisoprene, depending on the particular catalyst employed. For example, an AlR_3/VCl_4 system gives a 97 to 98% yield of *trans*-1,4-polybutadiene, but an AlR_3/TiI_4 catalyst yields 93 to 94% of the *cis*-1,4-polymer. The reasons for these effects are still a matter for speculation.

Finally, it is known that Ziegler–Natta catalysts can be used for the ring-opening polymerization of cycloolefins such as cyclobutenes, cyclooctadienes, cyclodecadienes, and so on. Cyclobutene itself gives a mixture of *cis*- and *trans*-polybutadiene in the presence of $Al(C_2H_5)_3/TiCl_4$ catalyst systems.

π-Allyl Complexes as Catalysts

The polymerization of diolefins, such as butadiene, is an important technological problem. It is of considerable interest, therefore, that transition metal π-complexes, such as $(\pi\text{-}C_4H_7)_2Ni$ (**10**), can catalyze the conversion of butadiene to high poly-mers. In fact, a broad range of related complexes, such as $(\pi\text{-}C_4H_7)_3Cr, (\pi\text{-}C_4H_7)_3Nb,$ $(\pi\text{-}C_4H_7)_4Ti,$ or $(\pi\text{-}C_4H_7)_4Zr$ behave in the same way. The yields, rates, and stereospecificity of the polymerizations are comparable to those found in Ziegler–Natta catalysis.

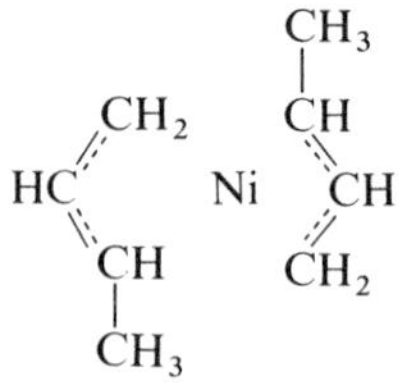

10

However, the mechanism of catalysis by π-allyl complexes is by no means fully established. It is known that the polymer formed in such reactions can be *cis*- or *trans*-polybutadiene and it is believed that the mode of attachment of the diene to the metal determines the stereochemistry of addition. Moreover, it is known that π-allyl complexes can be formed from, and may rearrange to, σ-complexes in which the ligand is covalently bound to the metal through one carbon atom (**11**). A transient rearrangement of this type (19) could strongly influence the pattern of diene insertion into the ligand–metal bond.

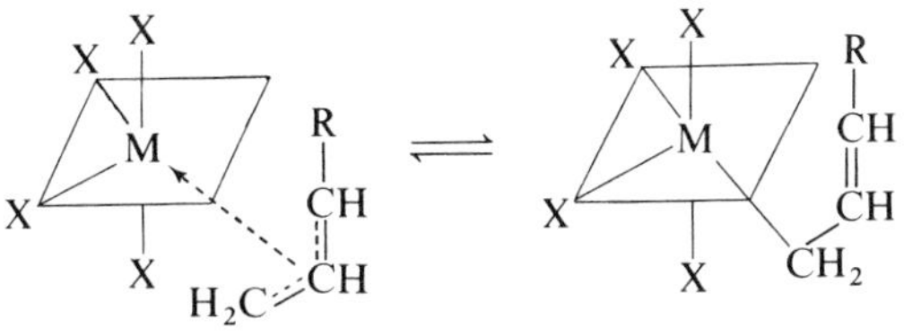

(19)

(a)   (b)

11

Other Anionic-Type Catalysts

A variety of transition metal oxides supported on alumina, silica, or charcoal induce the stereospecific polymerization of dienes. Typical catalysts include the oxides of vanadium, titanium, chromium, cobalt, nickel, and tungsten. The mechanisms of catalyst action are in most cases only partly understood. However, some type of π-coordination mechanism, similar to Ziegler–Natta catalysis, may exist.

 Dienes can also be polymerized stereospecifically by the use of Alfin catalysts. These are heterogeneous catalyst systems formed by the interaction of, for example, allylsodium, sodium isopropoxide, and sodium chloride. The latter component may function as a catalyst support. Alfin catalysts are used for the preparation of high-molecular-weight, stereoregular polyisoprene or poly(*trans*-1,4-butadiene). The mechanisms of these reactions are not fully understood and both anionic and free-radical mechanisms have been suggested.

Stereospecific Polymers

Any monomer molecule that possesses an asymmetric center at a skeletal atom has the capacity to form stereoregular polymers. Three primary possibilities exist with respect to the sequence in which the monomer units enter the chain. These are called *isotactic, syndiotactic,* or *atactic* polymerizations.

 Isotactic polymers are characterized by the presence of only one symmetry type of monomer residue in the chain. For example, a representation of isotactic polystyrene is shown in **12**. Or, more symbolically, this structure can be represented

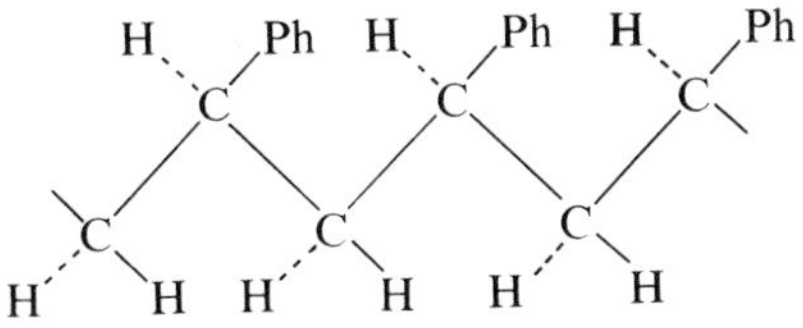

12

by **13**. One of the characteristic features of isotactic polymers is their ability to crystallize readily. This is a consequence of the fact that the regular disposition of

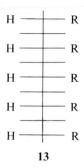

13

substituent groups along the chain permits the molecule to assume a regular helical conformation and allows adjacent chains to pack together in an ordered manner.

Syndiotactic polymers are characterized by *alternating* configuration of residues, as depicted in **14**. This structure is represented by the symbolism shown in **15**.

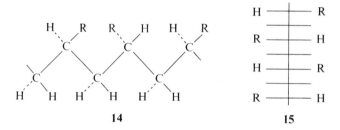

14   **15**

Syndiotactic polymers also tend to crystallize readily, again because of the opportunities that exist for the formation of helices and for efficient chain packing.

Atactic or heterotactic polymers contain no regular sequence of monomer residues along the chain. Because of this, the polymers are characterized by a low tendency for crystallization.

The simplest stereoregular systems are the ones described above in which the monomer has the structure $CH_2=CHR$. For monomers of structure, $CHR=CHR$ and $CHR=CHR'$, the following additional possibilities exist:

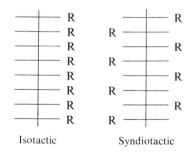

Isotactic   Syndiotactic

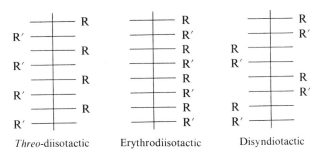

Threo-diisotactic   Erythrodiisotactic   Disyndiotactic

And for regular *copolymers* synthesized from ethylene and CHR—CHR, the following four alternative structures can be visualized:

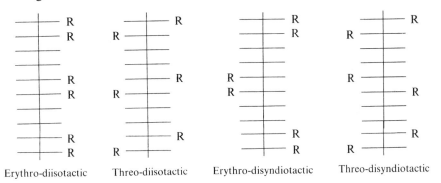

Erythro-diisotactic   Threo-diisotactic   Erythro-disyndiotactic   Threo-disyndiotactic

One further item of stereoregular nomenclature is necessary when a regular sequence of double bonds is present in the main chain as, for instance, in poly(1,4-butadiene). The configuration at each double bond may be *cis* or *trans*, as follows:

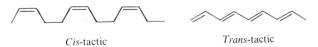

Cis-tactic   Trans-tactic

Finally, polymers can be formed that contain blocks of one type of stereoregular sequence followed by another. These are called *stereoblock* polymers.

In general, *ionic* catalysts, especially *anionic* catalysts, favor the formation of stereoregular polymers; thus, they also favor the synthesis of microcrystalline polymers. On the other hand, free-radical catalysts tend to favor the formation of atactic polymers with a corresponding low degree of crystallinity.

CATIONIC POLYMERIZATION

Catalysts for Cationic Polymerization

As mentioned previously, cationic polymerizations are those in which the growing chain end bears a positive charge. Table 4.4 lists representative compounds in the two main classes of compounds that are known to initiate cationic polymerizations. These two classes comprise the strong protonic acids and the Lewis acids.

TABLE 4.4 Cationic Catalysts

Catalyst Class	Example Formula
1. Strong acids	H_2SO_4
	$HClO_4$
	HCl
2. Lewis acids and their complexes	BF_3
	$BF_3:O(C_2H_5)_2$
	BCl_3
	$TiCl_4$
	$AlCl_3$
	$SnCl_4$

Most or perhaps all of the Lewis acids function as catalysts only if a *cocatalyst*, such as water or methanol, is present in an equimolar concentration or less. In such cases, the actual catalyst is a proton–cation complex formed by a process such as the one shown in reaction (20). An excess of the cocatalyst destroys the catalytic properties of the system.

$$F_3B:OH_2 \rightleftharpoons [F_3BOH]^- + H^+ \qquad (20)$$

Monomers for Cationic Polymerization

Monomers that polymerize under the influence of cationic catalysts usually contain *electron-supplying* substituent groups. Examples include isobutylene (**16**), 1,3-butadiene (**17**), vinyl ethers (**18**), *para*-substituted styrenes (**19**), α-methylstyrenes (**20**), and aldehydes (**21**).

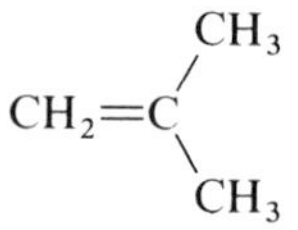

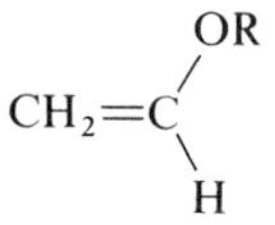

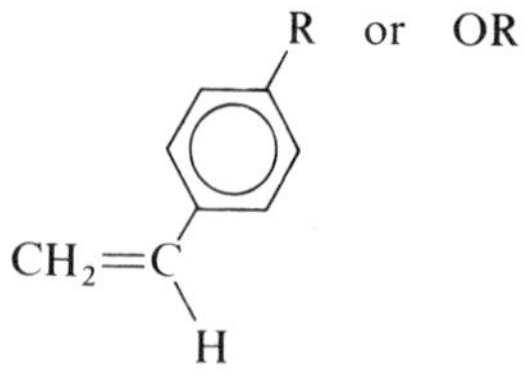

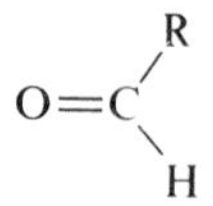

Experimental Conditions for Cationic Polymerizations

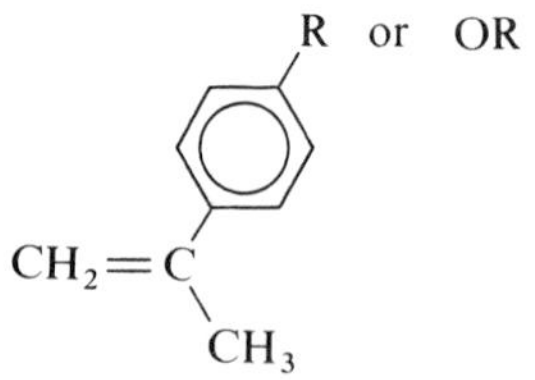

Low-temperature polymerization conditions are nearly always needed in order to suppress unwanted side reactions. Furthermore, the solvents used must be unreactive and very dry.

A classical example of a cationic reaction is provided by the polymerization of α-methylstyrene in the presence of boron trifluoride hydrate. Toluene may be used as a solvent after it has been dried over calcium hydride. Similarly, the α-methylstyrene should be distilled before use and dried over the same drying agent. The boron trifluoride can conveniently be obtained from a gas cylinder. The water needed as a cocatalyst will already be present in the system in spite of the precautions taken to exclude moisture. Two experimental approaches are available. In one (Figure 4.4) a beer or soft drink bottle is charged with a solution of the monomer, and air is removed with a stream of dry nitrogen. The bottle is capped with a rubber serum cap stopper and the bottle and contents are cooled to −78°C. The catalyst is introduced as a gas through the serum cap by means of a hypodermic syringe, and the "reactor" is then agitated for several hours at −78°C. When polymerization is complete (as indicated by the presence of a viscous reaction mixture), the cap is removed and the reaction products are poured into methanol to destroy the catalyst.

The second method is more suitable for following the rates of polymerization of gaseous monomers. The reaction is carried out in a glass vessel attached to a vacuum line (Figure 4.5). Monomer (e.g., isobutylene), solvent, catalyst, and cocatalyst are stored separately in gas storage bulbs or in flasks attached to the vacuum line, and quantities are measured as gases in bulbs of known volume. Stirring is by means of a magnetic stirrer, and the progress of the reaction is followed by observation of the monomer vapor pressure by means of a manometer. The reaction vessel must be cooled to dissipate the heat evolved. An apparatus such as this must be used for the study of the effects of different catalysts and cocatalysts on the reaction.

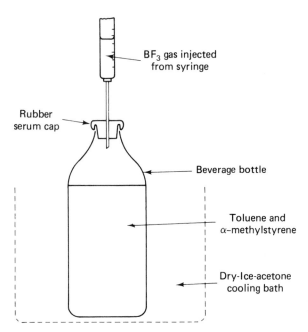

BF$_3$ gas injected from syringe

Rubber serum cap

Beverage bottle

Toluene and α-methylstyrene

Dry-Ice-acetone cooling bath

Fig. 4.4 Apparatus for the small-scale cationic polymerization of α-methylstyrene.

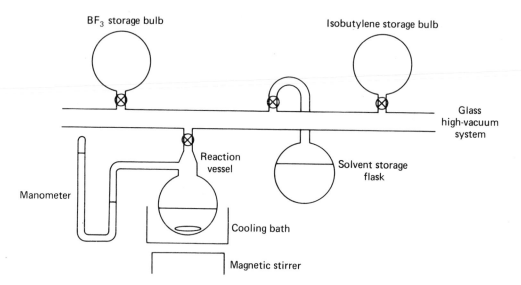

Fig. 4.5 Glass high-vacuum system designed for following the rate of polymerization of a volatile monomer under rigorously controlled conditions. (Based on a system designed by A. M. Eastham.)

The Mechanism of Cationic Polymerization

Again it is convenient to divide the overall mechanism into initiation, propagation, chain transfer, and termination steps.

1. *Initiation* actually involves two sequential steps—the generation of a proton and the addition of that proton to the monomer. Strong protonic acids in non-aqueous media liberate protons by the conventional ionization processes shown in (21), (22), and (23). Lewis acids interact with the cocatalyst to yield a proton donator by reactions such as (24), (25), or (26).

$$H_2SO_4 \rightleftharpoons H^+ + HSO_4^- \tag{21}$$

$$HClO_4 \rightleftharpoons H^+ + ClO_4^- \tag{22}$$

$$HCl \rightleftharpoons H^+ + Cl^- \tag{23}$$

$$BF_3 + H_2O \longrightarrow F_3B{:}OH_2 \rightleftharpoons F_3BOH^- + H^+ \tag{24}$$

$$BF_3 + CH_3OH \longrightarrow F_3B{:}OCH_3H \rightleftharpoons F_3BOCH_3^- + H^+ \tag{25}$$

$$SnCl_4 + H_2O \longrightarrow [Cl_4Sn{:}OH_2] \rightleftharpoons Cl_4SnOH^- + H^+ \tag{26}$$

Boron trifluoride and stannic chloride do not normally function as catalysts in the absence of a cocatalyst such as water or methanol. However, even when such cocatalysts are not deliberately added to the system, enough adsorbed water molecules are present on the inside surface of the apparatus to cocatalyze the reaction. It can, in fact, be shown that an optimum ratio of catalyst to cocatalyst exists (often 1:1) that gives a maximum reaction rate. Larger amounts of cocatalyst decrease the reaction rate by destruction of the catalyst. Lowering of the cocatalyst concentration below the optimum amount lowers the polymerization rate. As shown in Figure 4.6, the very lowest cocatalyst concentrations cannot be attained reproducibly, but extrapolation of the curve strongly suggests that the polymerization rate would be zero in the complete absence of a cocatalyst. Thus, for practical purposes, nearly all cationic catalysts can be considered as species of formula $H^{\oplus}X^{\ominus}$.

Initiation of a monomer molecule, in principle, then involves the addition of the catalyst ion pair across the double bond (27). The mode of addition across the

$$
H^{\oplus}X^{\ominus} + CH_2{=}\underset{\underset{R}{|}}{\overset{\overset{R}{|}}{C}} \longrightarrow HCH_2{-}\underset{\underset{R}{|}}{\overset{\overset{R}{|}}{C^{\oplus}}}X^{\ominus} \tag{27}
$$

double bond will be such that the proton will add to the carbon atom that bears the greatest electron density, thereby forming the most stable carbonium ion. If the side group, R, is an electron-supplying group, then the addition will take place as shown in (27).

2. *Propagation* takes place by successive insertion of monomer molecules into the cation–anion "bond." The first propagation step, to yield a dimer, is illustrated

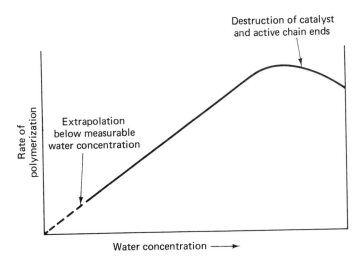

Fig. 4.6 Showing the falloff in the rate of a cationic polymerization as the amount of the cocatalysts (water) is reduced below the optimum ratio. See, for example, papers by A. G. Evans and G. W. Meadows, *Trans. Faraday Soc.*, **46**, 327 (1950), and A. M. Eastham, *J. Am. Chem. Soc.*, **78**, 6040 (1956).

in (28). At low temperatures, chain growth takes place rapidly without appreciable chain transfer.

$$CH_3-\overset{\overset{R}{|}}{\underset{\underset{R}{|}}{C}}{}^{\oplus}\ X^{\ominus} \qquad \longrightarrow \qquad CH_3-\overset{\overset{R}{|}}{\underset{\underset{R}{|}}{C}}-CH_2-\overset{\overset{R}{|}}{\underset{\underset{R}{|}}{C}}{}^{\oplus}\ X^{\ominus} \qquad (28)$$

$$CH_2=\overset{\overset{R}{|}}{\underset{\underset{R}{|}}{C}}$$

3. *Chain transfer* becomes important at temperatures near room temperature. An important transfer step involves the donation of a proton from a terminal side group to a monomer molecule (29). The newly initiated monomer molecule can,

$$\left[CH_3-\!\!\text{\tiny w}\!\!-\overset{\overset{CH_3}{|}}{\underset{\underset{R}{|}}{C}}\right]^{\oplus}\ X^{\ominus}\ +CH_2=\overset{\overset{R}{|}}{\underset{\underset{R}{|}}{C}} \longrightarrow CH_3-\!\!\text{\tiny w}\!\!-\overset{\overset{CH_2}{\|}}{\underset{\underset{R}{|}}{C}}\ +\ CH_3-\overset{\overset{R}{|}}{\underset{\underset{R}{|}}{C}}{}^{\oplus}\ X^{\ominus} \qquad (29)$$

of course, generate a new chain. If more chains are initiated in this way, the average chain length of polymer in the system will be reduced. Hence, a need exists to maintain low reaction temperatures if high-molecular-weight polymers are desired. (See also Chapter 13, equations 48 and 50.)

4. *Termination* of the polymer chain can occur by the transfer mechanism discussed above or by the loss of a proton to the X^- counterion, thereby terminating also the kinetic chain. Termination can also take place by the reaction of a growing chain end with traces of water or other protonic reagents (30). This mode of termination can be important even in rigorously dried glassware.

$$R-\!\!\text{\tiny w}\!\!-\overset{\overset{R}{|}}{\underset{\underset{R}{|}}{C}}{}^{\oplus}\ X^{\ominus} + H_2O \longrightarrow R-\!\!\text{\tiny w}\!\!-\overset{\overset{R}{|}}{\underset{\underset{R}{|}}{C}}OH + HX \qquad (30)$$

Special Characteristics of Cationic Polymerizations

The kinetic characteristics of cationic polymerization are considered in Chapter 13. Here, attention is drawn to the consequences of these kinetic features. Because chain transfer reactions can be important in cationic reactions, a curious relationship exists between the rates of chain transfer, propagation, and termination as the reaction temperature is lowered. If termination is more important than chain transfer, *the rate of polymerization will accelerate as the temperature is lowered.* This contrasts with the behavior of free-radical polymerizations. Furthermore, if the activation energy for termination and transfer is greater than that for propagation, the molecular weight also will increase as the temperature is lowered.

Cationic Polymerization of Aldehydes

Most of the discussion so far has been directed to the cationic polymerization of olefins. However, a second group of monomers—the aldehydes—are especially prone to cationic polymerization.

For example, formaldehyde can be polymerized in the presence of a boron trifluoride–water catalyst. The reaction mechanism is believed to resemble the one discussed above for olefin monomers. Initiation involves the protonation of the aldehyde oxygen atom (reactions 31 and 32).

$$F_3B:OH_2 \longrightarrow F_3^{\ominus}BOH + H^{\oplus} \quad (H^+X^-) \tag{31}$$

$$H^+X^- + O{=}CH_2 \longrightarrow HO{-}\overset{\displaystyle H}{\underset{\displaystyle H}{\overset{|}{\underset{|}{C}}}}{}^{\oplus} \; X^{\ominus} \tag{32}$$

$$\xrightarrow{\;O=CH_2\;} HO{-}CH_2{\left(O{-}CH_2\right)_{\!n}}O{-}\overset{\displaystyle H}{\underset{\displaystyle H}{\overset{|}{\underset{|}{C}}}}{}^{\oplus} \; X^{\ominus}$$

The prospect that boron trifluoride alone can function as a catalyst in this reaction has also been proposed (33). However, linear propagation by this mech-

$$BF_3 + O{=}CH_2 \longrightarrow F_3B{:}\overset{\ominus}{O}{-}\overset{\displaystyle H}{\underset{\displaystyle H}{\overset{|}{\underset{|}{C}}}}{}^{\oplus} \xrightarrow{\;O=CH_2\;} F_3B{:}\overset{\ominus}{O}{-}\overset{\displaystyle H}{\underset{\displaystyle H}{\overset{|}{\underset{|}{C}}}}{-}O{-}\overset{\displaystyle H}{\underset{\displaystyle H}{\overset{|}{\underset{|}{C}}}}{}^{\oplus} \quad \text{etc.} \tag{33}$$

anism would require the progressive separation of the anionic and cationic charges—an unlikely process. Charge separation could only be avoided by a "pseudo-cyclic" propagation mechanism, such as the one shown in **22**.

22

Formaldehyde also polymerizes in the absence of added catalyst, especially when cooled. Traces of formic acid impurity could function as catalytic species. Polyformaldehyde is a tough thermoplastic material. It is offered commercially under the trade name Delrin, and is particularly useful for the fabrication of gear wheels.

Acetaldehyde also polymerizes "spontaneously" below its freezing point, or when treated with protonic acids or Lewis acids. Amorphous polyacetaldehyde is a rubbery polymer with a molecular weight in the region of ~40,000. It readily depolymerizes back to acetaldehyde, particularly if the chain ends remain uncapped and active.

Higher aldehydes such as propionaldehyde or butyraldehyde can be polymerized only at low temperatures and high pressures. They depolymerize to the parent aldehyde when exposed to the atmosphere at ambient pressures. This depolymerization behavior represents a rather profound thermodynamic characteristic of many polymer systems, and it is considered in more detail in Chapter 10.

STUDY QUESTIONS

1. Given 100 g of styrene, and the appropriate apparatus and conditions to conduct the anionic polymerization of this monomer with *n*-butyllithium, calculate the expected average molecular weight of the resultant living polymer chains if you could introduce exactly 500 molecules of *n*-butyllithium. Assume no termination and total usage of the monomer and initiator.

2. Make a list of the main differences between anionic, cationic, free-radical, and condensation polymerizations. Which of these reaction types normally gives the lowest proportion of side reactions?

3. Suppose that, in an anionic or coordination polymerization, an opportunity exists to use several different initiators, RM, which contain the same R group, but different metals, M^1, M^2, M^3, etc. Speculate on the effects that might be observed if the R—M bond becomes progressively more covalent along the series R—M^1, R—M^3, etc.

4. What storage conditions would you choose if faced with the need to preserve an unterminated anionic polymer for a period of 10 years?

5. Design a laboratory apparatus or a large-scale reaction system that would allow an *n*-butyllithium-initiated vinyl polymerization to be carried out in a continuous-flow reactor. What advantages or disadvantages can you foresee for such a process? Repeat the exercise for the use of an Alfin catalyst or a heterogeneous Ziegler–Natta system.

6. Comment on the reasons why the following monomers are not normally polymerized by ionic processes: acrylic acid, allyl alcohol, acrylamide, vinyl chloride.

7. What effects might you predict in a BF_3-catalyzed polymerization if the solvent was changed from methylene chloride to benzene?

SUGGESTIONS FOR FURTHER READING

Anionic polymerization

BYWATER, S., "Anionic Polymerization," *Progress in Polymer Science* (A. D. Jenkins, ed.), **4**, 27 (1975).

BYWATER, S., "Polymerization Initiated by Lithium and Its Compounds," *Adv. Polymer Sci.*, **4**, 66 (1965).

GAYLORD, N. G., AND S. S. DIXIT, "One Electron Transfer Initiated Polymerization. II. Initiation through Monomer Anion Radicals," *J. Polymer Sci.* (*D*) (*Macromol. Rev.*), **8**, 51 (1974).

HALASA, A. F., D. N. SCHULZ, AND D. P. TATE, "Organolithium Catalysis of Olefin and Diene Polymerization," *Adv. Organometal. Chem.* (F. G. A. Stone and R. West, eds), **18** (1979).

HENDERSON, J. F., AND M. SZWARC, "The Use of Living Polymers in the Preparation of Polymer Structures of Controlled Architecture," *J. Polymer Sci.* (*D*) (*Macromol. Rev.*), **3**, 317 (1968).

HIROHARA, H., AND N. ISE, "On the Growing Active Centers and Their Reactivities in 'Living' Anionic Polymerizations of Styrene and Its Derivatives," *J. Polymer Sci.* (*D*) (*Macromol. Rev.*), **6**, 295 (1972).

MORTON, M., AND L. J. FETTERS, "Homogeneous Anionic Polymerization of Unsaturated Monomers," *J. Polymer Sci.* (*D*), (*Macromol. Rev.*), **2**, 71 (1967).

MULVANEY, J. E., C. G. OVERBERGER, AND A. M. SCHILLER, "Anionic Polymerization," *Fortschr. Hochpolym.-Forsch.*, **3**, 106 (1961).

SZWARC, M., *Carbanions, Living Polymers, and Electron Transfer Processes.* New York: Wiley–Interscience, 1968.

Coordination polymerization

BOOR, J., JR., "The Nature of the Active Site in the Ziegler-Type Catalyst," *J. Polymer Sci.* (*D*) (*Macromol. Rev.*), **2**, 115 (1967).

CARRICK, W. L., "The Mechanism of Olefin Polymerization by Ziegler–Natta Catalysts," *Fortschr. Hochpolym.-Forsch.*, **12**, 65 (1973).

COOVER, H. W., JR., R. L. MCCONNELL, AND F. B. JOYNER, "Relationship of Catalyst Composition to Catalytic Activity for the Polymerization of α-Olefins," *J. Polymer Sci.* (*D*) (*Macromol. Rev.*), **1**, 91 (1967).

FURUKAWA, J., "Stereoregular and Sequence-Regular Polymerization of Butadiene," *Accounts Chem. Res.* **13**, 1 (1980).

KEII, T., *Kinetics of Ziegler–Natta Polymerization.* London: Chapman & Hall, 1972.

KETLEY, A. D. (ed.), *The Stereochemistry of Macromolecules*, Vol. 1. New York: Dekker, 1967.

OLIVE, G. H., AND S. OLIVE, "Koordinative Polymerisation an löslichen Übergangsmetall-Katalysatoren," *Adv. Polymer Sci.*, 421 (1969).

REICH, L., AND A. SCHINDLER, *Polymerization by Organometallic Compounds.* New York: Wiley–Interscience, 1966.

ROHA, M., "The Chemistry of Coordinate Polymerization of Dienes," *Fortschr. Hochpolym.-Forsch.*, **1**, 512 (1960).

TSURUTA, T., "Stereoselective and Asymmetric-Selective (or Stereo-elective) Polymerizations," *J. Polymer Sci.* (*D*) (*Macromol. Rev.*), **6**, 179 (1972).

TSURUTA, T., AND K. F. O'DRISCOLL (eds.), *Structure and Mechanism in Vinyl Polymerization.* New York: Dekker, 1969.

WERBER, F. X., "Polymerization of Olefins on Supported Catalysts," *Fortschr. Hochpolym.-Forsch.*, **1**, 180 (1959).

YOUNGMAN, E. A., AND J. BOOR, JR., "Syndiotactic Polypropylene," *J. Polymer Sci. (D) (Macromol. Rev.)*, **2**, 33 (1967).

Cationic polymerization

BARKER, S. J., AND M. B. BRUCE, *Polyacetals.* New York: American Elsevier, 1970.

FURUKAWA, J., AND T. SAEGUSA, *Polymerization of Aldehydes and Oxides* (Polymer Reviews, Vol. 3). New York: Wiley–Interscience, 1963.

KENNEDY, J. P., *Cationic Polymerization of Olefins.* New York: Wiley, 1975.

KENNEDY, J. P., AND J. K. GILLHAM, "Cationic Polymerization of Olefins with Alkylaluminium Initiators," *Adv. Polymer Sci.*, **10**, 1 (1972).

KENNEDY, J. P., AND A. W. LANGER, JR., "Recent Advances in Cationic Polymerization," *Fortschr. Hochpolym.-Forsch.*, **3**, 508 (1964).

KENNEDY, J. P., AND S. RENGACHARY, "Correlation between Cationic Model and Polymerization Reactions of Olefins," *Adv. Polymer Sci.*, **14**, 1 (1974).

LYONS, A. R., "Polymerization of Vinyl Ketones," *J. Polymer Sci. (D) (Macromol. Rev.)*, **6**, 251 (1972).

PLESCH, P. H., *The Chemistry of Cationic Polymerization.* New York: Macmillan, 1963.

TSUKAMOTO, A., AND O. VOGL, "Cationic Polymerization," *Progr. Polymer Sci.* (A. D. Jenkins, ed.), **3**, 199 (1971).

Photolytic, Radiation, and Electrolytic Polymerization

The use of *chemical* catalysts and initiators provides only one method for the induction of olefin polymerization. Polymerization can also be induced by supplying the initiation energy through irradiation with visible or ultraviolet light, high-energy or ionizing radiation, or by the passage of an electric current. The conversion of a monomer to a polymer will occur through the normal propagation, termination, and transfer reactions (see Chapters 3 and 4). Only the initiation processes will be unusual.

PHOTOCHEMICAL POLYMERIZATION

Advantages and Uses

There are several advantages to be gained by the use of light for the initiation of vinyl polymerization. Perhaps the most obvious advantage in laboratory research is the avoidance of chemical contamination by initiator residues. Moreover, there is a marked convenience to photochemical reactions that appeals to many researchers.

From a practical standpoint, photopolymerization is used as a photographic process. In one type of system a plate or film is coated with a monomer such as acrylamide (CH_2=$CHC(O)NH_2$), together with a small amount of a divinyl compound. A photosensitizing dye may also be present. Exposure of parts of the film to light causes photopolymerization of the acrylamide, together with cross-linking through the divinyl residues. After the exposure, the film is "developed" by washing with water. Unpolymerized monomer is removed from those areas that were shielded from light, leaving a relief polymeric image corresponding to the exposed portions (Figure 5.1). A colored image is left if dyestuffs were incorporated into the matrix.

An alternative process makes use of the photoinduced crosslinking of polymers. For example, poly(vinyl cinnamate) or mixtures of cellulose cinnamate and poly(vinyl cinnamate) remain soluble if protected from light. However, exposure of a film of these materials to strong light (usually ultraviolet light) through a negative brings about crosslinking and insolubilization of those areas that lie beneath the transparent sections of the negative. The crosslinking mechanism involves cyclo-dimerization of cinnamoyl groups on different chains.

Images made either by photopolymerization or photocrosslinking are now used extensively in the preparation of letterpress and lithography plates, holography, and (coupled with etching processes) in the manufacture of printed and integrated circuits. Like conventional silver halide photography, photopoly-merization provides a method for *amplification* of the original photochemical event. In this case, the amplification involves the chain reaction of an addition polymerization. However, the fastest photopolymerization processes have only one thousandth of the speed and sensitivity of medium-speed silver halide systems, but it

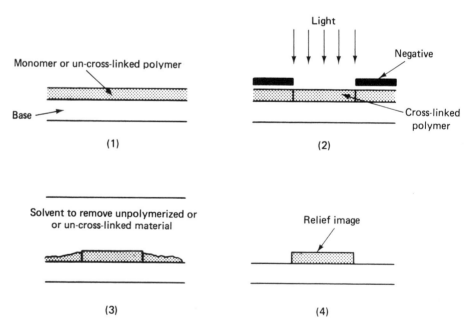

Fig. 5.1 Photopolymerization as a photographic process. Illumination of a monomer or a crosslinkable polymer forms an insoluble relief image.

is anticipated that future developments in this area could bring about the development of faster emulsions. Hopefully, this will occur before the use of silver in photography becomes drastically curtailed.

Monomers That Undergo Photopolymerization

Any monomer that will undergo chain reaction polymerization is susceptible to photopolymerization or photosensitized polymerization. The absorption of light simply produces free radicals or ions. Also, the chain-propagation steps and the termination reactions are generally not affected. The advantage of photopolymerization and photosensitized polymerization is that the initiation process may take place over a wider range of temperatures and with a greater specificity than is found in chemically initiated systems.

Some monomers, such as vinyl alkyl ketones and vinyl bromide, absorb 300-nm or longer wavelength light and dissociate directly to free radicals. These radicals initiate polymerization. Other monomers, such as styrene or methyl methacrylate, are susceptible to direct photopolymerization when exposed to 300-nm or shorter wavelength light. The detailed mechanism of the formation of the propagating radicals in this case is not completely understood, but it appears to involve the conversion of an electronically excited singlet state of the monomer to a long-lived excited triplet state.

Direct photopolymerization is not restricted to the initiation of free-radical chains. For example, the cationic chain polymerization of isobutylene has been initiated by irradiation with light of 117 and 123 nm wavelength (vacuum ultraviolet radiation)[1]. The direct absorption of such radiation by isobutylene produces positive ions and electrons, and the former initiate the polymerization. Presumably other monomers that are susceptible to ionic chain polymerization will also undergo direct photopolymerization when irradiated with vacuum ultraviolet radiation.

In spite of these facts, only a few unsaturated monomers are known which absorb light between 250 and 500 nm—the most convenient wavelength range for experimental work. For other monomers, a *photosensitizer* must be added to the system. Photosensitizers are compounds that absorb light in a convenient region of the spectrum and then dissociate into free radicals or transfer energy directly to the monomer. Photosensitizers are considered in more detail later.

Experimental Technique: A Photosensitized Polymerization of Styrene

A simple photopolymerization apparatus, constructed from Pyrex tubing, ordinary laboratory corks, or rubber stoppers, and a 15-W "black light"[2] is shown in

---

[1] E. W. Schlag and J. J. Sparapany, *J. Am. Chem. Soc.*, **86**, 1875 (1964).
[2] Available commercially from the General Electric Company.

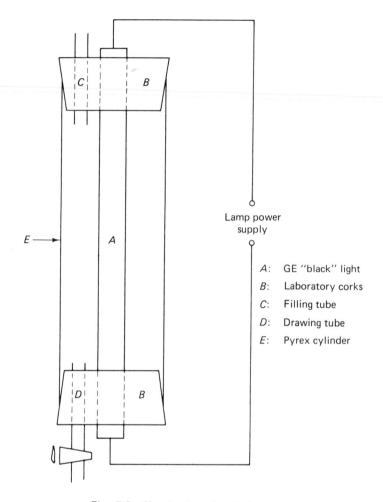

A: GE "black" light
B: Laboratory corks
C: Filling tube
D: Drawing tube
E: Pyrex cylinder

Fig. 5.2   Simple photochemical reactor.

Figure 5.2. The light emitted from this lamp has its maximum intensity at a wavelength of about 360 nm, and this matches well the near-ultraviolet absorption maximum of azoalkanes. The useful surface area of the lamp for emission is about 280 cm^2 and the light flux emitted may be taken as 1.0×10^{-8} einstein/cm^2-s. (An einstein is equivalent to 6.02×10^{23} light quanta.) The lamp A is held in place by holes drilled in the cork stoppers B, as are a filling tube C and a drain tube D.

Before assembly of the apparatus, the corks are thoroughly cleaned and then soaked for several hours in the solvent to be used. A 3-liter sample is prepared of a $2\,M$ solution of freshly distilled styrene and $0.2\ M$ azoisopropane in benzene. The azo compound is the photosensitizer. The solution is purged of oxygen by bubbling nitrogen gas through it for an hour. It is then introduced into the apparatus through tube C, and the irradiation is started. After 2 hours the irradiation is stopped, the solution is removed from the reactor, and the benzene solvent is removed by evaporation.[1] About 8 g of polystyrene of average molecular weight of 20,000 should remain as a residue.

To calculate the rate of initiation using equation (13), to be described later, it should be assumed[2] that the primary quantum yield, ϕ, is 0.3 and that the extinction coefficient, ε, is 10 liters-mol^{-1}-cm^{-1}.

Mechanism of Initiation by Direct Photolysis of the Monomer

Many monomers undergo free-radical chain polymerization when exposed to ultraviolet or visible light. The photopolymerization of a pure monomer occurs by the same chain *propagation* process that was discussed in Chapter 3. However, an all-inclusive mechanism has not yet been established for photolytic *initiation*.

Two different types of direct photoinitiation can be recognized. In the first, light absorption yields an electronically excited monomer molecule which subsequently decomposes to give radical fragments. Examples of monomers in this category include alkyl vinyl ketones and vinyl bromide, both of which dissociate when irradiated with ultraviolet light by the reactions shown in equations (1) to (3). Here R· denotes an alkyl radical. Following photodissociation of the monomer, the resultant monoradicals add to the monomer, and radical chain polymerization (Chapter 3) takes place.

$$
\overset{\displaystyle O}{\underset{\displaystyle \|}{}} \\
R-C-CH=CH_2 \quad \xrightarrow{h\nu} \quad R-C\cdot + \cdot CH_1=CH_2 \tag{1}
$$

$$
\overset{\displaystyle O}{\underset{\displaystyle \|}{}} \\
R-C\cdot \qquad \longrightarrow \quad R\cdot + CO \tag{2}
$$

$$
CH_2=CHBr \qquad \xrightarrow{h\nu} \quad Br\cdot + CH_2=CH\cdot \tag{3}
$$

The rate of initiation of the chains and the dependence of this rate on the monomer concentration and temperature is quite different in photochemical initiation from the situation found in thermal initiation. For example, when a thermal initiator, I, is used as a source of radicals, the rate of formation of the initiating radicals is

$$
\frac{d[R\cdot]}{dt} = 2k_d[I] \tag{4}
$$

where k_d is the rate constant for the first-order decomposition of the initiator and [I] is the initiator concentration. For photoinitiation using a pure monomer, the

---

[1] Evaporation of the benzene should be carried out in a rotary evaporator or in a fume hood, but *not* on the open bench. Benzene is toxic.

[2] J. G. Calvert and J. N. Pitts, *Photochemistry* (New York: Wiley, 1966), p. 464.

rate of radical formation averaged over the system may be written as

$$\frac{d[\text{R}\cdot']}{dt} = \frac{2\phi I_0 A}{V}(1 - e^{-\alpha[\text{M}]L}) = \frac{2\phi I_0 A}{V}\left(\alpha[\text{M}]L - \frac{1}{2}(\alpha[\text{M}]L)^2 + \cdots\right) \tag{5}$$

where I_0 is the number of einsteins per cm^2 incident on the system per second; α the absorption coefficient (an average, if the light is not monochromatic) in liters-mol^{-1}-cm^{-1}; [M] the monomer concentration in mol/liter; L the length of the light path in cm; A the area of the system illuminated in cm^2; V the volume in cm^3; and ϕ the fraction of the light quanta absorbed that result in the formation of a pair of radicals. If the system is to be irradiated uniformly, the fraction of the incident light that is absorbed must be kept small. This condition may be ensured by requiring that the term [M]L in (5) is kept sufficiently small that, in the expanded form of the exponential expression, all terms after the first may with negligible error be discarded. Under such conditions, and with replacement of α by the more familiar extinction coefficient, ε, (5) is transformed into (6). The geometric factor (LA/V) is unity, or nearly so, for most photolytic systems employed.

$$\frac{d[\text{R}\cdot']}{dt} = 4.6\phi I_0 \varepsilon[\text{M}]\frac{LA}{V} \tag{6}$$

Comparison of (5) and (6) with (4) illustrates the main differences between initiation by a thermal dissociation of an initiator and by the photodissociation of a monomer. The formation of the initiating radicals by a *thermal* dissociation process is strongly dependent on the temperature and is independent of the monomer concentration. On the other hand, the terms $\phi I_0 \alpha$ and $\phi I_0 \varepsilon$ in (5) and (6) are almost independent of temperature. Hence *the rate of initiating radical formation in a photoinitiation process is almost independent of the temperature but is proportional to the monomer concentration.*

The second type of initiation mechanism is exemplified by the photopolymerization of styrene or methyl methacrylate. Absorption of light in this case does not result in decomposition of monomer molecules. Instead, it has been suggested[1] that the absorption of light produces an excited singlet state of the monomer which may either fluoresce (7) or be converted to an excited (and long-lived) triplet excited state (8). The latter may be regarded as a diradical, that is, $\cdot CH_2 - \dot{C}(H)X$. Attack on the monomer by this diradical (9) ultimately yields two monoradicals (11),

$$\text{PhCH}=\text{CH}_2 + hv' \tag{7}$$

$$\text{Ph}-\text{CH}=\text{CH}_2 \xrightarrow{hv} (\text{PhCH}=\text{CH}_2)^* $$

$$\text{Ph}\dot{\text{C}}\text{H}-\text{CH}_3^{\cdot} \tag{8}$$

$$\text{Ph}\dot{\text{C}}\text{H}-\text{CH}_3^{\cdot} + \text{PhCH}=\text{CH}_2 \longrightarrow \text{Ph}\dot{\text{C}}\text{H}-\text{CH}_2-\text{CH}_2-\dot{\text{C}}\text{HPh} \tag{9}$$

[1] R. G. Norrish and J. P. Simons, *Proc. Roy. Soc.*, **A251**, 4 (1959).

which, in turn, initiate polymerization. An example of this complex initiation mechanism is illustrated for styrene photopolymerization by equations (7) to (11).

$$Ph\dot{C}HCH_2CH_2\dot{C}HPh \longrightarrow 2\,PhCH{=}CH_2 \tag{10}$$

$$Ph\dot{C}HCH_2CH_2\dot{C}HPh + PhCH{=}CH_2 \longrightarrow CH_3\dot{C}HPh + PhCH{=}CHCH_2\dot{C}HPh \tag{11}$$

It can be shown by the steady-state methods discussed in Chapter 11 that, at high monomer concentrations, essentially all the diradicals react by step (11) rather than by (10), and the rate of formation of the initiating radicals will be proportional to [M], the monomer concentration. On the other hand, if the monomer concentration is so low that the diradicals mainly revert to monomer by step (10), the rate of initiator formation will be proportional to $[M]^2$, one power of [M] coming from (7) and the second from (9). In both cases these rates are proportional to the light intensity and to the extinction coefficient of the monomer, as required by expression (6).

Photosensitized Polymerizations

In order for *direct* photopolymerization to occur it is essential that the monomer should absorb some of the light impinging on the system. However, even if such direct light absorption does not occur, polymerization can still be initiated if photosensitizers are present. Photosensitizers are substances that produce free radicals when they absorb ultraviolet or visible light. The same substances that are used for thermal initiation are often used for photosensitization. For example, azo compounds and peroxides are photosensitizers, and the photoinitiation reaction is the same as is the thermal initiation process. However, much lower reaction temperatures can be employed when light absorption is used to initiate radical formation. Moreover, many initiators can be used as photosensitizers even though they do not dissociate *thermally* at convenient rates or temperatures. Examples of some photosensitizers are given in Table 5.1.

For example, azoisopropane does not dissociate sufficiently rapidly below 180°C (Table 3.1) to be a useful thermal initiator. However, it photodissociates even at low temperatures when irradiated with near-ultraviolet light:

$$Me_2CHN{=}NCHMe_2 + h\nu(3000\text{Å} < \lambda < 4000\text{Å}) \longrightarrow 2\,Me_2\overset{\displaystyle H}{\underset{\displaystyle |}{C}}{\cdot} + N_2 \tag{12}$$

As might be expected from the previous discussion, the rate of radical formation from such photosensitizers is given by

$$\frac{d[\text{R}\cdot]}{dt} = 4.6\phi I_0 \varepsilon_s[\text{S}]\,\frac{LA}{V} \tag{13}$$

where ε is the extinction coefficient of the photosensitizer, [S] the concentration of photosensitizer, and the other symbols are as defined earlier.

TABLE 5.1   Photosensitizers for Photopolymerization

TYPE OF COMPOUND	EXAMPLE	MECHANISM OF POLYMERIZATION
Carbonyl compounds	Acetone	Radical formation
	Biacetyl	Radical formation
	Benzophenone	Radical formation
	Benzoin	
	α-Chloroacetone	Radical formation
Condensed ring aromatics	Anthracene	Energy transfer
Peroxides	*t*-Butyl peroxide	Radical formation
	Hydrogen peroxide	Radical formation
Organic sulfides	Diphenyl disulfide	Radical formation
	Dibenzoyl disulfide	Radical formation
Azo compounds	Azoisopropane	Radical formation
	Azobisisobutyronitrile	Radical formation
Halogen-containing compounds	Chlorine	Radical formation
	Chloroform	Radical formation
	Carbon tetrachloride	Radical formation
	Bromotrichloromethane	Radical formation
	Bromoform	Radical formation
	Bromine	Radical formation
Metal carbonyls	Manganese pentacarbonyl and carbon tetrachloride	Radical formation
	Rhenium pentacarbonyl and carbon tetrachloride	Radical formation
Inorganic ions	$FeOH^{2+}$	Radical formation
	$FeCl_4^-$	Radical formation

Condensed ring aromatic compounds, such as anthracene give rise to a more complex type of photosensitization process.[1] Photoirradiation generates the excited triplet state of anthracene. This excited molecule then interacts with the monomer ultimately to produce monoradicals. The monoradicals initiate the polymerization. With anthracene, denoted by A, and a monomer, $XCH=CH_2$, this initiation mechanism is depicted in steps (14) to (19).

$$A + h\nu \longrightarrow A^* \tag{14}$$

$$A^* \longrightarrow A + h\nu \quad \text{(fluorescence)} \tag{15}$$

$$A^* \longrightarrow {}^3A \quad \text{(triplet state)} \tag{16}$$

$$^3A + M \longrightarrow (AM)^* \quad \text{(intermediate diradical)} \tag{17}$$

$$(AM)^* \longrightarrow A + M \tag{18}$$

$$(AM)^* + A \longrightarrow XCH=CH-A\cdot + HA\cdot \tag{19}$$

[1] See page 100.

A complicated mathematical expression is needed to describe the rate of formation of the initiating monoradicals. However, if all the triplet-state anthracene molecules react with the monomer, the rate will be independent of the monomer concentration. In all cases, the rate of formation of the initiating radicals will be proportional to the intensity of the incident light. At sufficiently high sensitizer concentrations the rate will be proportional to [A], while at very low sensitizer concentration it will be proportional to $[A]^2$.

A somewhat different type of photosensitizer is exemplified by the metal carbonyl/carbon tetrachloride systems.[1] For example, photolysis of $Mn_2(CO)_{10}$ or $Re_2(CO)_{10}$ in the presence of small amounts of CCl_4 initiates the free-radical polymerization of vinyl monomers. Although the mechanism is not completely understood, the actual initiating radical is thought to be $\cdot CCl_3$. This radical is produced by the reaction of the metal carbonyl photolysis products with carbon tetrachloride.

Quantum Yields

The quantum yield of a simple photochemical reaction is defined as the number of molecules of the product formed, or reactant consumed, per quantum of light absorbed. In photopolymerization the quantum yield for initiation is defined by the number of *chains* initiated per quantum of light absorbed. This may be written in terms of rates as

$$\phi_i = \frac{\text{Rate of chain initiation}}{\text{Rate of light absorption}} \tag{20}$$

When a simple photoinitiation occurs and the absorbing compound dissociates directly to two monoradicals, ϕ_i must lie between zero and 2.

The *overall* quantum yield of photopolymerization refers to the number of molecules of *monomer consumed* per quantum absorbed. This may be written as shown in (21). Since the formation of a high polymer requires the existence of long-chain processes, ϕ_p is generally very large, typically being of the order of several hundred to several thousand.

$$\phi_p = \frac{\text{Rate of monomer consumption}}{\text{Rate of light absorption}} \tag{21}$$

RADIATION-INDUCED POLYMERIZATION

General Characteristics

Chain reaction polymerizations can also be induced by irradiation of a pure monomer (or a solution of the monomer) with high-energy or ionizing radiation. Alpha particles, beta rays (electrons), gamma rays, or high-velocity particles from

[1] C. H. Bamford, P. A. Crowe, and R. P. Wayne, *Proc. Roy. Soc.*, **A284**, 455 (1965).

an accelerator can all be used. A wide variety of monomers may be polymerized in this way—styrene, acrylonitrile, methyl methacrylate, vinyl chloride, butadiene, isoprene, or practically any other monomer that polymerizes by a free-radical mechanism. In addition, monomers such as isobutylene can be induced to undergo low-temperature cationic polymerization during irradiation.

In a practical sense, the rapid polymerization of monomers and the crosslinking of polymers by high-energy radiation have been considered as possible manufacturing processes. However, for reasons that will become clear from the following sections, radiation-induced reactions are less convenient to carry out on a large scale than are chemically induced reactions, although future developments may change this picture.

Experimental Methods

Monomers and solvents for radiation-induced polymerization must be purified in the same way as for conventional free-radical or ionic polymerizations. Oxygen is usually detrimental to the reaction, and the monomer is normally sealed in an evacuated reaction vessel or protected from oxygen by an atmosphere of nitrogen.

A variety of reaction vessels have been used, depending on the monomer, the type of radiation, and the scale of polymerization. Figure 5.3 shows a simple reaction cell that is convenient for small-scale laboratory experiments when gamma or X-rays are employed. It consists of a small, shallow glass cylinder to which is attached a glass side arm. The monomer is introduced into the cell through the side arm, air is removed by evacuation, and the tube is sealed. The cell is mounted in a heating or cooling bath with the upper flat cell face above the level of the heating or cooling fluid. The cell is then positioned below the X-ray or gamma-ray source and the contents are irradiated from above. The polymer can be removed either by dissolving it in a suitable solvent or by breaking the cell.

Alternatively, if the experimenter has access to a γ-ray nuclear irradiation facility, other procedures can be used. A particularly simple nuclear irradiation arrangement allows the sample to be lowered into position close to a ^{60}Co source. In a typical ^{60}Co irradiation facility, the ^{60}Co is encapsulated in many stainless

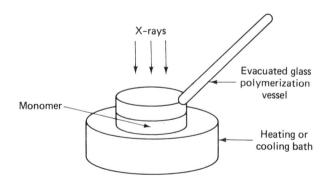

Fig. 5.3   Radiation-induced polymerization of a monomer in an evacuated glass ampoule.

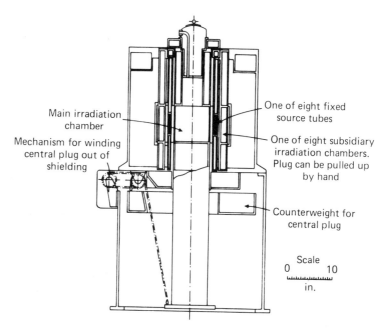

Main irradiation chamber

Mechanism for winding central plug out of shielding

One of eight fixed source tubes

One of eight subsidiary irradiation chambers. Plug can be pulled up by hand

Counterweight for central plug

Scale
0   10
in.

Fig. 5.4   Equipment for the irradiation of monomer or polymer samples with the use of a shielded cesium 137 source. [Reproduced from A. J. Swallow, *Radiation Chemistry of Organic Compounds* (New York: Pergamon Press, 1960); with permission from Pergamon Press.]

steel "pencils" which are themselves maintained beneath the surface of a "swimming pool" of water. The pencils can be arranged around a vertical aluminum tube (~ 7.5 cm diameter) that extends up to the surface of the pool. Various configurations of the pencils are used to supply differing dose rates to the sample. The sample to be irradiated is simply lowered into position and kept in place until the predetermined dose (total energy input) has been delivered. It is then removed for analysis. Typical γ-ray dose rates in a facility of this kind may be in the range of 0.1 to 1 megarad/h (10^7 to 10^8 ergs-g^{-1}-h^{-1}).

More compact nuclear irradiation sources generally use lead instead of water for shielding purposes. An example of this type is the ^{137}Cs irradiation unit shown diagrammatically in Figure 5.4. The ^{137}Cs is contained in fixed tubes surrounding a larger main irradiation chamber. Auxiliary irradiation chambers are located adjacent to each tube of ^{137}Cs. Again, samples are placed in the irradiation chambers for predetermined periods of time, and are then removed for examination.

The Absorption of Radiation

The initiation of polymerization by high-energy radiation is a far less selective process than is initiation by light. A wide variety of monomers can be induced to polymerize by almost all forms of high-energy radiation. In fact, the initiation of polymerization seems to result more from the gross "damage" sustained by the system than from the selective induction of specific chemical reactions. Because of this lack of selectivity, it is not possible to calculate a meaningful quantum yield. The energy yield in radiation-induced reactions is generally described by the G-value. This is defined by (22) for any substance A.

$$G(\pm A) = 100 \left(\frac{\text{Number of molecules of A formed or consumed}}{\text{Number of electron-volts of energy absorbed}} \right) \tag{22}$$

The high-energy irradiation of either a pure monomer or a solution of a monomer generates both free radicals and ions. Thus, either free-radical or ionic chain polymerizations may be induced. The preponderance of one or the other mechanism depends on the nature of the monomer, as well as on the temperature and the purity of the reagents. Some monomers, such as acrylic esters, vinyl esters, and vinyl fluoride, polymerize only by free-radical mechanisms. Isobutylene polymerizes only by a cationic chain mechanism. Others, such as styrene, acrylonitrile, or isoprene, apparently polymerize by both radical and ionic types of chain mechanisms. Ionic polymerizations are generally favored at low temperatures. Both types of mechanism are sensitive, in different degrees, to the presence of common impurities such as oxygen or water. Examples of both types of mechanism are discussed in later sections.

Gamma radiation is the most convenient type of high-energy radiation for the initiation of polymerization because the high penetrating power permits *uniform* irradiation of the system. Moreover, because gamma rays are absorbed to the same extent by solids as by liquids, solid monomers can be polymerized readily. This permits low-temperature polymerizations to be performed with many monomers.

The irradiation of a pure monomer will give rise to the simplest initiation reactions, because only one chemical species absorbs the radiation. However, systems that contain both monomer and a solvent undergo more complicated initial reactions because both species can absorb radiation energy to produce ions and free radicals. The details of these energy absorption processes are outside the scope of this book. However, as a very useful approximation, it is possible to assume that the fraction of the total energy absorbed by a particular species in solution is proportional to the electron fraction of that species. The *electron fraction* is defined as the number of electrons contained in that component divided by the total number of electrons in the whole solution.

For example, consider a 0.1 M solution of styrene in carbon tetrachloride. The concentration of carbon tetrachloride is about 10 M, and the number of electrons in styrene and carbon tetrachloride molecules are 56 and 74, respectively. Therefore, the electron fraction of carbon tetrachloride is calculated by equation (23):

$$\varepsilon_{CCl_4} = \frac{74\, M_{CCl_4}}{74\, M_{CCl_4} + 56\, M_{C_8H_8}} = \frac{740}{740 + 5.6} = 0.993 \tag{23}$$

Thus, during the irradiation of this solution, 99.3 % of the total energy absorbed will be absorbed by the solvent, and hence most of the initiating ions and free radicals will be derived from the solvent. In this example, the styrene polymerization will be initiated predominantly by species such as CCl_2^+, CCl_3^+, $Cl\cdot$, CCl_3, etc.

Three distinct phases may be identified in a radiation-induced polymerization. The interaction of the radiation with monomer and solvent molecules occurs within 10^{-16} to 10^{-15} s. In this brief interval the only significant motion is that of electrons, and the products of this first phase are electronically excited molecules, ions, and electrons. In the second phase, which takes place some 10^{-14} to 10^{-10} s after the initial interaction, the excited molecules, ions, and electrons dissociate or react with monomer to yield a set of initiating free radicals and ions. The third phase consists of the normal elementary reactions of initiation, propagation, transfer, and termination characteristic of chain reaction polymerization. It occupies the time regime of 10^{-10} to 10^{-1} s after the initial interaction.

Free-Radical Chain Initiation

As discussed earlier, a radiation-catalyzed free-radical polymerization differs from a chemically induced one mainly in terms of the initiation mechanism. The propagation, chain transfer, and termination processes are similar to those described in Chapter 3. Hence, the present comments will focus only on the initiation process.

The initiation step in a radiation-induced free-radical polymerization of a pure monomer can be symbolized by equation (24). The symbol $\longrightarrow\!\!\wedge\!\!\wedge\!\!\rightarrow$[1] means "under high-energy irradiation." M represents the monomer, and $R\cdot'$ denotes the initiating radicals. If a radiation-induced polymerization is carried out in solution, the initiation step depicted in (25) must also be taken into account. In (25), S denotes the

$$M \longrightarrow\!\!\wedge\!\!\wedge\!\!\rightarrow 2R\cdot' \qquad (24)$$

$$S \longrightarrow\!\!\wedge\!\!\wedge\!\!\rightarrow 2S\cdot \qquad (25)$$

solvent and $S\cdot$ represents initiating radicals derived from the solvent. Thus, even in these simple terms, two types of radicals can be generated. However, even in a well-defined initiation process, $R\cdot'$ and $S\cdot$ represent a whole *set* of initiating radicals and not just single species. Moreover, it should be kept in mind that the simple equations (24) and (25) are not to be considered as discrete chemical reactions. Rather, they represent the overall production of free radicals by a very fast *sequence* of elementary processes that follow the initial energy absorption. These processes include ion and excited molecule dissociations, ion and electron recombinations, and very fast collision reactions of ions and translationally "hot" atoms and radicals. For a discussion of the rapid elementary processes that follow radiation absorption the reader should consult texts and monographs on radiation chemistry such as those listed at the end of the chapter.

---

[1] This should not be confused with the similar symbolism that represents a polymer chain (see page 61).

MONOMER OR SOLVENT	FORMULA	$G(R \cdot')$ (radicals/100 eV)
Acrylonitrile	$CH_2=CHCN$	5.0
Isobutylene	$CH_2=C(CH_3)_2$	3.9
Methyl acrylate	$CH_2=CH-\overset{\overset{\textstyle O}{\|}}{C}-OCH_3$	6.3
Methyl methacrylate	$CH_2=C(CH_3)\overset{\overset{\textstyle O}{\|}}{C}-OCH_3$	6.1
Styrene	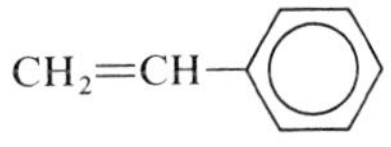 $CH_2=CH-\hexagon$	0.66
Vinyl acetate	$CH_2=CH-O-\overset{\overset{\textstyle O}{\|}}{C}-CH_3$	9.6
Benzene	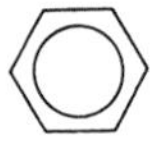 $\hexagon$	0.66
n-Hexane	$CH_3(CH_2)_4CH_3$	5.8
Toluene	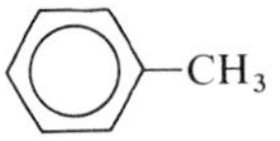 $\hexagon-CH_3$	2.4

The most important characteristic of the initiation step is the *yield* of radicals generated by the absorption of a given radiation dose. This characteristic is defined by the G-value or 100-eV yield of initiating radicals. This is referred to as $G_M(R \cdot')$ or $G_S(R \cdot')$, depending on whether the radiation energy is being absorbed by the monomer or by the solvent. A number of 100-eV yields of initiating radicals for some typical monomers and solvents have been determined by radical scavenging techniques. Some typical values are shown in Table 5.2. A knowledge of such values permits a calculation to be made of the rate of formation of the radiation-initiated radicals in the same way as was described earlier for thermal and photochemical initiation. Thus, for the radiation-induced polymerization of a pure monomer or a monomer in solution, the rate of formation of initiating radicals is described by (26) and (27) respectively, where the ε's are electron fractions and Q_A is the energy absorbed per unit volume.

$$\frac{d[R \cdot']}{dt} = \frac{G(R \cdot')}{100}\left(\frac{dQ_A}{dt}\right) \qquad \text{(Pure monomer)} \quad (26)$$

$$\frac{d[R \cdot']}{dt} = \left(\frac{\varepsilon_M G_M(R \cdot')}{100} + \frac{\varepsilon_S G_S(R \cdot')}{100}\right)\left(\frac{dQ_A}{dt}\right) \qquad \text{(Solution)} \quad (27)$$

The energy absorption per unit volume, Q_A, and the rate of energy absorption per unit volume, dQ_A/dt, are related to the dose and dose rates[1] by the density of the system.

Radiation chemists often express doses in a unit called the rad, which is defined as 100 ergs/g. A typical dose rate from a ^{60}Co γ-ray source would be 10^6 rads/h or 1 Mrad/h. The following example illustrates a calculation of the rate of formation of initiating radicals for a comparison with thermal and photochemical initiation.

Consider the irradiation of pure acrylonitrile (density $=0.81$ g/cm^3) at 20°C with γ-rays, with a dose rate of 1 Mrad/h. It is possible to calculate both the energy absorption and the rate of initiating radical formation. The energy absorption rate per unit volume, $dQ_A/dt = 1 \times 10^6$ rads/h $\times 10^2$ ergs/g-rad $\times 0.81$ g/cm$^3 \times 6.24 \times 10^{11}$ eV/erg $\times 2.78 \times 10^{-4}$ h/s $= 1.4 \times 10^{16}$ eV/cm^3-s. The rate of initiating radical formation may be derived from this value, equation (26), and the value listed in Table 5.2. Thus, $d[R\cdot]/dt = 0.050 \times 1.4 \times 10^{16} = 7.0 \times 10^{14}$ radicals/cm^3-s. This is equivalent to the rate of radical formation from 0.1 M benzoyl peroxide solution at 60°C.

Ionic Chain Initiation

The formation of ions from the monomer and solvent can be symbolized by (28) and (29). Again, the processes shown in (28) and (29) are not elementary reactions.

$$M \quad \longrightarrow\hspace{-1em}\text{\Large\char`\~}\hspace{-1em}\longrightarrow \quad R^+ + R^- \tag{28}$$

$$S \quad \longrightarrow\hspace{-1em}\text{\Large\char`\~}\hspace{-1em}\longrightarrow \quad R'^+ + R'^- \tag{29}$$

They represent the overall formation of initiating ions by very fast sequences of reactions that follow an initial ionization event. Thus, M^+, S^+, and e^- formed in the initial event are converted by very rapid reactions to the initiating ions denoted by R^+ and R^- in (28) and (29). The reader is again referred to texts and monographs on radiation chemistry for a detailed discussion.

The G-values for the *initial* formation of ion pairs in liquids are in the range 3 to 4. Although this is comparable to the G-values found for free-radical formation, the efficiency of ionic initiation is much lower than that of free-radical initiation. Most of the "gegenions" formed initially do not separate from each other, but instead undergo mutual charge neutralization. The radiation yield of "free" ions (ions capable of initiating polymerization) depends on the dielectric constant of the medium. In hydrocarbons, which are solvents of low dielectric constant (i.e., values of 2 to 4), the G-value for "free" ions is only 0.1, which means that only 1 out of every 30 or 40 ions that are formed goes on to initiate an ionic polymerization. Alcohols, which have dielectric constants in the range 20 to 40 show free-ion G-values of 0.6 to 1.5. Water, which has a dielectric constant of 78, has a free ion G-value of ~ 2.5.

---

[1] These terms are generally used by radiation chemists to express energy absorption per unit mass.

By analogy with free-radical polymerization, it is possible to use equation (30) to describe the radiation-induced rate of formation of initiating positive ions in the pure monomer, and equation (31) for a monomer in solution. As before, the ε's are

$$\frac{d[R^+]}{dt} = \frac{G(R^+)}{100}\left(\frac{dQ_A}{dt}\right) \tag{30}$$

$$\frac{d[R^+]}{dt} = \left(\frac{\varepsilon_M G_M(R^+)}{100} + \frac{\varepsilon_S G_S(R^+)}{100}\right)\frac{dQ_A}{dt} \tag{31}$$

electron fractions, Q_A is the energy absorbed per unit volume, and M and S refer to the monomer and the solvent, respectively. Identical expressions may be written for the rates of formation of negative ions. The rates of formation of the initiating ions can be estimated from the radiation dose rate using an analogous method to the one described above.

Ionic chain polymerizations are especially sensitive to traces of impurities, and these impurities can exert a strong inhibiting effect. For example, in the cationic polymerization of styrene, the propagating chains are effectively broken by proton transfer to a water molecule, as shown in reaction (32). It is very difficult to remove

$$R\!\left[CH_2\!-\!CH\right]\!CH_2\!-\!\overset{\overset{\displaystyle H}{|}}{C}{}^+ + H_2O \longrightarrow R\!\left[CH_2\!-\!CH\right]\!CH\!=\!CH + H_3O^+ \tag{32}$$

the last traces of water from a chemical system, and this experimental problem considerably delayed the recognition that radiation-induced polymerization can proceed by cationic chain mechanisms. Thus, styrene that has been dried by distillation from sodium–potassium alloy polymerizes under irradiation about 200 times faster than styrene that has merely been subjected to a single conventional distillation. Ionic polymerizations are generally much faster than free-radical polymerizations. If water is present, only the slower radical component of a radiation-induced polymerization can be observed because the cationic component is effectively inhibited.

Solid-State Radiation-induced Polymerization

General features

Some crystalline monomers or crystalline cyclic compounds can be induced to polymerize in the solid state. Typically, crystals of the monomer are irradiated with electrons, gamma rays, or X-rays, often at low temperatures. The source of radiation may be an accelerator, radioactive isotope, or an X-ray generator. Polymerization may occur either at the temperature of irradiation or during subsequent warming of the crystal to room temperature. The latter type of process is called *postpolymerization.* After polymerization is complete, the unchanged monomer can be dissolved away to leave the polymer behind, or the monomer can be removed by volatilization in vacuum or by reprecipitation into a nonsolvent for the polymer.

Monomers that have been polymerized in this way include acrylamide, acrylonitrile, styrenes, isobutylene, isoprene, butadiene, acrylic and methacrylic acids and their salts, formaldehyde, acetaldehyde, and acetone. Cyclic compounds that undergo solid-state radiation-catalyzed polymerization include trioxane, hexamethylcyclotrisiloxane, β-propiolactone, diketene, 3,3-bischloromethylcyclooxabutane, and hexachlorocyclotriphosphazene. The following examples illustrate the behavior of specific unsaturated monomers or cyclic trimers during solid-state polymerization.

Acrylamide

Acrylamide is a solid at room temperature (m.p. 84°C). The γ-ray-induced polymerization of this monomer was first achieved in 1954, an event which stimulated the initial development of the solid-state polymerization field. Irradiation at $-78°C$ results in no appreciable polymerization. However, rapid warming of the preirradiated polymer results in a violent polymerization process. A gradual increase in the temperature brings about a slow polymerization which may continue over a period of several months. Total conversion of monomer to polymer can be achieved during postpolymerization at 27°C.

It has been shown by electron spin resonance techniques that free radicals are formed during the irradiation process and that the radical concentration remains constant as the postpolymerization proceeds. Each radical is associated with one chain. Thus, if the polymerization is a free-radical process, it would appear that a radical termination step is absent, presumably because the radicals are trapped in a polymer matrix. However, the presence of free radicals does not prove a free-radical mechanism. Radical anions or radical cations can generate *ionic* polymerizations, even though radicals can be detected.

The polyacrylamide formed by this process is amorphous. Apparently, polymerization takes place at defects in the crystal lattice. If the crystal is scratched, polymerization proceeds rapidly along the scratch. Hence, it is assumed that defects and strains are generated at the crystal–polymer interfaces. Trapped radicals are perhaps released at these locations, thus favoring further polymerization.

Acrylonitrile

Continuous irradiation of solid acrylonitrile at $-196°C$ gives polyacrylonitrile, but the conversion to polymer never exceeds about 5%. However, if the 5% polymer–95% monomer mixture is allowed to melt and is then cooled and reirradiated, a higher conversion to polymer takes place. This effect may result from the crystallization of pure monomer crystals following melting, or it could be a consequence of the introduction of additional polymer–monomer interface defects. It has also been reported that irradiation of acrylonitrile at $-196°C$ with 7.8×10^5 rads, followed by subsequent heating to room temperature, results in an explosive polymerization before the melting point is reached. Large quantities of polymer are apparently formed in this process.

It is believed that the polymerization at $-196°C$ follows an ionic mechanism, but that free-radical processes predominate at temperatures near the melting point.

Vinyl carboxylates and derivatives

Salts of acrylic and methacrylic acids, M^+ $^-OOC—CH=CH_2$ or M^+ $^-OOC—C(CH_3)=CH_2$, can be polymerized by γ-irradiation in the solid state. However, changes in the cation, M^+, affect the rate of polymerization. At $-78°C$ the potassium salt polymerizes faster than the sodium salt, which in turn polymerizes faster than the lithium salt. These changes have been ascribed to differences in the crystal structures of the different salts. Similarly, different hydrates of acrylate salts polymerize at different rates. Methyl methacrylate resists polymerization in the solid state.

Aldehydes

Irradiation of solid formaldehyde with ionizing radiation at $-196°C$ can generate an almost explosive polymerization process. The violence of the reaction increases if irradiated solid formaldehyde is warmed or subjected to mechanical shock. Another curious feature of this reaction is the high G-value (5.4×10^6) for a 52% conversion and an average degree of polymerization of 10^4. This fact suggests that conventional ionic or free-radical initiation mechanisms are not operative. It has been proposed that irradiation generates a trapped formaldehyde excited state, such as $H_2C^+—O^-$, which initiates polymerization when the temperature is raised. The polymer formed by this technique has a fibrous appearance.

Trioxane

Crystals of trioxane, $(O—CH_2)_3$, undergo solid-state polymerization when irradiated with γ-rays or α-particles (see Chapter 6). The temperature during irradiation may be from $-78°C$ to $+55°C$. Radiation-catalyzed polymerization does not take place in the molten state or in solution. There is no induction period before polymerization begins, and the polymerization rate is apparently not influenced by free-radical scavengers or by air. The reaction rates are proportional to the radiation dose rate. For these reasons an ionic mechanism has been proposed, although this viewpoint is not universally accepted. The "in-source" polymerization of trioxane can be completely inhibited by the application of high pressures.

At $55°C$ the conversion of trioxane to polyoxymethylene does not exceed about 35%. It is believed that this limit corresponds to the point at which the irradiation process begins to cause decomposition of the polymer. The polymer matrix has the outward appearance of large crystals and, indeed, the polymer fibers obtained are the same length as the single crystals of the trimer from which they were formed. The crystallinity of these fibers is appreciable and the orientation is excellent. The polyoxymethylene formed by γ-irradiation of trioxane has a higher melting point and better heat stability than polymer formed by the conventional polymerization of formaldehyde.

X-ray diffraction analyses of both trioxane crystals and the polymer formed within them show that the trioxane crystal structure influences the route of the propagation reaction. Trioxane crystals are trigonal, with the molecular packing arrangement shown in Figure 5.5. The polymer crystals are twinned in such a way that two different helical chain axes can be discerned for the polyoxymethylene. These two chain orientations arise either by propagation down the c-axis of the trioxane crystal or by the linkage of molecules oriented along a diagonal axis oriented $76.1°$ to the c-axis. These two possibilities are illustrated in Figure 5.5, with the atomic separations (in Å) that must be bridged during polymerization indicated on the diagram.

α-Particle irradiation of trioxane crystals apparently generates a different twinned modification. In this case, polymerization originates from the damage streaks left by the α-particle tracks. Tetroxane, $(O—CH_2)_4$, also polymerizes under the influence of γ-rays to give a twinned polymer crystal. However, this twinned structure differs from that of trioxane because the crystal structure of tetroxane is different (monoclinic).

Other cyclic compounds

Hexachlorocyclotriphosphazene, $(NPCl_2)_3$, (see Chapter 7) polymerizes in the crystalline state when irradiated with X-rays. The maximum yield of polymer is about 10%. In fact, the rate of polymerization increases with temperature until the melting point ($114°C$) is reached, at which point the polymerization rate falls to zero. Presumably, the active sites are stabilized within the crystalline lattice but are destroyed rapidly in the liquid state.

Conclusions

A number of practical advantages are inherent in the radiation-catalyzed solid-state process. First, and most obviously, the method allows the polymerization of compounds which have low ceiling temperatures[1] and which cannot be polymerized thermally by conventional means. Second, solid-state reactions provide a method for polymerization in the absence of a potentially reactive solvent. Third, polymer order and crystallinity can be introduced by the use of the monomer crystal structure as a "template." Fourth, contamination by catalyst molecules or residues is avoided. Fifth, some "monomers" yield different polymers in the liquid and solid states. The conversion of diketene $(CH_2{=}\overline{C—O—C(O)}—CH_2)$ to the polymer $+C(=CH_2)—CH_2—C(O)—O+_n$ occurs only in the solid state.

Perhaps the most intriguing aspect of this field is the relationship between the crystal structure of the monomer (or cyclic oligomer) and the polymerization pathway. This subject provides a fertile area for future fundamental research.

---

[1] The ceiling temperature is the temperature above which the polymer cannot exist (see Chapter 10).

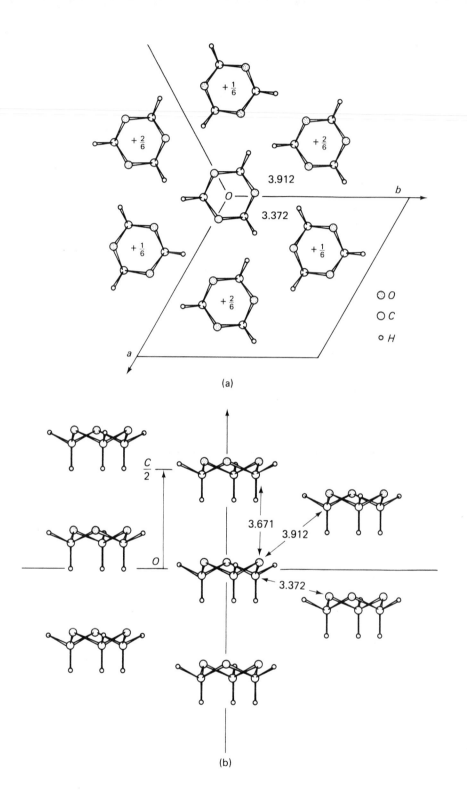

(a)

(b)

Fig. 5.5 (*Opposite*) (a) Packing of trioxane molecules between the planes $z = 0$ and $z = 1/2$, viewed down the c-axis. The z-height of each molecule is indicated. (b) Packing of the molecules along the contiguous threefold axis lying in the (110) plane. Polymerization may occur between molecules stacked above each other on one threefold axis or between molecules on different axes, displaced by $1/6c$. [The separations are indicated by the 3.671-Å and 3.372-Å distances shown in (b).] [From V. Busetti, M. Mammi, and G. Carazzolo, *Z. Krist.*, **119**, 310 (1963).]

ELECTROCHEMICALLY INITIATED POLYMERIZATIONS

The passage of an electric current through a liquid system takes place by the transport of electrons from the cathode to the anode. During conventional electrolysis, the current is carried through the solution by ions, as shown in Figure 5.6.

If an unsaturated monomer is present in solution, an electron can be transferred from the cathode to the monomer to generate a radical anion. At the anode, an electron can be removed from the unsaturated compound to generate a radical cation. These processes are illustrated by (33) and (34).

$$\underset{\overset{|}{H}}{\overset{\overset{R}{|}}{C}}{=}CH_2 \quad \xrightarrow{e^-} \quad \underset{\overset{|}{H}}{\overset{\overset{R}{|}}{\overset{\ominus}{C}}}{-}CH_2^{\cdot} \quad \text{At the cathode} \qquad (33)$$

$$\underset{\overset{|}{H}}{\overset{\overset{R'}{|}}{C}}{=}CH_2 \quad \xrightarrow{-e^-} \quad \underset{\overset{|}{H}}{\overset{\overset{R'}{|}}{\overset{\oplus}{C}}}{-}CH_2^{\cdot} \quad \text{At the anode} \qquad (34)$$

In practice each vinyl monomer has its own specific reduction potential below which electron transfer from the electrode to the monomer does not occur. Electron-withdrawing groups in the monomer facilitate reduction and thereby lower the reduction potential. Conversely, electron-supplying groups raise the reduction potential of a monomer but lower the oxidation potential at the anode.

It will be clear that electroinitiated monomers, such as those shown in (33) and (34), could generate anionic, cationic, or free-radical chain reactions. In fact,

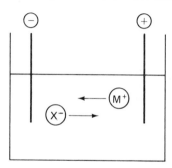

Fig. 5.6 Conventional electrolysis.

difficulty is often experienced in the elucidation of the exact polymerization mechanism. The following examples will illustrate a few of the possibilities and problems.

Acrylonitrile, methyl methacrylate, and styrene polymerize in dimethylformamide solvent under electrolytic conditions. Sodium nitrate is added as the supporting electrolyte. Apparently, acrylonitrile polymerizes by an anionic mechanism initiated by direct cathodic reduction of the monomer and dimerization of the radical anion (35). More than three polymer chains are formed for each electron

$$
\underset{\substack{| \\ H}}{\overset{\substack{C\equiv N \\ |}}{CH_2{=}C}} \xrightarrow{\ e^-\ } \underset{\substack{| \\ H}}{\overset{\substack{C\equiv N \\ |}}{\cdot CH_2{-}C^{\ominus}}} \longrightarrow \underset{\substack{| \\ H}}{\overset{\substack{C\equiv N \\ |}}{^{\ominus}C}}{-}CH_2{-}CH_2{-}\underset{\substack{| \\ H}}{\overset{\substack{C\equiv N \\ |}}{C^{\ominus}}} \qquad (35)
$$

transferred to the monomer, a result which suggests the existence of a chain transfer process. The polymer molecular weight is independent of the current density or the monomer concentration.

Electrolysis of solutions of styrene or methyl methacrylate in dimethylformamide in the presence of tetramethylammonium chloride leads to the formation of polystyrene or poly(methyl methacrylate) by an anionic mechanism. Copolymerizations are also possible. For example, copolymers of styrene and methyl methacrylate can be formed by the electroinitiation of mixtures of the monomers in dimethyl formamide, and this could indicate either an anionic or a free-radical process. However, in tetrahydrofuran, poly(methyl methacrylate) homopolymer is the principal product.

Electrochemical initiation of vinyl polymerization has also been attempted in heterophase monomer–aqueous systems. Polymers can be obtained from styrene, vinyl chloride, methyl methacrylate, and other monomers, but free-radical mechanisms appear to operate under these conditions. Cationic chain processes are believed to take place in nitrobenzene solvent when perchlorate salts are used as the supporting electrolyte.

STUDY QUESTIONS

1. What practical advantages exist for the use of radiation, photochemically, or electrolytically induced polymerizations compared with those initiated by the addition of chemical reagents? What are the disadvantages?

2. Under what circumstances would the high dielectric constant of water be an advantage if this solvent were to be used as a polymerization solvent?

3. Survey the X-ray crystallographic literature (start with *Chemical Abstracts*) and examine the crystal structures of as many vinyl monomers or cyclic compounds as you can find. In each case, speculate on whether or not the molecular arrangement in the crystal would be suitable for a solid-state polymerization.

4. A parallel beam of light of wavelength 366 nm and having an intensity of 3.0×10^{15} photons/cm^2-s is incident on the window of a cylindrical cell that is 0.500 in. in diameter and 1.50 in. in length. If the cell contains 0.0100 mol/liter of azoethane dissolved in benzene and

the extinction coefficient of azoethane at 366 nm is 9.00 liters/mol-cm., calculate the rate of absorption of photons per unit volume.

5. When the photolysis in Problem 4 was carried out for 15 min, it was found that the N_2 evolved from the solution exerted a pressure of 2.59 torr when contained in a vessel of 10-cm^3 volume at 27°C.

   (a) Calculate the primary quantum yield of decomposition of the azoethane.
   (b) Calculate the rate of formation of C_2H_5 radicals in the solution.
   (c) Assuming an efficiency of initiation of 0.4, calculate the rate of initiation of polymerization in a similar solution containing 3 mol/liter of styrene.

6. Given a light source that emits 10^{16} quanta/cm^2-s at 366 nm and some azobisisobutyronitrile ($\varepsilon = 9.5$ at 366 nm with a primary quantum yield of 0.43), design a system for the photosensitized polymerization of methyl methacrylate in benzene. The total volume of solution should be 50 cm^3 and the initiation rate should be 1×10^{-7} mol/liter-s. Assume the efficiency of initiation to be 0.50.

7. In a 15-min photopolymerization of methyl methacrylate using the system of Problem 6, 0.25 g of a polymer having an average molecular weight of 50,000 is obtained. Calculate the overall quantum yield of the polymerization, that is, the quantum yield for disappearance of monomer.

8. For γ-ray-induced polymerization of the following systems, calculate the fraction of energy absorbed by the monomer(s): (a) 5 M methyl acrylate in cyclohexane ($\rho = 0.779$ g/cm^3); (b) 0.1 M vinyl bromide in toluene ($\rho = 0.867$ g/cm^3); (c) an equimolar mixture of styrene ($\rho = 0.906$ g/cm^3) and methyl acrylate ($\rho = 0.950$ g/cm^3).

9. A certain ^{60}Co irradiator produces a dose rate of 1.5 Mrads/h. Using data in this chapter and assuming an initiation efficiency of 0.5, calculate the rate of initiation of free radical polymerization in the following liquid monomers: (a) styrene ($\rho = 0.906$ g/cm^3); (b) vinyl acetate ($\rho = 0.93$ g/cm^3); (c) acrylonitrile ($\rho = 0.806$ g/cm^3); (d) methyl acrylate ($\rho = 0.950$ g/cm^3).

10. A solution of 1 M acrylonitrile ($\rho = 0.806$ g/cm^3) in benzene ($\rho = 0.879$ g/cm^3) is irradiated with ^{60}Co γ-rays at a dose rate of 1.0 Mrads/h. Assuming that all radicals derived from either solvent or monomer initiate polymerization with an efficiency of 0.7, calculate the rate of initiation of free radical polymerization.

11. In Problem 10, the G-value for depletion of monomer is found to be 1500 molecules/100 eV. Calculate the rate of polymerization of acrylonitrile.

12. Suppose that you have available to you a ^{60}Co source with a dose rate of 0.50 Mrads/h. and you wish to polymerize methyl acrylate ($\rho = 0.95$ g/cm^3) in benzene solution ($\rho = 0.879$ g/cm^3) using an initiation rate of 10^{-7} mol/liter-s. Assuming an initiation efficiency of 0.5, calculate the weight of methyl acrylate that must be added to 100 cm^3 of benzene to prepare the solution.

SUGGESTIONS FOR FURTHER READING

Photochemical polymerization

COHEN, A. B., "Photopolymer Images," *Indust. Res.*, December, 39 (1976).

LABANA, S. S., "Photopolymerization," *J. Macromol. Sci.—Rev. Macromol. Chem.*, **C11**(2), 299 (1974).

Oster, G., "Photopolymerization and Photocrosslinking," *Encyclopedia of Polymer Sci. and Technol.* (N. M. Bikales, ed.), **10**, 145 (1969). See also G. Oster and N. Yang, *Chem. Rev.*, **68**, 125 (1968).

Radiation-induced polymerization

Carazzolo, G., S. Leghissa, and M. Mammi, "Polyoxymethylene from Trioxane by Solid State Polymerization," *Makromol. Chem.*, **60**, 171 (1963).

Chapiro, A., *Radiation Chemistry of Polymer Systems.* New York: Wiley–Interscience, 1962.

Chatani, Y., T. Uohida, H. Tadokoro, K. Hayashi, M. Nishii, and S. Okamura, "X-Ray Crystallographic Study of Solid State Polymerization of Trioxane and Tetraoxymethylene," *J. Macromol. Sci.—Phys.*, **B2**(4), 567 (1968).

Eastmond, G. C., "Solid State Polymerization," *Prog. Polymer Sci.* (A. D. Jenkins, ed.), **2**, 1 (1970).

Garratt, P. G., "Radiation-Induced Solid State Polymerization," *Polymer*, **3**, 323 (1962).

Henglein, A., W. Schnabel, and J. Wendenburg, *Einführung in die Strahlenchemie*, pp. 299–354. (Weinheim/Bergstr.: Verlag Chemie, 1969).

Herz, J. E., and V. Stannett, "Copolymerization in the Crystalline Solid State," *J. Polymer Sci.* (*D*) (*Macromol. Rev.*), **3**, 1 (1968).

Hsia Chen, C. S., "Polymerization Induced by Ionizing Radiation at Low Temperature: Polymerization of Substituted Styrenes," *J. Polymer Sci.* (*A*), **1**, 1293 (1963).

Magat, M., "Polymerization in the Solid State," *Polymer*, **3**, 449 (1962).

Okamura, S., K. Hayashi, and Y. Kitanishi, "Radiation-induced Solid State Polymerization of Ring Compounds," *J. Polymer Sci.*, **58**, 925 (1962).

Tabata, Y., "Solid State Polymerization," *Adv. Macromol. Chem.*, **1**, 283 (1968).

Williams, F., *Fundamental Processes in Radiation Chemistry* (P. Ausloos, ed.) Chap. 8. New York: Wiley–Interscience, 1968.

Wilson, J. E., *Radiation Chemistry of Monomers, Polymers and Plastics*, New York: Dekker, 1974.

Electrochemically initiated polymerizations

Breitenbach, J. W., O. F. Olaj, and F. Sommer, "Polymerisationsanregung durch Elektrolyse," *Fortschr. Hochpolym.-Forsch.*, **9**, 47 (1972).

Funt, B. L., "Electrolytically Controlled Polymerizations," *J. Polymer Sci.* (*D*) (*Macromol. Rev*), **1**, 35 (1967).

Yamazaki, N., "Electrolytically Initiated Polymerization," *Adv. Polymer Sci.*, **6**, 377 (1969).

6

Polymerization of Cyclic Organic Compounds

SCOPE

Two of the three general classes of polymerization processes—condensation and olefin polymerizations—have been discussed in Chapters 2 to 5. Here, we introduce the third category, the ring-opening polymerization of cyclic compounds, (1), by the overall process shown schematically in reaction (1).

$$
\begin{array}{c}
\text{A—B} \\
| \quad | \\
\text{(A—B)}_n
\end{array}
\quad \longrightarrow \quad \pm\text{A—B}\!\pm_{n+1}
\tag{1}
$$

1

Two main differences from the other two polymerization types can be emphasized. First, in contrast to condensation reactions, polymerization does not result in the loss of a small molecule. Second, ring-opening polymerization does not involve a loss of multiple-bonding enthalpy, whereas the loss of unsaturation is a powerful driving force for olefin polymerization. In fact, the cyclic compounds and the polymers formed from them are often remarkably similar in enthalpy per repeating segment. The consequences of this fact will be explored more fully in Chapter 10.

It should be noted that ring-opening polymerization is the principal method for the synthesis of inorganic polymers, a topic that is covered in Chapter 7. In this chapter we are mainly concerned with outlining the scope, experimental techniques, and general mechanisms for the polymerization of cyclic compounds, with a special emphasis on organic derivatives.

Cyclic organic compounds that have been polymerized include cyclic ethers, lactones (cyclic esters), lactams (cyclic amides), and imines (cyclic amines).

CYCLIC ETHERS

Trioxane

Trioxane (**2**) polymerizes under a variety of reaction conditions to yield polyoxymethylene (polyformaldehyde) (**3**), as shown in reaction (2). The polymerization

$$n\ H_2C \underset{\underset{\displaystyle CH_2}{O \qquad O}}{\overset{\overset{\displaystyle O}{\diagup \quad \diagdown}}{\qquad}} CH_2 \longrightarrow \left(CH_2 - O \right)_{3n} \qquad (2)$$

$$\mathbf{2} \qquad\qquad\qquad \mathbf{3}$$

takes place (a) in the presence of boron trifluoride or other Lewis acid catalysts, (b) during sublimation of the cyclic trimer, or (c) during γ-irradiation of the crystalline trimer.

The experimental procedure used for the Lewis-acid-catalyzed polymerization of trioxane is quite straightforward. A concentrated solution of trioxane in cyclohexane in a glass reaction vessel is purged of air by bubbling a stream of dry nitrogen through it. Boron trifluoride etherate catalyst is then added and the mixture is heated at 55 to 60°C and stirred out of contact with the air for several hours. Polyoxymethylene separates from the solution as a white powder. The polymer can be compression-molded at 180 to 220°C to give tough, translucent films that can be oriented by stretching. Polyoxymethylene has a marked tendency to depolymerize to formaldehyde at moderate temperatures. This process can be retarded by "end capping" of the chains by acylation or etherification, or by copolymerization with a small amount of an epoxide.

The partial polymerization of trioxane to polyoxymethylene during *sublimation* has been ascribed to the presence of traces of free formaldehyde which polymerizes, or to formic acid impurity which functions as a cationic catalyst.

The radiation-induced solid-state polymerization of trioxane has already been mentioned in Chapter 5. However, a solid-state polymerization can also be achieved by allowing solutions of Lewis acid catalysts to come into contact with trioxane crystals. Polymerization then proceeds inward from the crystal outer surfaces.

Tetroxane, $(O-CH_2)_4$, also polymerizes in the crystalline state, especially when irradiated with γ-rays.

Trithiane and Tetrathiane

Trithiane (**4**) and tetrathiane (**5**) are, respectively, the cyclic trimer and tetramer of thioformaldehyde. Both cyclic compounds can be converted to polythioformaldehyde (**6**). The cyclic trimer (**4**) is a stable, crystalline solid, m.p. 215 to 216°C. This

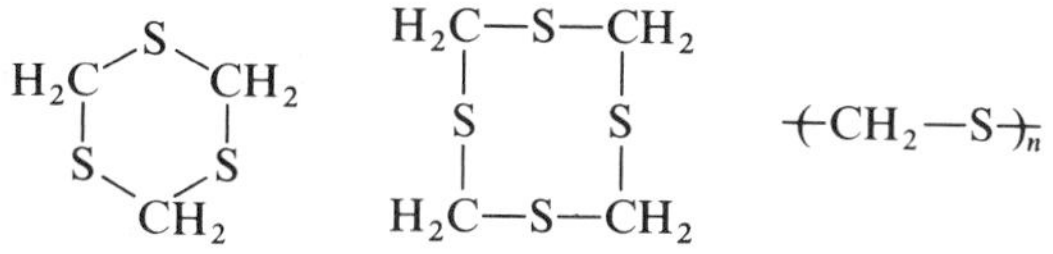

material polymerizes in the solid state when irradiated with γ-rays or when the molten material is treated with cationic-type catalysts, such as boron or antimony trifluorides. Tetrathiane behaves similarly, as does the cyclic pentamer. A cationic polymerization mechanism is believed to operate.

Tetrahydrofuran

Although tetrahydropyran (**7**) and 1,4-dioxane (**8**) are unreactive under polymerization conditions, tetrahydrofuran (**9**) can be induced to polymerize in the

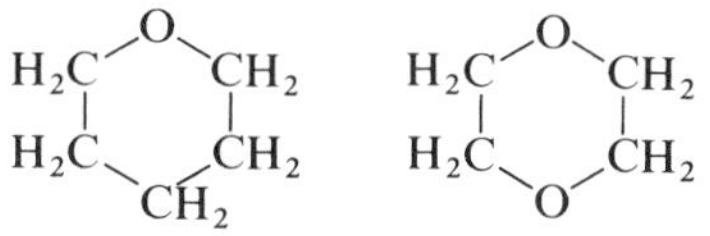

presence of phosphorus- or antimony pentafluorides or $[Ph_3C]^+[SbCl_6]^-$ as catalysts. The presence of the five-membered ring in tetrahydrofuran is apparently responsible for these differences.

In practical terms, tetrahydrofuran must be purified rigorously before the polymerization is carried out. The compound must be boiled at reflux over sodium hydroxide pellets, distilled in a nitrogen atmosphere, refluxed over lithium aluminum hydride, and then distilled immediately before use. A suitable polymerization catalyst is a coordination complex of tetrahydrofuran with phosphorus pentafluoride. Polymerization is effected during ~6 h at 30°C to yield poly(tetramethylene oxide) (**10**) with a molecular weight of about 300,000. The overall process is shown in reaction (3). The use of antimony pentachloride as a polymerization catalyst

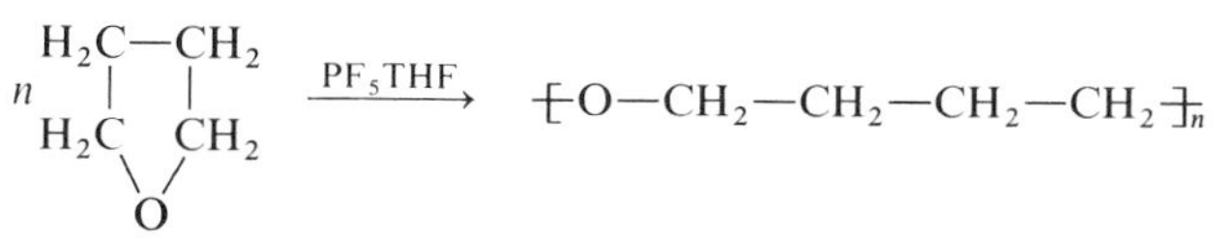

$$\tag{3}$$

yields a lower-molecular-weight polymer. Poly(tetramethylene oxide) is a tough, film-forming material, with a crystalline melting temperature of 45°C.

Oxetanes and Oxepanes

Substituted oxetanes, such as **11**, can be polymerized (sometimes violently) in the presence of Lewis acid catalysts such as phosphorus pentafluoride (reaction

$$
n \; ClCH_2{-}\underset{\underset{\displaystyle H_2C{-}O}{|}}{\overset{\overset{\displaystyle CH_2Cl}{|}}{C}}{-}CH_2 \quad \xrightarrow{PF_5} \quad \left[O{-}CH_2{-}\underset{\underset{\displaystyle CH_2Cl}{|}}{\overset{\overset{\displaystyle CH_2Cl}{|}}{C}}{-}CH_2 \right]_n \tag{4}
$$

$$\quad\quad\quad\quad \textbf{11} \quad\quad\quad\quad\quad\quad\quad\quad\quad \textbf{12}$$

4). Release of the ring strain in **11** almost certainly provides the driving force for polymerization. Polymer **12** is a crystalline, film-forming material that melts at 177°C. Oxetane itself (**13**) polymerizes readily at temperatures of 0°C or below to give high yields of polymer **14** (reaction 5).

$$
\underset{\underset{\displaystyle CH_2{-}O}{|}}{\overset{\overset{\displaystyle CH_2{-}CH_2}{}}{}} \quad \xrightarrow{BF_3} \quad (-O{-}CH_2{-}CH_2{-}CH_2)_{\overline{n}} \tag{5}
$$

$$\quad\quad \textbf{13} \quad\quad\quad\quad\quad\quad\quad\quad \textbf{14}$$

As shown in reaction (6), oxepane (**15**) polymerizes slowly in the presence of catalysts, such as $[(C_2H_5)_3O]^+(BF_4)^-$ or $[(C_2H_5)_3O]^+(SbCl_6)^-$, even though

$$
n \; \begin{matrix} CH_2{-}CH_2 \\ | \quad\quad | \\ CH_2 \quad\; CH_2 \\ | \quad\quad | \\ CH_2 \quad\; CH_2 \\ \diagdown \quad \diagup \\ O \end{matrix} \quad \xrightarrow{Et_3O^+BF_4^-} \quad \{O{-}(CH_2)_6\}_n \tag{6}
$$

$$\quad\quad\quad \textbf{15} \quad\quad\quad\quad\quad\quad\quad\quad \textbf{16}$$

only a minimal amount of ring strain must be released in this process. However, this polymerization is reversible, since depolymerization of **16** back to **15** takes place to yield an equilibrium mixture containing 2 to 3% of **15** and 97 to 98% **16** at 30°C. It has been shown that the polymerization reactivity falls in the order oxetane > tetrahydrofuran > oxepane.

Epoxides

Epoxide polymerization is a subject of considerable technological importance. Here again, the ease of ring-opening polymerization reflects a release of ring strain. Ethylene oxide (**17**), in particular, polymerizes readily to poly(ethylene oxide) (**18**) in the presence of both anionic- and cationic-type catalysts (reaction 7). Anionic catalysts that are suitable include alkoxide ions, hydroxides, metal oxides, and some organometallic derivatives. Cationic polymerizations are initiated by Lewis acids and protonic reagents.

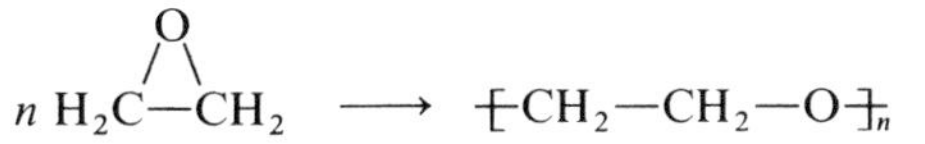

$$n \ H_2C\!\!-\!\!CH_2 \longrightarrow \{CH_2\!-\!CH_2\!-\!O\}_n \qquad (7)$$

$$\begin{array}{ccc} & 17 & & 18 \end{array}$$

The polymerization of ethylene oxide can be carried out at 50°C in a sealed polymerization tube in the presence of strontium carbonate as a catalyst. Initially, there is an induction period which is followed by a very rapid reaction, so rapid, in fact, that *explosions may occur*. The polymerization reaction is normally complete within 2 h. The polymer can then be cast from solution to give highly crystalline films.

Other epoxides, such as propylene oxide (**19**) can also be induced to undergo ring-opening polymerization, and alkylene sulfides behave similarly. On the other

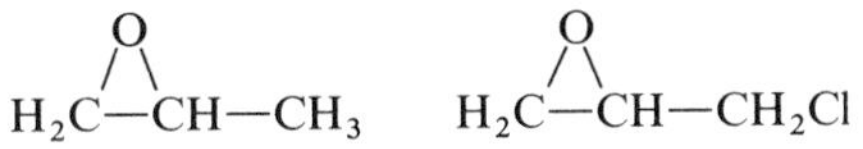

$$\begin{array}{cc} H_2C\!\!-\!\!CH\!-\!CH_3 & H_2C\!\!-\!\!CH\!-\!CH_2Cl \\ \textbf{19} & \textbf{20} \end{array}$$

hand, the well-known epoxy *resins* are usually prepared by the base-catalyzed reaction between an epoxide, such as epichlorohydrin (**20**) and a polyhydroxy compound, such as bisphenol A. The reaction yields a prepolymer by an initial base-catalyzed ring cleavage of the epoxide ring by the hydroxyl groups (**21**). The overall process is illustrated in reaction (8). The ultimate products contain both

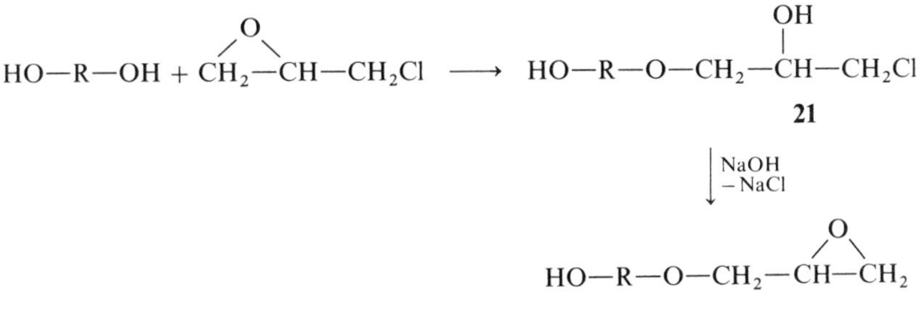

$$HO\!-\!R\!-\!OH + CH_2\!-\!CH\!-\!CH_2Cl \longrightarrow HO\!-\!R\!-\!O\!-\!CH_2\!-\!\overset{OH}{\underset{\textbf{21}}{CH}}\!-\!CH_2Cl$$

$$\Big\downarrow \begin{array}{l} NaOH \\ -NaCl \end{array} \qquad (8)$$

$$HO\!-\!R\!-\!O\!-\!CH_2\!-\!CH\!-\!CH_2$$

terminal epoxy groups and pendent hydroxyl groups. Crosslinking of the prepolymer is then effected by addition of reagents such as amines. Thus, epoxy resins are characterized more by ring cleavage and condensation than by simple ring-opening polymerization.

OTHER CYCLIC ORGANIC COMPOUNDS

Lactones (**22**), cyclic anhydrides (**23**), lactams (**24**), cyclic imines (**25**), and cyclopentene (**26**) will undergo ring-opening polymerization reactions as shown in reactions (9) to (13). Lactones (**22**) are polymerized to polyesters with the use of either anionic or cationic catalysts. Example initiators include alcohols, amines, organometallic compounds, and alcohol–titanium alkoxide mixtures. However, it should be noted that ring size has an important and rather curious influence on the polymerizability of lactones. γ-Butyrolactone, which contains a five-membered ring, apparently does not polymerize, although δ-valerolactone, with a six-membered ring, does polymerize.

$$n \underset{\mathbf{22}}{\overset{\overbrace{}}{O}{+}CH_2{+}_x C{=}O} \longrightarrow \left[O{-}(CH_2)_x{-}\overset{\overset{O}{\|}}{C}\right]_n \tag{9}$$

$$n \underset{\mathbf{23}}{O{=}\overset{\overbrace{O}}{C}{+}CH_2{+}_x C{=}O} \longrightarrow \left[\overset{\overset{O}{\|}}{C}{-}(CH_2)_x{-}\overset{\overset{O}{\|}}{C}{-}O\right]_n \tag{10}$$

$$n \underset{\mathbf{24}}{H{-}N{+}CH_2{+}_x C{=}O} \longrightarrow \left[\overset{\overset{H}{|}}{N}{-}(CH_2)_x{-}\overset{\overset{O}{\|}}{C}\right]_n \tag{11}$$

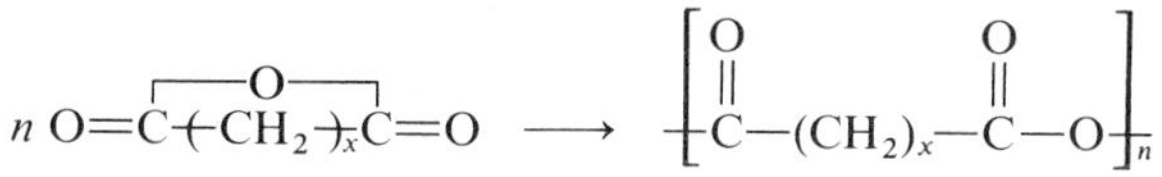

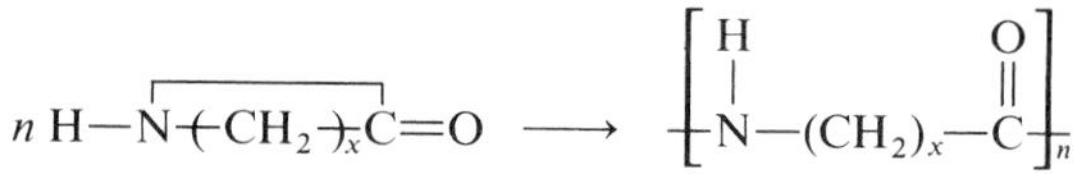

$$n \underset{\mathbf{25}}{\overset{\overset{\overset{H}{|}}{N}}{CH_2{-}CH_2}} \longrightarrow \left[CH_2{-}CH_2{-}\overset{\overset{H}{|}}{N}\right]_n \tag{12}$$

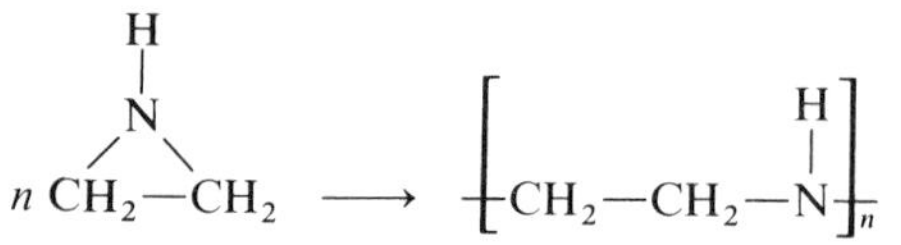

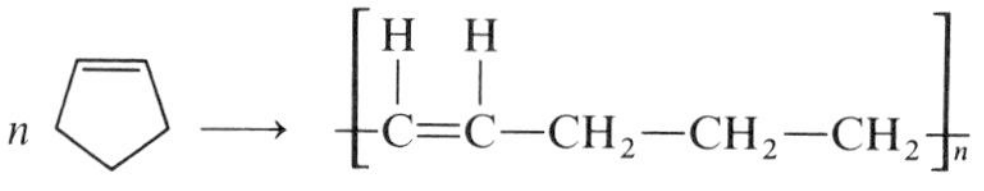

$$n \underset{\mathbf{26}}{\bigcirc} \longrightarrow \left[\overset{\overset{H}{|}}{C}{=}\overset{\overset{H}{|}}{C}{-}CH_2{-}CH_2{-}CH_2\right]_n \tag{13}$$

The polymerization of lactams (**24**), especially caprolactam (**27**), provides a valuable noncondensation route to the synthesis of nylons (reaction (14)). The

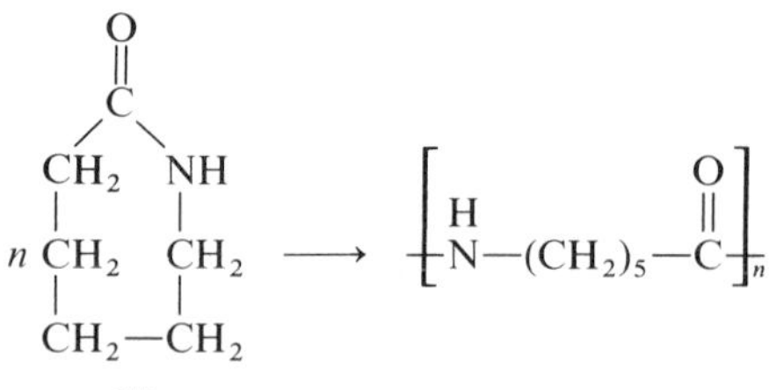

$$n \underset{\mathbf{27}}{\begin{matrix} \overset{\overset{O}{\|}}{C} \\ CH_2 \quad NH \\ | \qquad | \\ CH_2 \quad CH_2 \\ | \qquad | \\ CH_2{-}CH_2 \end{matrix}} \longrightarrow \left[\overset{\overset{H}{|}}{N}{-}(CH_2)_5{-}\overset{\overset{O}{\|}}{C}\right]_n \tag{14}$$

Nylon 6

polymerization can be initiated by reagents such as strong bases (metal hydrides, alkali metals, or metal amides), protonic acids, aromatic amines, or by water. The base-catalyzed initiation is often applied to N-acylated lactams, since these species are not subject to the long induction periods that characterize the polymerization of the parent lactams.

The water-catalyzed polymerization of caprolactam can be carried out on a laboratory scale provided that suitable safety precautions are taken.[1] A mixture of purified caprolactam and water in about a 50:1 weight ratio is sealed under nitrogen in a thick-walled polymerization tube. This tube is a potential bomb when heated, and intelligent precautions should be taken to provide shielding. The tube is then heated to 250°C for about 6 h, is then cooled, and the end of the tube is removed cautiously, again with the use of adequate shielding. The tube contents are now heated to 250 to 255°C as a stream of nitrogen is allowed to bathe the polymer surface. Most of the water will volatilize from the system, and the molten reaction mixture will undergo a viscosity increase. About 2 h of heating are usually sufficient to generate a polymer (nylon 6) of suitable molecular weight for melt spinning into fibers.

Ethylenimine (25) polymerizes very rapidly in the presence of cationic initiators, a result that probably reflects the release of ring strain.

MECHANISMS OF RING-OPENING POLYMERIZATIONS

General Mechanisms

Two general types of mechanisms have been proposed for ring-opening polymerizations. In the first, the catalyst is presumed to attack the ring initially, with concurrent or subsequent cleavage. The resultant ionic or zwitterionic end group then attacks another ring with concurrent ring cleavage, and so on. This overall process is illustrated in reaction (15).

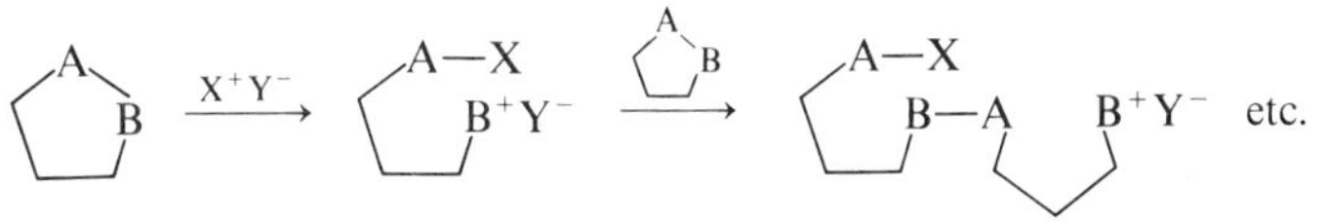

$$ (15) $$

The alternative mechanism supposes that an initial ring cleavage does not occur. Instead, the primary interaction of the catalyst with the cyclic monomer generates a coordination intermediate (usually an oxonium ion), which then functions as the true initiating species. This is illustrated in reaction (16). In many cases, the distinction between these two pathways is difficult to establish.

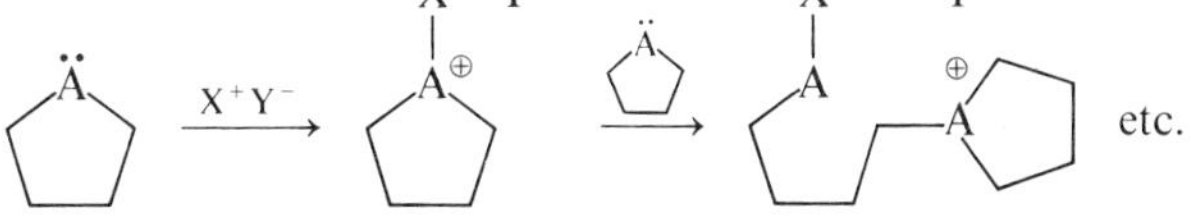

$$ (16) $$

[1] W. R. Sorenson and T. W. Campbell, *Preparative Methods of Polymer Chemistry*, 2nd ed. (New York: Wiley–Interscience, 1968), p. 344.

Mechanism of Trioxane Polymerization

Different reaction conditions apparently generate different polymerization mechanisms for trioxane. The general acid-catalyzed reactions are believed to involve the initiation and propagation steps depicted in (17) and (18). Thus, propagation

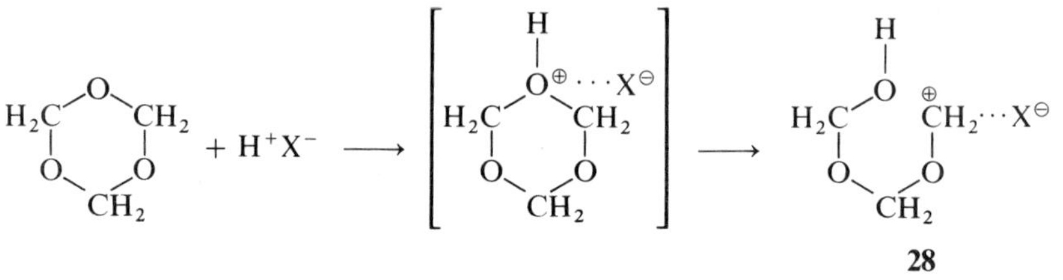

$$\text{(17)}$$

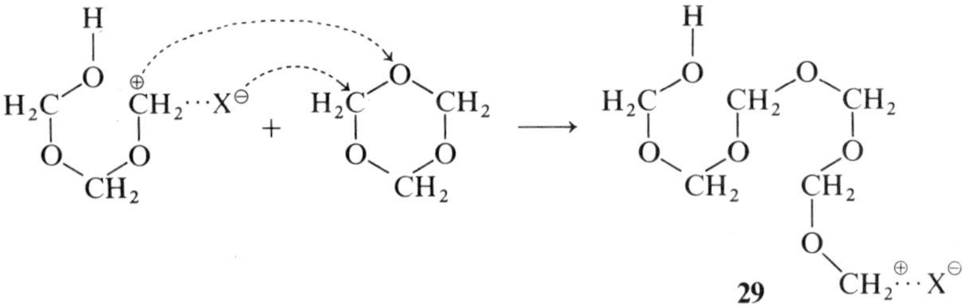

28

$$\text{(18)}$$

29

probably takes place by the insertion of the trimer molecules into the $CH_2^+ \cdots X^-$ ionic bond of **28**, **29**, and so on. Polymerization can also be accompanied by depolymerization and equilibration. These latter processes can regenerate trioxane or yield the monomer (formaldehyde). Moreover, tetroxane, $(O-CH_2)_4$, may also be generated by a "back-biting" process. Such processes are discussed in a more general sense in Chapter 10. Chain termination can occur either by reaction of the chain ends with anions, or by hydride abstraction by the terminal carbonium ion.

The polymerization of trioxane in methylene chloride solution under the influence of *anhydrous* Lewis acid catalysts, such as boron trifluoride, may follow a different mechanism. Apparently no cocatalyst (water) is required, and it has been proposed that polymerization involves the formation of a zwitterion (**30**). The sequence is illustrated in reactions (19) and (20). Because continued propagation

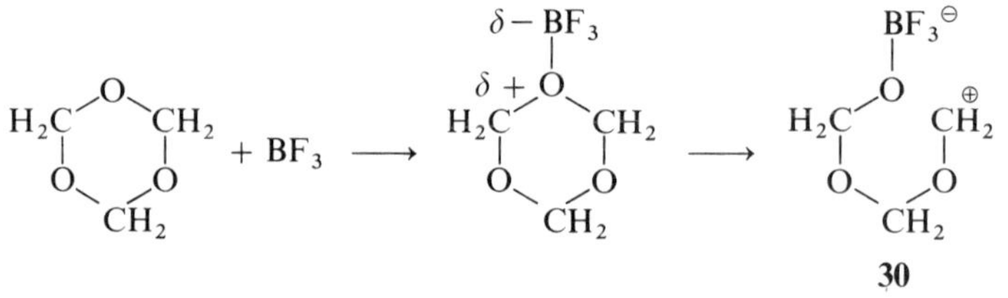

$$\text{(19)}$$

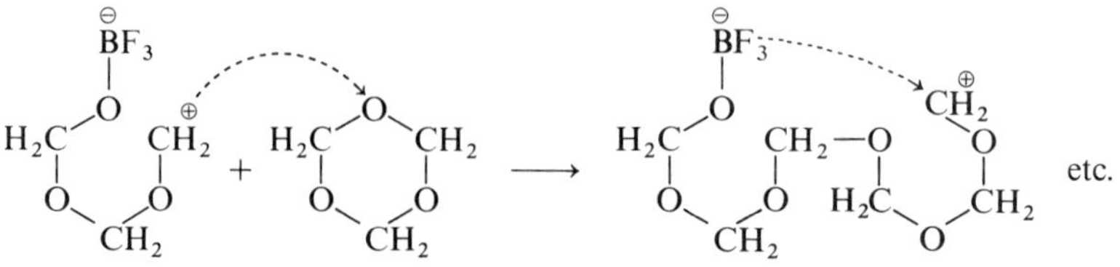

30

$$\text{(20)}$$

etc.

would progressively increase the number of skeletal atoms that separate the charges, it must be assumed that the chain ends remain in close proximity. If this is true, the polymerization can be viewed as the successive insertion of trimer molecules into a macrocyclic unit.

The mechanism of the radiation-induced *solid-state* polymerization of trioxane is still a subject for debate. An ionic mechanism has been proposed, but radical or cation–radical mechanisms must also be considered (see Chapter 5).

Mechanism of Tetrahydrofuran Polymerization

Lewis acids in the presence of traces of water may initiate the polymerization of tetrahydrofuran by protonation of the etheric oxygen atom (reaction 21). Propagation would then occur by insertion of tetrahydrofuran molecules into the ionic bond

$$
\begin{array}{c}
H_2C{-\!\!-\!\!-}CH_2 \\
| \qquad | \\
H_2C{\diagdown}_{\!\!O}{\diagup}CH_2
\end{array}
+ [PF_5OH]^{\ominus}H^{\oplus}
\longrightarrow
\begin{array}{c}
H_2C{-\!\!-\!\!-}CH_2 \\
| \qquad | \\
H_2C{\diagdown}_{\!\!O}\ \ \underset{\oplus}{CH_2}\cdots[PF_5OH]^{\ominus} \\
| \\
H
\end{array}
\qquad (21)
$$

31

of **31**. However, some tentative evidence exists that polymerization can occur even in the absence of water. If this is true, catalysts such as phosphorus pentafluoride or pentachloride may function as the ionic complexes $PF_4^{\oplus}PF_6^{\ominus}$ and $PCl_4^{\oplus}PCl_6^{\ominus}$. The latter formulation, in particular, is well known for PCl_5 in the solid state. If such ionic complexes are depicted symbolically as X^+Y^-, a polymerization mechanism of the type shown in (22) can be formulated, and propagation can be visualized as an insertion of tetrahydrofuran molecules into the ionic $-CH_2^{\oplus}\cdots Y^{\ominus}$ bond of species such as **32**.

$$
\begin{array}{c}
H_2C{-\!\!-\!\!-}CH_2 \\
| \qquad | \\
H_2C{\diagdown}_{\!\!O}{\diagup}CH_2
\end{array}
+ X^{\oplus}Y^{\ominus}
\longrightarrow
\begin{array}{c}
H_2C{-\!\!-\!\!-}CH_2 \\
| \qquad | \\
H_2C{\diagdown}_{\!\!O}\ \ CH_2^{\oplus}\cdots Y^{\ominus} \\
| \\
X
\end{array}
\qquad (22)
$$

32

When cocatalysts are absent, it is possible that Lewis acid/Lewis base complexes, such as **33**, function as the real catalytic species.

$$
\begin{array}{c}
CH_2{-}CH_2 \\
| \qquad\qquad {\diagdown} \\
\qquad\qquad \overset{\delta+}{O}\cdots\overset{\delta-}{BF_3} \\
| \qquad\qquad {\diagup} \\
CH_2{-}CH_2
\end{array}
$$

33

Mechanism of Epoxide Polymerization

The anionic polymerization of epoxides is initiated by alkoxides, hydroxides, metal oxides, and organometallic species such as zinc alkyls. Each of these catalysts can be depicted symbolically as X^+Y^-, for example, as M^+OR^-, M^+OH^-, and so on. The anionic initiation process then operates, as shown in (23). Since, for the catalysts

$$R-\overset{O}{\overset{\diagdown}{CH}}-CH_2 + X^+Y^- \longrightarrow R-\overset{\underset{|}{O^{\ominus}\cdots X^{\oplus}}}{CH}-CH_2-Y \qquad (23)$$

used, the $-O^{\ominus}\cdots X^{\oplus}$ bond is more ionic than the $-CH_2-Y$ bond, propagation occurs by insertion of monomer molecules into the $-O^{\ominus}\cdots X^{\oplus}$ bond, as depicted in (24).

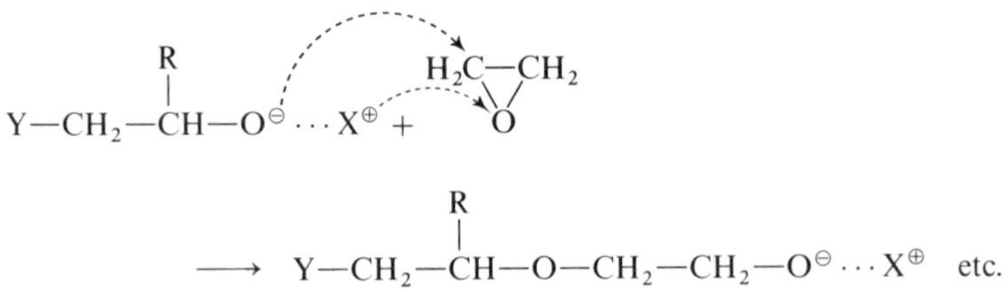

$$Y-CH_2-\overset{\underset{|}{R}}{CH}-O^{\ominus}\cdots X^{\oplus} + \overset{H_2C-CH_2}{\overset{\diagdown\diagup}{O}} \qquad (24)$$

$$\longrightarrow Y-CH_2-\overset{\underset{|}{R}}{CH}-O-CH_2-CH_2-O^{\ominus}\cdots X^{\oplus} \quad \text{etc.}$$

Termination may not occur unless protonic reagents are added. However, the polymer molecular weights are often low because of chain transfer. Chain transfer can occur by proton abstraction by the terminal anion from an alkyl group, R, with the concurrent formation of an allyl ether anion, (**34**), as shown in reaction (25).

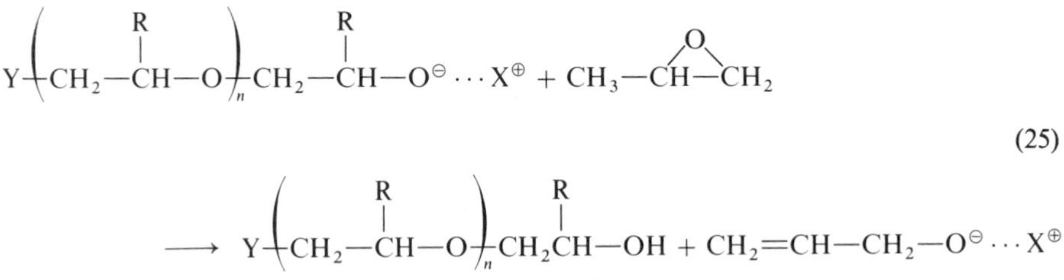

$$Y\left(CH_2-\overset{\underset{|}{R}}{CH}-O\right)_n CH_2-\overset{\underset{|}{R}}{CH}-O^{\ominus}\cdots X^{\oplus} + CH_3-\overset{O}{\overset{\diagdown}{CH}}-CH_2$$

$$(25)$$

$$\longrightarrow Y\left(CH_2-\overset{\underset{|}{R}}{CH}-O\right)_n CH_2\overset{\underset{|}{R}}{CH}-OH + CH_2=CH-CH_2-O^{\ominus}\cdots X^{\oplus}$$

$$\mathbf{34}$$

The cationic polymerization of epoxides probably proceeds through mechanisms that are similar to those described for trioxane and tetrahydrofuran.

Mechanisms of Lactam Polymerization

A variety of mechanisms have been proposed for lactam polymerization, depending on the type of initiator. Here we will consider only the commercially important processes—catalysis by bases and by water. Strong bases probably initiate polymerization by the replacement of hydrogen in the N—H residue of **35** by a cation to give **36**, as illustrated in scheme (26). However, the subsequent mechanism is

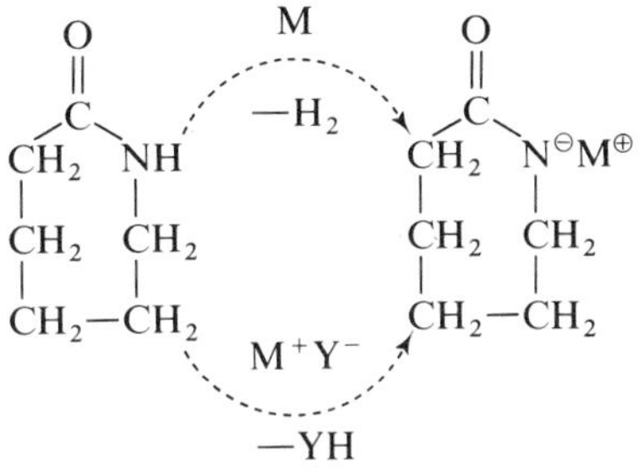

$$\tag{26}$$

35   36

complicated. Ring opening of the initiated monomer is presumed not to occur. Instead, the anionic center can attack the carbonyl carbon of another ring by a *slow* process, as shown in the formation of **37** (sequence 27).

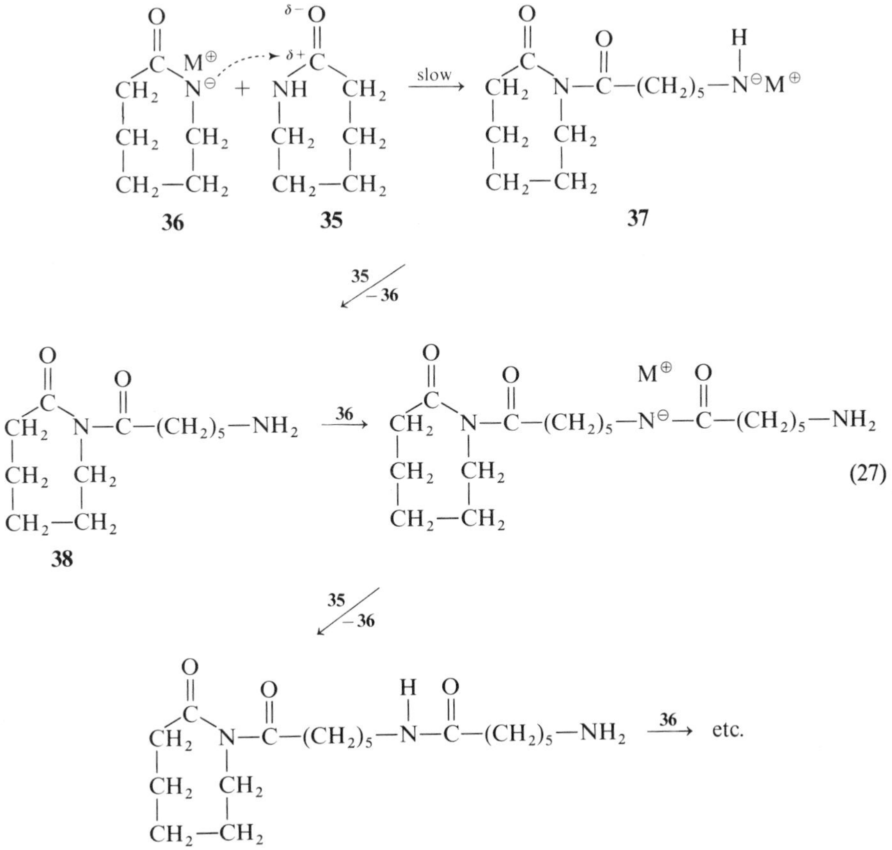

36   35   37

$$\tag{27}$$

38

39

The long induction periods observed for this polymerization are probably a consequence of the slowness of this step. However, compound **37** is assumed not to be the real initiating species. Instead, compound **38** is believed to fulfill that function by reaction with **36**. Propagation appears to take place by the unusual process of *insertion* of molecules of **36** into an —NH—CO— bond, followed by a remetallation of another monomer molecule by the polymer to yield, for example, **39**. The base-catalyzed polymerization is more conveniently applied to N-acyl-caprolactams, such as **40**, since the long induction periods are not encountered in these systems.

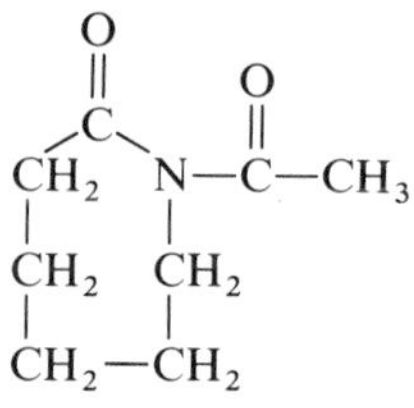

$$
\begin{array}{c}
\overset{\displaystyle O}{\underset{\displaystyle \Vert}{}} \\
\overset{C}{CH_2 \quad N-C-CH_3} \\
| \qquad | \\
CH_2 \quad CH_2 \\
| \qquad | \\
CH_2-CH_2
\end{array}
$$

40

The water-catalyzed reaction of caprolactam has a more straightforward mechanism. The primary step involves the hydrolysis of caprolactam to the amino acid, (**41**) (reaction 28). Propagation then involves either the direct, ring-opening

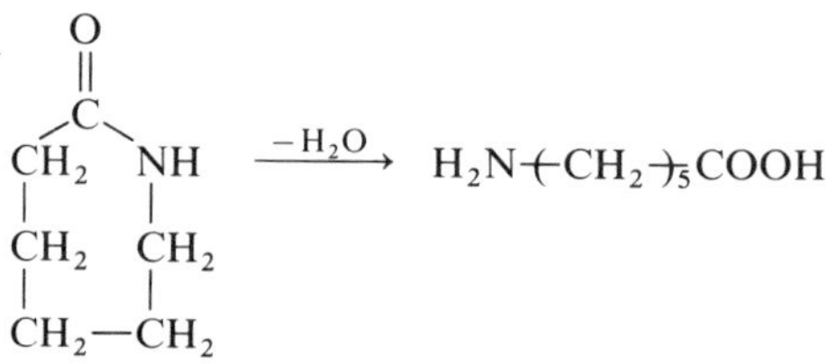

$$
\begin{array}{c}
\overset{\displaystyle O}{\underset{\displaystyle \Vert}{C}} \\
CH_2 \quad NH \\
| \qquad | \\
CH_2 \quad CH_2 \\
| \qquad | \\
CH_2-CH_2
\end{array}
\xrightarrow{-H_2O} \quad H_2N \text{+} CH_2 \text{)}_5 COOH
$$

(28)

41

attack of **41** on caprolactam, or a process in which the amino acid zwitterion, $H_3N^{\oplus}—(CH_2)_5C(O)O^{\ominus}$, undergoes a ring-opening attack on the cyclic monomer.

CYCLIC COMPOUNDS THAT RESIST POLYMERIZATION

Benzene (**42**), s-triazine (**43**), borazines (**44**), cyclohexane (**45**), tetrahydropyran (**7**), and 1,4-dioxane (**8**) have so far resisted all attempts to induce ring-opening polymerization. Both mechanistic and thermodynamic reasons can be put forward to rationalize this behavior. First, it is possible that suitable catalysts have not yet been found for the cleavage of skeletal bonds and the initiation of an ionic propagation. However, a more serious reason is apparently connected with the absence of ring strain in these compounds. Moreover, benzene, s-triazine, and borazine are especially stabilized by aromatic or pseudoaromatic π-bonding. Thus, these compounds constitute an "energy trap" in the polymeric series. Polymers can be made

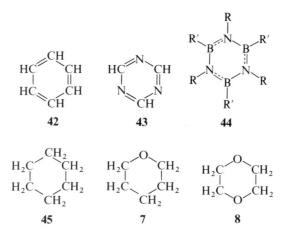

42   43   44

45   7   8

from the monomers (acetylene, nitriles, ethylene, etc.) provided that the "trap" can be avoided. However, once cyclotrimerization has occurred, further polymerization is essentially blocked. These and other influences on ring-opening polymerizations are discussed further in Chapter 10.

STUDY QUESTIONS

1. Speculate on the possibility that some of the ring-opening polymerizations discussed in this chapter might fall into the category of "living" polymerizations. Which structural factors would favor or prevent such a possibility?

2. The mechanisms discussed in this chapter are ionic mechanisms. Are free-radical mechanisms also plausible? If so, which cyclic monomer structures would be the most likely to participate in free-radical reactions? How might a free-radical mechanism be (a) initiated, and (b) detected?

3. Outline a cationic polymerization mechanism that might be applicable to epoxides.

4. Without referring to Chapter 10, speculate in detail on the reasons why polyethylene is not manufactured by the ring-opening polymerization of cyclohexane.

5. Some polymers (e.g., polyoxymethylene) can be prepared either from the monomer (formaldehyde) or from a cyclic species (trioxane). What are the main practical advantages or disadvantages to these alternative routes?

6. Glance through an organic chemistry textbook and compile a list of classes of organic cyclic compounds that are not mentioned in this chapter. Then suggest possible methods that might be used to induce their polymerization.

7. Suppose that you suspected that the polymerization of trioxane proceeded only by prior dissociation to formaldehyde, followed by polymerization of this monomer. Suggest ways in which you might distinguish between this mechanism and one that involved a prior trimerization of formaldehyde to trioxane, followed by a polymerization of trioxane.

8. Tetrahydrofuran is a common organic solvent that is often used in large quantities in the laboratory or in manufacturing. What reagents, other than PF_5, should you *not* bring into

contact with tetrahydrofuran if you wish to avoid a (possibly dangerous) polymerization process? How could you ensure that polymerization of tetrahydrofuran would be unlikely during normal laboratory use?

9. Speculate on the prospect that an epoxide could be copolymerized with caprolactam. What reaction conditions might you choose for this process? What complications do you foresee?

SUGGESTIONS FOR FURTHER READING

ALLCOCK, H. R., *Heteroatom Ring Systems and Polymers*. New York: Academic Press, 1967.

BAILEY, F. E., AND J. V. KOLESKE, *Poly(ethylene oxide)*. New York: Academic Press, 1976.

BAILEY, W. J., P. Y. CHEN, W. B. CHIAO, T. ENDO, L. SIDNEY, N. YAMAMOTO, N. YAMAZAKI, AND K. YONEZAWA, "Free-Radical Ring-Opening Polymerization," *Contemporary Topics in Polymer Sci.* (M. Shen, Ed.). **3**, 29 (1979).

DREYFUSS, P. AND M. P. DREYFUSS, "Polytetrahydrofuran," *Adv. Polymer Sci.*, **4**, 528 (1967).

EASTHAM, A. M., "Some Aspects of the Chemistry of Cyclic Ethers," *Fortschr. Hochpolym.-Forsch.*, **2**, 18 (1960).

GURGIOLO, A. E., "Poly(alkylene oxides)," *Rev. Macromol. Chem.*, **1**, 39 (1966).

LENZ, R. W., *Organic Chemistry of Synthetic High Polymers*. New York: Wiley–Interscience, 1967.

REIMSCHUESSEL, H. K., "Nylon 6: Chemistry and Mechanisms," *J. Polymer Sci. (D) (Macromol. Rev.)*, **12**, 65 (1977).

SORENSON, W. R., AND T. W. CAMPBELL, *Preparative Methods of Polymer Chemistry*, 2nd ed. New York: Wiley–Interscience, 1968.

STONE, F. W., J. J. STRATTA, L. C. PIZZINI, J. T. PATTON, R. A. BRIGGS, AND E. E. GRUBER. Three articles under the title "1,2-Epoxy Polymers," *Encyclopedia of Polymer Sci. and Technol.* (H. F. Mark, N. G. Gaylord, and N. M. Bikales, eds.), **6**, 103–175 (1967).

SZWARC, M., "The Kinetics and Mechanism of *N*-carboxy-α-amino Acid Anhydride (NCA) Polymerization to Poly(amino acids)," *Adv. Polymer Sci.*, **4**, 1 (1965).

VOGL, O., AND J. FURUKAWA, *Polymerization of Heterocyclics*. New York: Dekker, 1973.

WICHTERLE, O., J. SEBENDA, AND J. KRALICEK, "The Anionic Polymerization of Caprolactam," *Forschr. Hochpolym.-Forsch.*, **2**, 578 (1961).

7

Inorganic Polymers

REASONS FOR THE DEVELOPMENT
OF INORGANIC MACROMOLECULES

Nearly all the synthetic plastics and elastomers in use today are organic polymers. There are numerous reasons for this, including the ready availability of monomers from the petrochemicals industry and the fact that many years of synthetic organic research have led to the discovery of a vast array of compounds that can be used as polymer precursors.

The important differences between organic polymers can be traced to different glass transition temperatures, varying degrees of crystallinity, and specific melting characteristics. To a large extent, these differences are due to the presence of different substituent groups attached to the main chain or to the presence of elements other than carbon in the chain. Thus, by varying the side-group structure or by introducing oxygen or nitrogen into a carbon chain, a wide variety of polymers has been made available, with a very broad range of different properties.

However, the fact remains that many organic polymers have deficiencies. Except for some fluorocarbon polymers and a few aromatic ladder polymers (see Chapter 2), few organic polymers can be heated for prolonged periods above 150°C without either melting or decomposing. When decomposition occurs, it usually results from the reaction of the carbon atoms with atmospheric oxygen. Many organic polymers dissolve or swell in hot organic liquids or oils and for this reason cannot be used as seals or components in automobiles or aircraft engines. Moreover, few organic polymers remain flexible or rubbery over a wide enough temperature range for them to be useful at both low and high temperatures. In the arctic or in high-flying aircraft this hardness and brittleness of organic polymers can create serious hazards. There are other applications for which no suitable organic polymers have yet been found—for example, where prolonged resistance to ultraviolet radiation is essential, in the textile industry where flame-retardant fabrics are needed, and in medicine for the fabrication of artificial organs that will not lead to blood clotting or other undesirable effects. Almost incalculable advantages can be foreseen if electrically conducting polymers can be developed. Many polymer chemists now believe that the solution to problems such as these will require the development of radically different polymer systems. That is why, increasingly, attention is being focused on inorganic polymers.

SCOPE OF THE FIELD

The enormous variety of known organic polymers is a consequence of the ability of the carbon atom to catenate—that is, to form covalently bound chains—or to form macromolecules by bonding with oxygen, sulfur, or nitrogen atoms. However, carbon is not unique in its ability to form polymeric structures: most of the "main-group" elements can participate in the formation of rings or chains, as can a large number of the transition elements. The ability of transition elements to form metal–metal bonds and clusters offers additional prospects for polymer synthesis in the future.

In the following sections we will distinguish between macromolecules that possess a covalently bonded skeleton but an ionic side-group structure (such as mineralogical polymers) and those that are totally covalent, such as polymeric sulfur, siloxanes, phosphazenes, thiazenes, and so on. We will not discuss ionic crystal systems, such as those of alkali metal halides, because these materials possess none of the physical properties that are normally associated with useful macromolecules.

MINERALOGICAL-TYPE INORGANIC POLYMERS

In theory, any inorganic material which contains long-chain molecules should behave like an organic polymer. In other words it should be tough, flexible or elastomeric, easy to fabricate, and should dissolve in suitable solvents to give viscous solutions. Thus, the first question to be answered is why relatively few inorganic macromolecules are flexible or elastomeric.

A number of inorganic polymer systems have been in use for a long time. Glass is an important polymer made up of rings and chains of repeating silicate units. It is well known that glass fibers can be made by the extrusion of molten glass through spinarettes and that the fibers can be used for the fabrication of textiles, filters, or insulation mats. Similar chains of silicate units are found in pyroxene minerals (Figure 7.1). Ladder polymers, or double silicate chains, are present in amphibole minerals such as one form of asbestos (Figure 7.1).

In all these structures the side-group oxygen atoms bear negative charges and these are neutralized by positively charged metal ions, such as sodium, magnesium, or calcium ions. These charged side-group structures bind adjacent chains together to form a crosslinked matrix (1). Thus, instead of displaying rubbery elastomeric or

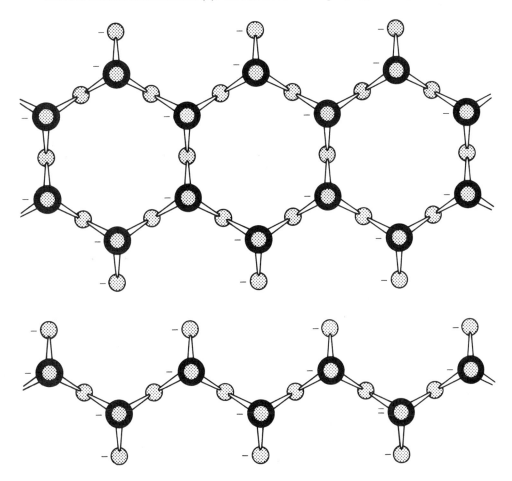

Fig. 7.1 Many mineralogical polymers contain silicate chains either as double-strand structures [as, for example, in the amphiboles (top)] or single-chain systems [as in pyroxenes (bottom)]. Connected to the skeletal silicon atoms are charged side-group oxygen atoms. These side groups are bound together in ionic crosslinks by di- or higher-valent metallic cations to generate a rigid, high-melting structure. [From H. R. Allcock, "Inorganic Polymers," *Scientific American*, **230**, 16 (March 1974). Copyright © 1974 by Scientific American, Inc. All rights reserved.]

flexible, film-forming properties, these polymers are hard and often brittle. They have high melting points, because only at high temperatures do the systems contain enough thermal energy to permit interchange of the counterions. In silica itself, the system is held together by a three-dimensional matrix of *covalent* silicon–oxygen bonds, and this structure generates a very high melting system.

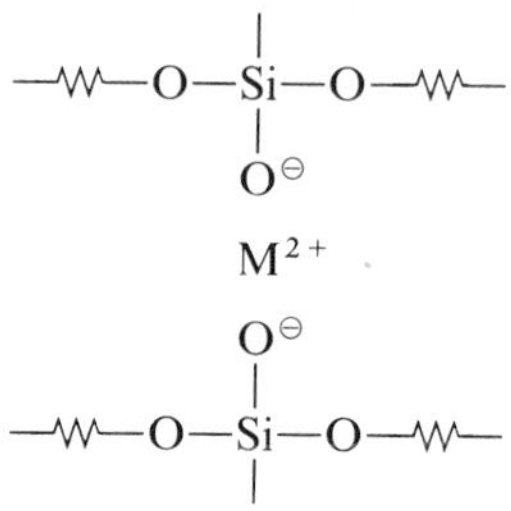

$$\begin{array}{c} \overset{|}{\underset{|}{-W-O-Si-O-W-}} \\ O^{\ominus} \\ M^{2+} \\ O^{\ominus} \\ \overset{|}{\underset{|}{-W-O-Si-O-W-}} \end{array}$$

1

A few other mineralogical-type polymers, as well as glass, have been fabricated into fibers by high-temperature techniques. Aluminum oxide monofilament can be made by melt extrusion. The fibers are stable up to 1400°C and can be used in thermal insulation or in filtration. Aluminosilicate fibers can be made from molten slag or rock. Chrystotile asbestos "monofilaments" are prepared by the wet spinning of an emulsion of pulverized asbestos in water. These continuous filaments are made up of intimately entangled asbestos fibrils. They can be used in textile machinery without exposure of the operators to the lung cancer hazards normally associated with the handling of asbestos dust. One widespread use of mineralogical fibers is as reinforcement fillers in crosslinked organic polymer matrices.

Phosphate polymers, with structures such as the one shown in **2** are also

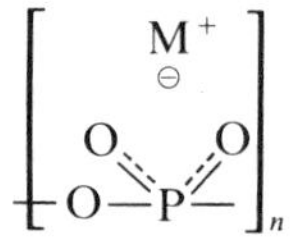

$$\left[\begin{array}{c} M^+ \\ \ominus \\ O \quad\quad O \\ \diagdown\diagup \\ -O-P- \end{array}\right]_n$$

2

known. They possess a backbone of alternating phosphorus and oxygen atoms and they formally resemble linear silicates. The crystalline high polymers of alkali phosphates have been known for many years as "Maddrell's salt" and "Kurrol's salt" (Figure 7.2). They are made by the thermal condensation of dihydrogen phosphates (**3**) according to the reaction shown in (1). Again the presence of the

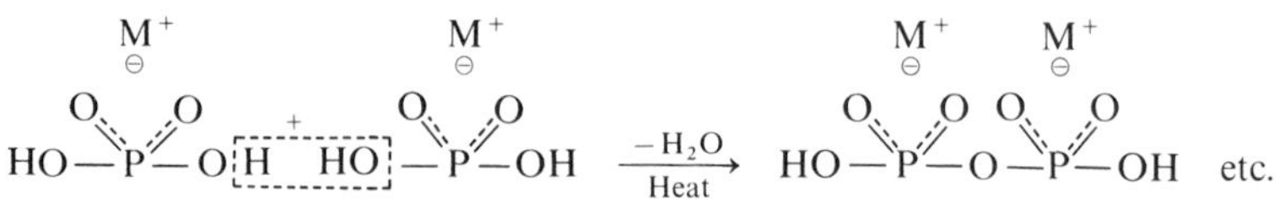

$$HO-\underset{\underset{O}{\overset{O}{\diagdown\diagup}}}{P}-O[H \quad HO]-\underset{\underset{O}{\overset{O}{\diagdown\diagup}}}{P}-OH \xrightarrow[\text{Heat}]{-H_2O} HO-\underset{\underset{O}{\overset{O}{\diagdown\diagup}}}{P}-O-\underset{\underset{O}{\overset{O}{\diagdown\diagup}}}{P}-OH \quad \text{etc.} \qquad (1)$$

3

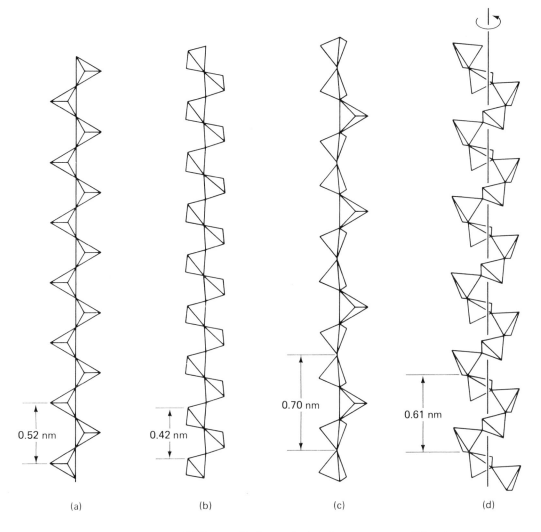

Fig. 7.2 Different alkali metal phosphates generate different conformational arrangements: (a) $(LiPO_3)_n$ low-temperature form; (b) $(RbPO_3)_n$; (c) $(NaPO_3)_n$ Maddrell's salt (high-temperature form); (d) $(NaPO_3)_n$ Kurrol's salt. [From E. Thilo, *Adv. Inorg. Chem. Radiochem.*, **4**, 1 (1962).]

charged side groups lowers the torsional mobility of the chains, with the consequent restriction of flexibility at normal temperatures. An additional problem encountered with some polyphosphates is that treatment with water causes breakage of the chains and brings about a conversion to low-molecular-weight products.

Most rocks, brick, concrete, and ceramics are three-dimensional inorganic polymers held together by a combination of covalent and ionic bonds. These substances have formed the fundamental materials for building construction, containers for food, and ceramic decoration since the dawn of civilization. However, their uses are limited for many modern applications by the difficulty of fabrication into useful objects except at very high temperatures. In general, they are not flexible, elastomeric, or resistant to impact. Although they have excellent stability in air at high temperatures, they do not fulfill many of the needs of modern technology.

COVALENT INORGANIC POLYMERS

For a number of years polymer chemists have realized that an answer to this problem lies in the middle ground between the organic polymers on the one hand and the mineralogical inorganic materials on the other. Basically, the need is for polymers that have a linear structure, with nonionic substituent groups to favor flexibility, and inorganic elements in the backbone to provide stability against heat and oxidation. The presence of inorganic elements in the polymer chain should also confer flexibility and elasticity over a wide temperature range because bonds to the heavier elements are longer than those involving carbon. Hence, the conformational mobility of the molecule should be higher. In theory, hundreds of different inorganic chain structures can be envisaged. In practice, certain restrictions must be considered.

A primary restriction is that the elements that make up the backbone should be linked together through covalent bonds. This restriction is employed to reduce the chances that the chains will be susceptible to attack by water or that ionic cross-links will exist. For that reason the primary search for useful inorganic polymers has concentrated on the covalent main-group elements, particularly boron, aluminum, silicon, germanium, tin, nitrogen, phosphorus, arsenic, antimony, oxygen, sulfur, selenium, and tellurium.

POLYMERIC SULFUR

One of the earliest covalent inorganic polymers to be made, and still one of the most fascinating, is polymeric sulfur. The stable form of sulfur at room temperature is rhombic sulfur, which contains cyclic molecules with eight sulfur atoms in a ring (Figure 7.3).

Rhombic sulfur is a brittle, crystalline material that melts at 113°C to form a yellow-red liquid. Liquid sulfur has a curious property that above 159°C its viscosity increases as the temperature is raised, contrary to the behavior of nearly all other liquids. The viscosity increase results from the opening of the eight-membered rings and their conversion to long chains by a free-radical polymerization process. Since long chains can become entangled more effectively than rings, the viscosity rises. As the temperature is raised above about 175°C, however, the

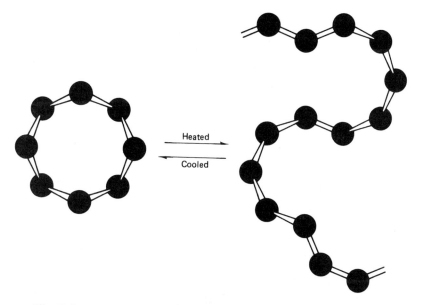

Fig. 7.3   Rhombic sulfur contains cyclic octameric rings of S_8. When heated above 140°C, the ring opens and polymerizes to polymeric sulfur. The process is reversed on cooling.

viscosity begins to decrease, an indication that depolymerization is now occurring to convert the chains back to rings (see Chapter 10).

The high-polymeric form of sulfur can be "quenched" or isolated in one of two ways. First, if the molten high polymer is "quick-quenched" by pouring into a Dry Ice–acetone cooling bath (−78°C), the polymer is isolated as a yellow, translucent glass. This material remains noncrystalline at temperatures below −30°C. When heated above the apparent glass transition temperature at −30°C, it becomes highly elastic. This form of the polymer is believed to consist of polymeric sulfur plasticized by S_8 rings. Removal of the S_8 molecules by extraction with carbon disulfide yields an unplasticized polymer that has a glass transition temperature of +75°C. At temperatures above −10°C, the crude, quenched polymer hardens as crystallization of the S_8 rings occurs.

Second, if the molten polymer is quenched to room temperature by pouring it into water (Figure 7.4), it forms a mixture of semicrystalline polymer and S_8 crystals, known as S_ω. This material undergoes very rapid reversion to rhombic sulfur when heated to temperatures above 90°C. However, at room temperature the rate of cyclization–depolymerization is apparently low. Because of this behavior, polymeric sulfur has only limited practical uses. Nevertheless, its thermal behavior illustrates a characteristic feature of many inorganic polymer systems; rings can be converted to high-polymeric chains, and at elevated temperatures this process tends to be reversed.

(a)   (b)

Fig. 7.4 Polymeric sulfur can be prepared by heating rhombic sulfur to a temperature between 140 and 170°C. If the polymer is quenched by pouring it into cold water (photograph at left), it will retain its flexible and elastomeric properties for a short time at room temperature. However, eventually the polymerized material will revert to its original cyclic oligomeric form. [From H. R. Allcock, "Inorganic Polymers," *Scientific American*, **230**, 16 (March 1974). Copyright © 1974 by Scientific American, Inc. All rights reserved.]

Polyselenium can be obtained from cyclic Se_8 by techniques that are similar to those described for the preparation of polysulfur.

MODIFIED SILICATES

If mineralogical polymers are rigid and difficult to fabricate because of the ionic or covalent crosslinking, a replacement of these ionic or covalent units by covalent "terminal" groups should bring about a modification of the properties. This has been attempted with minerals such as chrysotile asbestos, $Mg_3(OH)_4Si_2O_5$, olivine, $(Mg, Fe)_2SiO_4$, and aluminosilicates such as natrolite, $Na_2Al_2Si_3O_{10} \cdot 2H_2O$ by treatment with trimethylchlorosilane, $(CH_3)_3SiCl$.

Chrysotile asbestos reacts with trimethylchlorosilane in 2-propanol and concentrated hydrochloric acid. The modified mineral swells in organic solvents such as acetone, but retains its normal ribbonlike microstructure. The reaction appears to involve the removal of magnesium and hydroxyl ions and an introduction of $(CH_3)_3Si$ units on to the exposed oxygen atoms. The trimethylsilyl units generate a local solubility in the organic solvent, although each fiber is held together by covalent crosslinks.

POLY(ORGANOSILOXANES) (SILICONE POLYMERS)

Synthesis

We have seen earlier that one objective of inorganic polymer research has been the drive to synthesize polymers that have the stability of minerals but much more flexibility than glasses. Since the rigidity of glass and related polymers is due mainly to the presence of ionic side groups, it might be anticipated that replacement of these charged side units by uncharged groups should increase the flexibility of the chains.

This concept underlies the design of organosiloxane polymers (also known as "silicone polymers"). The most widely used silicone, poly(dimethylsiloxane), contains chains of alternating silicon and oxygen atoms, with two methyl groups attached to each silicon (4). Like polymeric sulfur, and like many other covalent

$$\left[\begin{array}{c} H_3C \quad CH_3 \\ O \text{——} Si \text{——} \end{array} \right]_n$$

4

inorganic polymers, polysiloxanes are made by polymerization of a low-molecular-weight cyclic analogue. In this case, the starting material is octamethylcyclotetra-siloxane (5), which itself is obtained by the hydrolysis of dimethyldichlorosilane (reaction 2). The cyclic tetramer (5) is a colorless, oily material. When heated above

$$(CH_3)_2SiCl_2 \xrightarrow[-HCl]{H_2O} \quad \begin{array}{c} CH_3 \quad CH_3 \\ | \qquad | \\ CH_3-Si-O-Si-CH_3 \\ | \qquad | \\ O \qquad O \\ | \qquad | \\ CH_3-Si-O-Si-CH_3 \\ | \qquad | \\ CH_3 \quad CH_3 \end{array} \xrightarrow[\text{or base}]{\text{Acid}} \left[\begin{array}{c} H_3C \quad CH_3 \\ O \text{——} Si \text{——} \end{array} \right]_n \quad (2)$$

5   **4**

100°C with a trace of acid or base, it polymerizes to form a highly viscous liquid or a gum. The molecular weight of the polymer may be as high as 2×10^6, which corresponds to over 25,000 silicon–oxygen repeating units per chain. "Silicone rubber" is made from the gum by crosslinking the chains by free-radical-type processes. Silicone stopcock grease contains lower-molecular-weight dimethylsiloxane polymers or oligomers plus silica as a filler.

The mechanism of siloxane polymerization follows an ionic process. Basic catalysts, such as alkali metal hydroxides or alkoxides, cleave silicon–oxygen skeletal bonds to yield linear species which can function as chain propagation sites.

Insertion of cyclic molecules takes place into the $-O^-K^+$ ionic "bond." The mechanism is illustrated by the sequence shown in (3).

$$
\begin{array}{l}
\underset{\substack{|\\CH_3}}{\overset{\substack{CH_3\\|}}{CH_3-Si-O-Si-CH_3}} \\
\overset{\displaystyle |}{\underset{\displaystyle |}{O}}\quad\overset{\displaystyle |}{\underset{\displaystyle |}{O}} \\
\underset{\substack{|\\CH_3}}{\overset{\substack{|\\}}{CH_3-Si-O-Si-CH_3}}\\
\qquad\qquad\qquad CH_3
\end{array}
\xrightarrow{\text{KOH}}
$$

$$
\text{CH}_3-\underset{\underset{\text{O}}{|}}{\overset{\overset{\text{CH}_3}{|}}{\text{Si}}}-\text{O}-\overset{\overset{\text{CH}_3}{|}}{\underset{}{\text{Si}}}{-}\text{OH}
$$

$$
\text{CH}_3-\underset{\underset{\text{CH}_3}{|}}{\overset{}{\text{Si}}}-\text{O}-\underset{\underset{\text{CH}_3}{|}}{\overset{}{\text{Si}}}-\text{O}^{\ominus}\text{K}^{\oplus}
$$

(3)

$$
\xrightarrow{[\text{O}-\text{Si}(\text{CH}_3)_2]_4}\quad
\text{HO}\left(\underset{\underset{\text{CH}_3}{|}}{\overset{\overset{\text{CH}_3}{|}}{\text{Si}}}-\text{O}\right)_7\underset{\underset{\text{CH}_3}{|}}{\overset{\overset{\text{CH}_3}{|}}{\text{Si}}}-\text{O}^{\ominus}\text{K}^{\oplus}\quad\text{etc.}
$$

The higher the catalyst concentration, the lower will be the average chain length. In practice, it is frequently necessary both to reduce the chain length and provide chain stabilization by the addition of an end-capping reagent to the polymerization system. The most convenient terminator in common use is hexamethyldisiloxane, $(CH_3)_3Si-O-Si(CH_3)_3$. A propagating siloxane chain can attack this reagent to generate an end-capped polymer and a new catalyst molecule, as shown in reaction (4). Thus, this reaction functions both as an end-capping procedure and as a chain

$$
\text{-\textbackslash\textbackslash\textbackslash-}\underset{\underset{\text{CH}_3}{|}}{\overset{\overset{\text{CH}_3}{|}}{\text{Si}}}-\text{O}^-\text{K}^+ + (CH_3)_3Si-O-Si(CH_3)_3
$$

$$
\longrightarrow\quad \text{-\textbackslash\textbackslash\textbackslash-}\underset{\underset{\text{CH}_3}{|}}{\overset{\overset{\text{CH}_3}{|}}{\text{Si}}}-\text{O}-\underset{\underset{\text{CH}_3}{|}}{\overset{\overset{\text{CH}_3}{|}}{\text{Si}}}-\text{CH}_3 + (CH_3)_3Si-O^{\ominus}K^{\oplus}\quad(4)
$$

transfer step, with the polymer chain length being inversely proportional to the amount of hexamethyldisiloxane added.

Catalysis by acidic reagents, such as hydrogen chloride, is less well understood. However, the acid is presumed to initiate polymerization by cleavage of a silicon–oxygen bond. A similar mixture of cyclic and polymeric homologues is ultimately formed irrespective of whether a basic or an acidic catalyst is employed.

Although the dimethylsiloxane structure forms the basis of most silicone polymers, other substituent groups have also been introduced. These include vinyl, ethyl, trifluoropropyl, p-cyanoethyl, phenyl, and biphenyl groups. The introduction of specific groups improves the oil resistance, strength, and toughness, flame resistance, or compatibility of the polymer. Cosubstituent groups can be introduced by one of two methods. First, cyclotetrasiloxanes that contain two or more different substituent groups can be synthesized by the cohydrolysis of two different chlorosilanes, such as $PhMeSiCl_2$ and Me_2SiCl_2. Second, copolymerization of two or more different cyclosiloxanes can be carried out. For example, $(O—SiPh_2)_3$, can be copolymerized with $(O—SiMePh)_3$ or $(O—SiMePh)_4$. Although such trimers react rapidly to form polymer, the polymers frequently *de*polymerize to cyclic tetramers as the reaction proceeds. This is a consequence of the thermodynamic problems mentioned in Chapter 10.

Properties of Poly (organosiloxanes)

Perhaps the most surprising feature of dimethylsiloxane high polymers is their flexibility and elasticity over a very broad temperature range. The glass transition temperature is $-130°C$. The temperature range of elasticity for silicone rubber is from -30 or $-40°C$ to $250°C$; the lower temperature marks the onset of crystallization. The flexibility of the bulk polymer is evidence of the ease with which the backbone bonds can undergo torsion. Indeed, organosiloxane polymers are among the most flexible macromolecules known.

Their high torsional mobility can be attributed to the lack of charge on the side groups and to the fact that the side groups are attached to *every other* skeletal atom instead of to every skeletal atom. In this way they differ structurally from many organic polymers. Thus, there are fewer opportunities for the side groups to "collide" with each other or even to attract or repel each other as the backbone bonds go through their torsional motions. It appears that the extreme flexibility of the siloxane backbone is responsible for the high permeability of silicone rubber to oxygen. Thus, silicone rubber films have been tested in "artificial gill" devices that would extract dissolved oxygen from water for diving purposes.

Poly(organosiloxanes) also repel water strongly. Partly because of this property, they are used in car polishes, in antistick formulations for cooking purposes, and in biomedical devices. For example, artificial heart valves and experimental heart bypass pumps are often fabricated from silicone rubber because the polymer has a lower tendency than most organic polymers to trigger the clotting of blood or to irritate tissues (see Chapter 22).

Organosiloxane Ladder Polymers

It has been indicated that one of the motivations for the development of polymers with inorganic backbones was the belief that these materials would be more stable at high temperatures than organic polymers. Although poly(organosiloxanes) are certainly resistant to oxidation at temperatures up to $200°C$, they suffer from one drawback—at temperatures above $250°C$ the siloxane chains break down to form rings, and the advantageous properties of the polymer are eventually lost. The depolymerization is similar to that observed with polymeric sulfur.

One solution to this problem is to design polymers that resemble the amphiboles, or double-chain silicates (Figure 7.1). Nonionic analogues of such polymers were first made with phenyl groups in place of the charged side oxygen atoms by the hydrolysis of phenyltrichlorosilane, $C_6H_5SiCl_3$. The resultant materials are called "silicone ladder polymers" (6) or poly(phenylsesquisiloxanes). As might be expected, the double-chain structure restricts the mobility of the silicon–oxygen bonds in the backbone and the polymers are high-melting, nonelastomeric materials. However, when dissolved in organic solvents they yield viscous solutions. Moreover, silicone ladder polymers swelled by the addition of small amounts of solvents can be stretched and oriented. Phenylsilicone ladder polymers remain stable up to a temperature of 300°C.

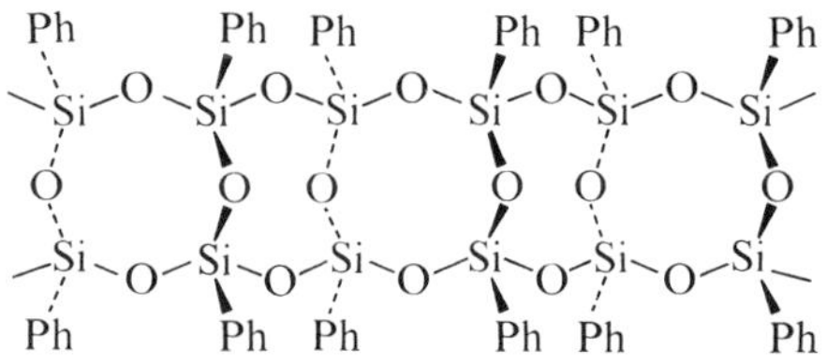

6

SILICONE–CARBORANE POLYMERS

The thermal instability of linear organosiloxane polymers has limited their uses for some high-performance applications. For this reason, attempts have been made to modify poly(organosiloxanes) in ways that will retain the advantageous chain flexibility and yet will raise the thermal stability. One approach has made use of the high thermal stability of carboranes.

Carboranes are cage-type molecules with a framework of boron and carbon atoms. *Meta*-carborane (shown in scheme 5) is an example of this class of compounds. Each boron (open circle) and carbon (black circle) atom bears a hydrogen atom.

Carboranes have a unique stability that results in part from the fact that they are "electron sinks." In other words, the bonding in carboranes is such that the electrons are free to move widely over the whole cage. As a consequence, a carborane cage can stabilize adjacent units bonded to it, such as, for example, a siloxane chain. For this reason, polymers have been developed that contain carborane cages linked together through one or more siloxane bridging units (7). Such materials are named Dexsil polymers. They can be synthesized by the route shown in scheme (5).

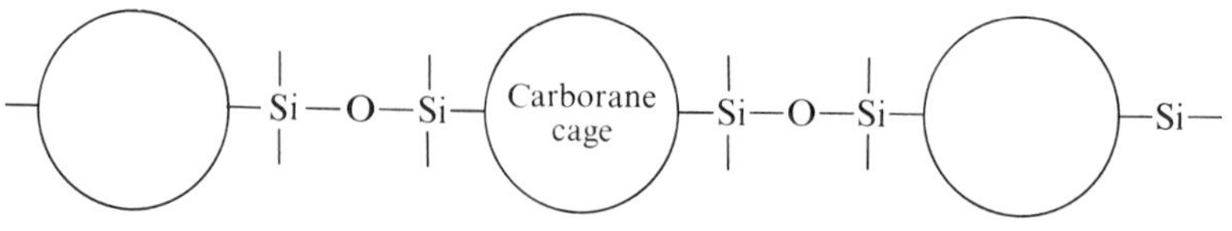

7

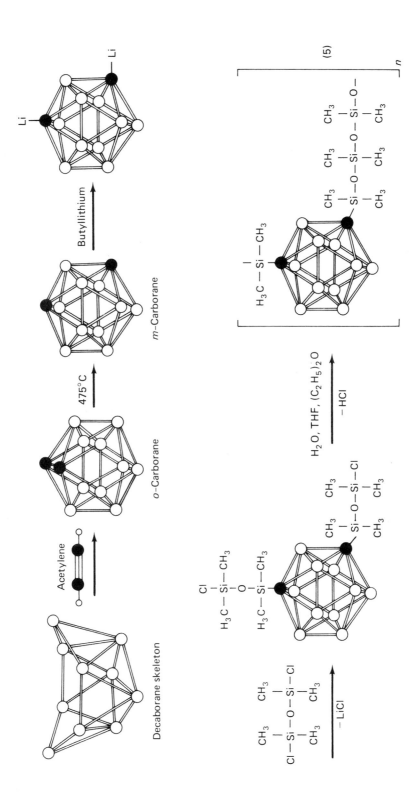

Decaborane skeleton

Acetylene

o-Carborane

475°C

m-Carborane

Butyllithium

Li   Li

$H_3C - Si - CH_3$

$Si - O - Si - O - Si - O -$

CH_3 CH_3 CH_3

CH_3 CH_3 CH_3

(5)

n

$H_2O, THF, (C_2H_5)_2O$

$- HCl$

CH_3 CH_3

$Si - O - Si - Cl$

CH_3 CH_3

$H_3C - Si - CH_3$

O

$Cl - Si - CH_3$

$H_3C - Si - CH_3$

CH_3 CH_3

$CH_3 - Si - O - Si - Cl$

CH_3 CH_3

$Cl - Si - O - Si - Cl$

$- LiCl$

Scheme 5

POLYORGANOSILANES

A relatively new class of silicon-containing polymers are the polycatenasilanes-compounds that contain silicon atoms only in the backbone. A generalized structure is shown in **8**. Such polymers are synthesized by the action of an alkali metal on a

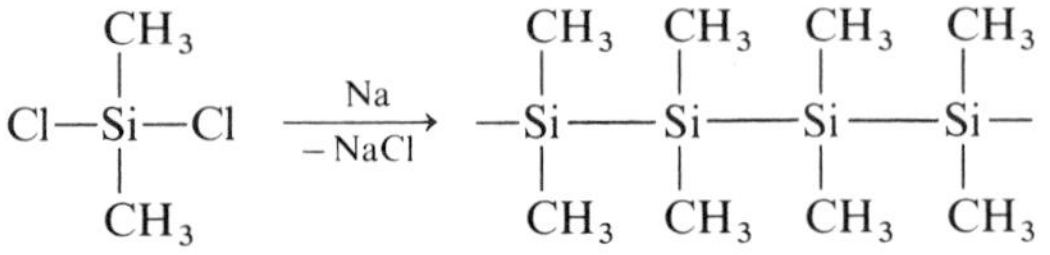

$$
\begin{array}{c}
\text{CH}_3 \\
| \\
\text{Cl}-\text{Si}-\text{Cl} \\
| \\
\text{CH}_3
\end{array}
\xrightarrow[-\text{NaCl}]{\text{Na}}
\begin{array}{c}
\text{CH}_3 \quad \text{CH}_3 \quad \text{CH}_3 \quad \text{CH}_3 \\
| \qquad | \qquad | \qquad | \\
-\text{Si}-\text{Si}-\text{Si}-\text{Si}- \\
| \qquad | \qquad | \qquad | \\
\text{CH}_3 \quad \text{CH}_3 \quad \text{CH}_3 \quad \text{CH}_3
\end{array}
$$

8

diorganodichlorosilane. Typically, cyclic oligomers, such as $(\text{SiR}_2)_6$, are formed most readily. Pyrolysis of these can lead to the formation of higher oligomers or silicon–carbide polymers. Because the backbone structure in these compounds appears to offer the prospect of extensive electron delocalization, the electrical properties of these polymers are of some interest.

PHOSPHAZENE POLYMERS

Some of the most rapid advances in covalent inorganic polymer research and technology are occurring in the field of phosphazene polymers. Polyphosphazene chains consist of an alternating sequence of phosphorus and nitrogen atoms, with two substituent groups attached to each phosphorus (**9**).

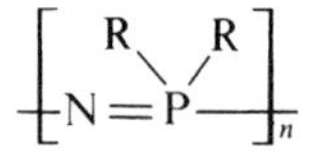

$$
\left[\begin{array}{c} R \quad R \\ \diagdown \diagup \\ N{=}P- \end{array} \right]_n
$$

9

Synthesis of Poly (dichlorophosphazene)

It has been known for more than a century that phosphorus pentachloride and ammonium chloride react to form a series of cyclic inorganic compounds of formula $(\text{NPCl}_2)_n$, where n is 3, 4, 5, 6, ... This reaction is illustrated in (5).

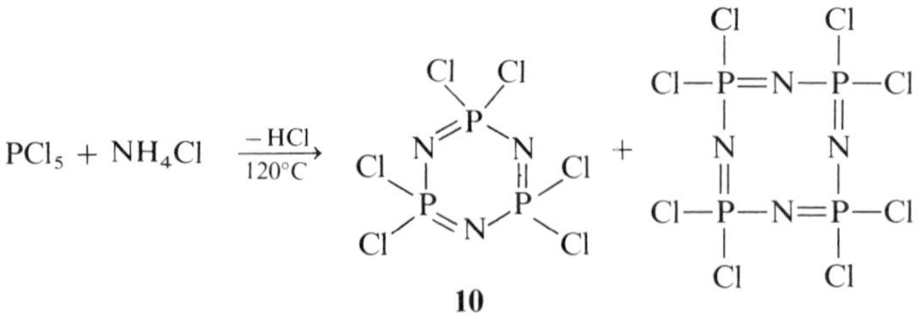

$$
\text{PCl}_5 + \text{NH}_4\text{Cl} \xrightarrow[120°\text{C}]{-\text{HCl}} \quad \mathbf{10} \quad + \quad \tag{5}
$$

10

The principal product is the cyclic trimer (**10**), which is called hexachlorocyclo-triphosphazene or "phosphonitrilic chloride trimer." This compound is a white, crystalline solid which melts at 114°C and is soluble in organic solvents. When heated at 230 to 300°C in an evacuated glass tube or reactor, it polymerizes to a transparent rubbery high polymer (11). The polymer is called poly(dichloro-

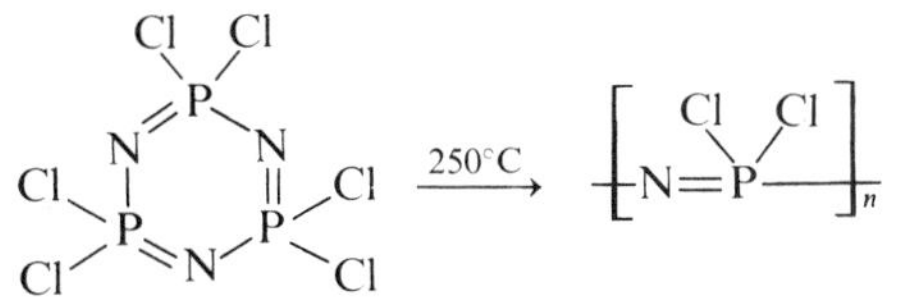

phosphazene). It is also known as "inorganic rubber." The polymer may contain 15,000 or more repeating units, with a molecular weight of over 2 million.

The mechanism of this polymerization is still not fully understood. Strong evidence exists that the mechanism is ionic and that the ionization of a phosphorus–chlorine bond takes place during initiation, as shown in (6). An ionized species, such as **12**, could, in effect, function as an ionic initiator by attacking the nitrogen

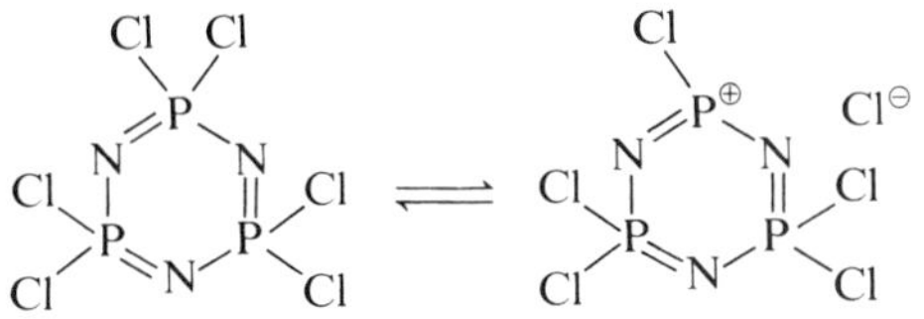

(6)

atom of another ring to initiate chain polymerization. However, it is also known that traces of water accelerate the polymerization rate, and it is now believed that the mechanism may be catalyzed by traces of moisture or other protonic reagents.

Considering that poly(dichlorophosphazene) is made from purely inorganic materials, it is a remarkable compound. In its stress-relaxation behavior, it is a more ideal elastomer than natural rubber. Perhaps more interesting is the observation that it remains rubbery at low temperatures and hardens only when the temperature falls to near the glass transition at −63°C. This behavior is indicative of a high degree of chain mobility.

The polymer formed under nonrigorous polymerization conditions is actually a crosslinked modification of poly(dichlorophosphazene). It swells in organic solvents, but it does not dissolve. It also shows a marked resistance to thermal degradation at temperatures up to 350°C. Polydichlorophosphazene would itself be a valuable technological material were it not for its tendency to react slowly with atmospheric moisture to yield phosphoric acid, ammonia, and hydrochloric acid. During this reaction the elastomer crumbles to a powder.

Synthesis of Poly(organophosphazenes)

The hydrolytic breakdown of poly(dichlorophosphazene) is initiated by a reaction of the phosphorus–chlorine bonds with water, rather than by primary degradation of the phosphorus–nitrogen skeleton. For this reason, the intriguing possibility existed that, if similar macromolecules could be prepared, but with nonhydrolyzable, organic substituent groups in place of chlorine atoms, the polymers should be hydrolytically stable. In the 1960s, it was apparent that a solution to this problem would provide a route to the synthesis of a wide variety of new and useful polymers. This expectation has now been largely fulfilled.

The main hurdle to be overcome was the need to devise a method for the synthesis of phosphazene high polymers with organic substituent groups attached to the main chain. This involved a search for methods for the direct replacement of the chlorine atoms in poly(dichlorophosphazene) by organic substituent groups. Attempts to perform substitution reactions on organic high polymers (see Chapter 9) are often disappointing because the coiling of polymer chains in solution tends to retard the reaction. If poly(dichlorophosphazene) is used as a reaction substrate, the situation is even more complicated because the normal polymer is crosslinked and insoluble. Complete replacement of halogen inside a crosslinked matrix would be virtually impossible. The solution to this problem required the synthesis of a poly(dichlorophosphazene) which was free from crosslinks and completely soluble in nonaqueous media. Such a polymer can be isolated from the mixture obtained by polymerization of hexachlorocyclotriphosphazene under rigorously controlled conditions. Crosslinking takes place in the final stages of polymerization, usually when 60 to 70% of the trimer has disappeared. Avoiding the crosslinking step is the key to the entire process.

Solutions of the uncrosslinked polymer react rapidly and completely with a wide variety of alkoxides or aryloxides, such as sodium ethoxide, sodium trifluoroethoxide, or sodium phenoxide, or with amines such as aniline or butylamine (7). Organometallic reagents, such as Grignard, dialkylmagnesium, or organolithium reagents can be employed to introduce alkyl or aryl groups bonded

$$\left[N=P \begin{matrix} Cl & Cl \\ \diagdown & \diagup \end{matrix} \right]_n$$

RONa / −NaCl   RNH$_2$ / −HCl   R$_2$NH / −HCl   (7)

$$\left[N=P \begin{matrix} RO & OR \\ \diagdown & \diagup \end{matrix} \right]_n \qquad \left[N=P \begin{matrix} RHN & NHR \\ \diagdown & \diagup \end{matrix} \right]_n \qquad \left[N=P \begin{matrix} R_2N & NR_2 \\ \diagdown & \diagup \end{matrix} \right]_n$$

directly to phosphorus. Note that the unusual principle behind this synthesis method is the use of a chemically unstable polymer as an *intermediate* for the preparation of a wide range of stable polymeric derivatives.

The organic-substituted polymers prepared in this way are stable elastomers or thermoplastics that are very resistant to hydrolysis. The polymers have high molecular weights, often with $\overline{M}_w$ values in the region of 1×10^6 to 4×10^6. The physical properties depend on the side groups present. For example, when methoxy or ethoxy side groups are introduced, the resultant polymers are rubbery elastomers. On the other hand, fluoroalkoxy, phenoxy, and some amino-substituted polymers are flexible, film-forming materials. It seems likely that almost any set of required properties can be designed into a phosphazene polymer by a judicious choice of the side groups. This almost unprecedented versatility is the most remarkable feature of the phosphazene system, and the number of different polymers that can, in principle, be synthesized by this method is comparable to that of the known synthetic organic macromolecules systems.

Two examples will serve to illustrate this point. When poly(dichlorophosphazene) reacts with sodium trifluoroethoxide in a tetrahydrofuran–benzene medium, the chlorine atoms are replaced by trifluoroethoxy groups to yield poly[bis(trifluoroethoxy)phosphazene] (13) (Figure 7.5). This polymer can be

$$\left[\begin{array}{c} OCH_2CF_3 \\ | \\ N=P \\ | \\ OCH_2CF_3 \end{array}\right]_n$$

13

solution-cast from acetone or methyl ethyl ketone solutions to give colorless, opalescent, flexible films, which superficially resemble polyethylene films in appearance. Solutions of the polymer can be extruded into a nonsolvent to yield flexible, slightly elastic fibers. The polymer has a low glass transition temperature ($-66°C$), and it remains flexible from this temperature up to its melting point at

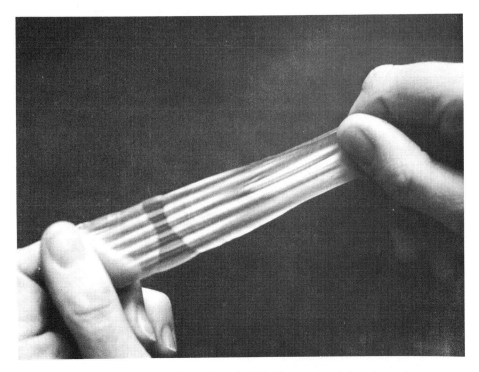

Fig. 7.5 Oriented film of poly[bis(trifluoroethoxy)phosphazene], $[NP(OCH_2CF_3)_2]_n$.

242°C. The presence of the fluorinated side groups makes the polymer highly water-repellent, more so in fact than Teflon or silicones. The high crystallinity of this polymer is responsible for the opalescent appearance. The crystallinity can be enhanced by stretching and orientation to yield strong fibers or films. In this form, the polymer yields excellent X-ray diffraction photographs (see Chapter 19).

A second example of a polyphosphazene structure is provided by the polymer prepared by the reaction of poly(dichlorophosphazene) with methylamine. The product is a clear, transparent, film-forming thermoplastic with the formula, $[NP(NHCH_3)_2]_n$. This compound differs from nearly all other synthetic polymers in being soluble in water. Amino substituent groups in general tend to generate hydrophilic properties rather than hydrophobic ones.

Other poly(organophosphazenes) can be synthesized by the polymerization of cyclic trimers or tetramers that possess organic side groups as well as halogen atoms attached to phosphorus.

Mixed-Substituent Organophosphazene Polymers

Polymer crystallinity results from the regular arrangement of substituent groups along a chain. The absence of crystallinity is often associated with rubbery or elastomeric properties. Thus, the random introduction of two or more different substituent groups should favor the appearance of elastomeric properties. The sequence shown in (8) to (11) shows three ways in which such polymers may be synthesized.

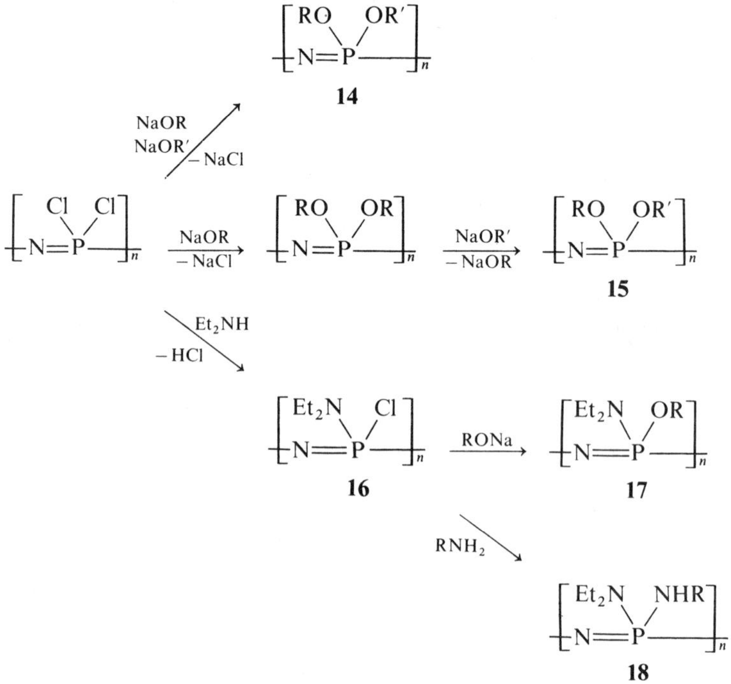

$$\left[\begin{array}{c} RO \quad OR' \\ \diagdown \diagup \\ N{=}P \\ \end{array}\right]_n \tag{8}$$

14

NaOR
NaOR' — NaCl

$$\left[\begin{array}{c} Cl \quad Cl \\ \diagdown \diagup \\ N{=}P \\ \end{array}\right]_n \xrightarrow[-NaCl]{NaOR} \left[\begin{array}{c} RO \quad OR \\ \diagdown \diagup \\ N{=}P \\ \end{array}\right]_n \xrightarrow[-NaOR]{NaOR'} \left[\begin{array}{c} RO \quad OR' \\ \diagdown \diagup \\ N{=}P \\ \end{array}\right]_n \tag{9}$$

15

Et$_2$NH
— HCl

$$\left[\begin{array}{c} Et_2N \quad Cl \\ \diagdown \diagup \\ N{=}P \\ \end{array}\right]_n \xrightarrow{RONa} \left[\begin{array}{c} Et_2N \quad OR \\ \diagdown \diagup \\ N{=}P \\ \end{array}\right]_n \tag{10}$$

16   **17**

RNH$_2$

$$\left[\begin{array}{c} Et_2N \quad NHR \\ \diagdown \diagup \\ N{=}P \\ \end{array}\right]_n \tag{11}$$

18

1. The simultaneous reaction of two different alkoxides with poly(dichloro-phosphazene) yields mixed substituent polymers (**14**). For example, cosubstitution with sodium trifluoroethoxide and sodium heptafluorobutoxide gives an elastomer with the basic composition $[NP(OCH_2CF_3)(OCH_2C_3F_7)]_n$. This material has valuable low-temperature properties and unusual solution behavior.

2. Mixed substituent polymers can be prepared by metathetical ligand exchange reactions (**9**). For example, the reaction of sodium trifluoroethoxide with the polymer $[NP(OCH_2C_3F_7)_2]_n$ yields a mixed trifluoroethoxy-heptafluorobutoxy-substituted phosphazene elastomer.

3. Polymers with two or more different substituents can be synthesized by a sequential nucleophilic substitution process which makes use of the steric hindrance associated with secondary amines. The reaction of poly(dichlorophosphazene) with diethylamine results in the replacement of only 50% of the chlorine atoms, and nmr spectra indicate that only one diethylamino group is attached to each phosphorus atom. Apparently, further substitution by diethylamine is inhibited by steric effects. However, treatment of the rubbery poly(diethylaminochlorophosphazene) (**16**) with a second, less bulky reagent, such as ammonia, methylamine, or sodium trifluoroethoxide, results in the replacement of the remaining chlorine atoms to yield polymers such as **17** or **18**. Terpolymers can be prepared by similar techniques.

Crosslinking

The usefulness of a polymer may be enhanced if the chains can be crosslinked. This is especially true if the material is an elastomer. One facile crosslinking process makes use of a metathetical ligand exchange process similar to reaction (9). The disodium salt of a diol will displace alkoxide ions from an alkoxyphosphazene high polymer and lead to crosslinking of the chains (12). Uncontrolled crosslinking may

$$
\begin{array}{c}
\underset{|}{\overset{OR}{}} \quad \underset{|}{\overset{OR}{}} \\
-N{=}P{-}N{=}P{-} \\
\underset{|}{\overset{|}{OR}} \quad \underset{|}{\overset{|}{OR}} \quad NaO \\
\end{array}
\;+\; \underset{|}{\overset{|}{R'}} \xrightarrow{-NaOR}
\tag{12}
$$

occur if a poly(organophosphazene) contains an appreciable number of residual phosphorus–chlorine bonds. In this case, water functions as the crosslinking agent by the formation of P—O—P bonds.

High-Temperature Depolymerization

Poly(organophosphazenes), in common with silicones, suffer from one defect. Above 200 to 250°C the polymer chains break down to yield small rings, particularly to form the cyclic trimer and tetramer. Because the useful properties of a polymer depend almost entirely on the presence of long chains, a depolymerization process such as this one is highly disadvantageous. Depolymerization places a serious upper limit on the temperatures at which many inorganic polymers can be used, and the blocking of thermal depolymerization constitutes one of the main challenges for research with inorganic polymers. This topic is considered in some detail in Chapter 10.

Structural Features of Polyphosphazenes

Two of the many questions that remain to be answered about polyphosphazenes are: (1) What is the electronic structure of the backbone, and (2) Why do polyphosphazenes show such unusual physical properties—low T_g values, elasticity, unusual crystalline transitions, and so on?

The electronic structure of the backbone is not well understood. If a normal covalent framework is assumed for these polymers, three electrons on each skeletal nitrogen atom and one electron on each phosphorus still remain to be accounted for. The most plausible current explanation (Figure 7.6) is that two of the electrons on each skeletal nitrogen form part of a "lone pair" that is responsible for the basicity

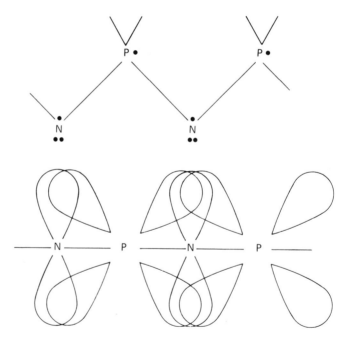

Fig. 7.6   Possible bonding arrangements in polyphosphazenes. In the upper diagram is shown the electrons that must be accounted for after the sigma bond framework has been generated from electron pairs. The lower diagram shows how a d_π-p_π system can be visualized as a stabilized excited state by utilization of one electron from each phosphorus and one from each nitrogen atom. The orbital overlap depicted is that of a phosphorus 3d orbital with a nitrogen 2p orbital.

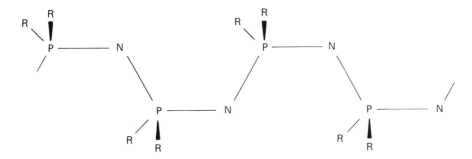

Fig. 7.7 The *cis–trans*-planar chain arrangement that is characteristic of a number of polyphosphazenes in the solid state. This conformation places the side groups at the greatest distance apart.

and coordination ability of these molecules. The remaining electrons on nitrogen and phosphorus then generate a π-system by overlap of a nitrogen *p*-orbital with a phosphorus *d*-orbital (Figure 7.6). This strengthens the backbone bonds while permitting free torsion of these bonds as the *p*-orbital on each nitrogen "switches" from one phosphorus *d*-orbital to another.

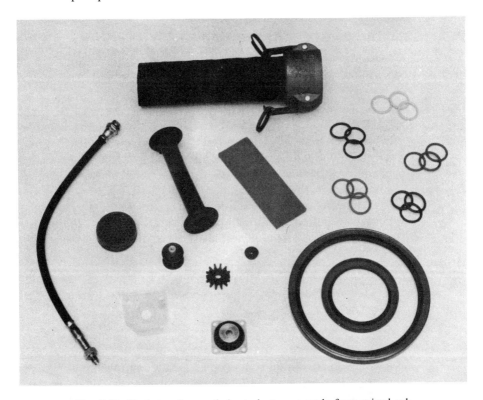

Fig. 7.8 Gaskets, pipe, and sheet elastomer made from mixed substituent fluoroalkoxyphosphazene elastomers. (By courtesy of the Firestone Tire and Rubber Company, Akron, Ohio.)

The unusual torsional freedom of the backbone bonds (and hence the flexibility and elasticity of many polyphosphazenes) appears to be connected with the wide bond angles at nitrogen, and with the absence of substituent groups on every other skeletal atom. Polyphosphazenes also show a tendency to assume a *cis–trans*-planar (0°, 180°) conformation (Figure 7.7), which represents the arrangement that places the side groups as far away from each other as possible.

Uses of Poly(organophosphazenes)

Organophosphazene elastomers have been developed for use as fuel lines, hoses, gaskets, and O-rings and as nonburning foam rubber articles (Figure 7.8). Most of the current technological applications depend on the oil resistance, nonflammability, and low glass transition temperatures of many phosphazene polymers. Other developments are taking place in the use of water-soluble carrier molecules for inorganic anticancer agents, metalloporphyrins, steroids, and so on, and as biomedical polymers that degrade in the body to phosphate, ammonia, and an amino acid (released from the side group). Organic dyestuffs have been covalently bound to polyphosphazene chains, as have a variety of biologically active agents.

POLY(SULFUR NITRIDE) (POLYTHIAZYL)

Synthesis and Appearance

One of the most remarkable inorganic polymer systems is poly(sulfur nitride), $(SN)_n$. This polymer is prepared by a sequence of reactions that start from cyclic "tetrasulfur tetranitride" (**19**). Tetrasulfur tetranitride itself is prepared from elemental sulfur and liquid ammonia, from SF_4 or S_2F_{10} and ammonia or, more commonly, by the reaction of S_2Cl_2 with ammonia. The cyclic tetramer (**19**) is an orange-yellow, crystalline solid, m.p. 178°C.

When heated to the sublimation temperature (85°C) in vacuum and when the vapor is passed through heated silver wool at 200 to 300°C, the cyclic tetramer (**19**) is converted to the potentially explosive cyclic dimer (**20**) which can be condensed

$$
\begin{array}{ccc}
\begin{array}{c} N{=}S{-}N \\ |\quad\ \ \| \\ S\qquad S \\ \|\qquad | \\ N{-}S{=}N \end{array}
& \xrightarrow[0.01\ \text{Torr}]{200\text{--}300°C}
& \begin{array}{c} S{=}N \\ |\quad | \\ N{=}S \end{array}
\xrightarrow{25°C} \ (\!S{=}N\!)_n \\[4pt]
\mathbf{19} & \mathbf{20} & \mathbf{21}
\end{array}
$$

as a white solid on a cold finger cooled in liquid nitrogen. The cyclic dimer is then purified by vacuum sublimation from the cold trap at 25°C to another trap at 0°C to form colorless crystals. A solid-state polymerization of the cyclic dimer then occurs at 25°C during 3 days, followed by heating in vacuum at 75°C for 2 h. During polymerization the crystals change from colorless (diamagnetic), to blue-black (paramagnetic), and then to lustrous gold (diamagnetic). The polymerization to (21) appears to begin at the surface of the crystals and proceed inward. The overall space-group symmetry of the crystals does not change during polymerization.

The solid-state polymerization process appears to take place by the cleavage of one bond in the cyclic dimer to form a diradical and a linkage of adjacent molecules to form the polymer (Figure 7.9). The polymer can be fabricated as thin films on glass, oriented poly(tetrafluoroethylene), or poly(ethylene terephthalate) by an epitaxial polymerization. The technique involves a depolymerization of $(SN)_n$ at 145°C, possibly to linear S_4N_4, which then repolymerizes on the glass or organic polymer surface. The $(SN)_n$ fibers are aligned parallel to each other along the orientation direction of the polymer support. The films are dark blue by transmitted light.

Crystals of $(SN)_n$ have a parallel fibrous structure, and long fibrous strands may be mechanically peeled from an individual crystal. The ends of the crystal always appear black rather than gold. The gold color constitutes a metallic-type reflectance which is indicative of a metallic-type arrangement of electrons. The crystals are soft and malleable and can be flattened at right angles to the fiber axis to form gold-colored sheets. Compression can cause detonation.

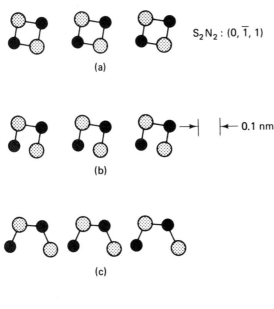

$S_2N_2 : (0, \bar{1}, 1)$

(a)

0.1 nm

(b)

(c)

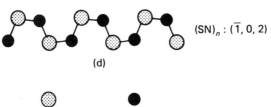

$(SN)_n : (\bar{1}, 0, 2)$

(d)

Sulfur   Nitrogen

Fig. 7.9 Diagramatic representation of the polymerization of $(SN)_2$ to $(SN)_n$ by lateral coupling of the open-chain diradical generated from the cyclic dimer. (From A. G. MacDiarmid, A. J. Heeger, and A. F. Garito, *McGraw - Hill Yearbook of Science and Technology—Polymer*, 1977.)

At 25°C the polymer is relatively inert to air and water and may be heated in air for short periods without tarnishing. However, the polymer decomposes (to a white-gray powder) during long exposure to air or water, and it breaks down to sulfur, nitrogen, and other species on long heating at elevated temperatures. It is insoluble in all solvents with which it does not react.

Electrical Properties and Structure

$(SN)_n$ is a new type of metal in that it has a metallic appearance and exhibits metallic-type electrical conductivity at ordinary temperatures. The electrical conductivity at 25°C is in the same range as for mercury, bismuth, or nichrome, with the conductivity being more pronounced along the direction of the polymer chains. The conductivity increases as the temperature is lowered until at 4.2°K it is 200 times as high as at 25°C. However, at 0.3°K the material apparently undergoes a change to a *superconductor*, at which point there is no resistance to electrical flow (Figure 7.10). The superconductivity is anisotropic (i.e., in three dimensions). If the polymer chains were totally "one-dimensional," the inevitable defects would presumably limit both the one-dimensional metallic properties and the super-conductivity. Indeed, defects in the structure are known to exist; hence, conductivity *between* chains probably occurs.

The electrical properties can be explained in terms of the polymer structure. The packing of the chains is illustrated in Figure 7.11. Individual chains occupy a *cis-trans*-planar conformation (see Chapter 18), with S—N bond lengths intermediate between those of single and double bonds. This suggests that a delocalized bonding arrangement exists along the chain via delocalized, half-filled π-orbitals, and this would permit the existence of a metallic conduction band. The interchain packing is such that electronic transmission could occur via orbital overlaps

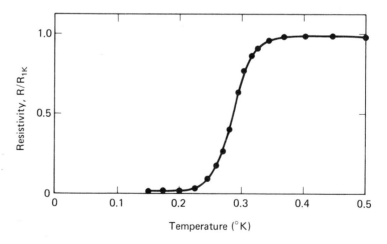

Fig. 7.10 $(SN)_n$ undergoes a change from a metallic conductor to a superconductor as it is cooled below 0.3°K. [From R. L. Greene, G. B. Street, and L. J. Suter, *Phys. Rev. Lett.*, **34**, 577 (1975).]

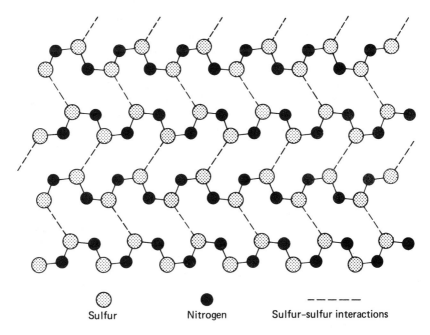

| Sulfur | Nitrogen | Sulfur-sulfur interactions |

Fig. 7.11 Packing of the $(SN)_n$ chains in a crystalline matrix. (From A. G. MacDiarmid, A. J. Heeger, and A. F. Garito, *McGraw-Hill Yearbook of Science and Technology—Polymer*, 1977.)

between S—S, N—N, or S—N pairs on adjacent chains. This would produce a system of pockets of electrons and holes (as in a semi-metal-like bismuth).

Halogenated Poly(*sulfur nitrides*)

$(SN)_n$ absorbs bromine vapor to yield a black polymer of composition, $(SNBr_{0.4})_n$. Similar materials can be prepared by the treatment of $(SN)_4$ with bromine. The polymer $(SNBr_{0.4})_n$ retains the fibrous structure of $(SN)_n$ and also exhibits electrical conductivity. Heating of $(SNBr_{0.4})_n$ in vacuum causes a loss of bromine and the formation of copper-colored crystals of $(SNBr_{0.25})_n$. These derivatives could be precursors of a range of derivatives based on the $(SN)_n$ structure.

POLYACETYLENE

Although polyacetylene is not an inorganic polymer, its relationship to poly(sulfur nitride) and carbon fibers (see the following section) makes it appropriate to describe its synthesis and properties here.

Acetylene can be polymerized under the influence of Ziegler–Natta-type catalysts to yield films of *cis-* or *trans*-polyacetylene that have a copper or silver, metallic appearance. These polymers are semiconductors. The silver *trans*-isomer is the thermodynamically stable form at room temperature.

Treatment of polyacetylene films with small amounts of electron acceptors, such as chlorine, bromine, iodine, or arsenic pentafluoride, or donors, such as sodium naphthalenide, results in marked increases in conductivity.

The electrical conductivity is electronic rather than electrolytic. Polyacetylene is presumed to possess a delocalized π-electron structure. However, the semi-conductivity of the undoped polymer appears to be due to the presence of impurities. The addition of the donors or acceptors may bring about the formation of charge-transfer π-complexes with the polymer, analogous to those formed between, say, halogens and ethylenic or aromatic species. This forms electrons or "holes" along the polymer chains. At low dopant levels, thermal activation of the bound electrons or holes generates carriers for electrical conductivity along the chains. However, at dopant levels above 1 mol %, the bound states are screened and the current carriers are free to move along the chains to give rise to metallic behavior.

Films of *cis*-polyacetylene doped with AsF_5 are sufficiently conducting that they can function as substitute "wires" in simple electrical circuits. Moreover, *p-n* junctions can be fabricated by mechanically pressing together sodium-doped and AsF_5-doped strips of polymer film, and these units exhibit typical diode properties.

CARBON FIBERS

Although carbon atoms form the principal building blocks of organic polymers, inorganic-type carbon polymers are also known. Diamond, charcoal, and graphite fall into this category. In recent years new developments have led to the preparation of fibers made from "inorganic" carbon. The starting point for the preparation of a carbon fiber is a conventional organic fiber, such as rayon or polyacrylonitrile, either in the form of a monofilament or a woven textile. Pyrolysis of the organic fiber results in the removal of hydrogen and the formation of polyaromatic structures.

For example, when polyacrylonitrile is pyrolyzed, internal addition takes place to yield a condensed polycyclic known as "black Orlon" (**22**). Further pyrolysis results in removal of the hydrogen atoms and the presumed generation of a polyaromatic structure (**23**). In general, pyrolysis reactions are carried out at

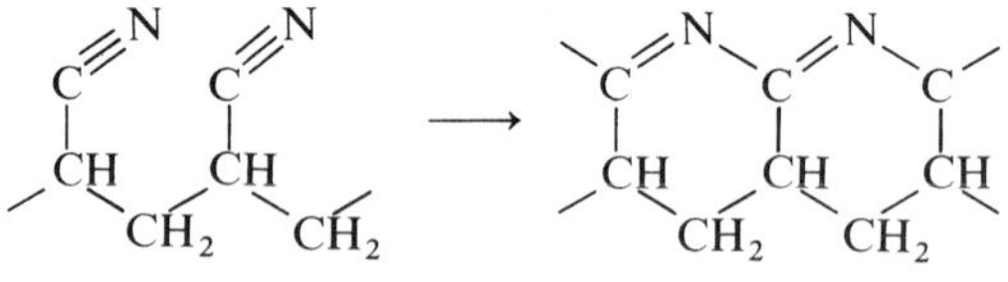

22

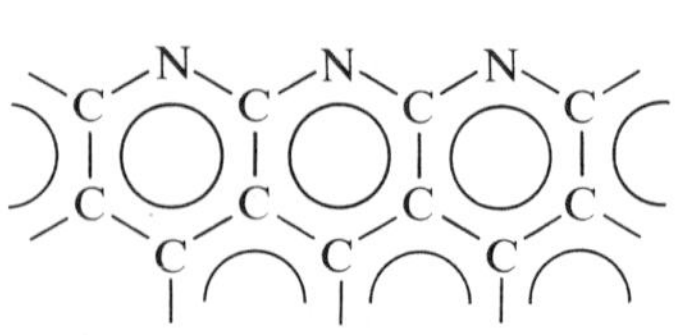

23

about 1000°C, but the ultimate "graphitization" may require temperatures as high as 2800°C.

Fibers made in this way are light in weight, chemically inert, and form electrical semiconductors or conductors. They have exceedingly high thermal stabilities. Carbon fiber fabrics can be heated to red heat with the flame from a propane torch without burning and without noticeable degradation taking place. Carbon fibers can be stiffer than glass fibers and they have found increasing use as reinforcement materials for plastics.

CYCLOLINEAR AND CYCLOMATRIX INORGANIC POLYMERS

In many cases the thermal stability of an inorganic system is a consequence of its ability to form six- or eight-membered rings. Such rings may not decompose at temperatures above 500°C. For this reason, attempts have been made to synthesize polymers that are made up of inorganic rings linked together by short chains. This principle has been introduced earlier with carborane–siloxane polymers.

A number of cyclolinear (24) and cyclomatrix (25) phosphazene polymers have been reported, formed by the reaction of chlorocyclophosphazenes with di- or even trifunctional linking agents, such as aliphatic or aromatic diols or diamines. For obvious reasons, these polymers are less flexible than linear polyphosphazenes, but they are more stable thermally. Cyclomatrix polymers such as 25 have been developed for use as high-temperature wire coatings or thermosetting resins.

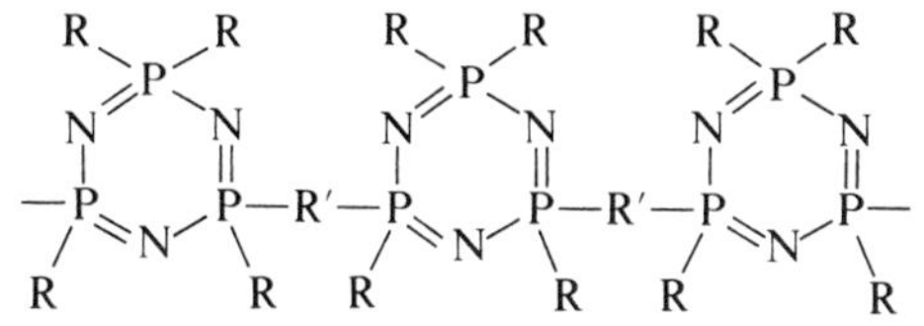

24

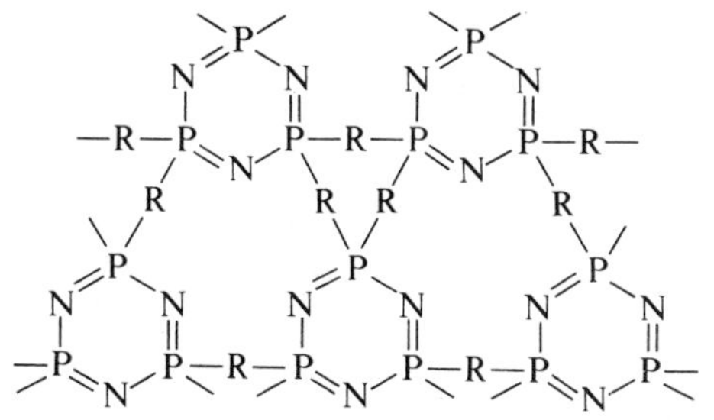

25

Similar types of products have been investigated in the field of boron–nitrogen chemistry. Boron nitride itself forms either polymeric sheets (26) or a diamond-type lattice; but cyclolinear polymers have also been made with the structure shown in **27**.

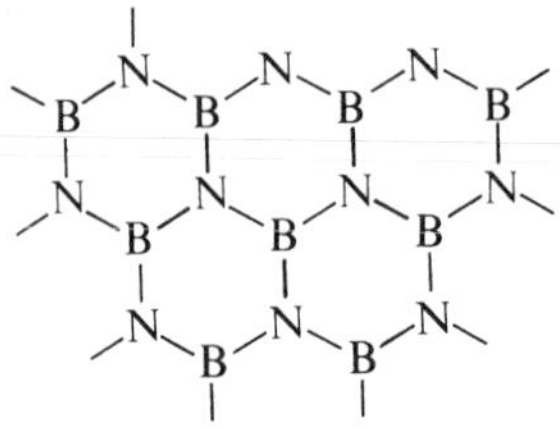

26

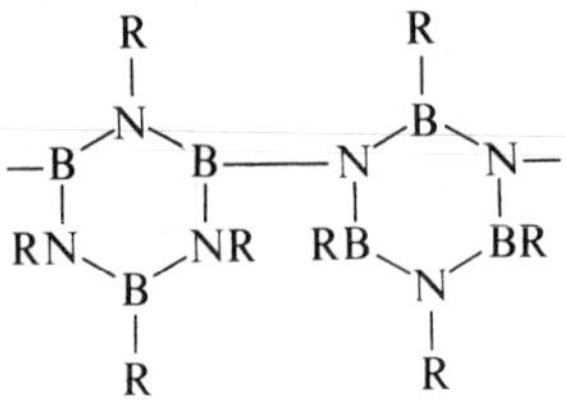

27

COORDINATION POLYMERS

A considerable amount of research has been carried out on the synthesis of polymers that contain metal atoms linked together by a coordinating organic ligand. A general structure for such polymers is depicted in **28**.

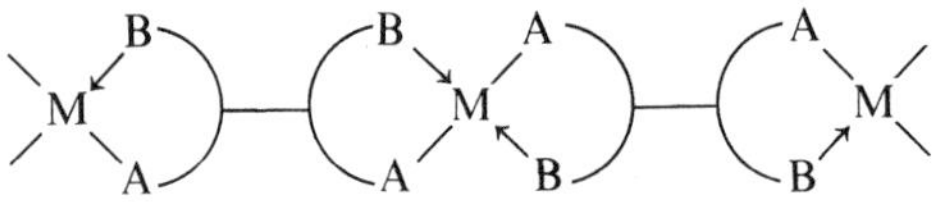

28

Organic ligands, such as Schiff bases, δ-quinolinol, and β-diketones, have been used in this way with metals such as copper, nickel, cobalt, zinc, manganese, palladium, and so on. Closely related to these polymers are those in which the chelating agent is a phosphinate structure, such as in the polymer depicted in **29**.

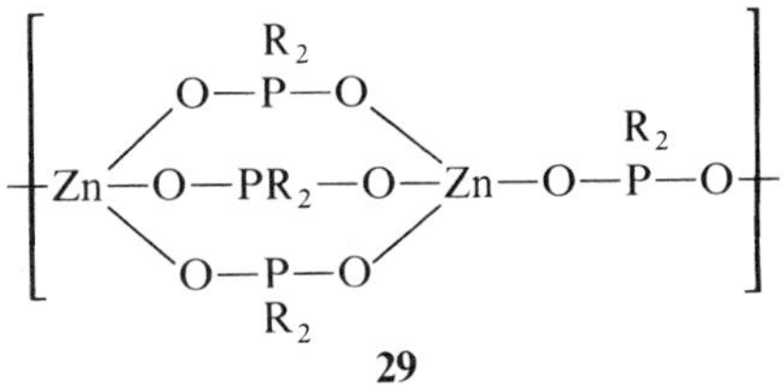

29

Other coordination polymers have been prepared in which metal phthalocyanine units (**30**) are linked together through the organic component. Related systems are also known in which a silicon atom occupies the central position in a phthalocyanine ring (M in **30**) with the silicon covalently bonded above and below the ring to a siloxane chain, and covalently bonded within the ring to two nitrogen atoms. The remaining two nitrogen atoms then coordinate to the silicon to form a six-coordinate structure. Such polymers are designed to lower the tendency for cyclization–depolymerization in organosiloxane polymers by inhibiting the occurrence of back-biting reactions.

Most of the known coordination polymers are of low molecular weight and are crystalline. Hence, they form brittle materials. However, they do display very high thermal stability and many are electrical semiconductors.

Synthesis and Reactions of Polymers / *Part I*

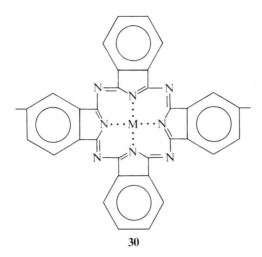

30

An equally intriguing, but newer type of coordination polymer, contains "sandwich" complexes of metals. Several types of metallocene polymers are known, but polyferrocenes, such as **31**, are typical. They can be prepared by reaction of a

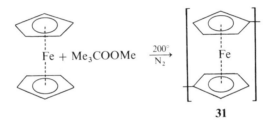

31

metallocene complex with a free-radical reagent, as in the formation of **31**, or by condensation or addition reactions of metallocenes that contain functionalized cyclopentadienyl rings.

ORGANIC POLYMERS WITH INORGANIC SUBSTITUENT GROUP

Finally, several classes of polymers are known in which an organic chain bears an inorganic unit in the side group. Typical of these polymers are those produced by polymerization of vinylsilicon, vinyltin, or vinylphosphorus compounds, such as the polymer shown in **32**. Such polymers closely resemble conventional organic

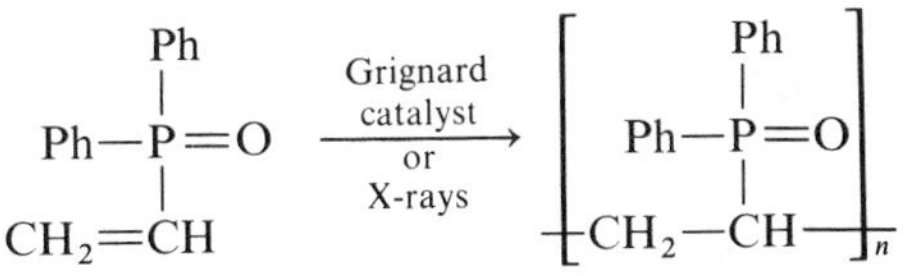

$$\begin{array}{c} \text{Ph} \\ | \\ \text{Ph}-\text{P}=\text{O} \\ | \\ \text{CH}_2=\text{CH} \end{array} \xrightarrow[\substack{\text{or} \\ \text{X-rays}}]{\substack{\text{Grignard} \\ \text{catalyst}}} \left[\begin{array}{c} \text{Ph} \\ | \\ \text{Ph}-\text{P}=\text{O} \\ | \\ \text{CH}_2-\text{CH} \end{array}\right]_n$$

32

polymers, but with added advantages, such as enhanced fire resistance, electrical semiconduction, or catalytic properties. Other polymers are known in which an organometallic unit is *coordinated* to the side group of an organic polymer. For example, rhodium compounds have been coordinated to phosphine residues that are linked to the aromatic rings of polystyrene. The synthesis of these systems is described in Chapter 9.

STUDY QUESTIONS

1. What options are available to the synthetic chemist who wishes to increase the thermal stability of those inorganic polymers that show a tendency to depolymerize to cyclic oligomers at elevated temperatures?

2. Relatively few polymer systems have been developed that contain backbone structures derived from the main-group elements other than carbon, nitrogen, oxygen, phosphorus, sulfur, or silicon. In an essay-type answer, suggest some of the reasons for this restriction and describe ways in which such polymers might be synthesized. What practical advantages might result from the incorporation of, say, tin or antimony into a polymer skeleton?

3. What are the main problems that might be encountered in the synthesis of macromolecules that contain transition metals coordinated to organic ligands in a backbone structure?

4. Poly(sulfur nitride) is an electrical conductor. Speculate on the underlying chemical and physical reasons for this phenomenon and suggest other high-polymeric systems that might behave in the same way.

5. Design synthesis routes to polyphosphazenes that would have the properties needed for use as (a) elastomers for use in artificial heart valves; (b) electrical wire insulators; (c) compounds to reduce turbulence in a recirculating water-cooling system; (d) polymers that are stable to ultraviolet light; (e) carrier molecules for chemotherapeutic drugs.

6. Large amounts of elemental sulfur are available at very low prices. What is the source of this material? How could it be converted to high-polymeric sulfur on a large scale? How might the polymer be stabilized against (a) reversion (depolymerization) and (b) crystallization? What potential uses can you foresee for polymeric sulfur if these two problems can be solved?

7. Discuss the prospect that polymers other than polyacrylonitrile or cellulose might be converted into carbon fibers. What advantages or disadvantages can you think of for the use of other polymers?.

SUGGESTIONS FOR FURTHER READING

Inorganic polymers (general)

ALLCOCK, H. R., *Heteroatom Ring Systems and Polymers.* New York: Academic Press, 1967.

GIMBLETT, F. G. R., *Inorganic Polymer Chemistry.* London: Butterworths, 1963.

HUNTER, D. N., *Inorganic Polymers.* Oxford: Blackwell, 1963.

LAPPERT, M. F., AND G. J. LEIGH (eds.), *Developments in Inorganic Polymer Chemistry.* New York: Elsevier, 1962.

STONE, F. G. A., AND W. A. G. GRAHAM (eds.), *Inorganic Polymers*. New York: Academic Press, 1962.

Mineralogical polymers

FRAZIER, S. E., J. A. BEDFORD, J. HOWER, AND M. E. KENNEY, "An Inherently Fibrous Polymer," *Inorg. Chem.*, **6**, 1693 (1967).

LENTZ, C. W., "Silicate Minerals as Sources of Trimethylsilyl Silicates and Silicate Structure Analysis of Sodium Solutions," *Inorg. Chem.*, **3**, 574 (1964).

LINSKY, J. P., T. R. PAUL, AND M. E. KENNEY, "Planar Organosilicon Polymers," *J. Polymer Sci. (A2)*, **9**, 143 (1971).

SKORIK, Y. I., E. V. KUKHARSKAYA, A. D. FEDOSEEV, AND K. P. KLIMOVA, "Modification of Crystotile Asbestos by Organopolysiloxanes in an Acoustic Field," *Zh. Prikl. Khim.*, **38**, 510 (1965).

VAN WAZER, J. R., AND C. F. CALLIS, "Phosphorus-Based Macromolecules," in *Inorganic Polymers* (F. G. A. Stone and W. A. G. Graham, eds.). New York: Academic Press, 1962.

WELLS, A. F., *Structural Inorganic Chemistry*, 4th ed. London: Oxford University Press, 1975.

Polymeric sulfur

GOETHALS, E. J., "Sulfur-containing Polymers," *J. Macromol. Sci.—Rev. Macromol. Chem.*, **C2**, 74 (1968).

SCHMIDT, M., "Sulfur Polymers," in *Inorganic Polymers* (F. G. A. Stone and W. A. G. Graham, eds.), p. 98. New York: Academic Press, 1962.

TOBOLSKY, A. V., AND W. J. MACKNIGHT, *Polymeric Sulfur and Related Polymers* (Polymer Reviews, Vol. 13), New York: Wiley–Interscience, 1965.

Poly (organosiloxanes)

BARRY, A. J., AND H. N. BECK, "Silicone Polymers," in *Inorganic Polymers* (F. G. A. Stone and W. A. G. Graham, eds.), p. 189. New York: Academic Press, 1962.

EABORN, C., *Organosilicon Compounds*. London: Butterworths, 1960.

HUGHES, J. S., "Some Recent Advances in Silicone Chemistry," in *Developments in Inorganic Polymer Chemistry* (M. F. Lappert and G. J. Leigh, eds.), p. 138. New York: Elsevier, 1962.

ROCHOW, E. G., *The Chemistry of the Silicones*, 2nd ed. New York: Wiley, 1951.

Polysilanes

INGHAM, R. K., AND H. GILMAN, "Organopolymers of Silicon, Germanium, Tin, and Lead," in *Inorganic Polymers* (F. G. A. Stone and W. A. G. Graham, eds.), p. 321. New York: Academic Press, 1962.

WEST, R., AND E. CARBERRY, *Science*, **189**, 179 (1975).

YAZIMA, S., K. OKAMURA, J. HAGASKI, AND M. OMORI, *J. Am. Ceramic Soc.*, **59**, 324 (1976).

Polyphosphazenes

ALLCOCK, H. R., *Phosphorus–Nitrogen Compounds*. New York: Academic Press, 1972.

ALLCOCK, H. R., AND R. L. KUGEL, *J. Amer. Chem. Soc.*, **87**, 4216 (1965).

ALLCOCK, H. R., "Poly(organophosphazenes)—Unusual New High Polymers," *Angew. Chem. (Intern. Ed. English)*, **16**, 147 (1977).

ALLCOCK, H. R., "Polyphosphazenes: New Polymers with Inorganic Backbone Atoms," *Science*, **193**, 1214 (1976).

ALLCOCK, H. R., R. W. ALLEN, AND J. P. O'BRIEN, "Synthesis of Platinum Derivatives of Polymeric and Cyclic Phosphazenes," *J. Am. Chem. Soc.*, **99**, 3984 (1977).

ALLCOCK, H. R., T. J. FULLER, D. P. MACK, K. MATSUMURA, AND K. M. SMELTZ, "Synthesis of Poly[(amino acid alkyl ester)phosphazenes], *Macromolecules*, **10**, 824 (1977).

ALLCOCK, H. R., "Small Molecule Phosphazene Rings as Models for High Polymeric Chains," *Accounts of Chem. Res.*, **12**, 351 (1979).

SINGLER, R. E., N. S. SCHNEIDER, AND G. L. HAGNAUER, "Polyphosphazenes: Synthesis— Properties—Applications," *Polymer Eng. Sci.*, **15**, 5, 321 (1975).

TATE, D. P., "Polyphosphazene Elastomers," *J. Polymer Sci., Polymer Symp.*, **48**, 33 (1974).

Poly(sulfur nitride)

AKHTAR, M., J. KLEPPINGER, A. G. MACDIARMID, J. MILLIKEN, M. J. MORAN, C. K. CHIANG, M. J. COHEN, A. J. HEEGER, AND D. L. PEEBLES, "A 'Metallic' Derivative of Polymeric Sulfur Nitride: Poly(thiazyl bromide), $(SNBr_{0.4})_x$," *J. Chem. Soc., Chem. Commun.*, 473 (1977).

BRIGHT, A. A., M. J. COHEN, A. F. GARITO, A. J. HEEGER, C. M. MIKULSKI, P. J. RUSSO, AND A. G. MACDIARMID, "Optical Reflectance of Polymeric Sulfur Nitride Films from the Ultraviolet to the Infrared," *Phys. Rev. Lett.*, **34**, 206 (1975).

COHEN, M. J., A. F. GARITO, A. J. HEEGER, A. G. MACDIARMID, C. M. MIKULSKI, M. S. SARAN, AND J. KLEPPINGER, "Solid State Polymerization of S_2N_2 to $(SN)_x$," *J. Am. Chem. Soc.*, **98**, 3845 (1976).

GOEHRING, M., "Sulphur Nitride and Its Derivatives," *Quart. Rev. Chem. Soc.*, **10**, 437 (1956).

GREENE, R. L., P. M. GRANT, AND G. B. STREET, "Low Temperature Specific Heat of Poly-sulfur Nitride," *Phys. Rev. Lett.*, **34**, 89 (1975).

MAURITZ, K. A., AND A. J. HOPFINGER, "Theory of Epitaxial Crystallization of S_2N_2 and $(SN)_x$ on Alkali Halide Substrates," *J. Polymer Sci. (Polymer Phys. Ed.)*, **14**, 1813 (1976).

MIKULSKI, C. M., P. J. RUSSO, M. S. SAVAN, A. G. MACDIARMID, A. F. GARITO, AND A. J. HEEGER, "Synthesis and Structure and Metallic Polymeric Sulfur Nitride, $(SN)_x$, and Its Precursor, Disulfur Dinitride, S_2N_2," *J. Am. Chem. Soc.*, **97**, 6360 (1975).

STREET, G. B., W. D. GILL, R. H. GEISS, R. L. GREEN, AND J. J. MAYERLE, "Modification of the Electronic Properties of $(SN)_x$ by Halogens; Properties of $(SNBr_{0.4})_x$," *J. Chem. Soc., Chem. Commun.*, 407 (1977).

WALATKA, V. V., M. M. LABES, AND J. H. PERLSTEIN, "Polysulfur Nitride—a One-Dimensional Chain with a Metallic Ground State," *Phys. Rev. Lett.*, **31**, 1139 (1973).

Coordination and miscellaneous inorganic polymers

ALLCOCK, H. R., AND R. L. KUGEL, "Ionic Polymerization of Diphenylvinylphosphine Oxide," *J. Polymer Sci.* (*A*), **1**, 3627 (1963).

ANDRIANOV, K. A. *Metalorganic Polymers*. New York: Wiley–Interscience, 1965.

BLOCK, B. P., "Coordination Polymers," in *Inorganic Polymers* (F. G. A. Stone and W. A. G. Graham, eds.), p. 447. New York: Academic Press, 1962.

BRADLEY, D. C., "Polymeric Metal Alkoxides, Organometalloxanes, and Organometalloxanosiloxanes," in *Inorganic Polymers* (F. G. A. Stone and W. A. G. Graham, eds.), p. 410. New York: Academic Press, 1962.

DAVISON, J. B., AND K. J. WYNNE, "Silicon Phthalocyanine–Siloxane Polymers: Synthesis and ^{1}H Nuclear Magnetic Resonance Study," *Macromolecules*, **11**, 186 (1978).

JONES, J. I., "Polymetallosiloxanes," Parts I and II, in *Developments in Inorganic Polymer Chemistry* (M. F. Lappert and G. J. Leigh, eds.), pp. 162, 200. New York: Elsevier, 1962.

KENNEY, C. N., "Metal Chelate Polymers," in *Developments in Inorganic Polymer Chemistry* (M. F. Lappert and G. J. Leigh, eds.), p. 256. New York: Elsevier, 1962.

LAPPERT, M. F., "Polymers Containing Boron and Nitrogen," in *Developments in Inorganic Chemistry* (M. F. Lappert and G. J. Leigh, eds.), p. 20. New York: Elsevier, 1962.

MCCLOSKEY, A. L., "Boron Polymers," in *Inorganic Polymers* (F. G. A. Stone and W. A. G. Graham, eds.), p. 159. New York: Academic Press, 1962.

8

Biological Polymers and Their Reactions

INTRODUCTION

Living things exist and reproduce because they contain macromolecules. Very early in the evolutionary process, the chemical (i.e., nonreplicative) synthesis of large molecules under natural conditions provided a mechanism for the reproductive processes on which even primitive life depends. Throughout the subsequent evolution of living organisms, polymers have been used as protective coatings, membranes, energy storage systems, skeletal systems, pathways for electrical conduction, and for countless other purposes.

There are three main classes of biological polymers. These are:

1. Polysaccharides.
2. Proteins and polypeptides.
3. Nucleic acid polymers.

The rest of this chapter will consider these three types of macromolecules. Polysaccharides are used mainly as skeletal reinforcement molecules in plants or as energy-storage molecules in plants and animals. Proteins function as protective or supportive molecules in animals but are also employed extensively as catalyst substrates, as oxygen-transport molecules, and in many other roles that accelerate and direct the chemical reactions of the cell or organism. Nucleic acids are used almost entirely for the purposes of information storage, cell replication, and protein synthesis.

It should be noted that all these macromolecules are *condensation* polymers that, in the simplest chemical sense, are produced by the elimination of water between diols, amino acids, or inorganic acid. These condensation reactions take place rapidly at only moderate temperatures in an aqueous environment. In this respect, the efficiency of the biological syntheses is a matter of considerable interest to the synthetic chemist as well as to the biologist.

POLYSACCHARIDES

General Composition of Polysaccharides

Polysaccharides are cyclolinear polyethers formed by the condensation reactions of sugars. Cellulose, starch, or glycogen are well-known examples, although a large number of other polysaccharides have been identified. Both homopolymers and copolymers are known in linear or branched sequences. Small amounts of noncarbohydrate components may also be present, including ester residues derived from phosphate, sulfate, malonate, or pyruvate units. The degree of polymerization may range from as little as 30 to as high as $\sim 10^5$.

A wide variety of sugar monomer residues are found in polysaccharides, but two of the most frequently occurring are the cyclic forms of glucose (**1** and **2**). These

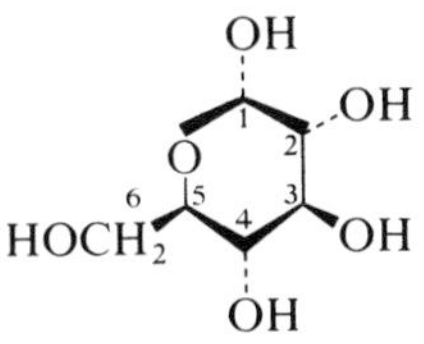

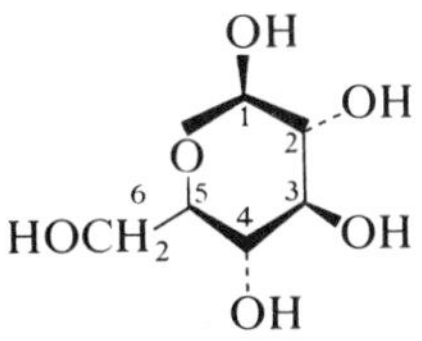

cyclic molecules are known as glucopyranoses (the pyran-type form of the sugar). The glucose molecule depicted in **1** is called α-D-glucose (or α-D-glucopyranose). The isomer shown in **2** is β-D-glucose. Starch, glycogen, and cellulose are homopolymers of glucose. Chitin, the structural polysaccharide found in insect exoskeletons, is a homopolymer of N-acetyl-D-glucosamine (**3**).

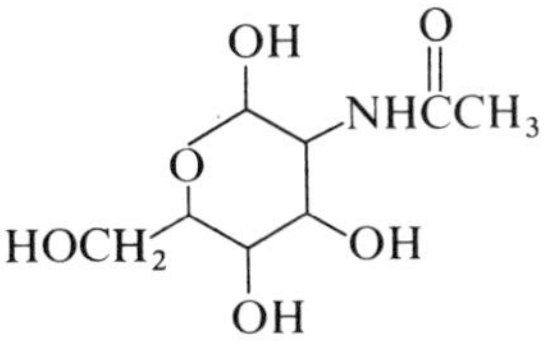

Specific Polysaccharides

Starch, glyogen, and cellulose are closely related homopolymers that are composed of glucopyranose residues (**1** or **2**) condensed via the hydroxyl groups at the 1:4 positions. Thus, the fundamental repeating structure in all three polymers is the one shown in **4**. The main differences between these three polymers are due to different monomer residue configurations or to different degrees of chain branching.

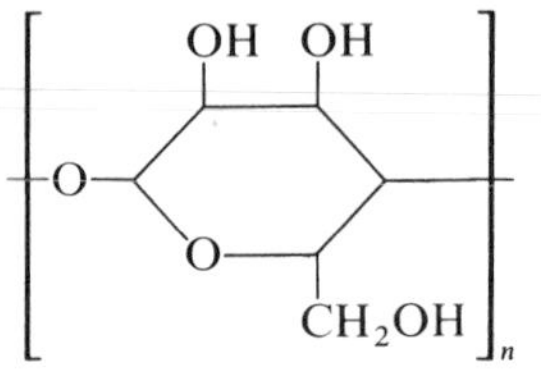

$$\left[O - \overset{OH \quad OH}{\underset{O \quad CH_2OH}{\bigcirc}} - \right]_n$$

4

Starch is the principal energy-storage polysaccharide of the photosynthetic plants. It is mainly a homopolymer of α-D-glucose (**1**). Most starches contain two structurally different components that are designated as the *amylose* and *amylopectin* fractions. The amylose fraction is a linear polymer, whereas the amylopectin fraction is highly branched. This branched component constitutes 70 to 80% of most starches.

The amylose component forms hydrated micelles in water, in which the polysaccharide chain twists into a helical conformation (Figure 8.1). The channel

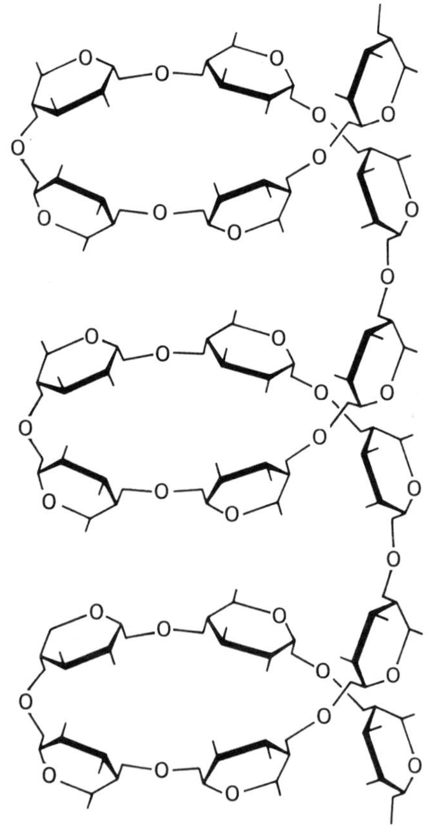

Fig. 8.1 Helical conformation assumed by the amylose fraction of starch. [From A. L. Lehninger, *Biochemistry*, (New York: Worth, 1970), p. 229.)

generated within this helix can accommodate iodine molecules to form the well-known deep blue starch–iodine complex. The molecular weights of the chains vary over a wide range, with $\overline{DP}$ values varying from 100 to 6000. However, some of this apparent molecular-weight variation may reflect hydrolytic chain cleavage that occurs during the isolation and fractionation of the polymers.

In amylopectin, a branch point exists at about every sixth to twelfth backbone residue, with each branch being roughly 12 to 15 glucose residues long. The branches involve the CH_2OH side groups—that is, they involve condensations at the 1, 4, and 6 carbon atoms of a backbone residue. Amylopectin gives a purple-brown color with iodine.

Glycogen is a storage polysaccharide found in animal tissues. It closely resembles starch in its general composition, but it is more highly branched than the amylopectin component of starch. It is estimated that an average of only three glucose residues separate each branch point in glycogen, although the molecular weights of glycogen and amylopectin are very similar ($\sim 10^7$). Glycogen yields a red-brown color with iodine.

Cellulose is the most abundant structural material used by plants. Cotton is nearly 100% cellulose and wood is roughly 50% cellulose. Like starch and glycogen, cellulose is a homopolymer of glucose (roughly 3500 residues per chain), but the glucose configuration is β rather than α (see structure **2**). This configurational difference is responsible for the different properties. Cellulose consists of fully extended chains that are hydrogen bonded into sheets to form a highly crystalline matrix (Figure 8.2). Hence, cellulose is insoluble in water and is an ideal material

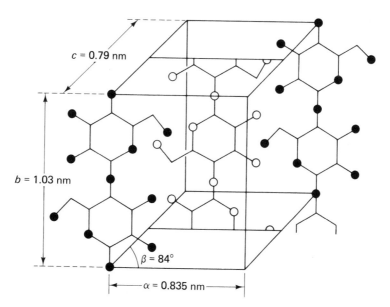

Fig. 8.2 Orientation of the polysaccharide chains in the unit cell of cellulose. Although three chains are shown, only two belong to each unit cell. [From K. H. Meyer and L. Misch, *Ber. Dtsch. Chem. Ges.,* **70B**, 266 (1937); *Helv. Chim. Acta*, **20**, 232 (1937).]

for structural reinforcement. Cellulose cannot be used as an energy source by primates, because the hydrolytic enzymes needed to degrade it to glucose or maltose are not present in the digestive tract. However, ruminant animals (such as cows) can utilize cellulose as a food via the action of symbiotic bacteria that exist in the gut. These secrete the enzymes required to hydrolyze cellulose to glucose.

Biological Synthesis of Polysaccharides

Starch and cellulose are available in such plentiful quantities from plant life that there has been virtually no incentive to devise laboratory syntheses of polysaccharides from sugars. However, the biological pathways by which starch, glycogen, and cellulose are formed from glucose are of considerable interest. The starting material is glucose 6-phosphate (5), which is first converted enzymatically to glucose

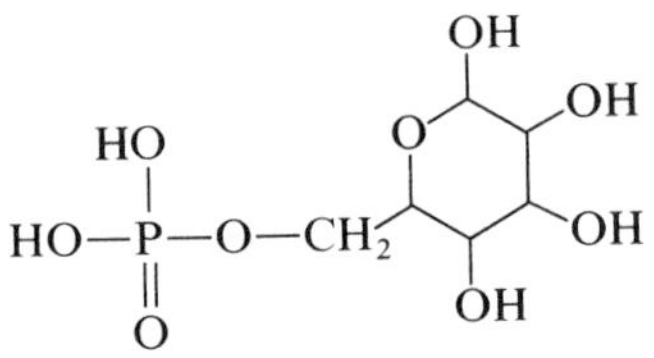

5

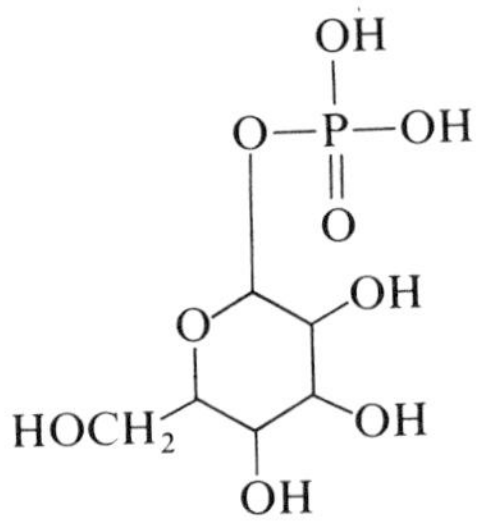

6

1-phosphate (6). The enzyme-induced reaction of glucose 1-phosphate with a nucleoside 5'-triphosphate, such as uridine 5'-triphosphate or adenosine 5'-triphosphate (ATP), yields a nucleoside diphosphate (NDP)-1-sugar plus pyrophosphate (reaction 1). The pyrophosphate is lost from the system by hydrolysis, and this

$$\text{NTP} + \text{sugar 1-phosphate} \; \rightleftharpoons \; \text{NDP-sugar} + \text{H}_2\text{O}_3\text{P}-\text{O}-\text{PO}_3\text{H}_2 \qquad (1)$$

provides a driving force for the reaction. Thus, it is the energy-yielding breakdown of a nucleoside triphosphate, such as ATP, to the energy-rich nucleoside diphosphate that enables condensation to occur even in a hydrolytic environment. Chain growth occurs by an enzyme-catalyzed loss of NDP from the 1-position of the glucose ring as that unit is condensed with the 4-position of a second sugar molecule (reaction 2). Chain branching by a condensation reaction at carbon 6 requires the presence of a special "branching enzyme."

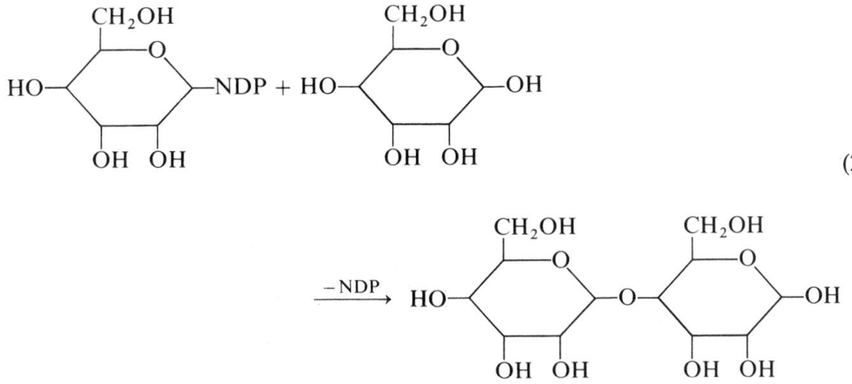

(2)

Reactions of Polysaccharides

Hydrolysis

Two enzymes, α-amylase and β-amylase, are capable of catalyzing the hydrolysis of the amylose fraction of starch. α-Amylase, produced in saliva and pancreatic juice, functions within the gastrointestinal tract to convert amylose first to medium-molecular-weight "dextrins," and then to glucose and maltose. β-Amylase, a plant enzyme found in malt, catalyzes the hydrolysis of amylose to maltose. The amylopectin fraction of starch, being highly branched, is hydrolyzed under the influence of α- and β-amylase with conversion of the outer *branches only* to glucose or maltose. However, the residual branched "core" (the limit dextrin) can be hydrolyzed by other "debranching" enzymes. Cellulose enzymes secreted by bacteria or fungi are responsible for the deterioration of cotton in warm, humid climates.

Polysaccharides can also be hydrolyzed to sugars by nonenzymatic processes. For example, cellulose undergoes skeletal hydrolysis reactions in aqueous acid which lead to a decrease in molecular weight and a loss of tensile strength. The crystalline regions of the polymer are apparently more resistant to hydrolysis than are the amorphous domains. The initial hydrolysis yields oligosaccharides, but glucose is the end product of hydrolytic degradation. The process is summarized in reaction (3).

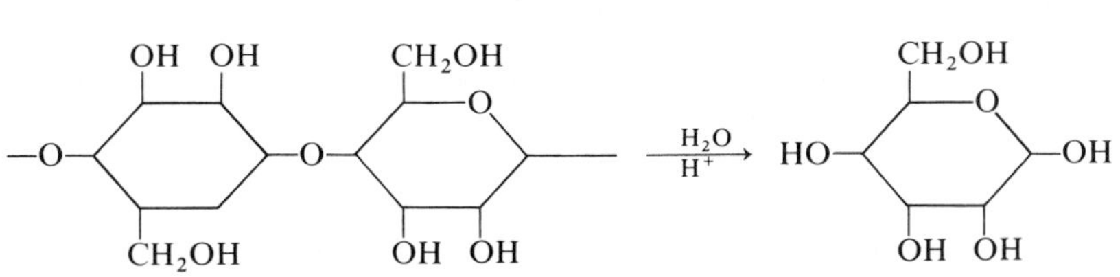

(3)

Esterification

Because each macromolecule in a polysaccharide bears pendent hydroxyl groups, substitution reactions can be performed on those hydroxyl sites. These reactions include esterification, etherification, crosslinking, and grafting processes. The first two types of reaction will be discussed here.

Esterification of cellulose can be effected with the use of inorganic or organic acids. Nitration occurs when cellulose is treated with a mixture of nitric and sulfuric acids at room temperature for about half an hour. The product is known as *cellulose nitrate*. In practice, some unreacted hydroxyl groups and some sulfate linkages will also be present. The actual nitrating agent is the nitronium ion, which reacts with both the crystalline and amorphous regions of the polymer. Cellulose nitrates that contained 12.5 to 13.4% nitrogen are explosive and are known as "gun cotton." Related polymers which contain 11 to 12% nitrogen have been used as lacquers, films, and plastics, although nowadays they have been largely displaced by newer, less flammable, synthetic polymers.

The reaction of cellulose with acetic acid or acetic anhydride yields *cellulose acetate*. The reaction usually requires a pretreatment of cellulose with acetic acid, which is followed by a reaction with acetic anhydride and sulfuric acid. The product from this reaction approximates in composition to a triacetate. It has a high melting point and a low solubility. Partial hydrolysis and replacement of sulfate ester groups is effected by treatment with aqueous acetic acid. This secondary process yields a polymer which has broad industrial use.

Rayon manufacture

Rayon is simply regenerated cellulose made from wood pulp. Celluloid is a film of the same material. Two processes exist for the manufacture of rayon—the xanthate method and the cuprammonium process. The xanthate method makes use of a cellulose esterification reaction. The process requires the treatment of the crude cellulosic material with aqueous sodium hydroxide and carbon disulfide to form a soluble ester by the process shown in (4).

$$
\underset{\underset{\text{H}}{|}}{\overset{\overset{\text{OH}}{|}}{\text{W—C—W}}} + \text{NaOH} + \text{CS}_2 \xrightarrow{-\text{H}_2\text{O}} \underset{\underset{\text{H}}{|}}{\overset{\overset{\text{O—}\overset{\overset{\text{S}}{\|}}{\text{C}}\text{—S}^{\ominus}\text{—Na}^{\oplus}}{|}}{\text{W—C—W}} \qquad (4)
$$

(Soluble in base)

Treatment of the xanthate ester with acid regenerates the cellulose. In practice, the cellulose may be precipitated as a fiber by extrusion of the xanthate solution (the "viscose") through spinnerettes into a bath of sulfuric acid that contains both sodium and zinc sulfates.

The cuprammonium process makes use of the fact that cellulose is soluble in solutions made from ammonium hydroxide and copper oxide. Spinning is then accomplished by extrusion of the solution into water. One problem with rayon and its derivatives as textile materials is the high degree of flammability. This necessitates the addition of flame retardants for certain uses.

Etherification

Ethers of cellulose can be formed by the treatment of alkaline cellulose compositions with an alkyl halide, such as methyl chloride. The reaction is summarized by equation (5).

$$\underset{\underset{H}{|}}{\overset{\overset{OH}{|}}{\text{W---C---W}}} \xrightarrow{\text{NaOH}} \underset{\underset{H}{|}}{\overset{\overset{O^-Na^+}{|}}{\text{W---C---W}}} \xrightarrow[-\text{NaCl}]{\text{CH}_3\text{Cl}} \underset{\underset{H}{|}}{\overset{\overset{O\text{---}CH_3}{|}}{\text{W---C---W}}} \tag{5}$$

Methyl iodide or dimethyl sulfate may also be employed as methylating agents. Commercial methylcellulose, as it is called, contains some residual hydroxyl groups, and is soluble in water. Ethyl cellulose is prepared commercially by the reaction of ethyl chloride with alkaline cellulose. Carboxymethylcellulose is isolated from the reaction of alkaline cellulose with chloroacetic acid or sodium chloroacetate (6).

$$\underset{\underset{H}{|}}{\overset{\overset{O^-Na^+}{|}}{\text{W---C---W}}} \xrightarrow[-\text{NaOH}]{\text{ClCH}_2\text{COOH}} \underset{\underset{H}{|}}{\overset{\overset{O\text{---}CH_2COOH}{|}}{\text{W---C---W}}} \tag{6}$$

Other reactions of cellulose

Finally, it should be remembered that a variety of reactions exist which permit the covalent binding of dyestuff molecules to cellulose. For example, colored aromatic vinyl sulfones undergo addition reactions to the hydroxyl groups of cotton (reaction 7). Alternatively, dyestuffs may be bound to s-triazines that con-

$$\underset{\underset{H}{|}}{\overset{\overset{O\text{---}H}{|}}{\text{W---C---W}}} + \text{CH}_2\text{==}CH\text{---}\overset{\overset{O}{\|}}{\underset{\underset{O}{\|}}{S}}\text{---Ar} \longrightarrow \underset{\underset{H}{|}}{\overset{\overset{O\text{---}CH_2\text{---}CH_2\text{---}\overset{O}{\overset{\|}{S}}\text{---Ar}}{|}}{\text{W---C---W}}}\,\,\underset{O}{} \tag{7}$$

tain active chlorine atoms. The chlorine atoms undergo reaction with the cellulosic hydroxyl groups (8).

$$\text{H---}\overset{\overset{\zeta}{|}}{\underset{\underset{\zeta}{|}}{C}}\text{---OH} + \text{Cl---C}\underset{\underset{\underset{C}{\|}}{N}}{\overset{\overset{N}{\diagdown}}{\diagup}}\text{C---Ar} \xrightarrow{-\text{HCl}} \text{H---}\overset{\overset{\zeta}{|}}{\underset{\underset{\zeta}{|}}{C}}\text{---O---C}\underset{\underset{\underset{C}{\|}}{N}}{\overset{\overset{N}{\diagdown}}{\diagup}}\text{C---Ar} \tag{8}$$

PROTEINS AND POLYPEPTIDES

General Composition of Proteins and Polypeptides

Proteins are complex polypeptide copolymers formed by the condensation reactions of amino acids (9). Synthetic polypeptide copolymers and homopolymers have also been made.

$$H_2N-\underset{\underset{H}{|}}{\overset{\overset{R}{|}}{C}}-COOH + H_2N-\underset{\underset{H}{|}}{\overset{\overset{R'}{|}}{C}}-COOH \xrightarrow{-H_2O} H_2N-\underset{\underset{H}{|}}{\overset{\overset{R}{|}}{C}}-\overset{\overset{O}{||}}{C}-NH-\underset{\underset{H}{|}}{\overset{\overset{R'}{|}}{C}}-COOH \quad \text{etc.}$$

(9)

Proteins exhibit an enormous variety of structures. The molecular weights of different proteins range from about 6000 to 1,000,000, with degrees of polymerization ranging from about 50 to over 8000. Some proteins consist of supramolecular agglomerates of two or more separate chains and, in such cases, the total molecular weight of the agglomerate may reach 40,000,000.

In addition to the polypeptide chains, some proteins contain nonproteinaceous components, known as "prosthetic groups." The prosthetic groups vary in structure from the iron porphyrins found in hemoglobin, myoglobin, and cytochrome c, to the ferric hydroxide, zinc, or copper in certain other metalloproteins. Phosphate prosthetic residues are present in casein. Ribonucleic acids constitute the prosthetic component of viruses and ribosomes. Proteins that contain prosthetic groups often fall into the category known as globular proteins—the materials that perform the chemical work of the living system. For example, the iron-porphyrin prosthetic group is responsible for the oxygen binding and transport function of hemoglobin and myoglobin. The metallocomponent of many enzymes is the actual site of the catalytic reactions.

Different proteins are different, not only because of the presence or absence of prosthetic groups, but also because of the different sequences of amino acid residues that can exist in polypeptide chains. There are only 20 commonly occurring amino acids (see Table 8.1), but the number of sequential permutations possible with 20 different residues is enormous. No less than 10^{300} different amino acid sequences could exist if only 12 different amino acids are present in equal amounts in a small protein that contains less than 300 total residues.

The amino acid sequence within a polypeptide chain is known as the "primary structure." This, in turn, governs the *conformation* of the polymer chain (see Chapter 18), also known as the "secondary structure," and determines the existence of "random coil" or helical segments. The location of the random coil and α-helix regions along a given chain, and the location of internal crosslinks or mutually attractive residues, determines the overall *shape* of the molecules— called the "tertiary structure." It follows that the secondary and tertiary structural characteristics will in turn influence the manner in which individual protein chains will agglomerate in supramolecular systems. This mode of agglomeration is called

TABLE 8.1   Naturally Occurring Amino Acids

Name	Structure	Symbol	Intramolecular Interaction Character
Alanine	CH_3 $H_2N{-}CH{-}COOH$	Ala (or A)	
Phenylalanine	CH_2Ph $H_2N{-}CH{-}COOH$	Phe (F)	
Valine	$CH(CH_3)_2$ $H_2N{-}CH{-}COOH$	Val (V)	
Leucine	$CH_2CH(CH_3)_2$ $H_2N{-}CH{-}COOH$	Leu (L)	
Isoleucine	$CH(CH_3)CH_2CH_3$ $H_2N{-}CH{-}COOH$	Ile (I)	Nonpolar, hydrophobic side groups
Methionine	$CH_2CH_2SCH_3$ $H_2N{-}CH{-}COOH$	Met (M)	
Proline	(ring structure) CH_2 CH_2 CH_2 $HN{-}CH{-}COOH$	Pro (P)	
Tryptophane	(indole ring structure) $CH_2{-}C$ $H_2N{-}CH{-}COOH$	Trp (W)	
Glycine	H $H_2N{-}CH{-}COOH$	Gly (G)	
Serine	CH_2OH $H_2N{-}CH{-}COOH$	Ser (S)	
Cysteine	CH_2SH $H_2N{-}CH{-}COOH$	Cys (C)	
Threonine	$CH(CH_3)OH$ $H_2N{-}CH{-}COOH$	Thr (T)	
Tyrosine	CH_2⟨ring⟩${-}OH$ $H_2N{-}CH{-}COOH$	Tyr (Y)	Polar, hydrophilic side groups
Asparagine	$CH_2C(O)NH_2$ $H_2N{-}CH{-}COOH$	Asn (N)	
Glutamine	$CH_2CH_2C(O)NH_2$ $H_2N{-}CH{-}COOH$	Gln (Q)	
Histidine	(imidazole ring structure) CH $HN{\diagup}{\diagdown}N$ $CH_2{-}C{=}CH$ $H_2N{-}CH{-}COOH$	His (H)*	

(*Continued*)

TABLE 8.1 (Continued)

Name	Structure	Symbol	Intramolecular Interaction Character
Lysine	$(CH_2)_4NH_2$ $H_2N-CH-COOH$	Lys (K)	Basic side groups
Arginine	$(CH_2)_3NHC(NH)NH_2$ $H_2N-CH-COOH$	Arg (R)	
Aspartic acid	CH_2COOH $H_2N-CH-COOH$	Asp (D)	Acidic side groups
Glutamic acid	CH_2CH_2COOH $H_2N-CH-COOH$	Glu (E)	

* A borderline amino acid which, at pH 7.0, functions more as a polar, hydrophilic unit than as a basic residue.

the "quaternary structure." Hence, the precise sequence of amino acids along a polypeptide chain is one of the main keys to the properties and function of that particular protein.

Amino Acids

As mentioned above, only 20 different monomer molecules are found in all the commonly occurring proteins, and it is the sequential "coding" of these residues along a chain that determines the shape and, to a large extent, the function of a protein. The 20 common amino acids are listed in Table 8.1.

The two most important differences between the different amino acids are the ability of a particular residue to generate an internal S—S crosslink (cystein residues) and, more important, the degree to which the side groups attached to different residues on the same chain have the ability to attract or repel each other and thereby determine the conformation of the chain. The amino acids listed in Table 8.1 are grouped according to the hydrophobic, hydrophilic, basic, or acidic character of their side groups. In an aqueous environment, hydrophobic groups will tend to attract each other in the same way that oil droplets tend to coalesce in water. Moreover, highly dipolar or charged side groups may be attracted to other dipolar or oppositely charged residues. Specific amino acids provide the binding sites for prosthetic groups. For example, the imidazole component of histidine is responsible for the coordinative binding between the protein and the Fe(II)-porphyrin unit in hemoglobin and myoglobin.

Different Proteins and Their Functions

Proteins can be divided into two large classes—fibrous and globular proteins. This classification is closely related to the conformational characteristics of polymers in the two groups, and also to the function served by the protein in a living system.

Fibrous proteins

Fibrous proteins are tough, insoluble materials found in the protective and connective tissues of animals, birds, and reptiles. They were among the first high polymers to be studied by X-ray diffraction techniques (see Chapter 19), and their secondary structure is much simpler than those of globular proteins. Fibrous proteins can be divided into three main classes: the α-keratins, the collagens, and the β-keratins.

α-Keratins are found in hair, wool, horn, nails, feathers, and leather. In general, these materials are nonelastic. They are tough or flexible substances that have been used by man since antiquity as technological raw materials. The molecular conformational structure found in α-keratins is the hydrogen-bonded α-helix. In many cases, a number of α-helical chains are wound together to form "ropes." This is illustrated in Figure 8.3. The α-helical arrays in the fibrous α-keratins are comprised of L-amino acid residues arrayed in right handed helices. The strength and toughness of α-keratin materials can be traced to the multiple hydrogen bonding which holds the helix in its conformation and to the presence of —S—S— crosslinks between cystein residues. Thus, reducing agents will soften α-keratin structures by cleavage of the crosslinks and the formation of S—H bonds.

Collagen is a tough, fibrous protein found in animal connective tissue and tendons. The normal α-helical arrangements are not found in this protein. Instead, each polypeptide chain forms a loose helix that is hydrogen-bonded *inter*molecularly to two other chains to form a triple-strand rope. This arrangement prevents slippage of the chains past each other.

The third fibrous protein structure that has been studied in detail is the β-keratin system or the silk fibroin arrangement. α-Keratins, when subjected to heat and moisture by "steaming," can be stretched to a considerable degree. This process involves a cleavage of intramolecular hydrogen bonds and the conversion of the α-helix coils to an extended zigzag conformation. The extended chains are now linked to their neighbors by *inter*molecular hydrogen bonds to form a pleated sheet arrangement.

Globular proteins

As mentioned earlier, globular proteins are soluble in aqueous media. They are responsible for organizing the chemical reactions that take place in a living system. Enzymes, antibodies, hemoglobin, myoglobin, serum albumin, and some hormones are all globular proteins. Globular proteins differ from the fibrous type because they are intricately folded molecules that contain both α-helix and nonhelical segments. Water molecules may occupy sites within the protein, and prosthetic groups are often present.

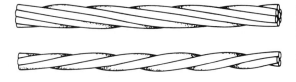

Fig. 8.3 Supercoiling of α-helical polypeptide coils to form "ropes." [From A. L. Lehninger, *Biochemistry*, (New York: Worth, 1970), p. 58.]

A detailed analysis of globular protein structures is beyond the scope of this chapter. Indeed, only a handful of globular protein structures are known in any detail at the present time. Three lines of attack are necessary before the structure and function of a globular protein can be partly understood. First, the sequence of amino acid residues along each chain must be established, usually by chemical means. Second, the conformation of the chain and the location of prosthetic groups must be determined, usually by X-ray diffraction techniques, although nmr, esr, and other spectroscopic methods may also be used. Third, attempts can be made to *rationalize* the secondary and tertiary structure in terms of the attractive or repulsive forces between different amino acid residues on the same chain. The degree of difficulty increases as one passes through these three successive steps. It is probably true to say that no globular protein is yet fully understood in all three ways.

Three examples only of globular proteins will be mentioned—myoglobin, cytochrome c, and lysosyme. Myoglobin is an iron-containing protein present in all animal tissues, but found in substantial quantities in the muscles of aquatic diving vertebrates, such as whales or seals. The function of the protein is to store oxygen. Myoglobin is a relatively simple protein, with a molecular weight of only 16,700, made up of 153 amino acid residues. It consists of one polypeptide chain to which is attached one Fe(II)-porphyrin or heme unit. One face of the heme unit is bound coordinatively through the iron to the imidazole component of a histidine residue, while a second histidine residue loosely occupies a similar position on the opposite face. Figure 8.4 illustrates this arrangement. The polypeptide chain is divided into eight helical regions separated by nonhelical bends. The whole tertiary structure forms a flexible box surrounding the heme component, and slight changes in the shape of the box and its central cavity, as the pH is changed, may explain why the molecule binds oxygen at pH 7.5 but releases it when carbon dioxide or lactic acid lowers the pH of the surrounding tissue. Hemoglobin contains four myoglobinlike molecules agglomerated into a supramolecular structure. Synthetic models and possible temporary substitutes for myoglobin and hemoglobin have been investigated (see Chapter 22).

Cytochrome c is a widely distributed protein in animals and plants. Its function is to act as an oxidation–reduction intermediary in the terminal oxidation chain. Cytochrome c resembles myoglobin in general composition. It possesses a single polypeptide chain of 104 amino acid residues and a heme unit. However, it contains no α-helical sections, but possesses instead many residues that occupy the extended β-conformation described earlier for β-keratins. Moreover, the heme unit of cytochrome c is not bound coordinatively to histidine residues as it is in myoglobin and hemoglobin. Instead, it is bound *covalently* through two side groups attached to the porphyrin ring that themselves are connected to two cystein residues on the main chain.

Lysozyme is an enzyme present in mucous or egg white. It kills bacteria by attacking the bacterial cell wall. Lysozyme contains no metal atom or other prosthetic group. It consists of only one polypeptide chain made up of 129 amino acid residues. The chain possesses four intramolecular cystein crosslinks (disulfide bridges), and only three α-helical segments. The tertiary structure is such that the hydrophobic amino acid residues are on the inside of the molecule, and the hydrophilic residues

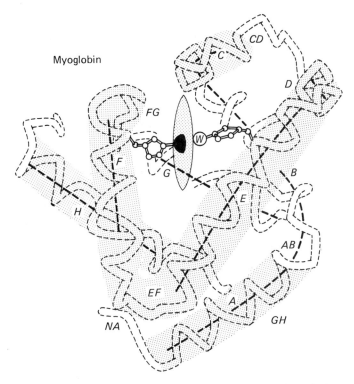

Myoglobin

Fig. 8.4   Location of the Fe(II)-porphyrin unit within the coiled protein chain of myoglobin. [From R. E. Dickerson and I. Geis, *The
Structure of Proteins* (New York: Harper & Row, 1969), p. 47.]

are on the outside. The structure is characterized by the presence of an obvious
crevice or cleft in the overall globular shape. It is in this cleft that the active polysaccharide cleavage site is believed to be located.

The Biological Synthesis of Proteins

In the living cell, proteins are synthesized from amino acids at ribosome sites under
the influence of enzymes and in sequences determined by the nucleic acid arrangement in messenger RNA. The energy required for the polymerization is supplied
by ATP.

Protein synthesis takes place in four stages. In the first step, amino acids become
enzymatically bound to a *transfer* RNA molecule by an esterification reaction.
Second, an *initiation* complex is formed between the amino acid-transfer RNA ester
and messenger RNA. Next, polypeptide synthesis begins by a sequential buildup of
amino acid residues according to the code specified by the messenger RNA. After
each amino acid residue has been introduced, both messenger RNA and the transfer RNA-peptide chain are moved along the ribosome to position the next coded
site. Finally, when the last coded instruction from the messenger RNA has been
followed, the completed protein separates from the ribosome. Construction of the

protein begins with a reaction between the *carboxylic acid end* of the first amino acid being condensed with the amino residue of the second amino acid, and so on.

Laboratory Synthesis of Polypeptides

Homopolymer synthesis

Homopolypeptides can be synthesized readily by conventional condensation reactions (Chapter 2) carried out with simple amino acids, such as glycine, alanine, or phenylalanine. Such polymers are not proteins—they are often only poorly soluble in water, may be highly crystalline, have polydisperse molecular weights, and do not adopt discrete folded conformations in solution or the solid state. However, homopolypeptides are of technological interest as fibers or absorbable surgical sutures, and they provide valuable models for estimating the role played by specific amino acid residues in determining the conformations of naturally occurring polypeptides. As such, they provide the raw data for conformational energy calculations on proteins, as discussed in Chapter 18.

Copolymer synthesis

Simple naturally occurring polypeptides have been synthesized by careful, sequential, copolymerization reactions. The main problem in protein synthesis is the prevention of condensation at *both* ends of the growing molecule. The polypeptide chain must be built up step by step, with the condensation normally taking place at one terminal group only. Such a process requires the use of "blocking groups" or "protective groups" to prevent unwanted, nonsequential reactions from occurring. The technique is illustrated in (10), where X and Y are protecting

$$
\begin{array}{ccccc}
\begin{array}{c}
\text{X} \\
| \\
\text{NH} \\
| \\
\text{H}-\text{C}-\text{R} \\
| \\
\text{COOH} \\
+ \\
\text{NH}_2 \\
| \\
\text{H}-\text{C}-\text{R}' \\
| \\
\text{C}=\text{O} \\
| \\
\text{Y}
\end{array}
&
\xrightarrow{-\text{H}_2\text{O}}
&
\begin{array}{c}
\text{X} \\
| \\
\text{NH} \\
| \\
\text{H}-\text{C}-\text{R} \\
| \\
\text{C}=\text{O} \\
| \\
\text{NH} \\
| \\
\text{H}-\text{C}-\text{R}' \\
| \\
\text{C}=\text{O} \\
| \\
\text{Y}
\end{array}
&
\xrightarrow[\text{both protecting groups}]{\text{Remove one or}}
&
\begin{array}{c}
\text{X} \\
| \\
\text{NH} \\
| \\
\text{H}-\text{C}-\text{R} \\
| \\
\text{C}=\text{O} \\
| \\
\text{NH} \\
| \\
\text{H}-\text{C}-\text{R}' \\
| \\
\text{C}=\text{O} \\
| \\
\text{OH}
\end{array}
\quad \text{etc.} \quad (10)
\end{array}
$$

groups). A commonly used dehydrating agent for the condensation step is dicyclohexylcarbodiimide (see page 34).

Normal laboratory procedures have been used to apply this method to the synthesis of simple peptide hormones containing 9 to 39 residues, as well as to the synthesis of insulin chains. However, the extension of such processes to the synthesis of complex proteins is a formidable task. One of the most significant breakthroughs in protein synthesis was the development between 1959 and

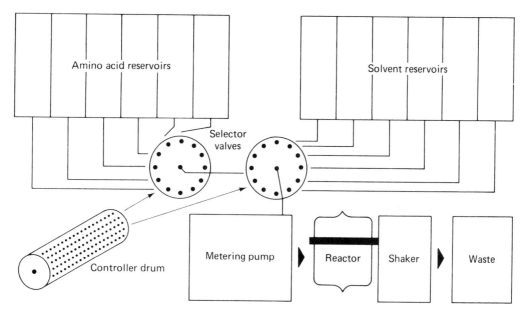

Fig. 8.5 Schematic diagram of a Merrifield polypeptide synthesizer. (From *Chem. Eng. News*, August 2, 1971, p. 26; Courtesy of the American Chemical Society.)

1964[1] of a machine that was capable of automatically performing the various sequential steps in high yield to generate complex polypeptide copolymers.

The fundamental idea underlying the whole procedure is this: the polypeptide chain is synthesized with an end of the molecule attached to an insoluble resin particle. This allows all side products, solvents, and so on, to be removed at each step in the process by the simple expedient of filtering off the solid resin particles. This procedure avoids the time-consuming and laborious techniques that were required earlier to purify each peptide at each step by crystallization. The machine consists of a series of amino acid, solvent, and reagent reservoirs connected via selector valves and metering pumps to a single reaction chamber. The insoluble resin is crosslinked polystyrene that has been chloromethylated by treatment with

$$\text{Polystyrene bead} - \bigcirc \xrightarrow[\text{SnCl}_4]{\text{CH}_3\text{OCH}_2\text{Cl}} \text{Polystyrene bead} - \bigcirc - \text{CH}_2\text{Cl} \quad (11)$$

chloromethyl ether and stannic chloride (reaction 11). The core of the machine is a rotating drum that bears pins which actuate microswitches as the drum rotates. The microswitches control the charges of reagents that enter the reaction chamber and the agitation and filtration of the contents. A schematic diagram of the apparatus is shown in Figure 8.5. The operator programs the sequence of chemical operations

[1] R. B. Merrifield, *Science*, **150**, 178 (1965).

by insertion of pins in the appropriate position on the drum. These chemical operations are as follows.

The amino acid starting materials are first blocked at the amino residue, as shown in **7**, and the first amino acid is coupled to the resin through the carboxylic acid group (**8**). Treatment of **8** with acid removes the blocking group as carbon

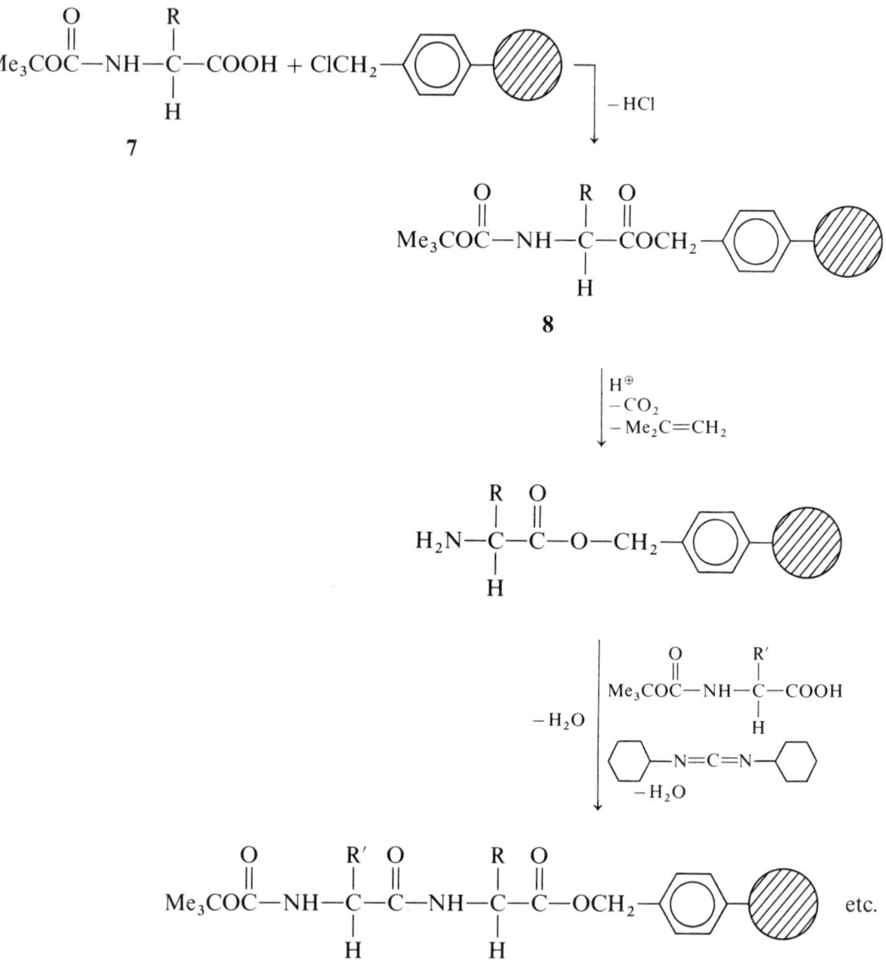

dioxide and isobutylene. The second amino-blocked amino acid can then be coupled to the first with the use of the powerful condensing agent—dicyclohexylcarbodiimide. These steps are repeated many times and the polypeptide is finally released from the resin by treatment with a reagent, such as HBr in trifluoroacetic acid.

The automatic nature of the sequence and the high yields at each step allowed the synthesis of the nonapeptide, bradykinin, in 27 h (3 h per peptide bond). The A-chain of insulin, with 21 residues, was synthesized in 8 days, and the B-chain (30 residues) in 11 days. Ribonuclease, with 124 residues, was synthesized in 6 weeks.

Reactions of Proteins

Denaturation and recoiling

The biological activity of proteins is a direct consequence of their structure and molecular conformation. This is especially true of globular proteins. Thus, any influence that breaks up the secondary or tertiary structure will destroy the biological activity. Heat, excessive cooling, pH changes, or the presence of many chemical reagents, will have this effect. This process is called *denaturation*, and it may be reversible or irreversible. The coagulation of the white of an egg when heated is an irreversible denaturation.

It should be noted that denaturation often does not change the chemical composition of the protein. The amino acid residues are present in the same sequence as before. Only the conformation of the chain has been changed from a discrete pattern of helix and bends to a more random coil arrangement. As we have seen, the tertiary structure depends mainly on hydrophobic and hydrophilic intramolecular interactions (plus the presence of disulfide linkages in some proteins). Hence, it might be expected that the biologically active conformation could reform once the perturbing conditions are removed. This can, in fact, be accomplished in the laboratory, outside a living cell. Refolding of the chains occurs spontaneously (if slowly) as the various attractive portions of the flexing, twisting chain find each other and pull the molecule back into the biologically active shape. The same phenomenon can be observed when the individual chains of supra-molecular proteins are first separated and then brought together. Hemoglobin rapidly reassembles itself from mixtures of the α and β chains. The heme residues may also be removed and then reintegrated into the structure. Moreover, the metal atoms of enzymes can be removed to cause a total loss of activity, but the reintroduction of the metal restores the catalytic powers. In some enzymes, *different* metal atoms may be introduced at the renaturation stage, sometimes to yield metallo-proteins that have a biological activity comparable to that of the original. Such metal substitutions are valuable for probing the mechanism of enzyme action.

Hydrolysis reactions

The hydrolytic conversion of proteins to oligomeric peptides and amino acids is a reaction that has been used by man since prehistory for the conversion of animal tissues to glues, or readily digestible foodstuffs such as gelatin. On a more subtle level, the selective hydrolysis of proteins can be used as a method of structural identification. The fundamental hydrolysis reaction is shown in (12). Acidic

$$\underset{}{\text{\tiny W}}\text{—CH—C—N—CH—}\text{\tiny W} \xrightarrow[\text{H}_2\text{O}]{\text{H}^+} \text{\tiny W}\text{—CH—C—OH} + \text{H}_2\text{N—CH—}\text{\tiny W} \qquad (12)$$

hydrolysis (for example, with 6 N HCl at 100 to 120°C for 10 to 24 h in a sealed system) will generate the hydrochloride salts of the free amino acids. Tryptophane is destroyed by this process and glutamine and asparagine are converted to glutamic and aspartic acids but, in general, the amino acid composition of the hydrosylate will reflect that of the protein. Alkaline hydrolysis does not cause decomposition of tryptophane, but destroys other amino acids.

The analysis of a protein is accomplished at two levels of detail. First, the gross amino acid composition of the hydrolylate may be determined by paper chromatography, electrophoresis, or ion-exchange chromatography. Second, the *sequence* of amino acid residues along the chain must be determined. This sequencing operation involves (1) an identification of the NH_2-terminal or COOH-terminal residues at the ends of the chain, and (2) cleavage of the protein into oligopeptides by the use of trypsin, followed by identification of the terminal residues in these fragments. This process is then continued until the sequence is known.

Identification of the NH_2-terminal amino acid can be accomplished by reaction of the protein with 2,4-dinitrofluorobenzene. This reagent reacts with terminal or pendent NH_2 groups, as shown in (13), to yield a linkage that will survive hydroly-

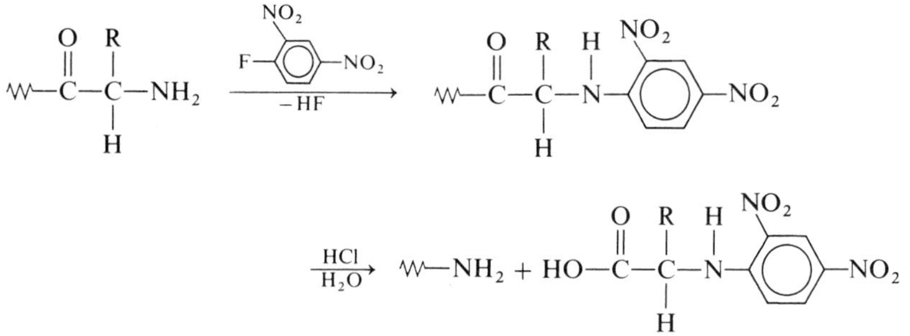

$$(13)$$

sis. Subsequent hydrolysis will yield a mixture of amino acids, only one of which bears the 2,4-dinitrofluorophenyl (DNP) groups. Phenyl isothiocyanate is also used as an $-NH_2$ end-group reagent, as illustrated in (14). The advantage of this reagent is that the hydrolysis to yield the phenylthiohydantoin (**9**) removes only

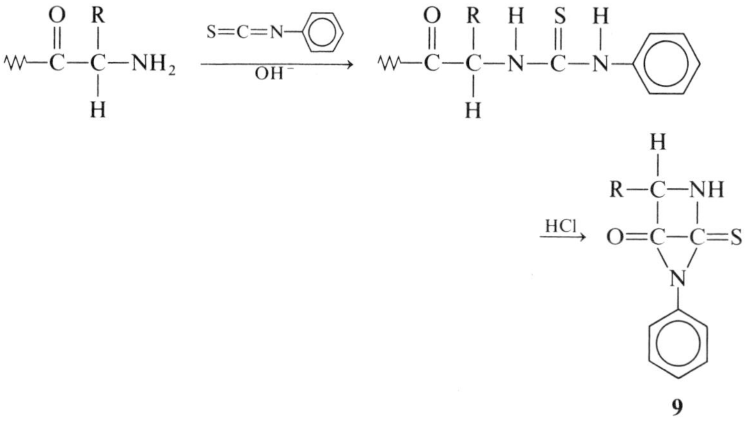

$$(14)$$

9

the terminal amino acid. Hence, the process can be continued repetitively to determine the actual amino acid sequence. The $-COOH$ terminal residue can be identified by reduction to the alcohol, followed by hydrolysis of the protein and identification of the amino alcohol, or by enzymatic cleavage of the terminal residue with carboxypeptidase.

Other reagents are known which selectively cleave the skeletal peptide bonds adjacent to specific amino acid residues. Bromine or *N*-bromosuccinimide selectively cleaves the tyrosine peptide bonds. Methionine peptide bonds are cleaved by iodoacetamide or cyanogen bromide. Many other selective cleavage processes are known.

Reactions of protein side groups

First, and perhaps most obvious, it is possible to exchange the hydrogen atoms of peptide linkages by treatment of the polymer with deuterium oxide or tritium oxide (15). However, the rate of exchange depends on the position and role of a

$$
\begin{array}{c} R \\ | \\ -C-C-N- \\ |\ \ ||\ \ | \\ H\ \ O\ \ H \end{array} \xrightarrow{\text{D}_2\text{O}} \begin{array}{c} R \\ | \\ -C-C-N- \\ |\ \ ||\ \ | \\ H\ \ O\ \ D \end{array} \tag{15}
$$

particular peptide link in the chain. Hydrophobic side groups (R) or hydrogen bonding involving the N—H group will retard exchange. Thus, deuterium-exchange studies can be used to examine the environment of a particular amino acid residue in an undenatured protein.

The different substituent groups (R) in proteins include those that contain —OH, –COOH, —NH$_2$, —NH—, —CONH$_2$, —SH, or —SCH$_3$ functional groups, and these can be induced to undergo reactions with appropriate reagents. Such reactions can be used for structure determination, or they can be carried out deliberately to modify the protein at specific sites in order to observe the resultant changes in physiological behavior. On an industrial level, side group reactions can be employed to improve the overall physical properties of a protein-containing product.

Halogen-containing reagents react with —SH, —OH, —NH$_2$, or —COOH groups. For example, iodoacetic acid, ICH$_2$COOH, reacts with —SH, or —NH$_2$ groups. Cysteine residues may be identified and modified by this procedure (16).

$$
\underset{\overset{|}{\lessgtr}}{\overset{\overset{\lessgtr}{|}}{CH}}-CH_2-SH + ICH_2COOH \xrightarrow{-HI} \underset{\overset{|}{\lessgtr}}{\overset{\overset{\lessgtr}{|}}{CH}}-CH_2-S-CH_2COOH \tag{16}
$$

Acyl halides or anhydrides react with —OH, —NH$_2$, —COOH groups, and so on (17). Acetic anhydride acetylates at least 90% of the pendent amino groups of

$$
\underset{\overset{|}{\lessgtr}}{\overset{\overset{\lessgtr}{|}}{CH}}-(CH_2)_4-NH_2 + Cl-\overset{\overset{\displaystyle O}{||}}{C}-R \xrightarrow{-HCl} \underset{\overset{|}{\lessgtr}}{\overset{\overset{\lessgtr}{|}}{CH}}-(CH_2)_4-NH-\overset{\overset{\displaystyle O}{||}}{C}-R \tag{17}
$$

wool. Carboxylic acid groups may be esterified with the use of alcohols in acidic media. Sulfuric acid reacts with —OH or —SH groups to generate sulfate esters or thioesters (18).

$$\overset{\displaystyle\gtrless}{\underset{\displaystyle\lessgtr}{CH}}-CH_2-OH \xrightarrow[-H_2O]{H_2SO_4} \overset{\displaystyle\gtrless}{\underset{\displaystyle\lessgtr}{CH}}-CH_2-O-SO_3H \qquad (18)$$

Treatment with nitric acid under mild conditions may be used to nitrate aromatic residues as, for example, in tyrosine units. Drastic nitration conditions cause decomposition of the protein. Iodination of a number of proteins in neutral or basic media leads to the conversion of tyrosine residues to 3,5-diiodotyrosine units.

Many proteins react with diazonium salts to yield diazo proteins (19). Alternatively, the protein itself may be diazotized with nitrous acid and then coupled, for example, to an amine (20).

$$\overset{\displaystyle\gtrless}{\underset{\displaystyle\lessgtr}{CH}}-R-\overset{\displaystyle R}{\underset{}{N}}-H + X^-N^+{=}N-R' \xrightarrow{-HX} \overset{\displaystyle\gtrless}{\underset{\displaystyle\lessgtr}{CH}}-R-\overset{\displaystyle R}{\underset{}{N}}-N{=}N-R' \qquad (19)$$

$$\overset{\displaystyle\gtrless}{\underset{\displaystyle\lessgtr}{CH}}-R-NH_2 \xrightarrow[-HCl]{HNO_2} \overset{\displaystyle\gtrless}{\underset{\displaystyle\lessgtr}{CH}}-R-N{=}N^+Cl^- \xrightarrow[-HCl]{RNH_2} \overset{\displaystyle\gtrless}{\underset{\displaystyle\lessgtr}{CH}}-R-N{=}N-R-NH_2$$

$$(20)$$

A number of unsaturated reagents undergo addition reactions with proteins. For example, as shown in reaction (21), formaldehyde adds to amino or amido

$$\overset{\displaystyle\gtrless}{\underset{\displaystyle\lessgtr}{CH}}-CH_2-\overset{\displaystyle O}{\overset{\displaystyle \|}{C}}-NH_2 \xrightarrow{CH_2O} \overset{\displaystyle\gtrless}{\underset{\displaystyle\lessgtr}{CH}}-CH_2-\overset{\displaystyle O}{\overset{\displaystyle \|}{C}}-NH-CH_2-OH \qquad (21)$$

groups to form a methylol derivative. As discussed in Chapter 2, methylol groups show a strong tendency to condense to generate crosslinks between the polymer molecules. Isocyanates can add to —OH, —SH, or —NH$_2$ groups. An example is shown in reaction (22).

$$\overset{\displaystyle\gtrless}{\underset{\displaystyle\lessgtr}{CH}}-CH_2-SH + O{=}C{=}N-Ph \longrightarrow \overset{\displaystyle\gtrless}{\underset{\displaystyle\lessgtr}{CH}}-CH_2-S-\overset{\displaystyle O}{\overset{\displaystyle \|}{C}}-NH-Ph \qquad (22)$$

The principal crosslink unit found in proteins is the disulfide bridge (**10**). Disulfide bridges are derived from cysteine residues by the oxidation process illustrated in equation (23). Interest in the crosslinkage of proteins stems from two

$$CH-CH_2-SH + HS-CH_2-CH \xrightarrow{O} CH-CH_2-S-S-CH_2-CH \quad (23)$$

10

sources: (1) the use of reagents to cleave disulfide bridges, and (2) the study of reactions that can introduce new crosslinks into a protein system.

Disulfide bridges can be cleaved by a variety of reagents, such as peracids (performic or peracetic acids), peroxides, or chlorine, which oxidize the bond, or reducing agent, such as borohydrides, bisulfites, or thiols. Water or hydroxide ion may cleave the disulfide bridges by hydrolysis. Even heat or ultraviolet light may break the bond. Of course, the destruction of native crosslinks increases the ability of the protein to swell in water or even to dissolve.

The introduction of additional crosslinks into proteins may confer significant technological advantages on the material. For example, the stability of leather, wool, or silk to acids, alkalis, oxidizing agents, or to attack by insects, is improved if additional crosslinking sites can be introduced. A wide variety of reagents have been used in attempts to achieve this end. Many of the reactions mentioned in the preceding section can be modified to generate crosslinks. The tanning of animal skins to form leather by treatment with tannins (phenolic tree bark extracts), chromium salts, or formaldehyde, takes place by crosslinking of collagen chains through hydrogen bonds, covalent linkages, or metal coordination binding.

POLYNUCLEOTIDES

General Composition of Polynucleotides

Polynucleotides are the macromolecules, found within living cells, that are responsible for the storage of genetic information, for its replication, and for providing templates that direct the synthesis of proteins. They are complex copolymers formed by the condensation of only a small range of closely related monomer molecules.

Two main types of polynucleotides are known: deoxyribonucleic acid (DNA) and ribonucleic acids (RNA). DNA molecules are the storage sites for genetic information within the chromosomal genes of the nucleus, and are responsible for the replication of that information during cell division. RNA molecules transmit information coded in their structure from the DNA molecules to the ribosomes and there direct the pattern of protein synthesis.

The monomer units from which polynucleotides are built are called mononucleotides. Each "monomer" is made up of three parts: a phosphoric acid residue plus a sugar residue in the main chain, and a heterocyclic organic base which forms a side group that is attached to the sugar (11). Different mononucleo-

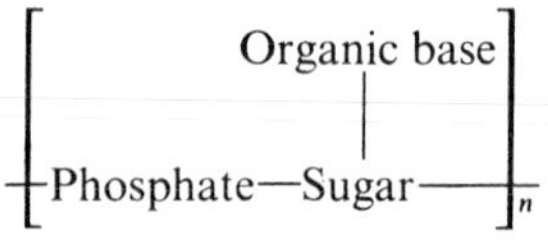

$$\left[\begin{array}{c} \text{Organic base} \\ | \\ \text{Phosphate—Sugar} \end{array} \right]_n$$

11

tide monomers contain one of two sugars—the cyclic, five-carbon ribose (12) or deoxyribose (13)—and one of five heterocyclic bases, as shown in 14 to 18. A

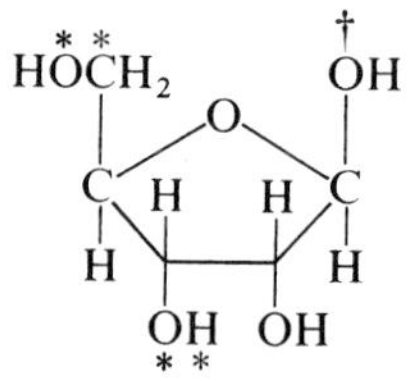

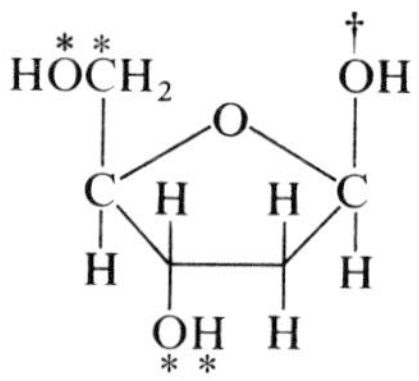

12	13	14
Ribose	Deoxyribose	Adenine (A)

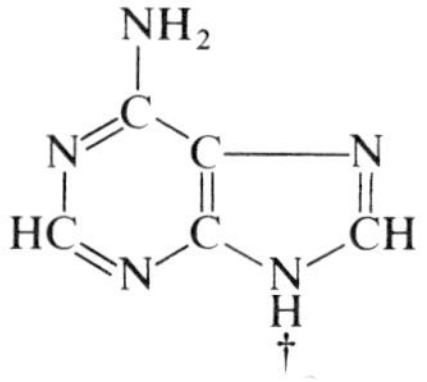

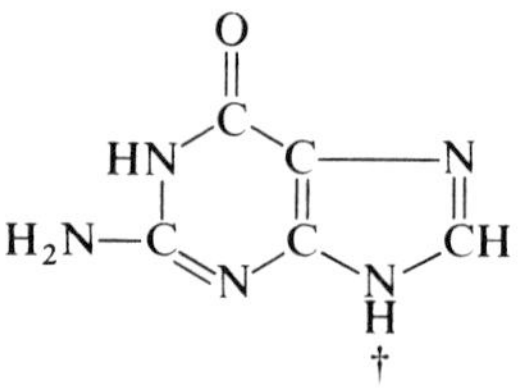

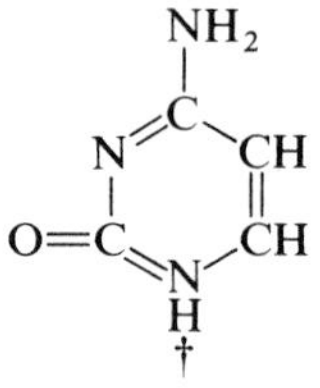

15	16	17	18
Guanine (G)	Cystine (C)	Uracil (U)	Thymine (T)

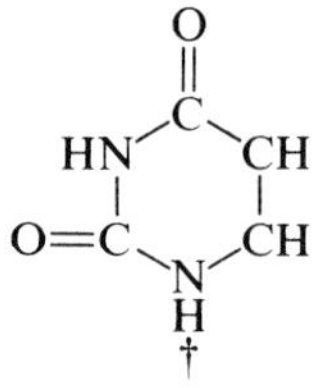

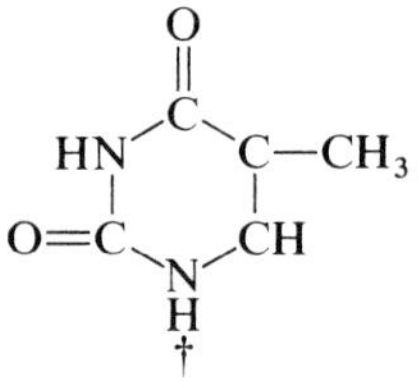

few other minor bases are found in some polynucleotides. The sites of attachment of the sugar to phosphate in the main chain are denoted by **, and the linkage sites between sugar and heterocyclic base are indicated by †. Thus, a typical nucleotide monomer residue would appear as shown in 19.

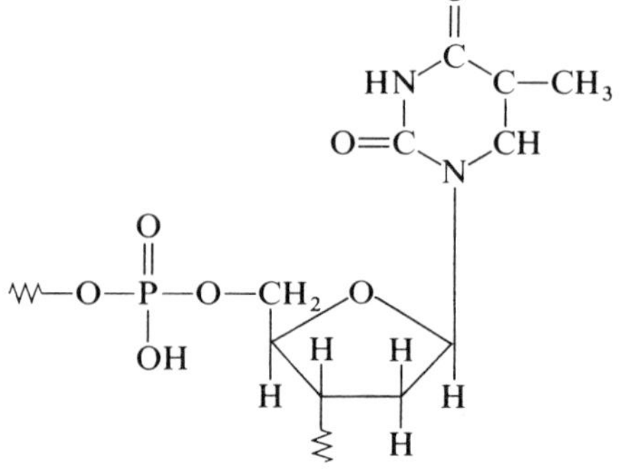

19

DNA contains only the sugar ribose and four bases, adenine, guanine, cytosine, and thymine, plus small amounts of methylated forms of these bases. The various forms of RNA contain only the sugar deoxyribose and four bases, adenine, guanine, cytosine, and uracil, together with minor amounts of less common bases that are often methylated derivatives of the major bases. In most cells, the amount of RNA is 5 to 10 times more than the amount of DNA.

Roles Played by Different Polynucleotides

DNA

Deoxyribonucleic acid provides a sequential polymeric transcription code that ultimately is responsible for the pattern of protein synthesis. Each DNA supra-molecular unit consists of two polymeric strands that are wound together in the form of the well-known "double helix." The double helix is held together by the coplanar stacking of base pairs within the interior of the helix (Figure 8.6). The only allowed base-pairing combinations are adenine–thymine and guanine–cytosine, each base pair being held together by hydrogen bonding linkages of the

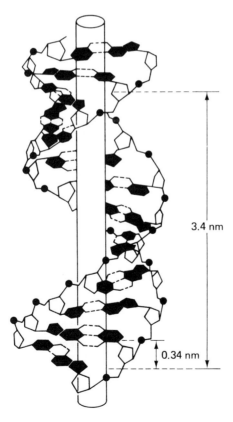

3.4 nm

0.34 nm

Fig. 8.6 DNA double helix. The pentagons represent sugar residues, the black dots show the location of the phosphate units, and the dotted lines depict the hydrogen bonds. [From J. N. Davidson, *Living Molecules*, The Royal Institute of Chemistry (Lecture Series), No. 1, 1963.]

types shown in **20** and **21**. This stacking arrangement places the bases on the inside of the helix and the sugar–phosphate main chains on the outside. Replication is accomplished by separation of the two strands of the double helix, each of which then functions as a template for the synthesis of a new complementary partner (Figure 8.7). Some DNA molecules are cyclic rather than linear.

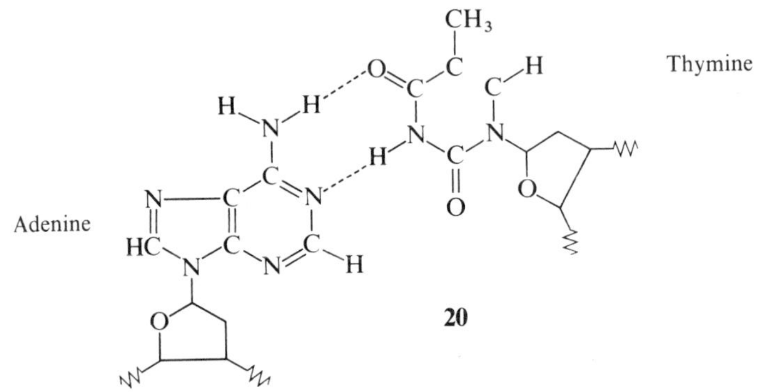

Messenger RNA

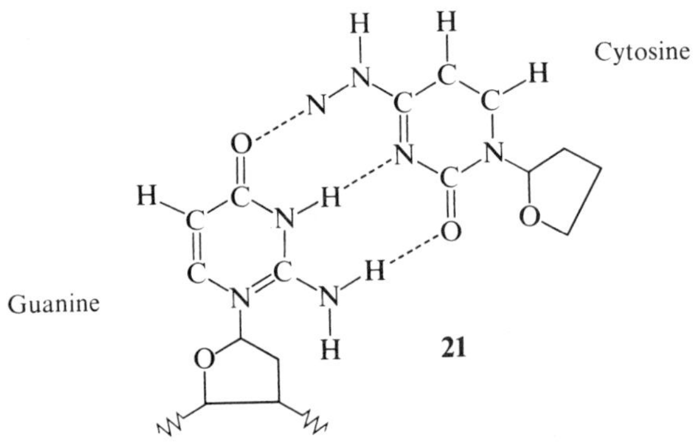

This form of RNA is a single-strand polynucleotide that originates in the cell nucleus. It functions as a template for protein synthesis at the ribosome sites. It is synthesized enzymatically on a single strand of DNA used as a template. Thus, messenger RNA contains bases that are complementary to those in the corresponding strand of DNA.

Transfer RNA

These molecules are short-chain polynucleotides (75 to 90 monomer units) that transport amino acids to the messenger RNA at the ribosome sites. It appears that each type of transfer RNA molecule is responsible for the transport of one type of amino acid, or several different transfer RNAs may be involved in the transfer of each type of amino acid. The amino acids appear to be bound to the polynucleotide

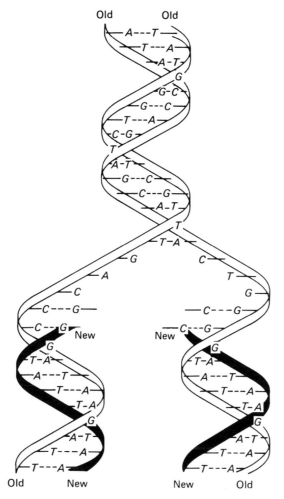

Fig. 8.7 Schematic representation of the replication mechanism of DNA. Both of the old strands of DNA can serve as a template to orient monomer units from a pool of triphosphates. Polymerization is effected under the influence of the enzyme DNA synthetase. [From R. J. Light, *A Brief Introduction to Biochemistry* (New York: Benjamin, 1968, Chap. VI.]

through ester linkages formed between the carboxylic acid residue and the 2'- or 3'-hydroxyl group of a *terminal* adenylic acid residue.

Ribosomal RNA

As the name suggests, this type of polynucleotide is found in the ribosomes, probably as a single-strand polymer. Its function is not understood. Ribosomes consist of RNA associated with protein.

Viruses

Viruses are supramolecular structures that contain a polynucleotide core surrounded by a protein coat. Depending on the type of polynucleotide present, they may be classified into RNA or DNA viruses. They function by invasion of a

Fig. 8.8 Drawing depicting part of the structure of the tobacco mosaic virus as determined by X-ray structure analysis. The RNA chain is depicted as the internal helix surrounded by protein subunits. [From A. Klug and D. L. D. Casper, *Adv. Virus Res.*, **7**, 225 (1960).]

living cell and displacement of the host messenger RNA molecules from the ribosomes. The virus polynucleotide then forms the template for the synthesis by the host cell of the protein coat for the virus and also induces the synthesis of more virus polynucleotide. Many viruses can be crystallized and studied by X-ray diffraction or electron microscopy (Figure 8.8).

Polymeric Coding in Polynucleotides

It is well known that the linear coding sequence of the bases along a DNA molecule is ultimately responsible for the linear sequence of amino acid residues in a particular protein. As we have seen, the base sequence in DNA is "transcribed" to a complementary base sequence along a chain of messenger RNA, which in turn acts as the real template on which the protein is constructed. The exact nature of coding along a polynucleotide chain required to specify a particular amino acid has been worked out. Only three base molecules in sequence are needed to specify a particular amino acid. Thus, the GCU sequence along a messenger RNA chain is the code for alanine, CUG for leucine, and so on. However, the system is "degenerate" in the sense that *several* similar nucleotide "words" are codes for *the same* amino acid. This is believed to provide a safety factor in case a mutation at one point along a polynucleotide chain interrupts the message. Triplet code messages exist for the termination of a protein chain and the start of the next chain, but no "punctuation" codes exist *between* the triplet codes for individual amino acids along a chain. Hence, an advantage exists if a mutation in one codon (a C, A, G, or U) of the triplet yields a sequence that is the code for the same amino acid or at least a similar one. Thus, following a mutation, the degeneracy of the system may prevent the synthesis of the wrong protein or, at the worst, it may provide a very minor change in the amino acid composition that could have an evolutionary survival value. It will be clear that investigations of the effects of deliberate mutational changes to the nucleotide sequence has had, and will continue to have, a profound influence on research in this area.

Synthesis of Polynucleotides

Synthesis within the living cell

DNA is synthesized within the living cell from nucleotide monomers under the influence of the remarkable enzyme, DNA polymerase. The process occurs only in the presence of preformed, single-strand DNA,[1] which acts both as a template and as a reaction "primer." DNA polymerase as isolated from *Escherichia coli* bacteria is a single-chain polypeptide containing about 1000 peptide residues. The enzyme can not only build polynucleotide chains at a rate of about 1000 residues per minute, but it can also catalyze the hydrolytic breakdown of polynucleotide chains when excess pyrophosphate is present. A second enzyme, DNA Ligase, is not capable of catalyzing the synthesis of complete DNA molecules, but serves to repair breaks in the polynucleotide chains or to cyclize linear bacterial polynucleotides to form circular DNA. Other enzymes appear to be capable of breaking the crosslinks that are formed when DNA chains are exposed to ultraviolet irradiation.

The RNA-type polynucleotide chains are synthesized from mononucleotide monomers by the enzyme. RNA polymerase. In this case, DNA serves as the template and chain primer in the presence of magnesium ion. Once again, the same enzyme can cleave the polynucleotide chains when high concentrations of pyrophosphate are present.

Laboratory synthesis of polynucleotides

The enzyme, polynucleotide phosphorylase, has been employed in laboratory experiments to synthesize both homopolymers and copolymers from mononucleotides. These synthetic polynucleotides were used in the critical experiments that led to the unravelling of the triplet code employed for protein synthesis.

The nonenzymatic construction of polynucleotide chains has been accomplished by the use of condensation reactions carried out under the influence of dicyclohexylcarbodiimide. One serious problem in syntheses of this type is the need to "block" those active positions on the monomer molecules that could participate in crosslinking. Short blocks of oligonucleotide units synthesized in this way can then be linked together to form longer chains.

---

[1] Double-helical DNA is apparently completely inactive as a polymerization template.

Reactions of Polynucleotides

Coiling and uncoiling reactions

The double-helix structure of DNA is held together largely by hydrogen bonds between the base pairs. Hydrogen bonds are relatively weak—they can often be broken by raising the temperature in an aqueous system, by the addition of reagents that can more readily coordinate to oxygen or nitrogen than can hydrogen, or by the addition of alcohols, ketones, or urea. For these reasons, the DNA double helix can unwind and separate into individual single strands in a denaturation process that is known as "melting." The melting phenomenon can be monitored by the sudden decrease in viscosity or the increase in ultraviolet absorption at 260 nm that occurs on denaturation. DNA samples from different cell sources have different melting temperatures.

Perhaps the most extraordinary feature of the coiling–uncoiling process is the ease with which *re*coiling occurs to form the double helix after the denatured DNA has been returned to its original environmental conditions. If the separation of the strands has proceeded to the point at which about 12 or more residues are still helically coiled, then recoiling of the remaining segments is extremely rapid. If, on the other hand, *total* separation of the strands has occurred, the reformation of the double helix is much slower. Presumably, random collisions between the chain ends of complementary strands are needed before the end structure can align itself to generate the duplex helical structure.

Metal ions have a profound effect on the coiling–uncoiling process. Metal ions that preferentially bind to the P—OH components of the backbone stabilize the helical structure and raise the melting point. This effect is a direct result of a neutralization of P—O$^\ominus$ charge. On the other hand, divalent metal ions, such as Cu(II) or Zn(II), that bind to the bases, generate spurious coordination crosslinks that assist unwinding at elevated temperatures, lower the melting point, but at the same time assist in the rewinding process when, for example, the temperature is again lowered. Effects such as these have been used to explain the toxicity, mutagenicity, or anticancer properties of some metals.

Mutations

Any change in the nucleotide sequence or in the chemical composition of a particular mononucleotide will upset the replication pattern and alter the structure of the protein derived from that template. Occasional errors in the normal replication process can generate defective DNA or RNA. Moreover, living cells are constantly exposed to environmental influences that can damage the polynucleotides. Cosmic-ray bombardment may cause chain cleavage. Peroxides formed within the cell, following cosmic ray, X-ray, or ultraviolet irradiation, may modify individual nucleotide residues or crosslink the chains. Many chemical reagents can react with DNA or RNA.

Simple cleavage or crosslinking of the polynucleotide chains may not constitute a permanent mutation since enzymes exist that can rejoin the ends of cleaved chains and can break crosslinks. Furthermore, the replacement in the DNA double helix of one nucleotide complementary pair by another may be a relatively benign change. Ultimately, it could result in the synthesis of a protein molecule that has one incorrect amino acid residue in the chain—possibly a harmless modification or one that could be beneficial in a changing environment.

However, if an *extra* nucleotide residue becomes incorporated into the DNA chain or if a nucleotide is missing, the DNA code will be severely misread beyond this point. Hence, proteins synthesized from such mutants will interfere with the biochemical mechanisms of the cell, causing death of the cell or malignant changes.

Hydrolysis of polynucleotides

DNA and RNA are degraded by specific enzymes, and such degradations are used to analyze the base sequence in polynucleotides. However, nonenzymatic hydrolysis also yields valuable structural information. For example, the hydrolysis of DNA with dilute acid causes removal of the side group bases without cleavage of the backbone bonds. DNA is stable to basic hydrolysis, but RNA can be hydrolyzed by base. The use of enzymatic and nonenzymatic hydrolysis as a sequencing technique for polynucleotides is a more complex problem than, for instance, the sequencing of a protein, because the amount of *detail* that can be derived from the hydrosylate is much less. There are only four principal bases involved in poly-nucleotide structures, compared to the 20 or so amino acids found in proteins.

STUDY QUESTIONS

1. Cellulose is one of the few polymers that form the starting point for a "substitutive" route to other derivatives. Suggest other reactions, not mentioned in this chapter, that might be used to modify cellulose and broaden its usefulness.

2. In your opinion, could starch be used as a starting point for the synthesis of useful, nonfood polymeric derivatives by the use of substitutive techniques?

3. Taking into account the helical structure of amylose (Figure 8.1) and its capacity to accommodate iodine molecules, what other small molecules or polymers might be induced to occupy the helical channels? What experimental conditions can you devise in order to make such adducts?

4. What advantages or disadvantages can you foresee for the introduction into a synthetic polyamide system of small amounts of histidine, lysine, aspartic acid, cysteine, or methionine units as copolymer residues?

5. Why does a conventional synthetic polymer, such as Nylon 66, occupy an extended zigzag conformation in the solid state, when α-keratins form a helical array?

6. Explain why the steam pressing of cotton or wool textiles is used to remove or introduce creases. Speculate on the reactions that take place when human hair is "permanently waved."

7. In your opinion, could synthetic macromolecules be constructed that would possess the intricately folded structures seen on globular proteins? If so, how would such polymers be synthesized? What uses can you anticipate for such materials?

8. What functions are served by the protein chains in myoglobin and hemoglobin? To what extent might these chains be modified without loss of the oxygen-carrying ability of the system? Could synthetic macromolecules be used in place of the protein chains?

9. What other polymers might be used as the stationary phase in the Merrifield synthesis?

10. A number of reactions of proteins are mentioned in this chapter. Discuss the possible uses of these reactions to modify synthetic polyamides.

11. Discuss the prospect that synthetic polymers might be prepared that could function as templates (possibly stereoregular templates) for the synthesis of new polymer molecules in a manner reminiscent of polynucleotide reactions.

SUGGESTIONS FOR FURTHER READING

ALLEN, T. C., AND J. A. CUCULO, "Cellulose Derivatives Containing Carboxylic Acid Groups," *J. Polymer Sci. (D) (Macromol. Rev.)*, **7**, 189 (1973).

ASPINALL, G. O., *Polysaccharides*. New York: Pergamon Press, 1970.

BLOOMFIELD, V. A., "Hydrodynamic Properties of DNA," *J. Polymer Sci. (D) (Macromol. Rev.)*, **3**, 255 (1968).

CHIEN, J. C. W., "Solid State Characterization of the Structure and Property of Collagen," *J. Macromol. Sci.—Rev. Macromol. Chem.*, **C12**, 1 (1975).

COLVIN, J. R., "The Structure and Biosynthesis of Cellulose," *CRC Crit. Rev. Macromol. Sci.*, **1**, 47 (1972–73).

DICKERSON, R. E., AND I. GEIS, *The Structure and Action of Proteins*. New York: Harper & Row, 1969.

EICHHORN, G. I., (ed.), *Inorganic Biochemistry*, Vols. 1 and 2, New York: Elsevier, 1973.

EISENBERG, H., *Biological Macromolecules and Polyelectrolytes in Solution*. Oxford: Clarendon Press, 1976.

GAL'BRAIKH, L. S., AND Z. A. ROGOVIN, "Chemical Transformations of Cellulose," *Adv. Polymer Sci.*, **14**, 87 (1974).

GREENWOOD, C. T., AND E. A. MILNE, *Natural High Polymers*. Edinburgh: Oliver & Boyd, 1968.

HERMANS, J., JR., D. LOHR, AND D. FERRO, "Treatment of the Folding and Unfolding of Protein Molecules in Solution According to a Lattice Model," *Adv. Polymer Sci.*, **9**, 299 (1972).

JONES, D. W., (ed.), *Introduction to the Spectroscopy of Biological Polymers*. New York: Academic Press, 1976.

KRASSIG, H. A., AND V. STANNETT, "Graft Co-polymerization to Cellulose and Its Derivatives," *Adv. Polymer Sci.*, **4**, 111 (1965).

LEHNINGER, A. L., *Biochemistry*. New York: Worth Publishers, 1970.

SCHMITT, F. O., *Macromolecular Specificity and Biological Memory*. Cambridge, Mass.: MIT Press, 1962.

SCHUERCH, C., "The Chemical Synthesis and Properties of Polysaccharides of Biomedical Interest," *Adv. Polymer Sci.*, **10**, 173 (1972).

SPIRIN, A. S., *Macromolecular Structure of Ribonucleic Acids*. New York: Reinhold, 1964.

STEINER, R. F. AND R. L. BEERS, JR., *Polynucleides: Natural and Synthetic Nucleic Acids*. New York: Elsevier, 1961.

TIMASHEFF, S. N., AND G. FASMAN (eds.), *Structure and Stability of Biological Macromolecules*. New York: Dekker, 1969.

VEIS, A. (ed.), *Biological Polyelectrolytes*. New York: Dekker, 1970.

VEIS, A., *The Macromolecular Chemistry of Gelatin*. New York: Academic Press, 1964.

WALTON, A. G., AND J. BLACKWELL, *Biopolymers*. New York: Academic Press, 1973.

WATT, I. C., "Copolymers of Naturally Occurring Macromolecules," *J. Macromol. Sci—Rev. Macromol. Chem.*, **C5**, 175 (1970).

WHISTLER, R. L. (ed.), *Industrial Gums, Polysaccharides and Their Derivatives*, 2nd ed. New York: Academic Press, 1973.

WHISTLER, R. L., AND C. L. SMART, *Polysaccharide Chemistry*. New York: Academic Press, 1953.

Reactions
of Synthetic Polymers

Synthetic high polymers undergo a variety of reactions either when they are brought into contact with chemical reagents or when heated to high temperatures. For convenience, these reactions will be divided into two classes—those which involve reactions of the polymer side groups, and those which involve additions to or cleavage of the main chain.

REACTIONS INVOLVING THE SIDE GROUPS

General Considerations

In Chapters 2 to 7 we considered the ways in which different polymers can be prepared by the polymerization of different monomers or cyclic compounds. However, there remains one further method that may be used to prepare different synthetic polymer structures. This involves the chemical modification of preformed high polymers. In theory, the vast arsenal of conventional organic and inorganic reactions could be employed to introduce new functional groups into a polymer. In practice, limitations exist. Some reactions which proceed rapidly and efficiently between small molecules cannot be applied effectively to high polymers.

There are two reasons for this. First, high-polymer molecules are randomly coiled in solution. This coiling may sterically inhibit the approach of reagent molecules to every monomer residue in the chain. Thus, reactions between a reagent and a polymer may be incomplete or unacceptably slow. Second, the introduction of one new substituent group may retard the introduction of a second group at adjacent monomer residue sites because of neighboring group polar or steric effects.

On the other hand, many polymer reactions are known which proceed at similar or even faster rates than the analogous nonpolymeric reactions. Enzymatic processes provide striking examples of this effect. In such cases it appears that the role of the polymer is to generate a more favorable collision efficiency between the reacting species so that the reaction becomes one between a highly mobile reagent and a relatively immobile substrate rather than one between two highly mobile reagents. As an analogy, it is easier to fire a bullet at a moving elephant than it is to fire a bullet at a moving bullet. Moreover, the initial introduction of a new substituent group may serve to catalyze the introduction of more new units at adjacent sites along the chain.

An enormous number of different reactions have been attempted with polymer molecules, and in this book we can review only a few examples. A few substitution reactions on polymers are discussed elsewhere in this volume (especially in Chapters 7 and 8). The following examples have been chosen to illustrate briefly additional possibilities.

Hydrolysis of Side-Group Structures

The hydrolyses of poly(vinyl esters) (1) or poly(vinyl amides) to a poly(vinyl carboxylic acid) (2) (reaction 1) are processes which differ in one important respect from the conventional hydrolyses of small molecule esters or amides to carboxylic

$$
\left[\begin{array}{c} H \\ | \\ \text{CH}_2-\text{C}- \\ | \\ \text{C}=\text{O} \\ | \\ \text{O} \\ | \\ \text{R} \end{array} \right]_n \xrightarrow[\text{Base}]{\text{H}_2\text{O}} \left[\begin{array}{c} H \\ | \\ \text{CH}_2-\text{C}- \\ | \\ \text{C}=\text{O} \\ | \\ \text{O}^{\ominus} \end{array} \right]_n \tag{1}
$$

$$\mathbf{1} \qquad\qquad \mathbf{2}$$

acids. Hydrolysis of a few ester or amide linkages yields a polymer in which the remaining uncharged groups are flanked by charged carboxylate groups (3).

$$
\begin{array}{ccccc}
 & H & & H & & H \\
 & | & & | & & | \\
-\text{CH}_2-\text{C}- & \text{CH}_2- & \text{C}- & \text{CH}_2- & \text{C}- \\
 & | & & | & & | \\
 & \text{C}=\text{O} & & \text{C}=\text{O} & & \text{C}=\text{O} \\
 & | & & | & & | \\
 & \text{O}^{\ominus} & & \text{OR} & & \text{O}^{\ominus}
\end{array}
$$

$$\mathbf{3}$$

The presence of these negatively charged groups would be expected to retard the approach of the reagent (OH^-) to the adjacent ester or amide groups, with a corresponding decrease in the reaction rate.

The hydrolysis of polymethacrylamide (4) appears to follow this pattern (reaction 2). This polymer undergoes hydrolysis in basic media to form a product

$$
\left[\begin{array}{c} CH_3 \\ | \\ -CH_2-C- \\ | \\ C=O \\ | \\ NH_2 \end{array}\right]_n \xrightarrow[OH^-]{H_2O} \begin{array}{ccc} CH_3 & CH_3 & CH_3 \\ | & | & | \\ -C-CH_2-C-CH_2-C- \\ | & | & | \\ C=O & C=O & C=O \\ | & | & | \\ O^{\ominus} & NH_2 & O^{\ominus} \end{array} \tag{2}
$$

$$4 \qquad\qquad\qquad 5$$

with carboxylate side groups. However, all the amide groups cannot be removed, and this effect has been ascribed to the inability of hydroxide ion to penetrate the polar field of the flanking carboxylate units (5).

However, several cases are known where the presence of a charged carboxylate ion actually *accelerates* the rate of hydrolysis of a neighboring ester function. For example, copolymers of acrylic acid and p-nitrophenyl methacrylate are hydrolyzed much more rapidly than are the p-nitrophenyl esters of small molecule carboxylic acids. This is because the carboxylate ion itself can attack the adjacent ester function (6). Such intramolecular effects should be dependent on geometric factors.

$$
\begin{array}{cc} R & H \\ | & | \\ -CH_2-C-CH_2-C- \\ | & | \\ O=C & C=O \\ | & | \\ O & O \\ | \\ R' \end{array}
$$

$$6$$

It has, in fact, been shown that isotactic poly(methyl methacrylate) undergoes hydrolysis more rapidly than do the syndiotactic or atactic modifications.

Substitutive Halogenation

Small molecule aliphatic compounds can be halogenated to yield fluorocarbons or chlorocarbons. High polymers undergo similar reactions. Polyethylene can be fluorinated in a heterophase system in the presence of fluorine gas diluted with nitrogen. Either polyethylene powder or a film of the polymer is used. Fluorination takes place at the polymer surface to yield a material that has fluorocarbon-type properties. Substitutive chlorination and bromination of polyethylene have also been reported, but polypropylene and polyisobutylene undergo chain degradation during chlorination. The photochlorination of poly(vinyl chloride) apparently yields poly(1,2-dichloroethane).

Nitration of Polystyrene

Just as benzene can be nitrated to form nitrobenzenes, so can polystyrene be nitrated by mixtures of nitric and sulfuric acids. The poly(nitrostyrene) (**7**) formed in this way can then be reduced to poly(aminostyrene) (**8**) by conventional tech-

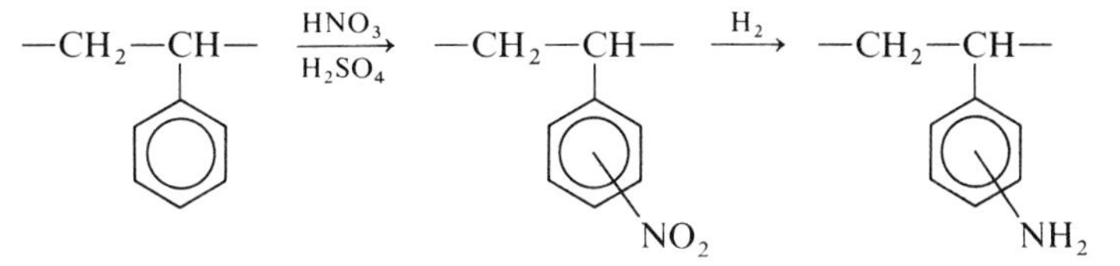

$$-CH_2-CH- \quad \xrightarrow[H_2SO_4]{HNO_3} \quad -CH_2-CH- \quad \xrightarrow{H_2} \quad -CH_2-CH- \tag{3}$$

NO₂   NH₂
7   **8**

niques (sequence 3). The sulfonation of polystyrene usually results in the formation of crosslinked materials, rather than simple sulfonic acid derivatives.

Chloromethylation of Polystyrene

Polystyrene is chloromethylated in the presence of chloromethyl ether and Friedel–Crafts catalysts such as aluminum chloride or zinc chloride. The overall reaction is shown in equation (4). Lightly chloromethylated polystyrene resins are used as a

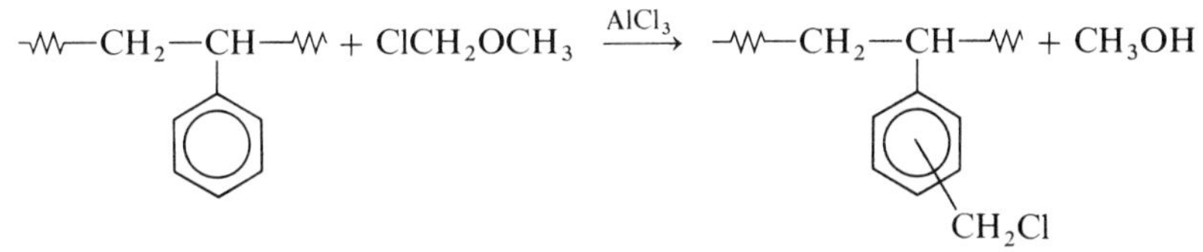

$$\text{-W-}CH_2-CH\text{-W} + ClCH_2OCH_3 \quad \xrightarrow{AlCl_3} \quad \text{-W-}CH_2-CH\text{-W} + CH_3OH \tag{4}$$

CH₂Cl

substrate in the controlled laboratory synthesis of proteins (see Chapter 8). The chloromethylation reaction can be forced to high conversions, with the ultimate product corresponding closely to poly(vinylbenzyl chloride) (**9**). Such polymers can be quaternized with tertiary amines to yield water soluble polymers (**10**) (reaction 5). Alternatively, as shown in reaction (6), the chloromethylated polystyrene can be treated with a phosphide to introduce phosphinic ligands (**11**) for

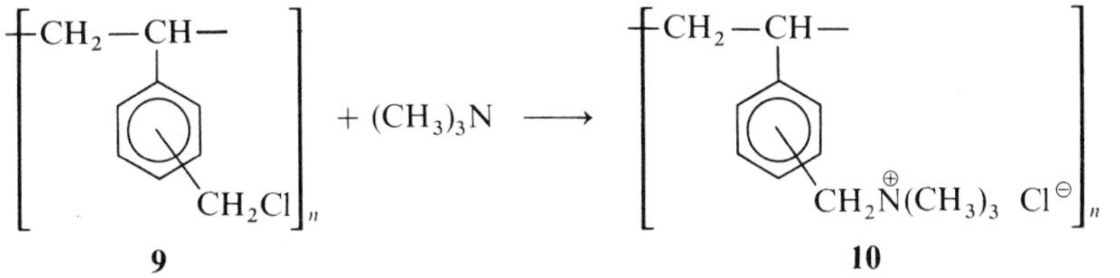

$$\left[\begin{array}{c} CH_2-CH- \\ \\ CH_2Cl \end{array}\right]_n + (CH_3)_3N \longrightarrow \left[\begin{array}{c} CH_2-CH- \\ \\ CH_2\overset{\oplus}{N}(CH_3)_3 \ Cl^{\ominus} \end{array}\right]_n \tag{5}$$

9   **10**

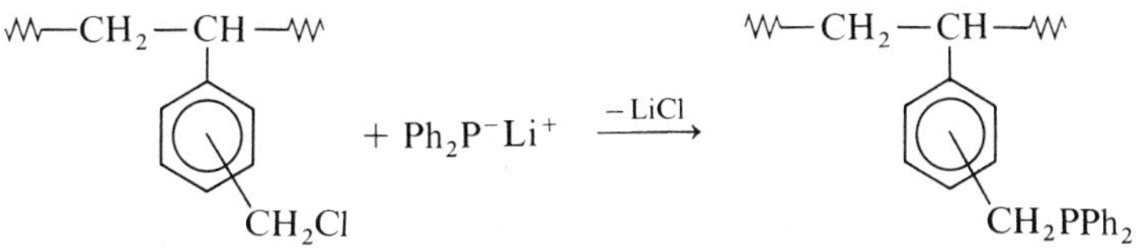

$$\text{W-}CH_2-CH\text{-W} \qquad\qquad \text{W-}CH_2-CH\text{-W}$$

$$+ Ph_2P^-Li^+ \quad \xrightarrow{-LiCl} \tag{6}$$

CH₂Cl   CH₂PPh₂
11

the binding of metal coordination complexes. These have been used as polymer bound catalyst systems, as discussed in the next section.

Polymer-bound Catalysts

Many important organic chemical reactions require the presence of homogeneous organometallic catalysts. However, large-scale manufacturing processes based on these reactions suffer from the extreme difficulty of recovery of the catalyst after the reaction is complete. Because such catalysts often contain expensive rare metals, the loss of the catalyst, or even a small part of it, constitutes an unacceptable cost.

Crosslinked polymers are insoluble in organic or aqueous media. Hence, if the catalyst can be bound to a crosslinked polymer substrate without loss of the catalytic activity, then recovery of the catalyst may be accomplished simply by removal of the polymer particles by filtration.

One of the commonest methods for binding a catalyst to a polymer is through a phosphine ligand that is itself bound to crosslinked polystyrene. The crosslinked polymer can be prepared by copolymerization of styrene with a small amount of divinylbenzene. The phosphine is attached to the polymer via chloromethylated polystyrene, as discussed in the last section. A typical system is shown in structure **12**, in which the Wilkinson hydrogenation catalyst, $(Ph_3P)_3RhCl$, is bound to a

$$-\text{\Large\textasciitilde}-CH_2-CH-\text{\Large\textasciitilde}$$

$$CH_2-\overset{\displaystyle Ph}{\underset{\displaystyle Ph}{P}} \cdots RhCl(PPh_3)_2$$

12

pendent phosphine ligand attached to the polymer substrate. A wide variety of catalysts containing iridium, cobalt, chromium, nickel, and titanium as well as rhodium have been bound to polystyrene not only through phosphine ligands, but also as π-complexes involving the aromatic rings.

A highly swelled but insoluble polystyrene particle will possess catalytic sites *within* the interior of each polymer "bead." Hence, reactant molecules will penetrate the swelled matrix, react, and diffuse out almost as readily as in a homogeneous system. To a large extent, the catalyst molecules will be separated from each other and will function by homogeneous rather than heterogeneous mechanisms. It should be noted that phosphine–metal coordination bonds are labile. Indeed, this may be a prerequisite for catalytic activity. Unfortunately, this characteristic means that small amounts of the catalyst may be lost from the substrate during many reaction cycles or in a continuous flow reactor.

It should also be noted that significant advances are occurring in the binding of enzymes to synthetic polymer substrates. The goal is to carry out biochemical reactions on a large manufacturing scale by allowing biological "feedstocks" to pass through columns of immobilized enzymes.

Phosphazene Polymer Substitution

The synthesis of organophosphazenes by the replacement of halogen in poly(dichlorophosphazene) by organic groups has been discussed in Chapter 7. Here it is sufficient to observe that these substitution processes are quite rapid compared to the substitution reactions of most organic polymers. Moreover, the substitution is essentially complete. Hence, it must be concluded that chain coiling in solution has only a minimal retarding influence on this particular substitution.

Graft Polymer Formation

Graft polymerization takes place when a preformed polymer provides initiation sites for the growth of branches formed from a new monomer. Thus, a graft copolymer can be represented by structure **13**.

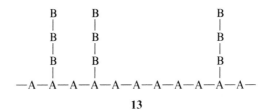

13

A number of different techniques can be used to form graft copolymers. Most of these fall into one of the following four categories.

1. A free-radical catalyst may be used to abstract a hydrogen atom from an organic polymer chain. The polymeric radical then initiates growth of a vinyl monomer.

2. A propagating vinyl polymer chain may abstract a hydrogen atom from a preformed polymer during a chain transfer step. The polymeric radical then initiates chain growth of the monomer.

3. The polymer can be irradiated by ultraviolet, gamma, or X-irradiation to generate free-radical sites along the chain. In the presence of a monomer, these free-radical sites initiate graft polymerization (see Chapter 3).

4. Two preformed polymers are blended together mechanically or subjected to high-speed stirring in solution. The chain backbones are mechanically cleaved by this process. Radical recombination and abstraction processes generate block and graft copolymers.

The techniques described in categories 1 to 3 generally lead to the formation of homopolymer from the monomer as well as to the synthesis of graft copolymers. It should also be noted that polymers which contain unsaturated substituent groups are particularly suitable for graft polymerization reactions. Experimentally, the likelihood of achieving a successful graft polymerization depends on the nature of the polymer, the monomer, and the catalyst. Some monomers will not polymerize from the radical sites of certain polymers. One example of a system that readily gives graft copolymers is the one formed from polystyrene and methyl methacrylate monomer in the presence of benzoyl peroxide.

Reactions Involving Decomposition of Substituent Groups

Polymers that contain nitrile, carboxylic acid, ester, or chloro side units decompose by processes that involve an initial reaction of the side groups at elevated temperatures. In a few cases, the reaction of the side group may yield a more stable polymer than the starting material. Polyacrylonitrile is a particularly good example. This polymer darkens in color when heated at 175°C. The change is attributed to intramolecular linkage of the nitrile groups to generate a conjugated ladder polymer (**14**) (reaction 7).

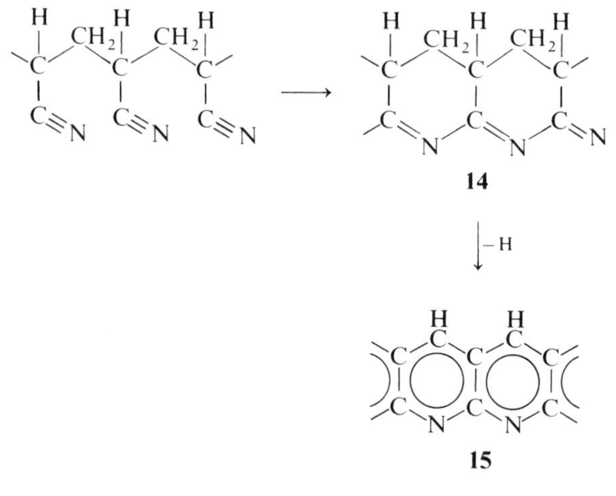

(7)

14

15

Pyrolysis at higher temperatures increases the aromaticity by oxidative removal of hydrogen atoms (**15**). Polymers that contain pendent carboxylic acid groups, such as poly(methacrylic acid), decompose to form anhydrides (**16**), as shown in reaction

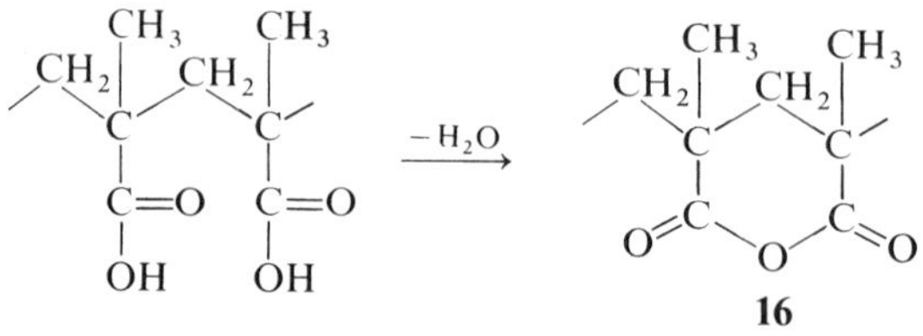

(8)

16

(8). Such polyanhydrides are more stable thermally than the polymers from which they are derived.

On the other hand, poly(vinyl acetate) (**17**) decomposes thermally to liberate acetic acid (9). The colored residue is believed to have a polyacetylene type of structure (**18**). Colored polymers are also formed when poly(vinyl chloride) is

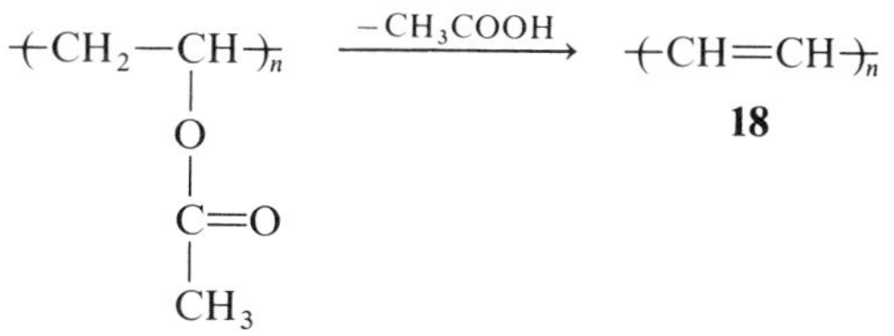

$$+CH_2-CH\rightarrow_n \xrightarrow{-CH_3COOH} +CH=CH\rightarrow_n$$

with structure **17** having the pendant group:

$$\begin{array}{c} | \\ O \\ | \\ C=O \\ | \\ CH_3 \end{array}$$

18 (9)

17

heated at temperatures near 130°C. Hydrogen chloride is liberated and, again, polyacetylene-like residues are formed (reaction 10). The loss of hydrogen chloride

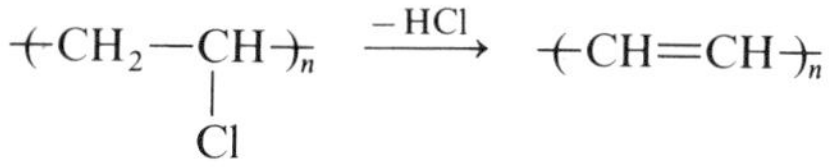

$$+CH_2-CH\rightarrow_n \xrightarrow{-HCl} +CH=CH\rightarrow_n$$

with pendant Cl on **17**:

$$\begin{array}{c} | \\ Cl \end{array}$$

(10)

is believed to take place first at the chain ends and then to be propagated by a free-radical process along the chain from these terminal sites. Exposure of the polyunsaturated polymer to the air causes a fading of the color. This has been attributed to air oxidation of the unsaturated structures.

REACTIONS INVOLVING THE MAIN CHAIN

Addition Reactions

Chlorine can be added across the double bonds of unsaturated polymers such as natural rubber or polybutadiene. Natural rubber is dissolved in carbon tetra-chloride, and chlorine is bubbled into the system until the polymer contains 66% or more of chlorine. The chlorinated rubber is nonflammable. Mechanistically, the reaction is very complex, with halogen addition, substitution, and skeletal cycliza-tion reactions taking place. By contrast, polybutadiene apparently reacts with chlorine almost exclusively by an addition mechanism, as shown in reaction (11).

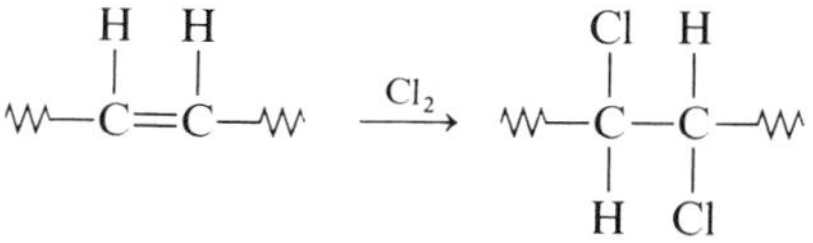

$$\begin{array}{ccc} H \ \ H & & Cl \ \ H \\ | \ \ | & Cl_2 & | \ \ | \\ \text{w}-C=C-\text{w} & \xrightarrow{\quad} & \text{w}-C-C-\text{w} \\ & & | \ \ | \\ & & H \ \ Cl \end{array} \qquad (11)$$

Hydrogen chloride adds across the double bonds of polyisoprene in chloroform solution to yield a material known as "rubber hydrochloride."

The hydrogenation of unsaturated polymers, such as polybutadiene, has been accomplished with the use of nickel or noble metal catalysts. Saturation of all the double bonds in the polymer does not normally occur. Increasing the degree of hydrogenation leads principally to an increase in the polymer crystallinity. However, drastic hydrogenation, especially at elevated temperatures causes skeletal cleavage. This has been suggested as a method for the conversion of old rubber tires to gasoline.

Hydrolytic Chain Cleavage

We have already seen, in Chapter 8, that polysaccharides and proteins undergo hydrolytic skeletal breakdown. Some synthetic polymers behave similarly.

Polyesters can be chain-cleaved by hydrolysis. Poly(ethylene terephthalate), for example, can be hydrolyzed in acidic, neutral, or basic media. In acidic or neutral media the overall reaction can be summarized in equation (12). Because the reaction

$$-\text{\footnotesize W}-R-\overset{\displaystyle O}{\overset{\|}{C}}-O-R'-\text{\footnotesize W} \xrightarrow{\text{H}_2\text{O}} \text{\footnotesize W}-R-\overset{\displaystyle O}{\overset{\|}{C}}-OH + HO-R'-\text{\footnotesize W}- \qquad (12)$$

is acid-catalyzed, the prospect exists that the neutral hydrolysis may be speeded up by the formation of carboxylic acid end groups. The basic hydrolysis of this polymer takes place even in heterogeneous systems, and this can generate problems when thin fibers of the polymer are passed through alkaline dye baths. Synthetic polyamides are also susceptible to hydrolytic cleavage, especially in acidic media.

Enzymatic Degradation of Synthetic Polymers

One of the main advantages of synthetic polymers over naturally occurring polymeric materials such as cellulose or leather is their resistance to bacterial or fungal attack. Hence, the synthetic materials are, in general, more permanent. However, a few synthetic polymers are susceptible to biological breakdown and it is clearly important, from an applications point of view, to know which polymers are the most labile in a biological environment. Furthermore, ecological considerations have focused attention on the need for polymers that are deliberately designed to degrade when discarded.

Polyurethanes in particular (see Chapter 2) appear to be susceptible to microbial attack. Polyether polyurethanes are more resistant to biological degradation than are polyester polyurethanes. The precise mechanisms of these degradations are not fully understood. However, susceptibility to biological degradation is the exception rather than the rule. Polyamides, fluorocarbons, polyethylene, polypropylene, polyfluorocarbons, polycarbonates, and many other polymer systems appear to be resistant to biological attack. The possibility always exists that mutant bacteria or fungi may arise or be developed that could attack most synthetic polymer systems, but that phenomenon has not yet been observed.

Oxidation Reactions

Many synthetic organic polymers are oxidized in contact with the atmosphere. At room temperature in the absence of light the reaction may be very slow. But at elevated temperatures or during exposure to ultraviolet light the rate of oxidation is often quite rapid. Appreciable decomposition of polyethylene occurs when the material is exposed to outdoor daylight for less than 2 years, and the preliminary effects of photooxidation are evident after only a few months. Polypropylene is even more susceptible to photoxidative breakdown. However, polyisobutylene is more stable under these conditions than is polyethylene, and the same is true for polystyrene. Mechanical stress or contact of the polymer with radical-producing reagents may accelerate the oxidation process. Oxidation of a polymer usually leads to increasing brittleness and a deterioration in strength. Clearly, the utility of a polymer for a particular application may depend on its resistance to oxidation.

The oxidative degradation of an organic polymer generally proceeds through free-radical reactions. Free radicals are formed by the thermal or photolytic cleavage of bonds. The radicals then react with oxygen to yield peroxides and hydroperoxides by processes such as those shown in reactions (13) to (19) (here $R\cdot$ represents a polymer radical). Such reactions lead to both chain cleavage and to crosslinking.

$$RH \longrightarrow R\cdot + H\cdot \tag{13}$$

$$RR \longrightarrow 2R\cdot \tag{14}$$

$$R\cdot + O_2 \longrightarrow ROO\cdot \tag{15}$$

$$ROO\cdot + RH \longrightarrow ROOH + R\cdot \tag{16}$$

$$ROOH \longrightarrow RO\cdot + HO\cdot \tag{17}$$

$$RO\cdot + RH \longrightarrow ROH + R\cdot \tag{18}$$

$$HO\cdot + RH \longrightarrow H_2O + R\cdot \quad \text{etc.} \tag{19}$$

Crosslinking can be visualized as resulting from the combination of radical sites on adjacent chains. Chain cleavage can occur either by primary homolytic skeletal cleavage or by a backbiting attack by a terminal radical unit on its own chain.

Polymers such as polyisoprene or polybutadiene, which contain unsaturated linkages, can be attacked by atmospheric ozone as well as by oxygen. Again, free-radical cleavage and crosslinkage processes are responsible for the loss of advantageous polymer properties following oxidation. Poly(vinyl chloride) is especially sensitive to oxidation and dehydrohalogenation reactions. On the other hand, fluorine-containing organic polymers are surprisingly stable.

Various compounds are added to polymers to retard free-radical-induced decompositions. These additives include ultraviolet absorbers such as substituted benzophenones, which reduce the rate of photolytic oxidation. Phenolic compounds are added as radical chain terminators. Carbon black functions both as an ultraviolet screening agent and as a chain terminator. Sulfur compounds are added as peroxide deactivation reagents. Still other additives are employed to inactivate traces of metals which can participate in radical formation.

Polymers that contain unsaturated organic groups can be deliberately oxidized to form epoxy polymers. For example, polybutadiene or polyisoprene yield polymeric expoxides when treated with hydrogen peroxide or aliphatic peracids. The overall reaction is depicted by (20).

$$-\overset{|}{C}=\overset{|}{C}- \quad \xrightarrow{\text{Peroxide}} \quad -\overset{|}{C}\overset{|}{\underset{\diagdown O \diagup}{-}}\overset{|}{C}- \tag{20}$$

High-Temperature Degradation Reactions

Most organic polymers decompose when heated to moderate or high temperatures. It is for this reason that few synthetic polymers can be used for long periods of time at temperatures above 150 to 200°C. This fact is largely responsible for the persistent use of metals and ceramics for many applications, even though synthetic polymers may be cheaper and, in some cases, stronger on a weight-for-weight basis.

The thermal instability of most polymers has perplexed many investigators, especially those who have attempted to predict thermal stabilities on the basis of bond strengths or by comparisons of polymers with low-molecular-weight model compounds. Polyethylene, poly(vinyl chloride), and many polyacrylates decompose at least 200°C below the corresponding decomposition temperatures of short-chain paraffins, chloroparaffins, or simple esters.

The thermal instability of organic high polymers can be traced to one or more of the following reasons: (1) degradation of a polymer to a low-molecular-weight compound is favored at high temperatures by entropy effects (see Chapter 10); (2) carbon–carbon bonds are relatively weak; (3) carbon–carbon bonds are oxidatively unstable; (4) structural abnormalities, such as branch points, exist along the chains; (5) terminal catalytic sites may initiate depolymerization; (6) a long *chain* of atoms may facilitate decomposition chain reactions, such as monomer "unzipping" processes.

The search for thermally stable high polymers has generally followed two lines of attack. The first involves the synthesis of polymers that contain inorganic elements in the backbone. This approach has been discussed in Chapter 7. The second strategy has been to study the thermal decomposition mechanisms of organic polymers in the hope that the data may suggest ways in which the decomposition mechanism might be inhibited. Here we deal with the second approach.

Three types of thermal decomposition mechanisms can be recognized: (1) depolymerizations—reactions that yield monomer from a vinyl-type polymer; (2) chain cleavage reactions, which yield random chain fragments; and (3) degradation reactions, which are initiated by decomposition of the side-group structures. Mechanism (3) often follows oxidation of the side-group structures, and this was discussed briefly in the preceding section. In practice, mechanisms (1) and (2) blend into one another, with many polymers showing evidence of both processes.

A number of olefin-type polymers decompose thermally to yield the olefin monomer (reaction 21). These depolymerization reactions constitute the reverse of

$$\left(\underset{\underset{\displaystyle CH_2-CH}{}}{\overset{\overset{\displaystyle R}{|}}{}}\right)_n \longrightarrow n\,CH_2\!\!=\!\!\overset{\overset{\displaystyle R}{|}}{CH} \tag{21}$$

the original polymerization process. In fact, such reactions can occasionally be used to "recycle" polymers by offering a method for the clean regeneration of the monomer. In practice, only a few polymers degrade in such a way that they are 100% converted to the monomer. Most polymers yield some monomer and some higher fragments, and a variety of situations are known, which range all the way from pure depolymerization to pure random fragmentation. The following examples will make this clear.

Poly(methyl methacrylate), poly(α-methylstyrene), and poly(tetrafluoroethylene) are three polymers which undergo 100% conversion to the monomer at elevated temperatures. However, the precise circumstances which lead to this effect are different in the three cases. Poly(methyl methacrylate) is a classical case which represents the extreme of a "pure" depolymerization. The depolymerization is a free-radical process that is initiated from the chain ends. Each initiated chain "unzips" rapidly to yield monomer. Thus, at any instant the system contains only unreacted polymer and monomer. Since whole chains apparently depolymerize in one rapid chain reaction, it is said that the "zip length" is large. The nature of the active chain ends is a question for debate. At moderate temperatures (220°C) only half the polymer chains unzip, and higher temperatures (350°C) may be needed to decompose the remaining polymer. Apparently chains which unzip at 220°C are terminated by unsaturated groups, whereas those which depolymerize only at higher temperatures have saturated end groups.

Poly(α-methylstyrene) also yields the monomer by an unzipping process. However, the chain reaction starts not from the chain ends, but from random fragmentation sites. Because of this, the rate of production of monomer depends on the molecular weight of the polymer. Presumably, the longer polymer chains are more likely to incur a random cleavage than are the shorter chains. Poly(tetrafluoroethylene) depolymerizes totally to monomer only at low pressures and high temperatures. At atmospheric pressure, the monomer molecules recombine to form dimer and other species. This polymer is one of the most thermally stable polyolefins known, but even so, it cannot withstand prolonged exposure to temperatures above about 350 to 400°C.

Polystyrene represents a case in which monomer is only one of several species formed by the thermal degradation process at 350°C. In fact, monomer, dimer, trimer, and tetramer are formed in the relative proportions of 40:10:8:1. The thermal breakdown process is believed to be initiated at weak links along the chain. Unsaturated linkages probably constitute the weak points. After the initial chain cleavage occurs at these sites, a free-radical mechanism leads to liberation of the monomer and to an intramolecular back-biting process. This latter process liberates dimer, trimer, and so on, by a mechanism such as the one shown in equation (22).

$$\text{W}-CH_2-CH-CH_2-CH-CH_2-CH\cdot$$
$$\quad\quad | \quad\quad\quad\; | \quad\quad\quad\; |$$
$$\quad\quad Ph \quad\quad\; Ph \quad\quad\; Ph$$

$$\longrightarrow \text{W}-CH_2-CH\cdot + CH_2{=}C-CH_2-CH-CH_2-CH_2$$
$$\quad\quad\quad\quad | \quad\quad\quad\quad\; | \quad\quad\quad | \quad\quad\quad\; |$$
$$\quad\quad\quad\quad Ph \quad\quad\quad Ph \quad\quad Ph \quad\quad\; Ph$$

(22)

Polyethylene yields virtually no monomer. Above about 300°C, the decomposition products form a continuous spectrum of unsaturated hydrocarbons which contain from 1 to at least 70 carbon atoms. Clearly, this suggests a random chain cleavage process. It is believed that the products represent the combined results of chain cleavage initiated at weak links (possibly oxygenated sites) followed by both inter- and intramolecular chain transfer. The existence of chain branch points may facilitate the transfer process. Polypropylene behaves in a very similar manner to polyethylene.

The foregoing comments have applied specifically to olefin addition polymers, but some observations on the thermal behavior of condensation polymers are also appropriate. Polyamides can decompose during melt spinning or molding procedures. Such decomposition, although slight, can affect the physical properties of the polymer. Apparently, the degradation process is initiated by free radicals formed by the homolytic cleavage of $-NH-CH_2-$ skeletal bonds. Water and carbon dioxide are also liberated. The water serves to hydrolyze amide linkages $(-NH-C(O)-)$ to further shorten the chains. Branches are also formed by reaction of terminal $\text{W}-NH_2$ groups with carbonyl units (reaction 23). Ultimately, the branches cause gelation of the molten polymer.

$$-\text{W}-C-\text{W}- \quad\xrightarrow{-H_2O}\quad \text{W}-C-\text{W}$$
$$\quad\quad\;\; || \quad\quad\quad\quad\quad\quad\quad\quad || $$
$$\quad\quad\;\; O \quad\quad\quad\quad\quad\quad\quad\quad N$$
$$\quad\quad\quad\quad\quad\quad\quad\quad\quad\quad\quad\quad\; |$$
$$\quad\quad NH_2 \quad\quad\quad\quad\quad\quad\; CH_2$$
$$\quad\quad\;\; | \quad\quad\quad\quad\quad\quad\quad\quad \text{W}\rfloor$$
$$\quad\quad CH_2$$
$$\quad-\text{W}\rfloor$$

(23)

Polyesters, such as poly(ethylene terephthalate), are fairly stable at temperatures just above the melting point. However, at temperatures between 300 and 550°C, decomposition of this polymer occurs to yield carbon dioxide, acetaldehyde, and terephthalic acid, together with smaller amounts of other decomposition species, such as water, methane, acetylene, and so on.

Electron Beam Depolymerization

Several of the polymers that depolymerize to monomer at high temperatures undergo the same reaction when irradiated with a beam of electrons. This phenomenon is made use of in the microetching of polymers in the preparation of micro-circuits.

CONCLUSIONS

A few reactions of synthetic high polymers are beneficial in the sense that they provide a means for the modification or improvement of the polymer properties. However, some of the most facile reactions, particularly those involving oxidation and chain cleavage are, in nearly all cases, detrimental to the polymer. For this reason, a considerable technology has developed around the techniques that retard the oxidation or thermal decomposition of polymers. Some of these techniques are based on fundamental thermochemical and mechanistic principles of the types discussed in Chapter 10.

STUDY QUESTIONS

1. Discuss in detail the validity of the "model compound" approach to the study of polymer reactions, that is, the study of the reactions of small molecules as a substitute for a direct study of the macromolecules.

2. Design a synthesis route starting from polyethylene or polystyrene that would allow the introduction of water-solubilizing side groups. Which steps in the process are likely to be the most difficult to perform, and why? Suggest possible uses for the final product of the reaction sequence.

3. Suggest applications in which the hydrolysis of a polyamide might be considered to be an advantage.

4. Discuss the possibility that the reactions of a polymer might be used to measure the stereo-regularity of that polymer.

5. Comment on the fact that polymer substitution reactions are not generally used for the modification of poly(dimethylsiloxane).

6. Devise a reaction scheme for the oxidative degradation of an aromatic ladder polymer (see Chapter 2) in the atmosphere at elevated temperatures.

7. What advantages or disadvantages can you foresee for the preparation of microcircuits (a) by electron beam decomposition of polymers, and (b) by photopolymerization of a monomer or photocrosslinking of a polymer?

SUGGESTIONS FOR FURTHER READING

AYREY, G., B. C. HEAD, AND R. C. POLLER, "The Thermal Dehydrochlorination and Stabiliza-tion of Poly(vinyl chloride)," *J. Polymer Sci. (D) (Macromol. Rev.)*, **8**, 1 (1974).

BASEDOW, A. M., AND K. H. EBERT, "Ultrasonic Degradation of Polymers in Solution," *Adv. Polymer Sci.*, **22**, 83 (1977).

BATTAERD, H. A. J., AND G. W. TREGEAR, *Graft Copolymers* (Polymer Reviews, Vol. 16). New York: Wiley–Interscience, 1967.

BOVEY, F. A., *The Effects of Ionizing Radiation on Natural and Synthetic High Polymers* (Polymer Reviews, Vol. 1). New York: Wiley–Interscience, 1958.

BRYDON, A., AND G. C. CAMERON, "Chemical Modification of Unsaturated Polymers," *Progr. Polymer Sci.* (A. D. Jenkins, ed), **4**, 209 (1975).

CAMERON, G. C., AND J. R. MacCALLUM, "The Thermal Degradation of Polystyrene," *J. Macromol. Sci.—Rev. Macromol. Chem.*, **C1**, 327 (1967).

CHARLESBY, A., *Atomic Radiation and Polymers.* New York: Pergamon Press, 1960.

CICCHETTI, O., "Mechanisms of Oxidative Photodegradation and of UV Stabilization of Polyolefins," *Adv. Polymer Sci.*, **7**, 70 (1970).

COLLMAN, J. P., L. S. HEGEDUS, M. P. COOKE, J. R. NORTON, G. DOLCETTI, AND D. N. MARQUARDT, "Resin-Bound Transition Metal Complexes," *J. Am. Chem. Soc.*, **94**, 1789 (1972).

CONLEY, R. T. (ed.), *Thermal Stability of Polymers*, New York: Dekker, 1970.

DAVIS, A., AND J. H. GOLDEN, "Stability of Polycarbonate," *J. Macromol. Sci.—Rev. Macromol. Chem.*, **C3**, 49 (1969).

DOLE, M. (ed.), *The Radiation Chemistry of Macromolecules.* New York: Academic Press, 1972–73.

DONARUMA, L. G., AND O. VOGL (eds.), *Polymeric Drugs.* New York: Academic Press, 1978.

FETTES, E. M., *Chemical Reactions of Polymers* (High Polymers, Vol. XIX). New York: Wiley–Interscience, 1964.

FINCH, C., *Polyvinyl Alcohol: Properties and Applications.* New York: Wiley, 1973.

FOX, R. B., "Photodegradation of High Polymers," *Progr. Polymer Sci.* (A. D. Jenkins, ed.), **1**, 45 (1967).

GRASSIE, N., "Degradation," *Encyclopedia of Polymer Sci. and Technol.* (H. F. Mark, N. G. Gaylord, and N. M. Bikales, eds.), **4**, 647 (1966).

HEITZ, W., "Polymeric Reagents. Polymer Design, Scope, and Limitations," *Adv. Polymer Sci.*, **23**, 1 (1977).

JELLINEK, H. H. G., "Depolymerization," *Encyclopedia of Polymer Sci. and Technol.* (H. F. Mark, N. G. Gaylord, and N. M. Bikales, eds.), **4**, 740 (1966).

KAPLAN, A. M., R. T. DARBY, M. GREENBERGER, AND M. R. ROGERS, "Microbial Deterioration of Polyurethane Systems," *Devel. Ind. Microbiol.*, **9**, 201 (1968).

KUNITAKE, T., AND Y. OKAHATA, "Catalytic Hydrolysis by Synthetic Polymers," *Adv. Polymer, Sci.*, **20**, 159 (1976).

MADORSKY, S. L., *Thermal Degradation of Organic Polymers* (Polymer Reviews, Vol. 7). New York: Wiley–Interscience, 1964.

MANO, E. B., AND F. M. B. COUTINHO, "Grafting on Polyamides," *Adv. Polymer Sci.*, **19**, 97 (1975).

MAYER, Z., "Thermal Decomposition of Poly(vinyl chloride) and of Its Low Molecular Weight Model Compounds," *J. Macromol. Sci.—Rev. Macromol. Chem.*, **C10**, 263 (1974).

MICHALSKA, Z. M., AND D. E. WEBSTER, "Supported Homogeneous Catalysts," *Chem. Tech.*, **5**, 118 (1975).

NIKITINA, T. S., E. V. ZHURAVSKAYA, AND A. S. KUZMINSKY, *Effects of Ionizing Radiation on High Polymers*. New York: Gordon and Breach, 1963.

O'DRISCOL, K. F., AND D. G. MERCER, "Gel Entrapped Multistep (Enzyme) Systems," *Contemporary Topics in Polymer Sci.* (M. Shen, Ed.), **3**, 319 (1979).

ONOZUKO, M., AND M. ASAHINA, "On the Dechlorination and Stabilization of Polyvinyl Chloride," *J. Macromol. Sci.—Rev. Macromol. Chem.*, **C3**, 235 (1969).

OVERBERGER, C. G., AND K. N. SANNES, "Polymers as Reagents in Organic Synthesis," *Angew. Chem., Intern. Ed. English*, **13**, 99 (1974).

PITTMAN, C. U., "Organic Polymers Containing Transition Metals," *Chem. Tech.*, **1**, 416 (1971).

PITTMAN, C. U., AND G. O. EVANS, "Polymer-Bound Catalysts and Reagents," *Chem. Tech.*, **3**, 560 (1973).

PRITCHARD, J. G., *Poly(vinyl alcohol)*. New York: Gordon and Breach, 1970.

REICH, L., AND S. S. STIVALA, *Elements of Polymer Degradation*. New York: McGraw-Hill, 1971.

REICH, L., AND S. S. STIVALA, "Uncatalyzed Uninhibited Thermal Oxidation of Saturated Polyolefins," *Rev. Macromol. Chem.*, **1**, 249 (1966).

SANDER, M., AND E. STEININGER, "Phosphorylation of Polymers," *J. Macromol. Sci.—Rev. Macromol. Chem.*, **C2**, 57 (1968).

SHIMIDZU, T., "Cooperative Actions in the Nucleophile-Containing Polymers," *Adv. Polymer Sci.*, **23**, 55 (1977).

SOHMA, AND M. SAKGUCHI, "ESR Studies and Polymer Radicals Produced by Mechanical Destruction and Their Reactivity," *Adv. Polymer Sci.*, **20**, 109 (1976).

STAMM, R. F., E. F. HOSTERMAN, C. D. FELTON, AND C. S. HSIA CHEN, "Grafting of Chloromethylstyrene to Propylene Fibers by the Use of Ionizing Radiation," *J. Appl. Polymer Sci.*, **7**, 753 (1963).

TROZZOLO, A. M., AND F. H. WINSLOW, "A Mechanism for the Oxidative Photodegradation of Polyethylene," *Macromolecules*, **1**, 98 (1968).

TSUJI, K., "ESR Study of Photodegradation of Polymers," *Adv. Polymer Sci.*, **12**, 131 (1973).

TURNER, D. T., "Role of Free Radicals in the Radiation Chemistry of Polymers," *J. Polymer Sci. (D) (Macromol. Rev.)*, **5**, 229 (1971).

WALL, L. A., "Polymer Decomposition: Thermodynamics, Mechanisms, and Energetics," *Soc. Plastics Eng. J.*, *810 (1960).

part **II**

THERMODYNAMICS
AND KINETICS
OF POLYMERIZATION

10

Polymerization
and Depolymerization
Equilibria

The synthesis of organic and inorganic polymers by the ring-opening polymerization of cyclic compounds was discussed in Chapters 6 and 7. Such polymers frequently break down at high temperatures to yield cyclic oligomers, often the same cyclic compounds from which the polymers were formed in the first place. Similarly, a number of addition polymers degrade at high temperatures to regenerate the original monomer. In many cases it can be shown that the polymerization and depolymerization steps are simply different aspects of an equilibration process. Whether polymerization or depolymerization occurs at a given temperature depends on the position of the equilibrium at that temperature.

An understanding of these processes is vitally important for those who are concerned about the practical problems of thermal stability in high polymers. It is also important for an understanding of why a substantial number of cyclic systems and unsaturated compounds have not yet been converted to high polymers.

It is worthwhile to approach this subject by first considering a number of questions. For example, why do certain unsaturated monomers or cyclic compounds polymerize to linear high polymers, whereas others do not? Why are some polymers more stable at high temperatures than others? Which molecular features in different systems might stabilize chains more than rings or monomers, or vice versa? What role does the *mechanism* of polymerization or depolymerization play in these processes? Do monomeric compounds participate in ring–chain interconversion, or do cyclic species play a role in monomer–polymer equilibrations?

In order to attempt to answer these questions, let us first consider some examples of systems which undergo monomer–polymer or ring-chain equilibration and some which do not. We can then formulate a number of general observations.

MONOMER–POLYMER EQUILIBRIA

At high temperatures poly(methyl methacrylate) depolymerizes to methyl methacrylate, and poly(tetrafluoroethylene) depolymerizes to tetrafluoroethylene monomer. This general phenomenon can be summarized by the process shown in **1**. Other polymers that behave in the same way include poly(α-methylstyrene),

$$\left[\begin{array}{cc} H & R \\ | & | \\ -C-C- \\ | & | \\ H & R' \end{array}\right]_n \longrightarrow n \begin{array}{cc} H & R \\ | & | \\ C=C \\ | & | \\ H & R' \end{array}$$

1

poly(methacrylonitrile), and poly(vinylidine cyanide). The monomers formed from these polymers cannot undergo α-hydrogen abstraction. Hence, hydrogen abstractive side reactions do not compete with simple depolymerization to the monomers.

Furthermore, it is known that although monomers, such as acetaldehyde, $MeCH=O$, propionaldehyde, $EtCH=0$, butyraldehyde, $PrCH=O$, or acetone, $Me_2C=O$, can be polymerized, the polymers depolymerize back to the monomer at only moderate temperatures. High-pressure, low-temperature reaction conditions are needed for polymerization, but release of the pressure and warming to room temperature results in depolymerization.

On the other hand, polystyrene undergoes thermal breakdown to yield not only styrene, but also products derived from random chain scission, α-hydrogen abstraction, and chain transfer processes. Thus, it is important to recognize that depolymerization (i.e., the reverse of polymerization) is only one process of several that are possible when a polymer is heated to high temperatures (see also Chapter 9).

EXAMPLES OF RING–POLYMER INTERCONVERSIONS

As discussed in Chapter 7, rhombic sulfur, S_8, can be polymerized to S_n at temperatures above 160°C, but depolymerization back to S_8 occurs at higher temperature. Selenium behaves similarly. Trioxane, $(OCH_2)_3$, and tetroxane, $(O-CH_2)_4$, polymerize to polyoxymethylene during γ- or X-irradiation or under the influence of cationic initiators. The polymer depolymerizes to the monomer (formaldehyde) or to cyclic oligomers above 100°C. Trithiane, $(S-CH_2)_3$, tetrathiane, $(S-CH_2)_4$, and higher cyclic homologues behave similarly.

Particularly interesting equilibria exist in the isocyanate series, where monomers (2), cyclic dimers (3), cyclic trimers (4), and high polymers (5) can all partici-

$$R-N=C=O$$

2 3 4 5

pate in an equilibration process. In the dimethylsiloxane system, the cyclic trimer,
$(O-SiMe_2)_3$, and tetramer, $(O-SiMe_2)_4$, polymerize readily with anionic or
cationic initiators to yield poly(dimethylsiloxane), $(O-SiMe_2)_n$. The polymer
depolymerizes at temperatures above 300 to 400°C to yield mainly the cyclic
tetramer.

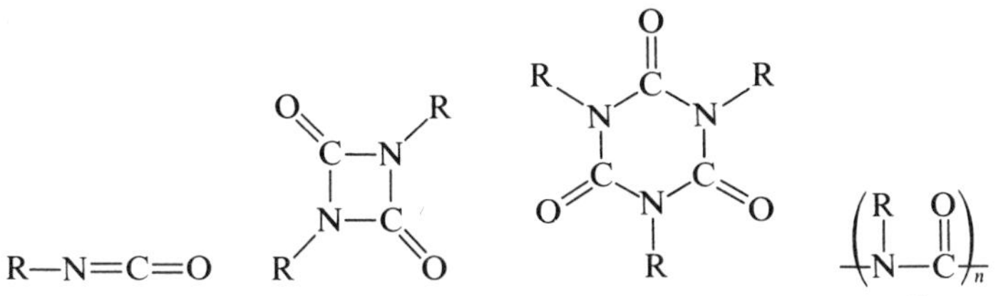

Halogen-substituted cyclophosphazenes, such as $(NPCl_2)_{3 \, or \, 4}$, $(NPF_2)_3$, or
$(NPBr_2)_3$, polymerize thermally, but organic-substituted cyclic derivatives, such
as $(NPPh_2)_3$ or $[NP(OCH_2CF_3)_2]_3$, do not. However, organic-substituted high
polymers, such as $[NP(OCH_2CF_3)_2]_n$ or $[NP(OPh)_2]_n$, prepared by an alternative route (see Chapter 7) depolymerize at elevated temperatures to the appropriate cyclic trimers or higher oligomers. A few other inorganic systems appear
to generate polymerization–depolymerization systems. For example, the cyclic
dimer (6), cyclic tetramer (7), and high polymer (8) of sulfur nitride may form an
equilibrating system.

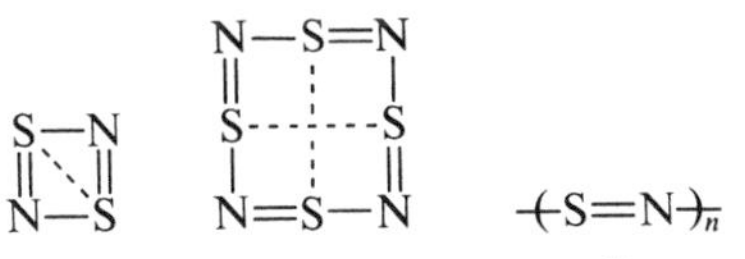

6 7 8

"UNPOLYMERIZABLE" COMPOUNDS

So many cyclic compounds and olefin derivatives polymerize that the fact is often
overlooked that a number of compounds which should polymerize have so far
proved highly resistant to polymerization. In Chapter 6 it was pointed out that
cyclic compounds, such as benzene, cyclohexane, s-triazines, 1,4-dioxane, tetrahydropyran, and borazines resist polymerization. Yet acetylene can be trimerized
to benzene or polymerized to polyacetylene. Benzene is an energy trap in the polymeric series (scheme 9). Similarly, s-triazines are the principal products formed
from the attempted polymerization of nitriles (scheme 10). Even some cyclic
dimers, such as tetramethylcyclodisilthiane (11), equilibrate to the cyclic trimer

(12), but do not yield high polymer. An analysis of the reasons for the unpolymerizability of systems such as these can provide valuable clues about the factors which influence polymerization-depolymerization equilibria in other systems. Some of these factors will be considered throughout this chapter.

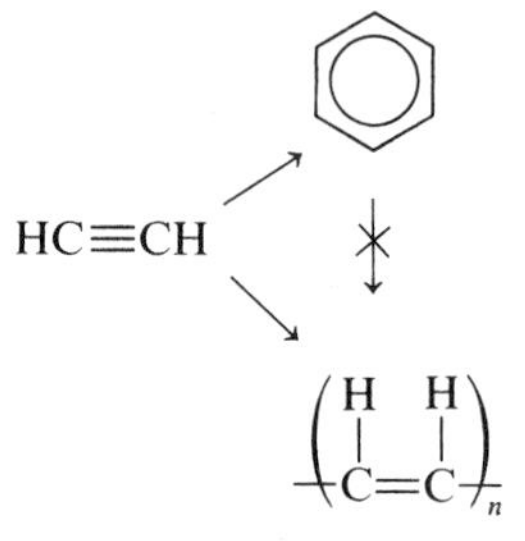

9

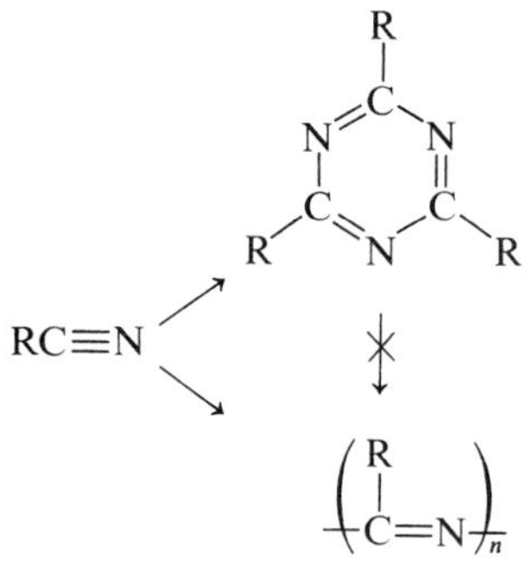

10

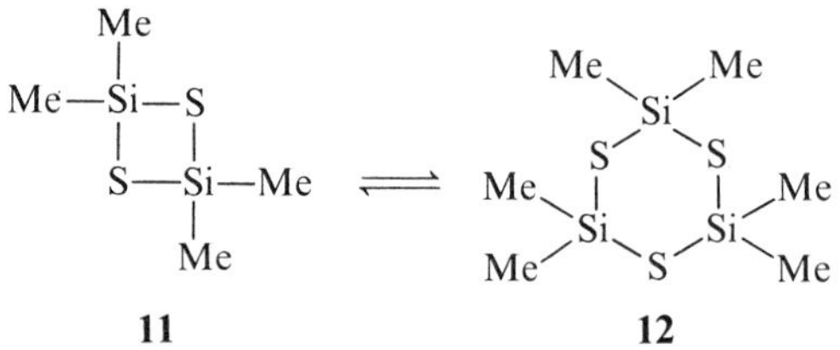

11

12

THE GENERAL THERMODYNAMIC PROBLEM

If we consider the polymerization of a monomer or a cyclic oligomer and the depolymerization of a polymer to be governed by thermodynamic factors,[1] some of the observations discussed in the preceding sections begin to make sense.

First, let us consider a hypothetical series of compounds which extends from the monomer, A=B, to the high polymer, $(A-B)_n$. It will be assumed initially

[1] As discussed later in this chapter, in practice, finite kinetic factors may override the thermodynamic effects.

that the lower oligomers in the series are cyclic and that the higher polymers are either macrocyclic or linear. If we make these assumptions, the series can be formulated as:

$$A{=}B \qquad \begin{matrix} A{-}B \\ | \quad | \\ B{-}A \end{matrix} \qquad \begin{matrix} A \\ B \diagup \diagdown B \\ | \qquad | \\ A \diagdown \diagup A \\ B \end{matrix} \qquad \begin{matrix} A{-}B{-}A \\ | \qquad | \\ B \qquad B \\ | \qquad | \\ A{-}B{-}A \end{matrix} \qquad (AB)_5 \qquad (AB)_6 \cdots (AB)_n$$

Let us also consider for a moment that equilibration can occur between all the members of this series. For this to happen, every compound in the series would need to possess a similar free energy. In other words, every member of the series can coexist in the equilibrium only if each compound has a similar stability to the others at a given temperature and pressure.

However, in practice, one or several of the compounds will usually be more stable than the others and will be present in greater amounts at equilibrium. In extreme cases, only one or two homologues will exist, with the total exclusion of all other species. Apparently, this is the situation found for benzene, s-triazines, borazines, and the other systems mentioned earlier. On the other hand, if an equilibrium does exist, it should, in principle, be possible to calculate an equilibrium constant for each interconversion, and this could be correlated with the free-energy change in the usual way.

SPECIFIC THERMODYNAMIC EFFECTS

The well-known thermodynamic expression

$$\Delta G = \Delta H - T\Delta S \tag{1}$$

can be used as a basis for understanding the polymerization–depolymerization behavior. In order for a polymerization to be thermodynamically feasible, the Gibbs free-energy change must be negative, that is, $\Delta G_p < 0$. If ΔG_p is positive, depolymerization will be favored. *Any factor that lowers the enthalpy or raises the entropy of a particular species in the system will shift the equilibrium to favor that species.* This elementary consideration allows us to understand many of the puzzling features of monomer–polymer or ring–polymer equilibria.

Polymerization is an *association* reaction in which many molecules come together to form one molecule. Regardless of the mechanism by which this occurs, this process results in a large loss in the number of translational and rotational degrees of freedom in the system. It is fairly obvious that the conversion of three monomer molecules to one cyclic trimer molecule is accompanied by a decrease in translational entropy (there are now fewer molecules in the system). Similarly, the polymerization of a thousand cyclic trimers to one linear polymer molecule reduces the translational entropy further. This major loss of entropy is not compensated by the small entropy increase associated with the torsional and vibrational degrees of freedom in a nonrigid polymer chain. The result is that the entropy change in the polymerization process is nearly always negative, that is, $\Delta S_p < 0$. This is true both for the conversion of olefinic monomers to high polymers and for the polymerization of cyclic compounds. Hence, on entropic grounds alone, depolymerization should always be favored over polymerization. Clearly, if polymerization is to predominate over depolymerization, it must do so under conditions where ΔH_p is sufficiently negative that it compensates for the entropy loss and yields a negative ΔG_p term. In practice, polymerization is often favored at low temperatures where the $T\Delta S_p$ term is small, but depolymerization often occurs at high temperatures where $T\Delta S_p$ is large. This is one of the main reasons for the thermal instability of many polymers at high temperatures.

On a slightly more detailed level, it is possible in many polymer systems to detect a *ceiling temperature* (T_c) above which no polymer can exist. This situation is found when the $T\Delta S_p$ term increases rapidly as the temperature is raised and sharply overtakes the ΔH_p term at the ceiling temperature. The ceiling temperature can be understood in a slightly different way by a consideration of Le Châtelier's principle, which states that a system will respond to a stress in such a way as to relieve that stress. It is clear from the discussion above that most known polymerizations are exothermic processes. Thus, in such cases, depolymerization will be endothermic. Therefore, as the temperature of the system is increased by supplying heat, depolymerization (which absorbs heat) becomes more probable than polymerization. At the ceiling temperature the thermodynamic tendencies for polymerization and depolymerization become equal. Above this temperature polymerization is thermodynamically unfavorable.

However, in other systems, a *floor temperature* (T_f) exists *below* which polymer cannot be detected. The exact behavior of any monomer–polymer or ring–polymer system as the temperature is raised depends on the relationship between ΔH_p and ΔS_p for that particular system. The real problem arises when one needs to know how different magnitudes of ΔH_p and ΔS_p will influence the system, or when it is necessary to make an intelligent guess about the way in which molecular structural changes will affect ΔH_p and ΔS_p. The first problem is discussed in the next two sections, and the latter problem will be considered later.

STANDARD ENTHALPIES, ENTROPIES, AND FREE ENERGIES OF POLYMERIZATION

The general process of polymerization may be described by the stoichiometric equation

$$M(l) \rightarrow \frac{1}{n} P_n(a) \tag{2}$$

This states that 1 mol of monomer in the liquid state is converted to $1/n$ mol of polymer with an average degree of polymerization, n, in the amorphous or slightly crystalline state. These states of aggregation of monomer and polymer are chosen because they are representative of many polymerizations. An actual polymerization of interest may in fact be carried out with the reactants and products in physical states other than the above (i.e., both in solution and in an inert solvent) and over a wide range of temperatures. However, for the purpose of tabulation and comparison, it is necessary to choose standard states such as those described above, together with a single temperature, which is conveniently taken to be 25°C. Standard thermodynamic methods generally permit straightforward conversions to be made between these standard states and other conditions of interest in actual polymerizations. It should also be noted that equation (2) assumes that the equilibrium lies so far to the side of the polymer that the system can be considered as an irreversible process.

We will now consider how experimental values are obtained for standard enthalpy, entropy, and free-energy changes for polymerization. Following that, we will discuss some specific examples of systems for which equilibrium considerations are important and some for which they are not.

Standard enthalpies of polymerization, ΔH_p°, are most generally obtained by direct calorimetric measurement of the amount of heat evolved when a known amount of monomer is converted to a known amount of polymer. Alternatively, enthalpies of polymerization may be determined by direct calorimetric measurement of heats of combustion of monomer and polymer to yield standard enthalpies of formation ΔH_f° for the monomer and polymer. The enthalpy of polymerization is then obtained, according to (2), by the well-known relationship

$$\Delta H_p^\circ = \frac{1}{n} \Delta H_f^\circ(P_n) - \Delta H_f^\circ(M) \tag{3}$$

Standard entropies of polymerization ΔS_p° are generally calculated from the absolute entropies of the monomer and polymer:

$$\Delta S_p^\circ = \frac{1}{n} S^\circ(P_n) - S^\circ(M) \tag{4}$$

The absolute entropies are determined from calorimetric measurements of the heat capacities of the monomer and polymer over a wide range of temperature using the expression

$$S^\circ(T) = \int_0^T \frac{C_p}{T} \, dT \tag{5}$$

where C_p is the molar heat capacity at constant pressure and T the absolute temperature. The integral is usually evaluated graphically from the experimental data. Much less information is available on ΔS_p than on ΔH_p.

The standard free energies of polymerization, ΔG_p°, are easily obtained by calculation, using (1), once ΔH_p° and ΔS_p° are known.

SPECIFIC MONOMER–POLYMER EQUILIBRIA

In practical terms, information is usually needed about the temperature range in which the monomer can be polymerized and about the ceiling temperature. Below the ceiling temperature (T_c), conversion of the monomer to a polymer involves a free-energy decrease, hence polymerization should occur. Above T_c, depolymerization involves a free-energy decrease and, therefore, depolymerization occurs. At T_c, the rates of polymerization and depolymerization should be equal. Thus, the question really revolves around the actual values of ΔH° and ΔS° for polymerization, since $T_c = \Delta H_p^\circ / \Delta S_p^\circ$.[1] Table 10.1 lists ΔH_p°, ΔS_p°, and calculated values of ΔG_p° for several olefin and aldehyde polymerization systems, together with calculated T_c values. The data in Table 10.1 indicate that the ΔH_p° values for a variety of monomer–polymer systems vary over a considerable range, from -5.1 kcal/mol for n-butyraldehyde through about -13 kcal/mol for the methacrylic esters, -17 kcal/mol for styrene, to the very high exothermicity of -37 kcal/mol for tetrafluoroethylene.

The decreasing exothermicity of polymerization through the series ethylene, propylene, and isobutylene can be correlated with the increasing steric hindrance toward polymerization that results from the successive replacement of hydrogen atoms on the same carbon atom by methyl groups. Very little effect is noted for replacement of the first hydrogen by a methyl, but a drastic reduction in exothermicity occurs for replacement of the second hydrogen to yield isobutylene. The same phenomena can be observed in the series ethylene, styrene, α-methylstyrene. However, substitution on the aromatic ring of styrene has very little or no effect on ΔH_p°, as may be seen by a comparison of styrene with 2,4,6-trimethylstyrene. Obviously, steric effects play an important role in the thermochemistry of polymerization.

Other significant influences on the magnitudes of ΔH_p° can be attributed to energy differences in the monomer and polymer that arise from the resonance stabilization that is lost on polymerization or from changes in bond hybridization that occur on polymerization. Differences in the extent of hydrogen bonding in the monomer and polymer can also have an influence on ΔH_p°.

On the other hand, the values for ΔS_p° show much less variation than do the values for ΔH_p°. The range for ΔS_p° values (Table 10.1) is from -19 to -28 cal/deg-mol. The reason for this narrow range is that the dominant factor in ΔS_p° is the loss of translational entropy brought about by the large reduction in the number of molecules present, and this factor is relatively constant from system to system.

For most of the monomer–polymer systems listed in Table 10.1, the equilibria at 25°C lie on the side of the polymer rather than the monomer. This fact can be verified by substitution of the ΔG_p° values from Table 10.1 into the well-known equation (6) that connects the equilibrium constant with the standard free-energy change:

$$\Delta G^\circ = -RT \ln K_{\text{eq}} \qquad (6)$$

[1] Or $T_c = \Delta H_0 / [\Delta S_0 + R \ln[M]/[M]_0]$.

TABLE 10.1 Standard Enthalpies, Entropies, Free Energies, and Ceiling Temperatures for Polymerization of Various Monomer–Polymer Systems at 25°C*

Monomer	$-\Delta H_p^\circ$ (kcal/mol)	$-\Delta S_p^\circ$ (cal/deg-mol)	$-\Delta G_p^\circ$ (kcal/mol)	Ceiling Temperatures (°C) $(T_c = \Delta H_p^\circ / \Delta S_p^\circ)$
Acetaldehyde	—	—	—	−35 to −40†
Butadiene	17.6	20.5	11.5	585
n-Butyraldehyde	5.1‡	22.3‡	—	−45‡
Chloral	8.0‡	28.0‡	—	13‡
Ethylene	21.2	24	14.0	610
Formaldehyde	7.4	19	1.7	116
Isobutylene	12.9	28.8	4.3	175
Isoprene	17.9	24.2	10.7	466
Methyl methacrylate	13.2	28	4.9	198
α-Methylstyrene	8.4	24.8	1.0	66
Styrene	16.7	25.0	9.2	395
Tetrafluoroethylene	37	26.8	29	1100
2,4,6-Trimethylstyrene	16.7	—	—	—

* Unless otherwise specified, data refer to standard states of pure liquid for the monomer and amorphous or slightly crystalline polymer for the polymer.

† Estimated by experimental observations.

‡ Data refer to standard state of 1 mol/liter in solution.

Source: C. T. Mortimer, *Reaction Heats and Bond Strengths* (New York: Pergamon Press, 1962), Chap. 5; R. M. Joshi and B. J. Zwolinski, *Vinyl Polymerization*, Part I (G. H. Ham, ed.), (New York: Dekker, 1967), Chap. 8; F. S. Dainton and K. J. Ivin, *Rev. Chem. Soc. (London)*, **22**, 61 (1958); T. Ohtsuka and C. Walling, *J. Am. Chem. Soc.*, **88**, 4167 (1966).

But as the temperature is raised, the equilibrium shifts to favor the monomer until, at sufficiently high temperatures, depolymerization predominates. The ceiling temperature (T_c) values in Table 10.1 reflect this change. An enormous spread exists in the T_c values. Thus, polyformaldehyde and poly(methyl methacrylate) depolymerize between 100 and 200°C. Poly(n-butyraldehyde) is unstable above −45°C, while poly(tetrafluoroethylene) should have an equilibrium ceiling temperature above 1100°C. In practice, of course, poly(tetrafluoroethylene) would undergo *fragmentation* reactions in a closed system well below this temperature. It is important to recognize that the ceiling-temperature concept applies only to *closed* systems at equilibrium. If the system is not closed, monomer can be lost by volatilization, and total depolymerization can occur well below the ceiling temperature. Thus, in actual practical use, few polymers demonstrate the thermal stability predicted from the ceiling temperature.

INFLUENCE OF ΔH AND ΔS ON RING-CHAIN EQUILIBRIA (NONRIGOROUS APPROACH)

An approach described by Gee[1] allows a qualitative prediction to be made of the effect on a ring–polymer equilibrium of changes in the magnitudes of ΔH and ΔS. This approach is perhaps the most useful one for systems where only minimal thermodynamic data are available.

[1] G. Gee, *Chem. Soc. (London), Spec. Publ.*, **15**, 67 (1961).

Consider the polymerization of a ring to a long polymer chain:

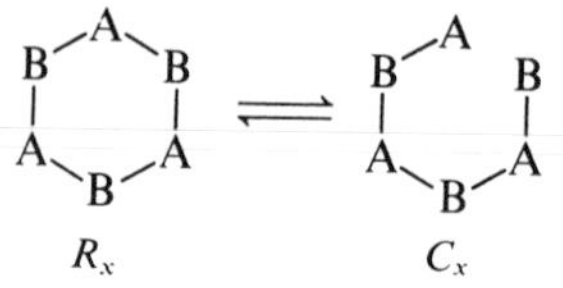

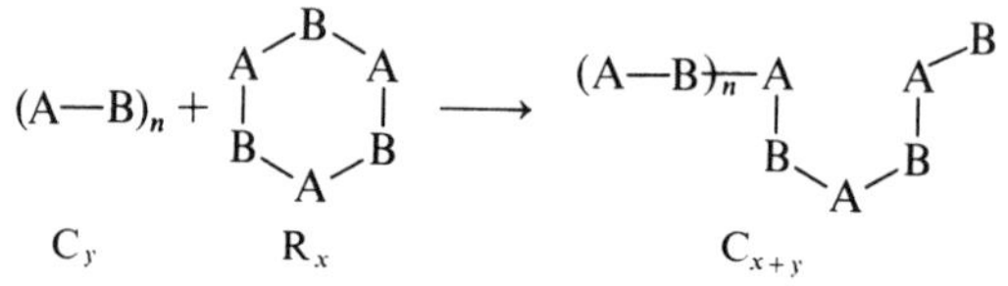

The mechanism of this process is not important since we are only concerned with the final position of an equilibrium.

The polymerization process can be viewed as a two-step sequence; (a) the opening of a ring (R_x) to yield a short chain (C_x), and (b) the ring-opening reaction of a ring with the end of an existing chain (C_y). The equilibrium state of the system is then given by

$$R_x \rightleftharpoons C_x \tag{7}$$

$$R_x + C_y \rightleftharpoons C_{x+y} \tag{8}$$

Reaction (7) involves scission of a bond. Hence, ΔH_7 will be much greater than zero, since bond cleavage requires the input of energy. The entropy change (ΔS_7) for this step will also be significantly greater than zero because the linear fragment C_x will have greater torsional and vibrational entropy than the cyclic species, R_x.

However, the enthalpy and entropy changes inherent in reaction (8) are more difficult to predict. One bond is broken and one bond is formed so that ΔH_8 could be close to zero. Furthermore, although a ring is opened in step (8) with a consequent increase in entropy, two "molecules" are combined into one so that there is a loss of three translational degrees of freedom. This means that ΔS_8 is usually less than zero. We can say with certainty, that, if reaction (8) takes place spontaneously, then the free energy, ΔG_8, must decrease in the process and therefore, ΔH_8 must be less than $T\Delta S_8$.

Let us consider an equilibration in which rings have been partly converted to chains. For reaction (8) we can then define an equilibrium constant K_8 given by

$$K_8 = \frac{[C_{x+y}]_{eq}}{[C_y]_{eq}[R_x]_{eq}}$$

where the subscript "eq" refers to *equilibrium* concentrations. If equilibration has progressed to the point where rings are being added to the ends of *long* chains, then $[C_y]$ will be approximately the same as $[C_{x+y}]$. Therefore, K_8 will become

$$K_8 \approx \frac{1}{[R_x]_{eq}}$$

The relationship between the *standard* free-energy change, ΔG_8°, and the equilibrium constant is given by

$$\Delta G_8^\circ = -RT \ln K_8$$

and by substitution for K_8 we obtain

$$\Delta G_8^\circ = RT \ln [R_x]_{eq}$$

Now, to define the state of polymerization or depolymerization of this system we need to know (a) the weight fraction of molecules that are in the form of chains (denoted by ϕ), and (b) the degree of polymerization of the chains. If $[R_x]_0$ is the concentration of rings in the pure ring compound (i.e., in the starting material), it is possible to relate ΔG_8° to ΔG_8 by the standard methods of thermodynamics. Thus, for reaction (8),

$$\Delta G_8 = \Delta G_8^\circ + RT \ln \frac{[C_{x+y}]}{[C_y][R_x]}$$

If now the approximation that $[C_{x+y}] \approx [C_y]$ can be made at all stages of the equilibration, including the initial stages, then

$$\Delta G_8 = \Delta G_8^\circ - RT \ln [R_x]_0$$

and by substitution of the expression for ΔG_8°,

$$\Delta G_8 = RT \ln \frac{[R_x]_{eq}}{[R_x]_0}$$

If no solvent is present, $[R_x]_{eq}/[R_x]_0 = 1 - \phi$, and substituting for ΔG_8 in $\Delta G_8 = \Delta H_8 - T\Delta S_8$ gives

$$-R \ln (1 - \phi) = \Delta S_8 - \frac{\Delta H_8}{T} \tag{9}$$

This approximate equation should yield reasonably accurate values for the weight fraction of chains (ϕ) provided that the calculated values for ϕ lie between 0 and 1. If ϕ falls outside this range, the approximate treatment obviously fails and must be replaced by a more rigorous approach. The results of substituting various values for ΔH_8 and ΔS_8 into this equation and assuming that they are independent of temperature are shown in Figure 10.1.

In general terms, there are three main equilibrium situations that can be anticipated.

1. If step (8) is exothermic (ΔH_8 is less than zero), then substitution in the preceding equation indicates that when ΔS_8 is zero or positive, the polymer concentration (as defined by ϕ) will fall with increasing temperature. This situation is summarized by case (b) in Figure 10.1. However, if ΔS_8 is also negative, no polymer

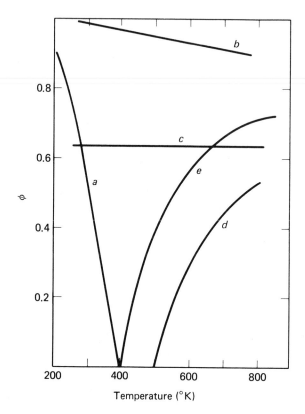

Fig. 10.1 Equilibrium fraction of chain polymer (theoretical). [From G. Gee, *Chem. Soc.* (*London*), *Spec. Publ.*, **15**, 67 (1961).]

Curve	a	b	c	d	e
ΔH_2 (kcal/mol)	-2	-2	0	2	2
ΔS_2 (cal/deg-mol)	-5	2	2	4	5

can exist above a *ceiling temperature* defined by the value of $\Delta H_8/\Delta S_8$. This possibility is depicted by curve (a) in Figure 10.1.

2. If ΔH_8 is zero, the polymer concentration will be independent of temperature, and given by the term $-R \ln (1 - \phi) = \Delta S_8$. This situation is depicted by curve (c) in Figure 10.1. Significant values of ϕ are obtained only if ΔS_8 is negative. If ΔS_8 is zero or positive, the polymer will always be unstable relative to rings.

3. If ΔH_8 is positive and if ΔS_8 is also positive, the polymer concentration *increases* with increasing temperature and no polymer can exist below a *floor temperature*, T_f (curves d and e, Figure 10.1). The value of the floor temperature is given by $\Delta H_8/\Delta S_8$, and equation (9) can be written as

$$- \ln (1 - \phi) = \frac{\Delta H_8}{R} \left(\frac{1}{T_f} - \frac{1}{T} \right) \tag{10}$$

However, if ΔS_8 is equal to or less than zero, no polymer can exist at any temperature.

The presence of a solvent will favor the formation of rings at the expense of polymer. This is because an additional term, $R \ln c$ (where c is the weight fraction of total solute in the solvent) must be added to the right-hand side of equation (9).

If two or more sizes of ring are in simultaneous equilibrium with the same polymer chain, equation (9) must be modified to replace $(1 - \phi)$ by the weight fraction ϕ_x of the ring under consideration. Thus, if two rings (R_{x1} and R_{x2}) are present, two simultaneous equations are generated:

$$-\ln \phi_{x1} = \frac{\Delta H_{81}}{R} \left(\frac{1}{T_{\phi 1}} - \frac{1}{T} \right) \qquad (11)$$

$$-\ln \phi_{x2} = \frac{\Delta H_{82}}{R} \left(\frac{1}{T_{\phi 2}} - \frac{1}{T} \right) \qquad (12)$$

Since $\phi_{x1} + \phi_{x2}$ must remain less than 1, this treatment is restricted to the situation where the two rings and the polymer are simultaneously present in significant amounts.

This approach gives particularly favorable results for the $S_8 \rightleftharpoons S_n$ equilibrium, where a floor temperature is known to exist at 159°C. It is also a valuable approach for use in systems where accurate thermodynamic data are not available. If an *estimate* of ΔH_8 and ΔS_8 can be made for a new ring–chain equilibration system, then the existence or absence of ceiling or floor temperatures can perhaps be predicted.

THE SULFUR EQUILIBRIUM (RIGOROUS APPROACH)

A more rigorous approach to ring–chain equilibration calculations was developed by Tobolsky,[1-3] and this method also has been very successfully applied to the sulfur system. The derivation is as follows. Because the polymerization of S_8 is a diradical reaction, the process can be considered in terms of three steps:

(a) $S_8 \xrightarrow{K} \cdot SSSSSSSS \cdot$ (or S_8^*) (Initiation)

(b) $S_8^* + S_8 \xrightarrow{K_3} S_{16}^*$ (Initial propagation)

(c) $S_{8n}^* + S_8 \xrightarrow{K_3} S_{8(n+1)}^*$ (General propagation)

It is assumed that the equilibrium constants for the initial and general propagation steps are identical for this reaction, but that these are different from K for the

[1] A. V. Tobolsky, *J. Polymer Sci.*, **25**, 220 (1957); **31**, 126 (1958).
[2] A. V. Tobolsky and A. Eisenberg, *J. Am. Chem. Soc.*, **81**, 780, 2303 (1959); **82**, 289 (1960).
[3] A. V. Tobolsky and W. J. MacKnight, *Polymeric Sulfur and Related Polymers*, H. F. Mark and E. H. Immergut, eds. (New York: Wiley–Interscience, 1965).

initiation step. Thus if S_8 (ring) is designated as M, S_8^* as M_1^*, and S_{8n}^* as M_n^*, the equilibrium constants for steps (a), (b), and (c) are:

(a) $\quad K = \dfrac{[M_1^*]}{[M]} \quad$ or $\quad [M_1^*] = K[M]$

(b) $\quad K_3 = \dfrac{[M_2^*]}{[M_1^*][M]} \quad$ or $\quad [M_2^*] = K_3[M_1^*][M] = K[M](K_3[M])$

(c) $\quad K_3 = \dfrac{[M_3^*]}{[M_2^*][M]} \quad$ or $\quad [M_3^*] = K_3[M_2^*][M] = K[M](K_3[M])^2$

Therefore,

$$[M_n^*] = K[M](K_3[M])^{n-1}$$

This expression allows the calculation of K and K_3 at any temperature if [M] and the number-average degree of polymerization, P, are known. Since $P = [W]/[N]$, where [W] is the total concentration of S_8 units in the polymer mixture, and [N] is the total concentration of polymer molecules, [W] and [N] can be obtained from

$$[N] = [M_1^*] + [M_2^*] + [M_3^*] + \cdots$$
$$[N] = K[M]\{1 + K_3[M] + (K_3[M])^2 + \cdots\}$$
$$[N] = \dfrac{K[M]}{1 - K_3M}$$
$$[W] = [M_1^*] + 2[M_2^*] + 3[M_3^*] + 4[M_4^*] + \cdots$$
$$[W] = K[M]\{1 + 2(K_3[M]) + 3(K_3[M])^2 + \cdots\}$$
$$[W] = K[M]/(1 - K_3[M])^2$$

whence

$$P = \dfrac{1}{1 - K_3[M]}$$

The total concentration of S_8 units, $[M_0]$, in monomer and polymer is given by

$$[M_0] = [M] + [W] = [M] + \dfrac{K[M]}{(1 - K_3[M])^2}$$

$$= [M](1 + KP^2)$$

[M] can be eliminated, and the final equation becomes

$$[M_0] = \dfrac{P - 1}{PK_3} + \dfrac{KP(P - 1)}{K_3}$$

where $[M_0]$ is the number of moles of S_8 units per kilogram of sulfur and has the value of 3.90 mol/kg at all temperatures.

Thus, determination of K and K_3 at two temperatures allows the calculation of ΔH°, ΔS°, ΔH_3°, and ΔS_3° from the assumed van't Hoff equation for each equilibrium constant:

$$\ln K = -\left(\frac{\Delta H^\circ}{RT}\right) + \left(\frac{\Delta S^\circ}{R}\right)$$

$$\ln K_3 = -\left(\frac{\Delta H_3^\circ}{RT}\right) + \left(\frac{\Delta S_3^\circ}{R}\right)$$

and the linearity of plots of $\ln K$ against $1/T$ and $\ln K_3$ against $1/T$ allows K and K_3 and, hence, $[M]$ and $[P]$ to be evaluated at all temperatures. As shown in Figures 10.2 and 10.3, there is a striking correspondence between the experimental and calculated values, both above and below the floor temperature.

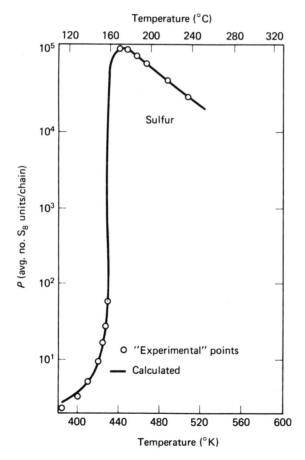

Fig. 10.2 Plot of P versus T for liquid sulfur. [From A. V. Tobolsky and A. Eisenberg, *J. Am. Chem. Soc.*, **81**, 780, 2303 (1959); **82**, 289 (1960), © by Amer. Chem. Soc., and A. V. Tobolsky and W. J. MacKnight, *Polymeric Sulfur and Related Polymers*, H. F. Mark and E. H. Immergut, eds., (New York: Wiley–Interscience, 1965, © by Amer. Chem. Soc.).]

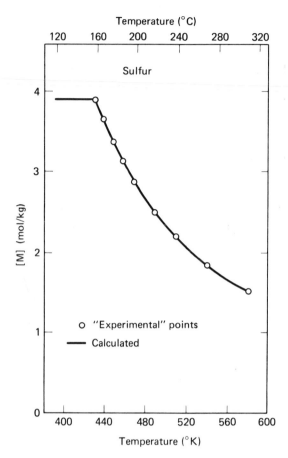

Fig. 10.3 Plot of [M] versus T for liquid sulfur. [From A. V. Tobolsky and A. Eisenberg, *J. Am. Chem. Soc.*, **81**, 780, 2303 (1959); **82**, 289 (1960), © by Amer. Chem. Soc.; and A. V. Tobolsky and W. J. MacKnight, *Polymeric Sulfur and Related Polymers*, H. F. Mark and E. H. Immergut, eds., (New York: Wiley–Interscience, 1965, © by Amer. Chem. Soc).]

Similar expressions have been derived for ring–chain equilibria which proceed through other mechanisms,[1,2] but insufficient experimental data are available for most systems to allow this approach to be more widely applied.

It will be clear from the preceding discussion that techniques are available for the interpretation of ring–chain equilibria in terms of thermodynamic data only for systems which have been subjected to considerable experimental investigation. Unfortunately, these approaches generally do not permit detailed predictions to be made about the expected thermal behavior of new or superficially studied polymers.

THE STATISTICAL INFLUENCE

A ring–chain equilibrium can conveniently be viewed from the statistical point of view. Two approaches to this aspect have been developed. In one, the concentration of any cyclic species in equilibrium with its open-chain homolog is considered in

---

[1] A. V. Tobolsky, *J. Polymer Sci.*, **25**, 220 (1957); **31**, 126 (1958).
[2] A. V. Tobolsky and A. Eisenberg, *J. Am. Chem. Soc.*, **81**, 780, 2303 (1959); **82**, 289 (1960).

terms of the *probability* of ring closure. In the other, equilibrations are viewed as scrambling reactions between monomer and end-capping units. Both of these viewpoints will now be discussed.

Consider a growing polymer chain

$$AB \xrightarrow{AB} ABAB \xrightarrow{AB} ABABAB \xrightarrow{AB} ABABABAB \xrightarrow{AB} etc.$$

At each step in the formation of open-chain species, there is a possibility that cyclization can take place. Although not indicated in the diagram, cyclization often involves the loss of the end-capping catalyst components. The probability that cyclization will occur from an open-chain homologue depends on the ease with which the chain ends can come together to form a ring, and this, in turn, is a function of the bond angles, bond lengths, and torsional flexibility of the skeletal bonds. Because many skeletal bond angles are in the region of 109 to 120°, cyclic trimers and tetramers are often heavily represented among the cyclic oligomers but, as the degree of polymerization increases, linear chains appear in greater concentrations at the expense of cyclic species.

The polymerization of rings and the depolymerization of chains are simply different aspects of the same backbone scrambling process. This is one reason why the phenomenon is frequently found in inorganic backbone systems, because the polar skeletal bonds can scramble easily at elevated temperatures.

Several investigators have examined the probability of cyclization in terms of chain length and molecular parameters. In one approach,[1] a model is used which assumes that the distribution of polymer end-to-end distances, r, in a randomly coiled polymer is given by a Gaussian function. For polymers of the type discussed here, the following equation has been derived:

$$R_n = BVx^n n^{-5/2}$$

where R_n is the number of rings with a degree of polymerization, n; $B = (3/2\pi v)^{3/2}/2b^3$; V is the volume of the system; x is the fraction of unreacted end groups in the chains; v is the number of skeletal atoms per monomer unit; and b is the "effective link length" of the chain, which equals the individual bond lengths times $(1 + \cos \alpha)/(1 - \cos \alpha)$, α being the skeletal bond angle. A plot of $\log R_n$ against $\log n$ should be a straight line with a slope of $-\frac{5}{2}$ if this model is valid.

The poly(dimethylsiloxane) system has been studied sufficiently that an appreciable amount of experimental ring–chain equilibration data has been accumulated. Thus, this system is particularly suitable for an analysis in terms of the theory

[1] H. Jacobson and W. H. Stockmayer, *J. Chem. Phys.*, **18**, 1600 (1950).

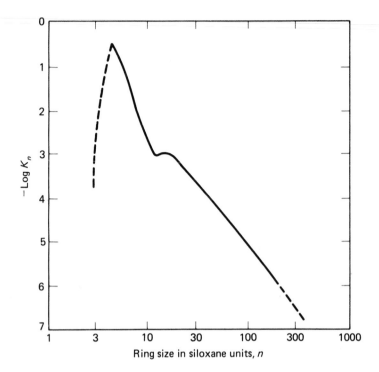

Fig. 10.4 Dependence of the molar cyclization constant, K_n, on the number of siloxane units per ring, n, for polydimethylsiloxanes in toluene at 110°C. The dashed sections of the curve are considered less reliable than the remainder. [From J. F. Brown and G. M. J. Slusarczuk, *J. Am. Chem. Soc.*, **87**, 931 (1965), © by Amer. Chem. Soc.).]

discussed above. Experimental data are available[1] which provide the ring size of the species present at equilibrium. These data are shown in Figure 10.4, in which K_n is equivalent to R_n. The cyclic trimer, $(Me_2Si\!-\!O)_3$, is always present in smaller amounts than the tetramer. The expected distribution is found from $n = 35$ at least up to $n = 200$, but deviations from the theory occur when n is 3 to 35. A small trough in the concentration versus n curve is found between $n = 12$ to $n = 16$, which indicates a smaller proportion of these species present than is predicted by the theory.

A major weakness of the theory is the presumed free torsion of the skeletal bonds. An alternative approach has been used in which a threefold rotational isomeric model for the chain is postulated.[2] This model gives a satisfactory explanation of the relative concentrations of trimer and tetramer at equilibrium, but not of the higher homologues. However, equilibrium ratios of the $n = 15$ to 200 homologues have been predicted correctly by statistical weighting of the calculation in favor of a *trans*-planar conformation.[3]

[1] J. F. Brown and G. M. J. Slusarczuk, *J. Am. Chem. Soc.*, **87**, 931 (1965).
[2] J. B. Carmichael and J. B. Kinsinger, *J. Polymer Sci.*, **A1**, 2459 (1963).
[3] P. J. Flory and J. A. Semlyen, *J. Am. Chem. Soc.*, **88**, 3209 (1966).

It seems clear that the deviations from theory result mainly from specific molecular effects which predominantly influence cyclization of the lower-molecular-weight cyclic homologues. Ring strain, intramolecular repulsions and attractions, and electrostatic effects associated with the polar chain ends must all affect the ease of cyclization, particularly in the range $n = 3$ to $n \approx 20$. Unfortunately, comparable data are not yet available for other skeletal systems, and valid extrapolations of the theory cannot be made.

The second statistical approach is concerned with the random scrambling of chain end units and middle units to create rings or chains.[1] An equilibrating system can be viewed as a scrambling mixture of chain units, end units, and ligands. For example, in terms of the terminology used previously, a simple equilibration between rings of formula $(A—B)_n$ and chains of formula $X(A—B)_n Y$ might be resolved into statistical scrambling reactions between $A—B$, XY, $X(A—B)_n Y$, and so on, units.

An analysis of the equilibrium constants, assuming random scrambling, suggests that pure chain-chain equilibria should be unaffected by dilution, but that ring-chain equilibria should be shifted toward small rings and the XY species at high dilutions.[2] Inorganic polymer systems are especially suited to this kind of analysis, and experimental correlations have been made for systems that contain $Si—O$, $Si—S$, and $S—O$ skeletal bonds.

MOLECULAR STRUCTURAL EFFECTS (QUALITATIVE APPROACH)

So far in this chapter, we have considered the general thermodynamic factors which influence polymerization–depolymerization equilibria. Unfortunately, for many organic systems and for nearly all inorganic systems, insufficient thermodynamic data are available to draw accurate conclusions or, indeed, to make even crude predictions about the thermal equilibration behavior. However, it is fortunate that certain qualitative trends can be recognized that enable molecular structural features to be correlated with polymerization or depolymerization behavior. These trends provide a nonrigorous, but useful, method of interpretation. Specific molecular factors that can be interpreted in this way include (1) an analysis of skeletal bond energies, (2) the influence of skeletal bond angles, (3) the presence or absence of pseudoaromaticity, and (4) the effect of side-group interactions. Each of these factors will now be considered.

SKELETAL BOND ENERGIES

In any polymerization system, it is possible to visualize a complete series of compounds which would contain all species from the monomer through the cyclic or linear oligomers to high polymers. As we have already seen, the amount of each

[1] J. R. Van Wazer, *J. Macromol. Sci.*, **A1**(1), 29 (1967).

[2] J. R. Van Wazer, *J. Macromol. Sci.*, **A1**(1), 29 (1967).

species present at equilibrium will depend on the free energy of that species. Although statistical influences are important, a major factor which influences the amount of each species present will be the enthalpy. This, in turn, will depend largely on the skeletal bond energy.

Many systems are known in which the skeletal bond energy varies very little from cyclic oligomers to high polymers, and it is in these systems that statistical factors tip the balance in favor of one species or another. However, significant bond energy differences usually exist between monomers or cyclic dimers and the higher homologues in the series. Cyclic dimers are often destabilized by ring strain, and this factor will be discussed in the next section.

However, an olefin-type monomer will have a higher enthalpy than the cyclic oligomers or high polymers derived from it because of the energy stored in the double bond. Thus, most unsaturated monomers are expected to polymerize exothermically and with a negative entropy change (see Table 8.1). The degree to which the polymerization is exothermic depends on the particular energy of the π-bond in the monomer, and this, in turn, is a function of the types of substituent group present.

The type of multiple bonding is also important. No stable unsaturated monomers are known of structure, $R_2Si = O$, $R_2Si = NR'$, $R_2P \equiv N$, $S \equiv N$, and so on, and species such as these either do not participate in the equilibration process or at best exist only as transient high-energy intermediates. Silicon, in particular, shows a strong tendency to avoid the formation of $p_\pi - p_\pi$ bonds. Thus, the depolymerization products from poly(organosiloxanes) or poly(organophosphazenes) are cyclic oligomers rather than monomers.

SKELETAL BOND ANGLES

The phenomenon of ring strain is well known. It occurs when the formation of cyclic compounds requires the distortion of bond angles from their preferred values. Thus, the destabilization of ring-strained species can be ascribed to an increased enthalpy term, although the probability of cyclization occurring to form such species would also be reduced. We may describe the "preferred" bond angle as the bottom of an energy well, with angular distortions from this value represented by higher enthalpy values.

It is obvious that if, for example, all the skeletal bond angles have a minimum energy at a 120° angle, then the six-membered ring will probably be a favored homologue during equilibration, whereas a four-membered ring will be a higher-energy species and will either be present at equilibrium in very small amounts or may not be found at all. Some ring strain may also be present in cyclic tetramers, pentamers, or hexamers, but the conformational mobility of higher homologues should provide for a release of the strain.

Nevertheless, some bonds have a much greater angular flexibility than others, and the bond angle may vary over as much as a 30° range before molecular destabilization becomes evident. In general, angular flexibility is often associated with skeletal atoms which can rehybridize easily, which have a large atomic radius, or which form bonds with a high ionic character. Thus, bonds to tetrahedral carbon (109.5° angle) or tetrahedral silicon often have little angular flexibility, whereas bonds to dicoordinate skeletal oxygen, nitrogen, or sulfur can vary over a wide range before appreciable strain enthalpy becomes manifest.

To illustrate the powerful destabilizing influence of ring strain, it is only necessary to note the nonexistence of a formaldehyde dimer, $(H_2C-O)_2$, in the oxymethylene equilibrate, and the absence of cyclic dimers in the siloxane, phosphazene, or vinyl-type equilibrations. Those cyclic dimers which do exist, such as $(S=N)_2$ or $[(CF_3)_2C-S-]_2$, polymerize readily (and sometimes violently). Ring strain can also be a factor in reducing the concentration of some cyclic trimers in an equilibrium mixture. For example, it has been suggested that the low equilibrium concentration of $(Me_2Si-O)_3$ in the dimethylsiloxane equilibrate may reflect 3 to 9 kcal/mol of ring strain. Higher cyclic oligomers can usually relieve bond-angle strain by puckering of the ring.

AROMATICITY AND DELOCALIZATION

The concepts of aromaticity and delocalization are well known. Delocalization of π-bonding electrons over several centers generally results in a lowering of the enthalpy of the molecule. If a polymeric series of compounds contains certain species in which the delocalization per repeating unit is particularly favored, then that molecule may be stabilized sufficiently to be a predominant species at equilibrium.

The most obvious example is benzene. The delocalization in this molecule is overwhelmingly favored by the 120° skeletal bond angles and by the planarity of the ring. By contrast, cyclobutadiene is unstable and cyclooctatetraene is puckered and nonaromatic. Furthermore, the well-known $4n + 2$ π-electron rule (Hückel's rule) provides an additional rationalization for the particular stability of benzene. A similar situation is found with s-triazines and borazines.

However, the same arguments do not apply if the skeletal multiple bonding is of the $d_\pi-p_\pi$ type. In phosphazenes (13), for example, the "unsaturation" results

13

from the use of phosphorus 3d orbitals. Because $d_\pi-p_\pi$ bonds have different symmetry requirements from $p_\pi-p_\pi$ bonds, the usual aromatic restrictions do not apply to phosphazenes, and no one member of the series appears to be stabilized by this effect. Similarly, although siloxanes are also believed to be stabilized by $d_\pi-p_\pi$ skeletal bonding, the trimer is not especially stable when compared to the other homologues.

Electron delocalization phenomena may also stabilize certain vinyl monomers (e.g., styrene) more than others, and this would be expected to become manifest as a lower than expected heat of polymerization.

SIDE GROUP INTERACTIONS

Bulky side groups attached to a polymer chain can destabilize a polymer relative to the monomer or cyclic oligomers. This effect is so far-reaching that it can overpower many of the other factors that have already been discussed.

When an equilibrium exists between a cyclic oligomer and a high polymer, the position of the equilibrium is usually determined by the relative enthalpies of the two species. If the side groups attached to the skeleton are small, then the intramolecular repulsions in the oligomer and the polymer will be comparable. However, if bulky side groups are present, polymerization of a cyclic oligomer (a trimer or tetramer, for example) to the polymer will be accompanied by an enthalpy increase—a result of intramolecular steric repulsions within the polymer. This can be illustrated by the polymerization schemes shown in Figure 10.5 and by the energy profile shown in Figure 10.6.

The side group-side group and side group-chain distances are always shorter in polymers than in cyclic dimers, trimers, or tetramers, irrespective of the preferred conformation of the polymer chain. The shorter intramolecular distances in the polymer nearly always result in more serious van der Waals repulsions, and the ultimate outcome is the destabilization of the polymer relative to the cyclic dimer, trimer, or tetramer. When such a situation exists, polymerization of the cyclic oligomers may be thermodynamically impossible. Such reasons are believed to underlie both the failure of many cyclic compounds to polymerize, and the facile depolymerization of hindered polymers prepared by alternative routes (see Chapter 7). It is, in fact, possible to predict roughly which side-group units cause destabilization for a particular skeletal system with the use of molecular models or hard-sphere van der Waals radii approaches or by a rough calculation of the energies involved within a short polymer segment.

Much difficulty has been reported with the preparation of high-molecular-weight polysiloxanes which contain side groups larger than methyl groups. For example, although equilibration of dimethylsiloxanes yields a mixture of 87% linear polymers and approximately 13% cyclic tetramer plus pentamer, the presence of one trifluoropropyl group per silicon atom in $(F_3CC_2H_4Si(CH_3)—O)_n$ markedly shifts the equilibrium to favor the cyclic oligomers.[1] Thus, the $n = 3$ to 6 cyclic content is 86.5% at equilibrium. The same type of behavior has been reported when ethyl, aryl, and trialkylsilyl side groups are present, although mechanistic influences may be partly responsible in these cases.

Evidence exists that there is a rough correlation between the ease of depolymerization of poly(organophosphazenes) and the dimensions of the substituent group. For example, poly(dichlorophosphazene) appears to resist depolymeriza-

---

[1] E. D. Brown and J. B. Carmichael, *J. Polymer Sci.*, **B3**, 473 (1965).

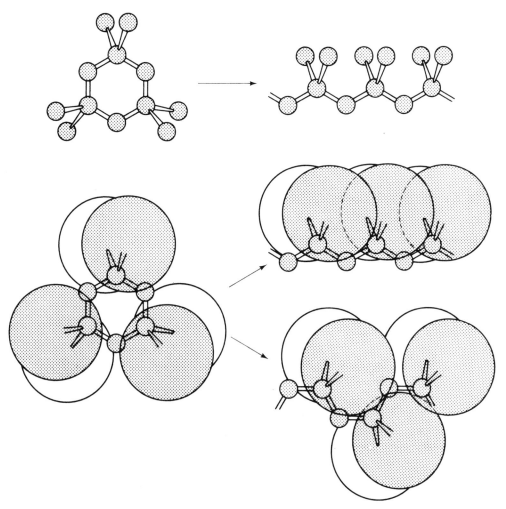

Fig. 10.5 Chain stability is favored by small side groups (top), which do not interfere with each other when a ring is polymerized to a chain. If a ring contains bulky side groups (bottom), polymerization becomes difficult if not impossible because the side groups repel each other. Moreover, chains with bulky side groups tend to depolymerize at lower temperatures than chains with small ones.

tion at temperatures below 350°C, whereas $[NP(OCH_2CF_3)_2]_n$ and $[NP(OPh)_2]_n$ depolymerize at lower temperatures. Linear organophosphazene polymers, such as $[NP(OCH_2CF_3)_2]_n$ or $[NP(OPh)_2]_n$, suffer appreciable intramolecular steric hindrance between the side groups themselves and between the side groups and the chain atoms. Van der Waals radii-type molecular models for these polymers can be constructed only with difficulty. The destabilization inherent in this crowding must be especially serious at elevated temperatures. On the other hand, *cyclic* trimers or tetramers with the same substituent groups suffer negligible intramolecular steric repulsions because the side groups are oriented away from each

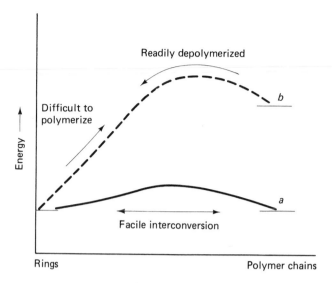

Fig. 10.6 Those rings which can be polymerized yield polymer chains that are comparable in energy (per repeating unit) to the original cyclic molecules (*a*). However, if bulky substituent groups are present (*b*), the energy of the polymer may be higher than that of the small rings. In such a case, at high temperatures the polymer will be converted easily over a small activation energy barrier to rings, but the rings cannot be polymerized to chains.

other and away from the nearby skeletal atoms. Thus, a strong argument can be made that many organosubstituted linear polyphosphazenes are thermodynamically less stable than the related cyclic trimers or tetramers. At moderate temperatures the polymers are kinetically stabilized against depolymerization, but at high temperatures a trimer-favoring equilibrium will be established. The failure of many organo*cyclo*phosphazenes to polymerize at high temperatures can be readily understood in these terms.

If an unsaturated *monomer* can be formed by depolymerization, the side-group influence may be even more serious. Conversion of a cyclic oligomer or polymer to an unsaturated monomer not only relieves the repulsions illustrated in Figure 10.5, but it also relieves the repulsions between two side groups attached to the same skeletal atom (**14**). The relatively low enthalpy of polymerization of

$$\left[\begin{array}{c} R \\ \diagdown \diagup 109.5° \\ -A-B- \\ \diagup \diagdown \\ R \end{array} \right] \longrightarrow \begin{array}{c} R \\ \diagdown \diagup \\ A=B \diagdown 120° \\ \diagup \diagdown \\ R \end{array}$$

14

isobutylene (-12.6 kcal/mol) may reflect the influence of methyl-group hindrance in the conversion of the monomer to the polymer. For example, conversion of a polyaldehyde to an aldehyde changes the hybridization at carbon from sp^3 to sp^2. This means that the R—C—R bond angle changes from 109.5° to 120°. Hence, the repulsions from the presence of bulky R groups can be relieved especially easily by depolymerization.

Although formaldehyde can be polymerized to trioxane, tetroxane, and poly-oxymethylene, aldehydes or ketones with side groups larger than hydrogen polymerize only with difficulty. For example, equilibration of acetaldehyde at 15°C yields 94.3 wt % of the cyclic trimer, $(MeHC—O)_3$, and at 0°C the equilibrate contains 5.6% monomer, 91% cyclic trimer, and 3.2% cyclic tetramer. High polymers are formed only below the ceiling temperature of $-40°C$, and they de-polymerize at room temperature. The thermal instability of this polymer contrasts strikingly with that of poly(oxymethylene), and the differences must be ascribed to steric repulsions by the methyl side groups. When the side group is larger than methyl, the facile depolymerization of the polymer becomes even more noticeable. Aldehydes with ethyl, n-propyl, isopropyl, n-butyl, and cyclohexyl groups can be polymerized only at high pressures and often at low temperatures. Once formed, the polymers depolymerize to monomer under ambient conditions. Polyacetone depolymerizes readily at room temperature, as does poly(hexafluorothioacetone), $[(F_3C)_2C—S]_n$.[1]

THE MECHANISTIC ASPECT

Polymerization of an unsaturated monomer or a cyclic oligomer, or depolymer-ization of a high polymer, depends not only on the free-energy change, but also on the activation energy for the reaction. It is, in fact, quite possible for a poly-merization or depolymerization to be energetically feasible, yet kinetically in-hibited. If the activation energy is high, the rate of equilibration will be infinitely slow at moderate temperatures. This raises the interesting point that, if the activa-tion energy for polymerization is high, and if depolymerization is thermodynami-cally preferred at high temperatures, then it may be impossible to find conditions for the uncatalyzed polymerization of a monomer or cyclic oligomer. Thus, cata-lysts play an important role in facilitating most monomer–chain and ring–chain equilibration processes. It must also be emphasized that serious experimental difficulties are involved in the separation of mechanistic influences from thermo-dynamic factors. For example, in many systems the observed change in concentra-tion of one homologue as the temperature is raised results not from changes in the equilibrium constant but from side reactions. Thus, extreme care must be exercised in the interpretation of equilibration data.

Most ring–chain interconversions are strongly influenced by the presence of a catalyst or end-capping species. Thus, the initiation for a heterolytic ring cleavage generally involves a reaction with the initiator. A general mechanism can be de-picted as

---

[1] W. J. Middleton, H. W. Jacobson, R. E. Putnam, H. C. Walter, D. G. Pye, and W. H. Sharkey, *J. Polymer Sci.*, **A3**, 4115 (1965).

where X^+Y^- is the initiator and B is the most electronegative element in the skeleton. Propagation then involves attack by the BX or AY bonds on another oligomer molecule.

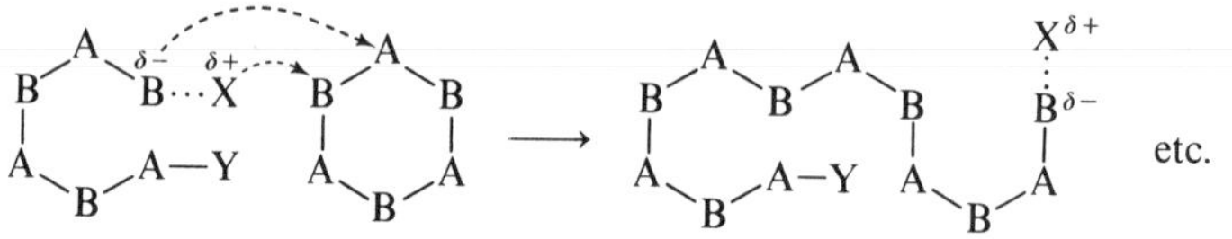

$$\text{(reaction scheme)} \longrightarrow \text{etc.}$$

If the chain is short, cyclization can occur by the process

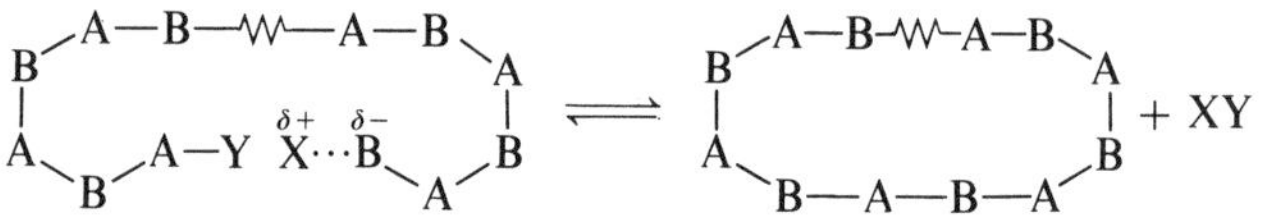

$$\text{(cyclic structure)} \rightleftharpoons \text{(cyclic structure)} + XY$$

But, if the chain is long, cyclization may occur only infrequently, and end-capped high-molecular-weight chains may be present in appreciable quantities. Of course, if the chain ends are held together by electrostatic forces, cyclization will always be preferred. However, the presence of high-molecular-weight chains with active end groups is a prime reason for the thermal instability of many polymers, since facile cyclization–depolymerization reactions occur at moderate or high temperatures, often by a "back-biting" mechanism such as

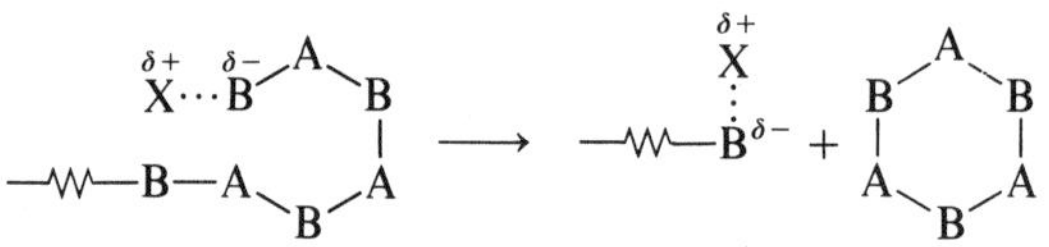

$$\text{(back-biting reaction scheme)}$$

It will be recognized that this general mechanism operates in the polymerization and depolymerization of trioxane (Chapter 6), siloxanes, and phosphates (Chapter 7). Since the mechanism provides a low-energy pathway for depolymerization, polymers susceptible to this type of mechanism often depolymerize at moderate temperatures. For this reason, the thermal stability of a high polymer can frequently be improved by total removal of the catalyst or by replacement of the end groups by nonionic substituents.

Free-radical cleavage processes of carbon–carbon bonds occur readily at 200 to 300°C. However, it is important to distinguish between the *random* fragmentation of polymer chains (which does not represent an equilibration process) and the "unzipping" of monomer molecules from the chain ends that takes place with depolymerization. The latter process can be understood readily in terms of the discussion in this chapter.

STUDY QUESTIONS

1. Discuss possible reasons why a cyclic compound such as trioxane can be induced to polymerize, but benzene cannot.

2. A cyclic trimer, $(A-B)_3$, polymerizes to form a linear macromolecule, $(A-B)_n$, in the melt at 100°C. At equilibrium, 40 wt % of the total species present is high polymer. Calculate ΔG

and ΔH for the addition of a trimer unit to the end of a growing chain. Assume that $\Delta S = 28$ cal/deg-mol.

3. In what way does a catalyst influence a polymerization–depolymerization equilibrium, and why?

4. Define the terms "ceiling temperature" and "floor temperature." Why do these definitions generally apply only to closed systems?

5. Assume that the polymerization of S_8 proceeds through the formation of a zwitterion $^+S—(S_6)—S^-$, instead of through the diradical, $\cdot S—S_6—S\cdot$. How would this be expected to affect the overall character of the equilibrium, T_c, T_f, ΔS_p, and ΔH_p?

6. What effects on the equilibration behavior of organosiloxanes would be expected from the replacement of methyl side groups by ethyl, phenyl, chloro, fluoro, cyano, or trimethylsilyl groups?

7. A certain ring-opening polymerization reaction (cf. equation 8) of a cyclic compound is exothermic by 7.0 kcal/mol and occurs with a negative entropy change of -21 cal/deg-mol.
   (a) Using the Gee approach, construct a plot of the weight fraction of molecules that exist as chains as a function of temperature.
   (b) Evaluate the ceiling temperature for the polymerization.
   (c) At what temperature is the weight fraction of chains equal to 0.95?
   (d) Specify any implicit or explicit assumptions that you have made in your calculations.

8. In a polymerization of rhombic sulfur from the monomeric form of S_8 rings, the following data were found at equilibrium [A. V. Tobolsky and A. Eisenberg, *J. Am. Chem. Soc.*, **81**, 780, 2303; **82**, 299 (1960)]:

	267°C	167°C
Average degree of polymerization	1.0×10^4	9.5×10^4
Monomer concentration (mol/kg)	1.82	3.63

   (a) Using the approach of Tobolsky, evaluate $\Delta H°$, $\Delta S°$, $\Delta H_3°$, and $\Delta S_3°$, as these quantities are defined on page 231, for the sulfur equilibrium.
   (b) Construct van't Hoff plots (i.e., $\ln K$ versus $1/T$) for both K and K_3.
   (c) Calculate the average degree of polymerization and the monomer concentration in this system at equilibrium at 200°C.

SUGGESTIONS FOR FURTHER READING

DAINTON, F. S., AND K. J. IVIN, "Some Thermodynamic and Kinetic Aspects of Addition Polymerisation," *Quart. Rev. Chem. Soc.* (*London*), **22**, 61 (1958).

GRASSIE, N., "Degradation," *Encyclopedia of Polymer Sci. and Technol.*, **4**, 647 (1966).

JELLINECK, H. H. G., "Depolymerization," *Encyclopedia of Polymer Sci. and Technol.* (H. F. Mark, N. G. Gaylord, and N. M. Bikales, eds.), **4**, 740 (1966).

SAWADA, H., "Thermodynamics of Polymerization. II. Thermodynamics of Ring-Opening Polymerization," *J. Macromol. Sci.—Rev. Macromol. Chem.*, **C5**, 151 (1970).

SAWADA, H., "Thermodynamics of Polymerization. V. Thermodynamics of Copolymerization, Part 1," *J. Macromol. Sci.— Rev. Macromol. Chem.*, **C10**, 293 (1974).

SAWADA, H., "Thermodynamics of Polymerization. VI. Thermodynamics of Copolymerization, Part 2," *J. Macromol. Sci.— Revs. Macromol. Chem.*, **C11**, 257 (1974).

SEMYLEN, J. A., "Ring-Chain Equilibria and the Conformations of Polymer Chains," *Adv. Polymer Sci.*, **21**, 41 (1976).

TOBOLSKY, A. V., *Properties and Structure of Polymers*, New York: Wiley, 1960.

11

Kinetics of Condensation (*Step-Growth*) Polymerization

GENERAL CONSIDERATIONS

As discussed in Chapter 2, condensation (or step-growth) polymerization occurs by consecutive reactions in which the degree of polymerization and average molecular weight of the product increase as the reaction proceeds. Usually, although not always, the reactions involve the elimination of a small molecule such as water, and hence the name *condensation polymerization* is used as a general term. We may represent the reactions occurring in a water-elimination polymerization by processes (1) to (6).

$$\text{Monomer} + \text{monomer} \longrightarrow \text{Dimer} + H_2O \qquad (1)$$

$$\text{Monomer} + \text{dimer} \longrightarrow \text{Trimer} + H_2O \qquad (2)$$

$$\text{Monomer} + \text{trimer} \longrightarrow \text{Tetramer} + H_2O \qquad (3)$$

$$\text{Dimer} + \text{dimer} \longrightarrow \text{Tetramer} + H_2O \qquad (4)$$

$$\text{Dimer} + \text{trimer} \longrightarrow \text{Pentamer} + H_2O \qquad (5)$$

$$\text{Trimer} + \text{trimer} \longrightarrow \text{Hexamer} + H_2O \qquad (6)$$

etc.

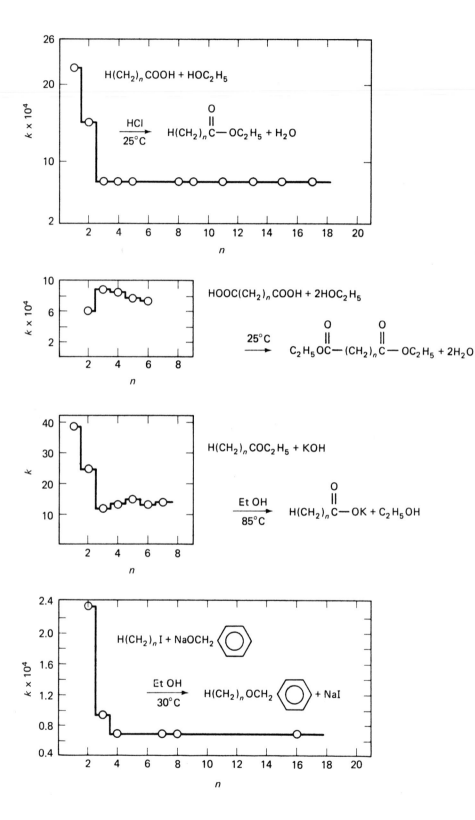

Fig. 11.1 (*Opposite*) Dependence of rate constants of functional group reactions on molecular size. [After P. J. Flory, *Principles of Polymer Chemistry* (Ithaca, N.Y.: Cornell University Press, 1953), pp. 70–71.]

Generally, the reactions are reversible, so that the eliminated water must be removed if a high-polymeric product is to be formed. To describe the course of these reactions (which produce, consecutively, stable dimers, trimers, tetramers, pentamers, ..., high polymers) in terms of reaction kinetics would seem at first sight to be a very complicated task. However, fortunately, it is possible to introduce a simplifying approximation that makes the kinetic problem tractable.

For example, consider a polyesterification in which a dibasic acid condenses with a glycol. No matter what the degree of conversion or the molecular weight of the reactants happens to be, the chemical interaction in each step is the same and may be written as in (7). The simplification that renders the kinetic problem soluble

$$\text{\textbackslash\!\textbackslash\!\textbackslash—COOH + HO—\textbackslash\!\textbackslash\!\textbackslash} \longrightarrow \text{\textbackslash\!\textbackslash\!\textbackslash—}\overset{\displaystyle O}{\overset{\displaystyle \|}{C}}\text{—O—\textbackslash\!\textbackslash\!\textbackslash} + H_2O \tag{7}$$

is the assumption that the rate constant of (7) is independent of the size of the molecules to which the functional groups are attached. The effect of molecular size on the rate constants of several small molecule reactions that are quite analogous to stepwise polymerization is shown in Figure 11.1. It may be seen in this figure that, except for the very small reactants ($n \le 2$), the rate constants or reactivities are independent of molecular size. Thus, as a reasonable approximation, each and every condensation reaction in a polyesterification may be assumed to proceed with the same rate constant. Only in the very early stages, that comprise only a minor part of the polymerization, is this assumption of questionable validity.

THE REACTIVITY OF LARGE MOLECULES

The assumption that the reaction rate may be considered to be constant in spite of changes of molecular size of the reactants is, at first sight, startling. Therefore, it is necessary to inquire into a physical rationalization for the validity of this assumption. There appear to be two major reasons:

1. Consider a molecule undergoing a polyesterification reaction to be a long chain with many possible conformations and having a COOH functional group on the end. The center of mass of the molecule is far removed from the COOH group. It is true that the larger (and heavier) the molecule, the slower will be the diffusion of the center of mass through the solution. However, the rate of movement of the COOH and OH functional groups through the solution may be very different from the rate of movement of the center of mass, and in the final analysis it is the encounter of functional groups that results in chemical reactions. Thus, through changes in the polymer chain conformations, the functional groups COOH and OH may encounter each other at much higher rates than is suggested by the masses

of the molecules of which they are a part. This enhanced encounter rate, which is due to changes in the polymer conformation, would be expected to be approximately independent of molecular size.

2. According to the liquid cage theory of Rabinowitch and Wood,[1] a colliding pair of particles (or functional groups) will be surrounded by a confining cage of solvent molecules. Before the colliding pair can escape from this solvent cage by diffusion away from each other, they will collide frequently. Slower diffusion rates are found in liquids, as compared with gases, and this means that two potential reactants will not come together rapidly, but once brought together will not separate rapidly either. Thus, each encounter leads to multiple collisions. Both the average number of collisions per encounter and the time between encounters increase as the molecular size increases (i.e., as the diffusion rate decreases).

We may represent this in a kinetic sense for two reactants, A and B, by

$$A + B \underset{k_{-8}}{\overset{k_8}{\rightleftarrows}} (A + B) \tag{8}$$

$$(A + B) \xrightarrow{k_9} P \tag{9}$$

where $(A + B)$ represents the pair of reactants trapped in the liquid cage and P is a product molecule (i.e., polymer). k_8 and k_{-8} represent diffusion rate constants of the reactants into and out of the liquid cage, respectively, while k_9 is the rate constant for chemical reaction. If we assume a steady state for the concentration of trapped pairs, the observed rate of product formation, $d[P]/dt$, is easily shown to be given by

$$\frac{d[P]}{dt} = \frac{k_8 k_9}{k_{-8} + k_9} [A][B] \tag{10}$$

There are two cases to consider:

(a) Diffusion is much more rapid than chemical reaction, $k_{-8} \gg k_9$, and the observed rate constant is given by

$$\text{Case (a):} \quad \frac{d[P]}{dt} = \frac{k_8}{k_{-8}} k_9 [A][B] \tag{11}$$

Since k_8 and k_{-8} represent diffusion into and out of the liquid cage, they are affected in the same way by increases in the size of the reactants. Therefore, the effect of the size of the reactants on the reaction rate will be determined by the effect of size on k_9. The rate constant for a reaction of two functional groups *contained in the same solvent cage* should be independent of the size of the molecule to which they are attached. Hence, for case (a), namely when diffusion is fast compared to chemical reaction, the observed rate constant should be independent of molecular size.

[1] E. Rabinowitch and W. C. Wood, *Trans. Faraday Soc.*, **32**, 1381 (1936).

(b) If the chemical reaction is very fast relative to diffusion, $k_9 \gg k_{-8}$, and (10) reduces to

$$\text{Case (b):} \quad \frac{d[P]}{dt} = k_8[A][B] \tag{12}$$

In this case the observed rate constant will be the same as the diffusion rate constant, and this itself will depend on molecular size.

It is apparent from (10) to (12) that, on the basis of liquid cage theory, the lack of dependence of the reaction rate constant on molecular size requires that the chemical reaction be slow compared to diffusion. Those reactions that are characteristic of condensation polymerization are typically the slow reactions of organic chemistry, with appreciable activation energies of at least 12 kcal/mol. It may therefore be concluded that the rate constants of these chemical reactions will, in general, be significantly lower than the diffusion rate constants. Hence, according to (11), the rate constants observed in condensation polymerization will be independent of molecular size. In other words, it is a valid approximation to look upon polyesterification, polyamidation, or polyurethane-forming reactions, and so on, as general reactions of the forms shown by (13) to (15) respectively, in which the rate constants k, k', and k'' do not depend on the sizes of the molecules to which the functional groups are attached.

$$\text{W—COOH + HO—W} \xrightarrow{k} \text{W—}\overset{\overset{\displaystyle O}{\parallel}}{C}\text{O—W + H}_2\text{O} \tag{13}$$

$$\text{W—}\overset{\overset{\displaystyle O}{\parallel}}{C}\text{—Cl + H}_2\text{N—W} \xrightarrow{k'} \text{W—}\overset{\overset{\displaystyle O}{\parallel}}{C}\text{—NH—W + HCl} \tag{14}$$

$$\text{W—N=C=O + HO—W} \xrightarrow{k''} \text{W—NH}\overset{\overset{\displaystyle O}{\parallel}}{C}\text{O—W} \tag{15}$$

RATES OF POLYCONDENSATION REACTIONS

Since the rate constant of a condensation polymerization reaction (or reactivity of two functional groups) is independent of molecular size, it is possible to measure the rate of reaction simply by determining the concentration of functional groups as a function of time. For example, this may be done easily by titration of the unreacted carboxylic acid groups during a polyesterification reaction. Thus, in a polyesterification, the general reaction at any time t is as shown by (13) and, as the reaction proceeds, the functional groups —COOH and —OH disappear at the same rate. Therefore, samples can be removed from the reaction mixture at various intervals and the concentration of carboxylic acid groups can be determined. The rate of the reaction is then defined as

$$\text{Reaction rate} = -\frac{d[\text{COOH}]}{dt} \tag{16}$$

KINETICS OF POLYESTERIFICATION

It is well known in organic chemistry that esterification reactions are catalyzed by acids. Polyesterification is no exception. The rate law may then be written

$$-\frac{d[COOH]}{dt} = k[COOH][OH][acid] \tag{17}$$

To proceed further it is necessary to distinguish between those systems in which no acidic catalyst is added and those to which a catalyst has been added and in which its concentration remains constant throughout the polymerization.

Case 1: No Acidic Catalyst Added

In this situation the carboxylic acid groups themselves must function as the acid catalyst and (17) becomes

$$-\frac{d[COOH]}{dt} = k_3[COOH]^2[OH] \tag{18}$$

where the subscript to k denotes a third-order rate constant. Assume now that stoichiometric quantities of the reactants were present initially. In other words, at $t = 0$,

$$[COOH]_0 = [OH]_0 = 2[HOOC-R-COOH]_0 = 2[HOR'OH]_0 \tag{19}$$

As shown by the general reaction, (13), COOH and OH groups disappear at the same rate. Therefore, at all times $[COOH] = [OH]$ and the rate equation (18) for the uncatalyzed reaction becomes

$$-\frac{d[COOH]}{dt} = k_3[COOH]^3 \tag{20}$$

Integration of (20) leads immediately to

$$\frac{1}{[COOH]^2} = \frac{1}{[COOH]_0^2} + 2k_3 t \tag{21}$$

It is convenient to express the conversion, or extent of reaction, in terms of the fraction of COOH groups (or OH groups) that have reacted. Thus, if P is the fraction of COOH groups reacted, then

$$P = 1 - \frac{[COOH]}{[COOH]_0} \tag{22}$$

or

$$[COOH] = [COOH]_0(1 - P) \tag{23}$$

GLYCOL	A (kg/equiv)^2 min^{-1}	E (kcal/mol)	k at 202°C (kg/equiv)^2 min^{-1}
HO—(CH$_2$)$_2$OH	—	—	~0.005
HO—(CH$_2$)$_{10}$OH	4.8×10^4	14	0.0175
HO—(CH$_2$)$_{12}$OH	—	—	0.0157
HO—(CH$_2$)$_2$O—(CH$_2$)$_2$OH	4.7×10^2	11	0.0041

Substitution of (23) into (21) gives the result

$$\frac{1}{(1 - P)^2} = 1 + 2[\text{COOH}]_0^2 k_3 t \tag{24}$$

When plots are made of experimental values of $(1/1 - P)^2$ versus time, it is found that (24) is not obeyed from $P = 0$ up to about $P = 0.80$ (i.e., for the first 80% of the esterification of —COOH and —OH groups). After 80% conversion the integrated rate expression (24) is obeyed very well. The deviations below 80% conversion are not unique to *poly*esterifications, however, because they are also observed for the simple esterifications that result when the dicarboxylic acid is replaced by a monocarboxylic acid.

Apparently, the major reason for the nonadherence to (24) below about 80% conversion is that the reaction medium is changing from one of pure reactants initially to one in which the ester product is the solvent. The prevalent[1] (although not universal) view among polymer chemists is that the kinetics of condensation polymerization have meaning only for the last 20% of the reaction when the reaction medium has become essentially invariant. Hence, the *true* reaction rate constants are to be obtained from the linear portion of plots of $1/(1 - P)^2$ versus time. For example, typical plots of approximately the last 20% of the uncatalyzed polyesterification of adipic acid and 1,10-decanediol, namely,

$$n\,\text{HOOC}—(\text{CH}_2)_4—\text{COOH} + n\,\text{HO(CH}_2)_{10}\text{OH}$$

$$\longrightarrow \text{HO} \left[\overset{\overset{\displaystyle O}{\|}}{\text{C}}—(\text{CH}_2)_4 \overset{\overset{\displaystyle O}{\|}}{\text{C}}—\text{O(CH}_2)_{10}\text{O} \right]_n \text{H} + (2n - 1)\text{H}_2\text{O} \tag{25}$$

are shown in Figure 11.2. The rate constants obtained from the slopes and initial concentrations of [COOH] are shown in Table 11.1, along with those from some other esterifications of adipic acid. Note that the concentration units of the rate constants are in terms of "equivalents per kilogram"; this is a more convenient measure of concentration than the usual "moles per liter" because the volume of the system decreases significantly. The Arrhenius parameters A and E of the equation $k = Ae^{-E/RT}$ are also tabulated in Table 11.1 for those reactions that have been studied kinetically at more than one temperature.

---

[1] D. H. Solomon, *Step-Growth Polymerizations*, D. H. Solomon, ed. (New York: Dekker, 1972), Chap. 1.

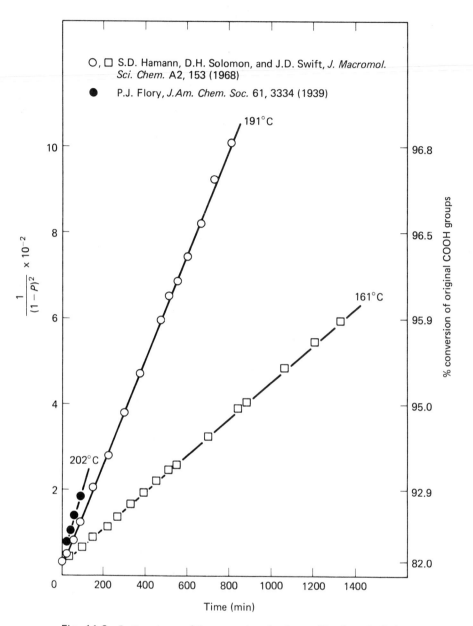

Fig. 11.2 Latter stages of the uncatalyzed polyesterification of adipic acid and 1,10-decamethylene glycol.

Case 2: Acid-Catalyzed Polyesterification

If an acid catalyst is added to a polyesterification system (which contains equal quantities of COOH and OH), the general equation, (17), becomes

$$-\frac{d[\text{COOH}]}{dt} = [\text{COOH}]^2(k_3[\text{COOH}] + k_{\text{cat}}[H^+]) \tag{26}$$

where k_3 is the rate constant for the uncatalyzed reaction and k_{cat} is the rate constant for the catalyzed process. By the definition of a catalyst, the $[H^+]$ does not change throughout the course of the reaction and generally $k_{\text{cat}}[H^+] \gg k_3[\text{COOH}]$. As a result, (26) usually can be approximated by (27), in which the second-order rate constant k_2 is related to k_{cat} by the expression $k_2 = k_{\text{cat}}[H^+]$.

$$-\frac{d[\text{COOH}]}{dt} = k_2[\text{COOH}]^2 \tag{27}$$

Integration of (27) and substitution of (23) leads to

$$\frac{1}{1 - P} = 1 + k_2[\text{COOH}]_0 t \tag{28}$$

which is a description of the dependence of the conversion on reaction time for a catalyzed polyesterification. Second-order plots according to (28) are shown in Figure 11.3 for the last 20% of the polyesterifications of adipic acid by 1,10-decanediol and diethyleneglycol catalyzed by *p*-toluene sulfonic acid. The increase in reaction rate resulting from the presence of the acid catalyst can be seen by comparing the respective conversions as a function of time in Figures 11.2 and 11.3. The second-order rate constants for catalyzed polyesterifications are obtained from the slopes of such plots and the initial concentration of carboxyl groups in accordance with (28).

Second-order rate constants for some acid-catalyzed polyesterifications and polyamidations obtained in this way are shown in Table 11.2.

TABLE 11.2 Rate Constants for Some Acid-Catalyzed Polyesterifications and Polyamidations

Monomer System	Catalyst	T (°C)	k_2 (kg/equiv-min)	A (kg/equiv-min)	E (kcal/mol)
HO(CH$_2$)$_2$O(CH$_2$)$_2$OH + HOOC(CH$_2$)$_4$COOH	0.4% *p*-toluene sulfonic acid	109	0.013	—	—
HO(CH$_2$)$_{10}$OH + HOOC(CH$_2$)$_4$COOH	0.4% *p*-toluene sulfonic acid	161	0.097	—	—
H$_2$N—(CH$_2$)$_{16}$COOH	*m*-Cresol (solvent)	175	0.012	1.7×10^{12}	29
H$_2$N—(CH$_2$)$_{10}$COOH	*m*-Cresol (solvent)	176	0.011	1.4×10^{13}	31

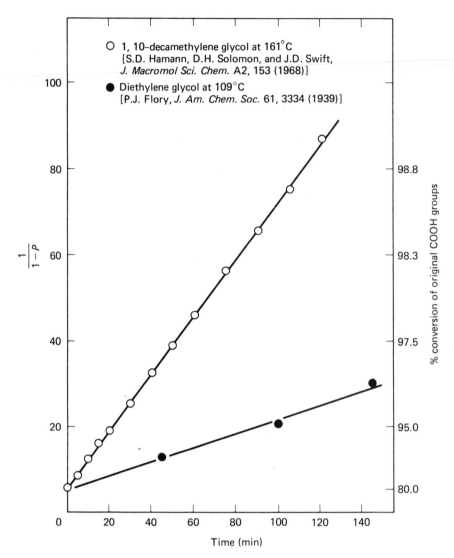

Fig. 11.3 Later stages of the polyesterification of adipic acid catalyzed by *p*-toluene sulfonic acid. (The reaction time of zero corresponds to ~82% esterification of the original COOH groups present.)

TIME DEPENDENCE OF THE AVERAGE DEGREE OF POLYMERIZATION AND THE AVERAGE MOLECULAR WEIGHT

Consider a polyesterification of bifunctional monomers in which initially equal amounts of dibasic acid and glycol are present. Under such conditions, the initial number of COOH groups is equal to the total number of molecules present initially in the system. As a consequence of each esterification reaction, namely (7),

$$\text{w—COOH} + \text{HO—w} \longrightarrow \text{w—}\overset{\overset{\displaystyle O}{\|}}{C}\text{—O—w} + H_2O \tag{7}$$

one COOH group disappears, but the total number of molecules present is unchanged. However, if the water formed in the reaction is removed (and this must be done to obtain high polymer), then, for each COOH group lost, one molecule is removed from the system. Thus, with an efficient removal of water, the number of COOH groups present is equal to the number of molecules present, not only initially, but throughout the reaction. If N is the total number of molecules in the system and V is the volume, it is possible to write

$$\frac{N}{V} = [\text{COOH}] = [\text{COOH}]_0(1 - P) \tag{29}$$

The *repeating unit* of each polyester molecule formed from the monomers HOOC—R—COOH and HOR′OH, namely

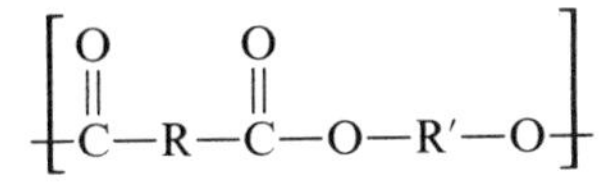

$$\begin{bmatrix} \overset{\overset{\displaystyle O}{\|}}{C}\text{—R—}\overset{\overset{\displaystyle O}{\|}}{C}\text{—O—R′—O} \end{bmatrix}$$

contains one *structural unit* from the glycol, namely, —O—R′—O—, and one *structural unit* from the dibasic acid, namely $-\overset{\overset{\displaystyle O}{\|}}{C}\text{—R—}\overset{\overset{\displaystyle O}{\|}}{C}-$. Structural units are never removed from the system. Therefore, the total number of structural units present at all times is a constant and is equal to the initial number of molecules. Hence, in view of (29) it is possible to write

$$\frac{N_{\text{structural units}}}{V} = [\text{COOH}]_0 \tag{30}$$

The average degree of polymerization of the system, $\overline{DP}$, is defined as the average number of structural units per molecule. Therefore, in view of (23), (29), and (30), the $\overline{DP}$ can be defined by

$$\overline{DP} = \frac{[\text{COOH}]_0}{[\text{COOH}]} = \frac{1}{1 - P} \tag{31}$$

Note that, for condensation polymers prepared from two reactants, the average number of *repeating units* per molecule is one-half of the average degree of polymerization.

If $\overline{M}_0$ is the average molecular weight of the structural units that make up a repeating unit, then the number-average molecular weight of the polyester is given by

$$\overline{M}_n = \frac{\overline{M}_0}{1 - P} + 18 \qquad (32)$$

where 18 is added to account for unreacted groups at the ends of each polyester chain.

As an example of the dependence of the average molecular weight on the conversion, consider the polyesterification of adipic acid, $HOOC-(CH_2)_4-COOH$, and 1,10-decanediol, $HO-(CH_2)_{10}-OH$. The average molecular weight of a structural unit is: $\overline{M}_0 = (112 + 172)/2 = 142$. Hence, the dependence of the number-average molecular weight on conversion P is, from (32),

$$\overline{M}_n = \frac{142}{1 - P} + 18$$

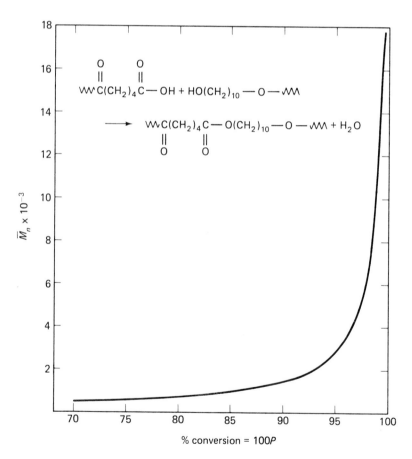

Fig. 11.4   Number-average molecular weight of poly-1,10-decanediol-adipate as a function of conversion.

A plot of $\overline{M}_n$ versus conversion, P, for this system is shown in Figure 11.4. This illustrates the fact that very high conversions are required to produce useful polymers having molecular weights above 10,000.

By combining (32) with (24) and (28), the dependence of the molecular weight on reaction time for uncatalyzed and catalyzed polyesterifications, respectively, is obtained as shown by

$$\overline{M}_n = \overline{M}_0(1 + 2[COOH]_0^2 k_3 t)^{1/2} + 18 \tag{33}$$

$$\overline{M}_n = \overline{M}_0(1 + [COOH]_0 k_2 t) + 18 \tag{34}$$

The kinetic expressions (24) and (28) are obeyed for conversions above $\sim 80\%$. For conversions of this magnitude it is generally true that the values of t are sufficiently large that unity in the parentheses of (33) and (34) may be neglected. The approximate equations

$$\overline{M}_n(\text{uncatalyzed}) \approx \overline{M}_0[COOH]_0(2k_3)^{1/2}t^{1/2} \tag{35}$$

$$\overline{M}_n(\text{catalyzed}) \sim \overline{M}_0[COOH]_0 k_2 t \tag{36}$$

are then obtained. These equations may be used with the rate-constant data of Tables 11.1 and 11.2 to construct curves of number-average molecular weights as a function of time.

While most of the kinetic relationships derived in this section have referred to polyesterification reactions between a dicarboxylic acid and a glycol, an extension to other step-growth polymerizations of bifunctional monomers is straightforward.

MOLECULAR-WEIGHT DISTRIBUTIONS
OF LINEAR CONDENSATION POLYMERS

The preceding section has covered the average molecular weight and its dependence on reaction time. However, it is also of interest to determine the *distribution* of molecular weights and the dependence of this distribution on the reaction time. An evaluation of the molecular-weight distribution is readily accomplished with the general assumption that the reactivity of functional groups is independent of the size of the molecule to which they are attached.

Consider, for example, the polyamidation of an amino acid of structure $H_2N-R-COOH$. With a single-component system, the equality of reacting functional groups can be guaranteed at all times (i.e., $[NH_2] = [COOH]$) and only one structural unit is present (i.e., $-NH-R-C-$). Of course, the analysis will also be valid for polycondensations or step-growth polymerizations that involve the interaction of two bifunctional monomers, and an extension of the argument to such cases is straightforward.

What is the probability that a molecule selected randomly from the polymerizing mixture will be found to contain *exactly* x structural units? In other words, if the terminal NH_2 group of this randomly selected polymer molecule is considered, what is the probability that it is connected to exactly x structural units? To answer this question, recall that P is the fraction of COOH groups that have reacted in time t. Then, $1 - P$ is the fraction of COOH groups remaining at time t. Expressed in another way, P is the probability that at time t a given COOH group will have reacted; $1 - P$ is then the probability that at time t a given group will *not* have reacted. It is instructive at this stage to consider Table 11.3, describing the progress of the polyamidation.

Examination of Table 11.3 shows that in the x-mer (i.e., the randomly selected polymer molecule containing exactly x structural units), $(x - 1)$ reacted COOH groups and one unreacted COOH group will form part of the residue that is connected to each NH_2 group. Beginning at the terminal NH_2 group of a randomly selected polymer molecule, the probability that $(x - 1)$ reacted COOH groups and one unreacted COOH group will be found is the probability that the molecule will contain exactly x structural units, and this is given by

$$\text{Prob}(x) = P^{x-1}(1 - P) \tag{37}$$

The chance that a randomly selected polymer molecule contains exactly x structural units is given by (37) and is equal to the fraction of molecules that is composed of x-mers. Hence, the number of x-mers, N_x, in a system of N molecules is given by

$$N_x = NP^{x-1}(1 - P) \tag{38}$$

The total number of structural units in the system, as discussed earlier, is equal to the initial number of molecules present. Furthermore, if the water produced by

TABLE 11.3 Progress of Polyamidation

MOLECULE	NUMBER OF STRUCTURAL UNITS PRESENT	NUMBER OF REACTED COOH GROUPS
$H-NHR\overset{O}{\overset{\|}{C}}-OH$	1	0
$H-NHR\overset{O}{\overset{\|}{C}}-NHR\overset{O}{\overset{\|}{C}}-OH$	2	1
$H-NHR\overset{O}{\overset{\|}{C}}-NHR\overset{O}{\overset{\|}{C}}-NHR\overset{O}{\overset{\|}{C}}-OH$	3	2
$H-NHR\overset{O}{\overset{\|}{C}}-NHR\overset{O}{\overset{\|}{C}}-NHR\overset{O}{\overset{\|}{C}}-NHR\overset{O}{\overset{\|}{C}}-OH$	4	3
$\vdots$		
$H\left[NHR\overset{O}{\overset{\|}{C}}\right]_{x-1}NHR\overset{O}{\overset{\|}{C}}-OH$	x	$x - 1$

the reaction is removed, the number of COOH groups is at all times equal to the number of molecules present. Therefore, we may write

$$N_{COOH} = N = N_0(1 - P) \tag{39}$$

which on substitution into (38) gives the number of x-mers present in terms of the initial number of molecules, N_0, as

$$N_x = N_0(1 - P)^2 P^{x-1} \tag{40}$$

The distribution in (40) shows that, for any given conversion, P, (or reaction time t), the highest probability is that low-molecular-weight polymers will be present. However, the distribution becomes broader and the average molecular weight increases as the conversion increases. This is shown clearly in Figure 11.5,

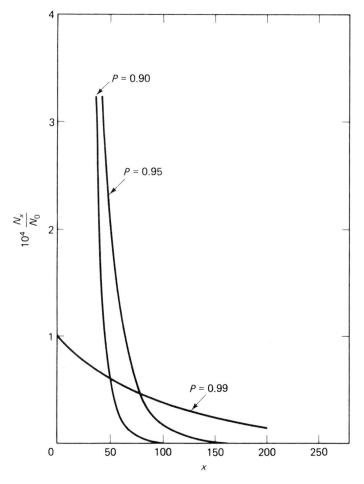

Fig. 11.5 Numerical distribution of the number of structural units in a condensation polymer for conversions of 90%, 95%, and 99% (see equation 40).

in which the fraction N_x/N_0 is plotted as a function of x for conversions of 90%, 95%, and 99%. According to the definition of N_x/N_0, the area under each curve is $1 - P$.

The number-average molecular weight may be obtained from the probability function in (37) and the usual definition of an arithmetic average. Neglecting the weight of water contained in the terminal groups, the molecular weight of an x-mer is given by $M_{x\text{-mer}} = xM_0$, where M_0 is the molecular weight of the structural unit. Therefore, we may write

$$\overline{M}_n = \sum_{x=1}^{N_0} xM_0 \text{ Prob}(x) \tag{41}$$

Substitution of (37) into (41) yields

$$\overline{M}_n = M_0 \sum_{x=1}^{N_0} x(1 - P)P^{x-1} = M_0(1 - P) \sum_{x=1}^{N_0} xP^{x-1} \tag{42}$$

The weight fraction of x-mers, W_x, is the weight of molecules containing exactly x structural units divided by the total weight of the polymer, as shown by

$$W_x = \frac{N_x M_x}{\sum_{x=1}^{\infty} N_x M_x} = \frac{M_0 x N_x}{M_0 \sum_{x=1}^{\infty} xN_x} = \frac{x(1 - P)^2 P^{x-1}}{\sum_{x=1}^{\infty} x(1 - P)^2 P^{x-1}} \tag{43}$$

As will be shown later, $\sum_{x=1}^{\infty} xP^{x-1} = (1 - P)^{-2}$, so that from (43) we obtain the weight fraction distribution of x-mers as a function of conversion P. This distribution is given by (44) and is plotted for several values of the conversion in Figure 11.6.

$$W_x = x(1 - P)^2 P^{x-1} \tag{44}$$

The *weight-average molecular weight*, $\overline{M}_w$, is defined by (45), which is, again, simply the concept of an average.

$$\overline{M}_w = \sum_x W_x M_x = M_0 \sum_x xW_x \tag{45}$$

Substitution of (44) into (45) then gives

$$\overline{M}_w = (1 - P)^2 M_0 \sum_{x=1}^{N_0} x^2 P^{x-1} \tag{46}$$

The summations appearing in the expressions for number-average molecular weight (42) and weight-average molecular weight (46) are well known for values of $P < 1$. Because P is less than unity, by definition, the summations may be written as

$$\sum_{x=1}^{N_0} xP^{x-1} = 1 + 2P + 3P^2 + 4P^3 + \cdots + nP^{n-1} + \cdots = \left(\frac{1}{1 - P}\right)^2 \tag{47}$$

$$\sum_{x=1}^{N_0} x^2 P^{x-1} = 1 + 4P + 9P^2 + 16P^3 + \cdots + n^2 P^{n-1} + \cdots = \frac{1 + P}{(1 - P)^3} \tag{48}$$

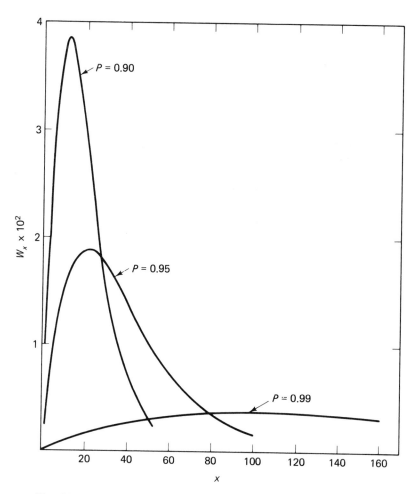

Fig. 11.6 Distribution by weight of the number of structural units in a condensation polymer for conversions of 90%, 95%, and 99% (see equation 44).

Substitution of (47) and (48) into (42) and (46), respectively, then gives (49) and (50) for the average molecular weights:

$$\overline{M}_n = \frac{M_0}{1 - P} \tag{49}$$

$$\overline{M}_w = M_0\left(\frac{1 + P}{1 - P}\right) \tag{50}$$

The weight-average molecular weight is seen to be always greater than the number-average molecular weight. In fact, for polymers produced by step-growth (condensation) mechanisms, (49) and (50) combine to show this relationship as

$$\overline{M}_w = \overline{M}_n(1 + P) \tag{51}$$

EFFECT OF NONSTOICHIOMETRIC REACTANT RATIOS ON LINEAR CONDENSATION POLYMERS

To produce a high polymer (i.e., $\overline{M} > 10^4$) in condensation polymerization it is necessary to allow the reaction to proceed to a very high degree of conversion or, in other words, to a product that contains a very small number of *chain ends*. In general, this will be possible only when equal concentrations of reactive functional groups are maintained throughout the course of the reaction. This means that not only must the reaction be initiated with stoichiometric reactant ratios, but the system must also be free of impurities that contain the same functional groups. The system must also be free from side reactions that might selectively consume functional groups and thereby destroy the equality of the functional-group concentrations.

Consider a condensation polymerization of a monomer that has two functional groups of type A (i.e., COOH) with a monomer having two functional groups of type B (i.e., OH). The general reaction is an esterification reaction, and it may be written as

$$A\!-\!\!\text{W}\!-\!A + B\!-\!\!\text{W}\!-\!B \longrightarrow A\!-\!\!\text{W}\!-\!ab\!-\!\!\text{W}\!-\!B + a'b' \qquad (52)$$

Let N_A° and N_B° be the respective number of functional groups that are present initially, and assume that their ratio, r, is less than unity (i.e., $r = N_A^\circ/N_B^\circ < 1$). Because one structural unit is present for every two functional groups, the *total* number of structural units, N_{su}, is given by

$$N_{su} = \tfrac{1}{2}(N_A^\circ + N_B^\circ) = \tfrac{1}{2}N_A^\circ\left(\frac{1+r}{r}\right) \qquad (53)$$

Let P be the fraction of A groups that have reacted at time t. Because the functional groups A and B destroy each other on a 1 : 1 basis (i.e., $\Delta N_A = \Delta N_B$), the fraction of B groups that have reacted at time t is rP as shown by

$$\frac{\Delta N_B}{N_B^\circ} = \frac{\Delta N_A}{N_B^\circ} = \frac{r\Delta N_A}{N_A^\circ} = rP \qquad (54)$$

At any time t, the number of chain ends, N_{ce}, must be equal to the sum of the numbers of *unreacted* A and B groups, or

$$N_{ce} = (1 - P)N_A^\circ + (1 - rP)N_B^\circ \qquad (55)$$

which is easily converted to (56), since $N_B^\circ = N_A^\circ/r$.

$$N_{ce} = N_A^\circ\left[2(1 - P) + \frac{1 - r}{r}\right] \qquad (56)$$

For bifunctional monomers and linear polymers, the number of chain ends, N_{ce}, is at all times equal to twice the total number of molecules present (i.e., $N = \frac{1}{2}N_{ce}$). Following substitution of this result into (56), the total number of molecules is given by

$$N = N_A^\circ\left(1 - P + \frac{1+r}{2r}\right) \tag{57}$$

As discussed previously, we express the average degree of polymerization $\overline{DP}$ in terms of the numbers of structural units and of molecules as in (58):

$$\overline{DP} = \frac{N_{su}}{N} \tag{58}$$

Substitution of (53) and (57) into (58) leads to (59), which describes the average degree of polymerization in terms of the conversion, P, and the ratio, r.

$$\overline{DP} = \frac{1+r}{2r(1-P)+1-r} \tag{59}$$

The reader may verify by inspection that (59) reduces to (31) in the limit of $r = 1$ (i.e., at the stoichiometric reactant ratio).

For $r < 1$, an excess of B groups will be present and the reaction is, therefore, complete when all the A groups have reacted. Hence, the maximum average degree of polymerization possible corresponds to a complete conversion of the A groups (i.e., to $P = 1$). According to (59), the maximum $\overline{DP}$ is then given by

$$(\overline{DP})_{max} = \overline{DP}(P = 1) = \frac{1+r}{1-r} \tag{60}$$

Equation (60) illustrates clearly the necessity for the use of very pure reagents, and for ensuring exact stoichiometric functional-group concentrations. Similarly, an absence of side reactions that destroy the functional groups selectively must be guaranteed if high-molecular-weight polymers are to be formed in good yield. As an example, consider the effect of a 1% monocarboxylic acid impurity in the adipic acid on the polyesterification of adipic acid with ethylene glycol. The general reaction is

$$\text{W}-\overset{\overset{\text{O}}{\|}}{\text{C}}-(CH_2)_4-COOH + HO-CH_2CH_2O-\text{W}$$

$$\longrightarrow \text{W}-\overset{\overset{\text{O}}{\|}}{\text{C}}-(CH_2)_4-\overset{\overset{\text{O}}{\|}}{\text{C}}-OCH_2CH_2O-\text{W} + H_2O$$

For this polymer $\overline{M}_0 = 86$ g/mol. A 1% impurity in the acid corresponds to a value of r of

$$r = \frac{[\text{Acid}]_0}{[\text{Glycol}]_0} = 0.99$$

Then, according to (60), the maximum average degree of polymerization and average molecular weight are

$$(\overline{DP})_{max} = \frac{1.99}{0.01} = 199$$

$$(\overline{M}_n)_{max} = (199)(86) = 17,114 \text{ g/mol}$$

BRANCHED AND CROSSLINKED CONDENSATION POLYMERS

In the preceding sections the reactions of *bifunctional* monomers to yield *linear* condensation polymers have been described. However, if at least one of the reactants is a tri- or multifunctional species, the polymerization will generate a branched polymer.

For example, if glycerol, $HOCH_2CH(OH)CH_2OH$, is allowed to react with a dicarboxylic acid or its anhydride, each glycerol molecule that has reacted completely will generate a branch point. Let B be an unreacted OH group on glycerol and b a reacted OH group on glycerol. Similarly, let A and a be unreacted and reacted COOH groups on the acid. Then a reaction in which 9 mol of acid has reacted with 4 mol of glycerol may be written as in (61). The product molecule **1**

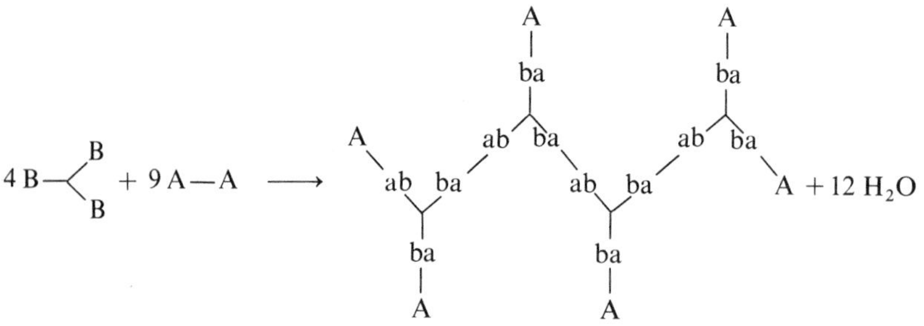

$$4\,B{-}\!\!\!\begin{array}{c}B\\\\B\end{array} + 9\,A{-}A \longrightarrow \quad \cdots \quad A + 12\,H_2O \qquad (61)$$

1

has six unreacted acid groups which may condense with more glycol. Hence, such branched polymer molecules can grow to very high molecular weights and may ultimately form an infinite network. If internal coupling occurs (reaction of a hydroxyl group and an acid function from branches of the same or different molecule), then the polymer will become crosslinked. In practice, extensive branching and crosslinking cause "gelation" of the polymer. In this state the polymer is swellable by solvents, but it does not dissolve. Highly crosslinked polymers are totally unaffected by solvents.

If a diol B—B were present, condensation of A—A with B—B would produce linear chains. Clearly, the degree of branching or crosslinking in (61) can be controlled by the relative amounts of triol and diol added to the system. It is possible to calculate the conditions needed to avoid or ensure the reaching of the gel point by use of the Carothers equation,

$$P = \frac{2}{F_{av}} - \frac{2}{\overline{DP}F_{av}}$$ (62)

where P is the extent of reaction, F_{av} is the average functionality of the system, and $\overline{DP}$ is the average degree of polymerization. Gelation is presumed to occur when average degree of polymerization becomes infinitely large, in which case, (62) reduces to

$$P_c = \frac{2}{F_{av}}$$ (63)

For example, suppose that we wish to compare the behavior of two polyester systems, one of which contains 2 mol of glycerol and 3 mol of a diacid (system X) and another which contains 2 mol of glycerol, 1 mol of a diol, and 4 mol of diacid (system Y). In system X, 12 functional groups are present for every 5 monomer molecules. Hence, F_{av} is $\frac{12}{5}$, or 2.4. Thus, for system X, gelation should occur at the critical point (P_c) of 83.3% reaction. On the other hand, in system Y, there are 16 functional groups per 7 monomer molecules and F_{av} is 2.29. Gelation should occur in this system at 87.3% reaction. In practice, this approach overestimates the reaction point at which gelation occurs, because polymer molecules exist that have molecular weights higher than the average value. These will reach the gelation point before those which have the average value of the molecular weight.

An alternative approach to the prediction of gelation points is based on a statistical treatment. The fundamental argument behind this approach is that the presence of three or more branch points in one molecule is sufficient to cause gelation. Thus, the problem revolves around a calculation of the probability that one of the four terminal branches in the segment in **2** will eventually yield another branch point.

2

If the functionality of the branch-point monomer is f, then the residual functionality of this unit at the chain end will be $f - 1$. Thus, the probability that one of the four branches will generate another branch point is given by $1/(f - 1)$. Because this also represents the probability that gelation will occur, we can write

$$\alpha_c = \frac{1}{f - 1}$$

where α_c is the critical branching coefficient required for gel formation.

Several assumptions are needed to simplify the treatment to a manageable degree. First, it is assumed that all functional groups are equally reactive. This is, in fact, an erroneous assumption even for glycerol, where the secondary OH group is known to be less reactive than the primary ones. Second, it is assumed that all condensation steps take place between different molecules and that internal condensations to give species such as **2** do not occur. The deviations from these assumptions probably account for the numerical discrepancies between experiment and theory.

Again an important practical problem is to calculate the point at which gelation should occur for different ratios of tri- (or multi-) and difunctional reagents. When a trifunctional monomer is present, gelation will occur if α (the branching coefficient) is greater than $\frac{1}{2}$. The value $\alpha = \frac{1}{2}$ therefore constitutes the critical condition for the formation of an infinite network. In a more general sense, the critical value of α is defined, as mentioned previously, by the term: $\alpha(f - 1) = 1$ or $\alpha_c = 1/(f - 1)$.

Thus, an estimate of α is the key requirement. This estimate may be made using a similar statistical approach to that discussed earlier. The result is

$$\alpha = \frac{P_B^2 p}{r - P_B^2(1 - p)} = \frac{rP_A^2 p}{1 - rP_A^2(1 - p)} \tag{64}$$

As discussed previously, P_A and P_B represent the fractions of the original A and B functional groups that have reacted. Similarly, r is the ratio of the initial number of A groups to the initial number of B groups. The term p is the fraction of original A groups that are contained in the tri- (or higher) functional monomer. It can be shown[1] that α_c can be attained only when $1/(1 + p) < r < 1 + p$.

There are three interesting special cases of (64).

1. No bifunctional monomer that contains A groups is present. In this case $p = 1$ and (64) simplifies to

$$\alpha = rP_A^2 = \frac{1}{r} P_B^2 \tag{65}$$

2. The initial numbers of A and B groups are equal. In this case $r = 1$ and $P_A = P_B = P$ and (64) simplifies to

$$\alpha = \frac{P^2 p}{1 - P^2(1 - p)} \tag{66}$$

3. If the initial number of A and B groups are equal *and* no bifunctional monomer that contains A groups is present, (64) simplifies to

$$\alpha = P^2 \tag{67}$$

[1] P. J. Flory, *J. Am. Chem. Soc.*, **63**, 3083 (1941).

As an example, let us apply the expressions above to predict the gel points of the hypothetical systems mentioned earlier:

For system X, with 2 mol of glycerol and 3 mol of bifunctional acid, $r = 1$ and $p = 1$. The critical branching coefficient for this trifunctional system is $\alpha_c = \frac{1}{2}$. Thus, from (67), the gel point will occur at a degree of reaction of

$$P = \sqrt{\tfrac{1}{2}} = 0.707$$

or a conversion of 70.7%.

For system Y with 2 mol of glycerol, 1 mol of diol, and 4 mol of acid, $r = 1$ and $p = \frac{3}{4}$. Again $\alpha_c = \frac{1}{2}$, so that from (66) the gel point will occur at

$$P = \sqrt{\tfrac{8}{14}} = 0.756$$

or at a conversion of 75.6%.

STUDY QUESTIONS

1. Three samples of monodisperse poly(1,10-decanediol adipate) are mixed together as follows: 10 g of A, having a molecular weight of 40,000; 5 g of B, having a molecular weight of 100,000; and 3 g of C, having a molecular weight of 200,000. Calculate:
   (a) The number-average molecular weight of the polymer mixture.
   (b) The weight-average molecular weight of the polymer mixture.

2. A sample of a condensation polymer has the following experimental distribution of molecular weights:

Range of $M \times 10^{-3}$	Weight (mg)	Range of $M \times 10^{-3}$	Weight (mg)
0–20	0.35	120–140	14.43
20–40	2.52	140–160	14.70
40–60	5.70	160–180	13.77
60–80	8.75	180–200	12.16
80–100	11.52	200–220	8.40
100–120	13.31	220–240	5.75

Determine the weight-average and number-average molecular weights of the polymer.

3. The hydroxy acid $HO-CH_2CH_2CH_2CH_2COOH$ undergoes a condensation polymerization to form the polymer

$$H\left[OCH_2CH_2CH_2CH_2\overset{\overset{\displaystyle O}{\|}}{C}\right]_n OH$$

It is found that in a certain polymerization the product has a weight-average molecular weight of 18,400 g/mol. Calculate:

(a) The percentage of carboxyl groups that have esterified.
(b) The number-average molecular weight of the polymer.
(c) The average number of structural units in the polymer molecules.
(d) The probability that a polymer molecule chosen at random will contain twice the average number of structural units.

4. It has been reported [Zhubanov et al., *Izv. Akad. Nauk Kaz. SSR Ser. Khim.* ,**17**, 69 (1967)] that aminoheptanoic acid, H_2N—CH_2—$(CH_2)_4 CH_2 COOH$, in *m*-cresol solution undergoes condensation polymerization to a polyamide. The reaction was found to be second-order in the amino acid concentration, with the following rate constants:

t (°C)	150	187
k (kg/mol-min)	1.0×10^{-3}	2.74×10^{-2}

(a) Write a balanced chemical reaction that describes the conversion of monomer to a polyamide of molecular weight equal to 12,718 g/mol.
(b) Calculate the activation energy of the reaction and the preexponential factor of the Arrhenius expression for the rate constant.
(c) What percent conversion of monomer is necessary to produce a polyamide with a number-average molecular weight of 4.24×10^3 g/mol?
(d) What percent conversion is necessary to produce a polyamide with a weight-average molecular weight of 2.22×10^4 g/mol?

5. A solution of aminoheptanoic acid (see Problem 4) in *m*-cresol, having a concentration of 3.3 mol of amino acid per kilogram of solution, is prepared and quickly brought to a temperature of 187°C.

(a) Derive an expression for the degree of polymerization (i.e., the average number of structural units per polymer molecule) as a function of reaction time. Modify your expression to apply to a system in which caproic acid is present at a level of 0.65 % of the amino acid.
(b) Calculate the time required to form polyamide with a number-average molecular weight of 6340 g/mol.
(c) If the polymerization is carried out in a solution weighing 1 kg and for the reaction time calculated in (b), what would be the weight of polyamide formed that has a molecular weight of 12,718 g/mol?

6. Derive expressions that relate the mole fraction of *x*-mer to the reaction rate constant, initial concentration of carboxyl groups, and reaction time for (a) an uncatalyzed polyesterification, and (b) an acid-catalyzed polyesterification. Plot the mole fractions of polymer with $x = 10$ and $x = 100$ as a function of reaction time.

7. The following data was obtained by Hamann, Solomon, and Swift [*J. Macromol. Sci.— Chem.*, **A2**, 153 (1968); reproduced by permission of Marcel Dekker, Inc.] for the polyesterification of an equal molar mixture of

$$HO—(CH_2)_{10}—OH + HOOC—(CH_2)_4—COOH$$

The kinetics were studied by taking time zero to correspond to 82 % esterification of the original COOH groups. $[COOH]_0$ may be taken as 1.25 equiv/kg of mixture.

Further Polymerization of Poly(1,10-decanediol adipate)

(a)		(b)		(c)	
				Temp. 161°C	
				Catalyzed by *p*-Toluene Sulfonic Acid (0.004 mol per Mole of Polymer)	
Temp. 190°C		Temp. 161°C			
Time (min)	% Reaction	Time (min)	% Reaction	Time (min)	% Reaction
0	0	0	0	0	0
30	20.6	20	9.1	5	34.6
60	39.0	40	16.0	10	54.7
90	50.2	100	31.6	15	65.5
150	61.2	150	41.1	20	70.8
225	66.8	210	47.9	30	77.9
300	71.5	270	52.5	40	82.9
370	74.4	330	57.0	50	85.7
465	77.2	390	60.0	60	87.9
510	78.2	450	62.6	75	90.1
550	78.8	510	64.6	90	91.5
600	79.6	550	65.5	105	92.6
660	80.6	700	69.2	120	93.6
730	81.7	840	71.9		
800	82.5	880	72.4		
		1060	74.8		
		1200	76.2		
		1320	77.2		

(a) Determine the rate constants and the activation energy for the uncatalyzed reaction.
(b) Determine the rate constant for the catalyzed polymerization.

8. From the rate constant obtained in Problem 7 for the catalyzed polyesterification of 1,10-decanediol and adipic acid, calculate and plot the mole fraction and weight fraction distribution of *x*-mer [N_x/N versus x, and W_x versus x] at 120 min and at 240 min.

9. Suppose that in the polyesterification reaction in Problem 7, the adipic acid had an impurity of 0.85 mol %, of which you were unaware. What are the maximum values of the number-average and weight-average molecular weights that you could obtain in the reaction?

10. The following data were obtained [P. J. Flory, *J. Am. Chem. Soc.*, **61**, 3334 (1939)] for the polyesterification of diethylene glycol with adipic acid at 166°C:

Time (min)	% Conversion	Time (min)	% Conversion
6	13.79	321	86.72
12	24.70	398	88.37
23	36.75	488	89.74
37	49.75	596	90.84
59	60.80	690	91.63
88	68.65	793	92.20
170	78.94	900	92.73
203	81.61	1008	93.03
235	83.49	1147	93.54
270	85.00	1370	94.05

(a) Determine the order of the reaction with respect to the concentration of carboxyl groups.

(b) Determine the rate constant of the reaction in appropriate units. (*Hint*: Take the concentration units to be moles of COOH per kilogram of reactant mixture and require that $[OH]_0 = [COOH]_0$.)

(c) Calculate and plot the weight-fraction distribution of molecular weights obtained at reaction times of 270 and 1370 min.

11. Calculate the % conversion of acid at which gelation will occur in a mixture of 50 mol % adipic acid, 40 mol % ethylene glycol, and 10 mol % glycerol.

SUGGESTIONS FOR FURTHER READING

ALLEN, P. E. M., AND C. R. PATRICK, *Kinetics and Mechanisms of Polymerization Reactions*, Chap. 5. New York: Wiley, 1974.

FLORY, P. J., *Principles of Polymer Chemistry*, Chap. 3. Ithaca, New York: Cornell University Press, 1953.

LENZ, R. W., *Organic Chemistry of Synthetic High Polymers*, Chap. 3. New York, New York: Wiley–Interscience, 1967.

ODIAN, G., *Principles of Polymerization*, Chap. 2. New York: McGraw–Hill, 1970.

SOLOMON, D. H., (ed.), *Step-Growth Polymerizations*. New York: Dekker, 1972.

12

Kinetics of Free-Radical Polymerization

GENERAL CONSIDERATIONS

The speed of a polymerization reaction, the molecular weight of the product, the composition of a copolymer, and the formation of unwanted side products are factors that have an enormous significance both in laboratory experiments and in the manufacture of polymers. Before the course of a chemical reaction can be predicted, it is necessary to understand at least some of the underlying principles that determine the influence of different reaction conditions on the yields and products.

The study of polymerization mechanisms and reaction kinetics aims to uncover these underlying principles in order to permit predictions to be made. This chapter deals with free-radical polymerizations, about which a great deal is known. Chapters 11 and 13 cover condensation and ionic-type processes.

The elementary reactions involved in a free-radical polymerization were discussed in Chapter 3. In that discussion the following mechanism was used to describe a polymerization under conditions where the conversion of monomer to polymer is sufficiently low that the polymeric product molecules do not undergo reactions with the free radicals:

$$\text{Initiator} \xrightarrow{k_i} 2R'$$ Formation of free radicals

$$R' + M \xrightarrow{k'} R_1$$ Initiation of chains

$$R_1 + M \xrightarrow{k_{1p}} R_2$$ Propagation of chains

$$R_2 + M \xrightarrow{k_{2p}} R_3$$ Propagation of chains

- -

$$R_n + M \xrightarrow{k_{np}} R_{n+1}$$ Propagation of chains

- -

$$R_n + R_m \xrightarrow{k_{tnmc}} P_{n+m}$$ Termination of chains

$$R_n + R_m \xrightarrow{k_{tnmd}} P_n + P_m$$ Termination of chains

In this mechanism, M represents a monomer molecule, R' is an initiating radical, R_n is a propagating radical of degree of polymerization n, P_n is a polymer molecule of degree of polymerization n, and so on. The rate constants for each elementary reaction are shown above the arrows. For simplicity at this stage we omit chain transfer reactions from the mechanism. An extension of this simple mechanism to include chain transfer to the monomer, to the solvent, and to added chain transfer agents will be made in a later section of this chapter.

APPROXIMATIONS

A rigorous kinetic treatment of this mechanism leads to immense complexities and, in order to derive useful and tractable results, it is necessary to introduce some simplifying assumptions and approximations.

Kinetic Chain Length

It is assumed that the kinetic chain lengths are very large. This means that the number of monomer molecules consumed in the chain initiation process (i.e., that react with R') is negligible compared with the number consumed in the chain propagation reactions. Since we are, by definition, dealing with products that are high polymers containing many monomer units, this approximation is a very good one.

The Direction of Radical Addition to the Monomer

Only one type of radical is assumed to be present as a chain carrier. This will be true if each radical addition to the monomer occurs in the same way. As discussed in Chapter 3, the head-to-tail addition greatly predominates over head-to-head addition. It is assumed therefore, that head-to-tail addition is the sole type occurring. This means that, in the polymerization of $CH_2=CHX$, the structure of all the propagating radicals is of the type shown in **1**, where the zigzag line denotes a polymeric chain of $-(CH_2CHX)-$ units.

$$\text{W}-CH_2-\underset{\underset{X}{|}}{\overset{\overset{H}{|}}{C}}\cdot$$

1

Radical Reactivity and Size

It is assumed that the reactivity of the propagating radicals is independent of the size or degree of polymerization of the radical. The result of this assumption is that the rate constants for propagation, that is, $k_{1p}, k_{2p}, k_{3p}, \ldots, k_{np}, \ldots$ are taken to be equal and are written simply as k_p. Similarly, for termination reactions, the rate constants k_{nmc} and k_{nmd} are assumed to be independent of n and m and are written simply as k_{tc} and k_{td}.

The application of this assumption to the propagation reactions may be rationalized in terms of simple collision theory. According to this theory, the rate constant for propagation, k_p, may be written as

$$k_p = \xi_p \sigma_p \left(\frac{RT}{\pi\mu}\right)^{1/2} e^{-E_p/RT} \tag{1}$$

where ξ_p is the steric factor, σ_p is the cross section of the collision, μ is the reduced mass of the colliding pair, E_p is the activation energy for propagation, R is the gas constant, and T is the absolute temperature. Since the only effective collisions will be those of monomer M with the growing end of the propagating radical R_n, the product $\xi_p \sigma_p$ should be roughly independent of the degree of polymerization of the radical. Because the chemical nature of the reactive end of the radical is independent of the degree of polymerization, E_p should also be independent of the degree of polymerization of the radical. The reduced mass is defined by

$$\mu = \frac{M_R M_M}{M_R + M_M} \tag{2}$$

where M_R and M_M are the masses of the propagating radical and monomer, respectively. For most of the propagation reactions, M_R is much larger than M_M. Therefore, to a good approximation, equation (2) may be written

$$\mu \approx M_M \tag{3}$$

and the conclusion may be drawn that k_p should be a constant that is independent of radical size. This will generally be a valid approximation except in the initial stages of the propagation.

However, the assumption that k_t^1 is independent of size is usually less valid. This is because, for most termination reactions, the rate constant is determined by the collision frequency of radicals (diffusion-controlled reactions), and the reduced mass does not become independent of radical size as in (3). Nevertheless, this assumption must be made in order to obtain tractable results. The reader should keep in mind the limitation of this assumption and recognize that in most cases an *average* termination rate constant is being used.

The Steady-State Approximation

All free radicals present in the system are assumed to be at steady-state concentrations. This means that the total concentration of propagating radicals (which is the sum of the concentrations of propagating radicals of all degrees of polymerization) is also a steady-state value. Expressed algebraically this means that

$$\frac{d[R']}{dt} = \frac{d[R_n]}{dt} = \frac{d[R]}{dt} = 0 \tag{4}$$

where $[R] = \sum_n [R_n]$ and the other symbols are as described earlier.

Justification of this assumption may be seen as follows. Let $[M]_0$ be the initial concentration of the monomer. Monomer molecules that have reacted must be contained either in the propagating radicals or in the polymer (i.e., product molecules). Therefore, the stoichiometry requires that, at all times, equation (5) must hold.

$$[M]_0 = [M] + \sum_n n[P_n] + \sum_n n[R_n] \tag{5}$$

If (5) is differentiated with respect to time and then rearranged, the expression shown in (6) is obtained.

$$-\frac{d[M]}{dt} = \sum_n n\left(\frac{d[P_n]}{dt}\right) + \sum_n n\left(\frac{d[R_n]}{dt}\right) \tag{6}$$

Experimentally, it is found that, except in the very earliest (and generally negligible) stages of the reaction, the loss of monomer is, in fact, accounted for quantitatively by the appearance of the polymeric product. Thus, we can write equation (6) in the terms shown in (7).

$$-\frac{d[M]}{dt} = \sum_n n\left(\frac{d[P_n]}{dt}\right) \tag{7}$$

Equation (7) is generally valid only if (4) is also valid.

---

[1] $k_t = k_{tc} + k_{td}$.

STEADY-STATE CONCENTRATIONS OF THE PROPAGATING RADICALS

The steady-state approximation (4) is probably the most important single assumption required to find a tractable solution to the rate equations needed for the mechanism described on page 272. This approximation enables us to express free-radical concentrations in terms of the concentrations of *stable* substances by the solution of simple *algebraic* equations. Without this approximation, a set of nonlinear simultaneous differential equations would have to be solved.

In each propagation step of the mechanism, one radical is destroyed but another is created. Therefore, the entire sequence of chain propagation reactions has no effect on the *total* concentration of propagating radicals. If this total concentration is to be in a steady state, as assumed by (4), the rate of initiation of propagating radicals must be equal to their rate of termination, or

$$\frac{d[R]}{dt} = r_i - r_t = 0 \tag{8}$$

where r_i and r_t are the rates of initiation and termination, respectively, of the propagating radicals.

According to the mechanism on page 272, and the assumption that the initiating radicals are at a steady state, the rate of initiation is given by

$$r_i = k'[R'][M] \propto 2k_i[I] \tag{9}$$

Actually, as discussed in Chapter 3, some loss of initiating radicals always occurs through side reactions that are not shown in the mechanism. To recognize this fact, the rate of initiation is usually written as

$$r_i = 2fk_i[I] \tag{10}$$

where f represents the fraction of initiating radicals produced that actually add to monomer to form R_1.

In considering the rate of termination, r_t, it is necessary to distinguish between the reactions of *like* radicals and those of *unlike* radicals. Thus, for both combination and disproportionation, we must distinguish between the two termination processes

$$R_n + R_n \xrightarrow{k_{tnn}} \text{Polymer} \tag{11}$$

$$R_n + R_m \xrightarrow{k_{tnm}} \text{Polymer} \tag{12}$$

In a chemical sense, all the propagating radicals are the same: they differ only in size. Therefore, because we have assumed that reactivity is independent of size, the rate constants k_{tnn} and k_{tnm} can differ only by a factor of 2. This is a consequence of the relative collision frequencies of like and unlike species (i.e., $k_{tnm} = 2k_{tnn}$).

This well-known result may be seen as follows. Consider the case in which two types of radicals, R_n and R_m, are present. The total rate of collisions may be written

$$\text{Collision rate} = \alpha[R]^2 \tag{13}$$

where α is the rate constant. Substitution of the total radical concentration $[R] = [R_n] + [R_m]$ into (13) gives the collision rate as

$$\text{Collision rate} = \alpha[R_n]^2 + 2\alpha[R_n][R_m] + \alpha[R_m]^2 \tag{14}$$

The first and last terms in (14) are the rates of collisions of like radicals, while the second term gives the collision rate of unlike species. The rate constant for unlike collisions, namely 2α, is twice that for like collisions. In terms of chain termination rate constants, then, $k_{tnm} = 2k_{tnn}$.

Application of the mass action law to (11) and (12) gives the termination rate for radicals of degree of polymerization n:

$$r_t(n) = 2k_{tnn}[R_n]^2 + [R_n]\sum_{m \neq n} k_{tmn}[R_m] \tag{15}$$

The factor 2 in (15) arises because two radicals of degree of polymerization n are consumed by the process shown in (11), while only one is consumed in (12). Since $k_{tnm} = 2k_{tnn}$, (15) becomes

$$r_t(n) = 2k_{tnn}[R_n]\left\{[R_n] + \sum_{m \neq n} [R_m]\right\} \tag{16}$$

or

$$r_t(n) = 2k_{tnn}[R_n]\sum_m [R_m] = 2k_t[R_n][R] \tag{17}$$

where k_t is the rate constant for termination of like radicals and $[R]$ is the total radical concentration. The subscript nn has been dropped because of the assumption on page 273. The total rate of termination of all radicals is obtained by summation over n as

$$r_t = \sum_n r_t(n) = 2k_t[R]\sum_n [R_n] = 2k_t[R]^2 \tag{18}$$

Substitution of (10) and (18) into (8) gives, for the steady-state concentration of propagating radicals,

$$[R] = \left(\frac{r_i}{2k_t}\right)^{1/2} = \left(\frac{fk_i[I]}{k_t}\right)^{1/2} \tag{19}$$

In the case of photolytic or radiation-induced polymerization, $2k_i[I]$, which is the rate of formation of initiating radicals, may simply be replaced by the appropriate initiating radical formation rates given in Chapter 5.

RATE OF POLYMERIZATION

The rate of polymerization, r_p, is defined as the instantaneous decrease of monomer concentration, $[M]$, with respect to time, t, as shown by (20). However, polymerization rates,

$$r_p = -\frac{d[M]}{dt} \tag{20}$$

generally depend on the monomer concentrations and, since the monomer concentration is known most accurately at zero time, it is useful to work with *initial* rates of polymerization, as given by (21). Unless otherwise specified

$$r_p^0 = -\left(\frac{d[M]}{dt}\right)_0 = \lim_{\Delta t \to 0} \frac{[M]_0 - [M]}{\Delta t} \tag{21}$$

all rates mentioned in the following sections will be initial rates, and the superscript and subscript of r_p and $(d[M]/dt)$, respectively, in (21) will be dropped.

It is easy to determine the rate of polymerization since, according to (21), it is merely necessary to measure $[M]$ as a function of t and then determine the *initial* slope of a plot of $[M]$ versus t. Moreover, such plots are usually sufficiently linear up to about 10 to 15% monomer depletion that only one or two experimental points are necessary. Despite the simplicity of the measurements, rates of polymerization measured under a variety of reaction conditions provide a basis for the determination of a large amount of practical and theoretical kinetic information.

Experimental Measurement of Rates of Polymerization

In order to determine the rate of polymerization it is necessary to measure the monomer concentration $[M]$ as a function of the reaction time. The most obvious and straightforward way to do this is to stop the reaction at some predetermined time (e.g., by a sudden chilling of the reaction mixture) and then separate the monomer reactant from the polymeric product. The amount of monomer remaining or the amount of polymer formed at this reaction time may then be determined simply by weighing. This process can then be repeated for other reaction times until sufficient information has been obtained to permit a plot of $[M]$ as a function of t. The initial slope, $\lim_{\Delta t \to 0} \Delta[M]/\Delta t$, is the initial rate, r_p.

However, the measurement of $[M]$ by this method is time-consuming and requires the preparation of a new reaction mixture for each experimental point on the concentration–time plot. The method is useful as a standard method since

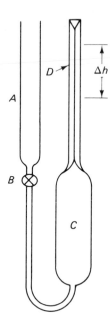

Fig. 12.1 Simple dilatometer.

absolute concentrations are obtained from it. But it is much faster and more convenient to measure some *physical property* of the reaction mixture that changes as the polymerization proceeds and which may be related to the concentration of the monomer. Although a number of suitable physical methods of analysis are available, *dilatometry*[1] is most often used in the measurement of polymerization rates. In this method a dilatometer is employed to measure the volume contraction that occurs as the monomer is converted to high polymer.

The principle of operation of a dilatomer may be seen from the simple apparatus shown in Figure 12.1. The monomer or a solution of the monomer in a solvent are introduced, along with an initiator, into the apparatus through filling tube A until the liquid is drawn well up into the capillary tube D. The stopcock B is then closed and the apparatus is placed in a thermostatic bath maintained at the desired reaction temperature and regulated to $\pm0.002°C$. The diameter and total height of the capillary depend on whether the polymerization is carried out in pure monomer or in a solution of the monomer. During the reaction the change in the height of the liquid in the capillary, Δh, is measured with a cathetometer (a rigidly mounted, vertically sliding telescope).

The total volume of the reaction system is given by

$$V = V_{BC} + \pi r^2 h \tag{22}$$

where V_{BC} is the volume contained in the space from the stopcock B through the bulb C to the entrance to the capillary tube D, r is the radius of the capillary, and h is the height of the liquid in the capillary above some arbitrary reference point.

[1] L. C. Rubens and R. E. Skochdopole, *J. Appl. Poly. Sci.*, **9**, 1487 (1965).

The change in volume is proportional to the change in h, as in

$$\Delta V = V_0 - V = \pi r^2 \Delta h \tag{23}$$

where V_0 is the initial volume of the system and V is the volume at time t.

The change in volume of the system may be related to the monomer concentration, reaction yield, and rate of reaction as follows. The total volume of the system, V, at any time t is given by

$$V = w_m \bar{v}_m + w_p \bar{v}_p + w_s \bar{v}_s \tag{24}$$

where w_m, w_p, and w_s are the weights and $\bar{v}_m$, $\bar{v}_p$, and $\bar{v}_s$ are the partial specific volumes[1] of monomer, polymer, and solvent, respectively. To a very good approximation, $w_p = w_m^\circ - w_m$, where w_m° is the initial weight of monomer. Solving (24) gives the weight of monomer at time t, as in

$$w_m = \frac{V - w_m^\circ \bar{v}_p - w_s \bar{v}_s}{\bar{v}_m - \bar{v}_p} \tag{25}$$

Because no polymer is present initially, the initial volume is given by

$$V_0 = w_m^\circ \bar{v}_m + w_s \bar{v}_s \tag{26}$$

and, provided that all the monomer is converted to polymer at the completion of the reaction, the final volume is

$$V_\infty = w_m^\circ \bar{v}_p + w_s \bar{v}_s \tag{27}$$

Equations (26) and (27) may be used to eliminate $\bar{v}_m - \bar{v}_p$ and $w_s \bar{v}_s$ from (25) and, when this is done, equation (28) is obtained.

$$w_m = \frac{V - V_\infty}{V_0 - V_\infty} w_m^\circ \tag{28}$$

The yield of the reaction on a wt % basis, Y, may then be written as

$$Y = 100 \frac{w_m^\circ - w_m}{w_m^\circ} = 100 \frac{V_0 - V}{V_0 - V_\infty} \tag{29}$$

or

$$Y = \frac{100 \Delta V}{V_0 - V_\infty} = \frac{100 \Delta h(t)}{\Delta h(t = \infty)} \tag{30}$$

---

[1] G. N. Lewis and M. Randall, *Thermodynamics*, revised by K. S. Pitzer and L. Brewer (New York: McGraw–Hill, 1961), p. 208.

TABLE 12.1 Polymerization of Methyl methacrylate at 50°C; Initiator: 1% Benzoyl peroxide

REACTION TIME (s)	YIELD (wt %)	$r_p \times 10^4$ (mol-liter^{-1} s^{-1})
0	0	—
480	0.77	1.5
1200	2.17	1.64
1920	3.50	1.65
4080	7.27	1.62
5880	10.38	1.60
8280	14.60	1.60
9600	17.16	1.62

Similarly, it is possible to proceed from (28) and utilize the definition of the initial rate of polymerization (21) to obtain the expression

$$r_p = \frac{Y}{100} \frac{[M]_0}{t} = \frac{\Delta h(t)}{\Delta h(t = \infty)} \frac{[M]_0}{t} \tag{31}$$

Some polymer yields from the polymerization of methyl methacrylate at 50°C [measured dilatometrically and calculated by equation (30)] and the corresponding polymerization rates calculated from (31) are shown in Table 12.1. The con-

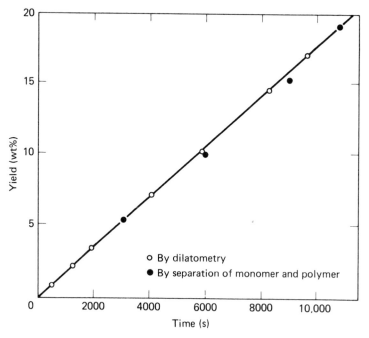

Fig. 12.2   Reaction yields in polymerization of methylmethacrylate at 50°C. [From G. V. Schulz and G. Harborth, *Angew. Chem.*, **59A**, 90 (1947).]

stancy of r_p in Table 12.1 shows that, up to the point of 17% conversion, depletion of the reactant does not significantly affect the measurement of the initial rate.

The validity of this simple method may be seen in Figure 12.2, in which the yields are plotted versus time for values determined both dilatometrically and absolutely (by the separation of polymer and monomer).

Despite the success of the dilatometric method with many polymerization systems, it is not an *absolute* method and complications can arise. The applicability of the method to an untried system cannot be assumed but must first be tested against an absolute method. Once the dilatometric method has been shown to be appropriate for the particular system under study as, for example, by data of the type shown in Figure 12.2, it provides probably the simplest and fastest method for the determination of polymerization rates. It is largely on the basis of such experimental rates, determined under a variety of conditions, that the free-radical mechanisms of polymerization have been developed.

Theoretical Rates of Polymerization

Application of the mass action law to the mechanism given on page 272 for the rate of disappearance of monomer leads to the expressions shown in (32) and (33) for the *theoretical* rate of polymerization.

$$r_p = k'[R'][M] + k_{1p}[R_1][M] + k_{2p}[R_2][M] + \cdots \tag{32}$$

or

$$r_p = [M]\left\{k'[R'] + \sum_n k_{np}[R_n]\right\} \tag{33}$$

According to the "assumption of long chains" made earlier, $k'[R']$ is very much less that $\sum_n k_{np}[R_n]$ and may be neglected in the bracketed term of (33). Similarly, according to the assumption made regarding radical size and reactivity, $k_{1p} = k_{2p} = \cdots k_{np} = k_p$. Incorporation of these two approximations into (33) yields

$$r_p = k_p[M] \sum_n [R_n] = k_p[M][R] \tag{34}$$

where [R] is the total concentration of propagating radicals. We have already evaluated the steady-state radical concentration, (19), in terms of the rate of initiation, and substitution of (19) into (34) gives the steady-state rate of polymerization:

$$r_p = \left(\frac{k_p^2}{2k_t}\right)^{1/2} r_i^{1/2}[M] = \left(\frac{k_p^2}{2k_t}\right)^{1/2} (2fk_i)^{1/2}[I]^{1/2}[M] \tag{35}$$

where the last term in (35) refers specifically to initiation by the thermal dissociation of an initiator.

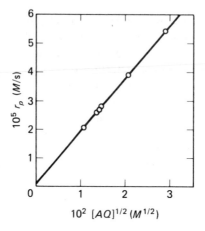

Fig. 12.3 Variation in initial rates of polymerization with concentration of anthraquinone in THF at 30°C ([MMA] = 4.68). [Reproduced from A. Ledwith, G. Ndaalio, and A. R. Taylor, *Macromolecules*, **8**, 1 (1975); with permission of the American Chemical Society, Washington, D.C.]

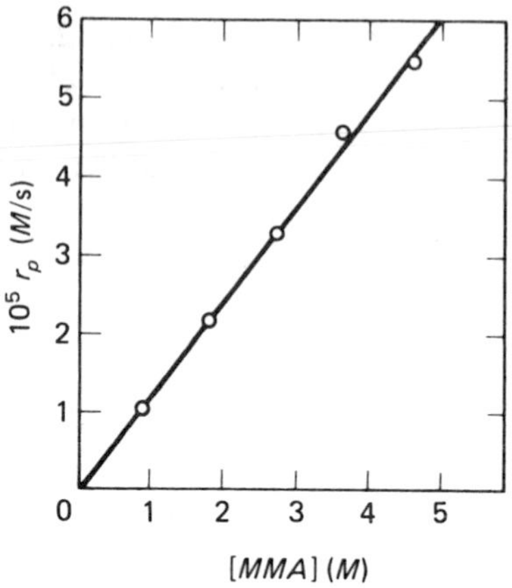

Fig. 12.4 Variation in initial rates of polymerization with concentration of MMA in THF at 30°C ([AQ] = 2.91 × 10⁻⁴ M). [Reproduced from A. Ledwith, G. Ndaalio, and A. R. Taylor, *Macromolecules* **8**, 1 (1975); with permission of the American Chemical Society, Washington, D.C.]

Equation (35) predicts that *the rate of polymerization at a given temperature should vary with the square root of the initiation rate or initiator concentration and with the first power of the monomer concentration.* The validity of this equation has been verified many times. An example of the dependence of r_p on the initiator concentration is shown in Figure 12.3, while an example of the dependence of r_p on monomer concentration is given in Figure 12.4.

According to (35), the slope of the straight line resulting from a plot of r_p versus [M], such as in Figure 12.4, is $(k_p^2/2k_t)^{1/2}r_i^{1/2}$. If r_i is known, which is usually the case, then plots such as the one shown in Figure 12.4 lead to numerical values of the ratio $(k_p^2/2k_t)^{1/2}$. Values of this important ratio for the polymerization of a number of monomers at 60°C are shown in Table 12.2.

TABLE 12.2   Rate Constant Ratios and Temperature Dependence for Free-Radical Polymerization

Monomer	$\left(\dfrac{k_p^2}{2k_t}\right)^{1/2}$ (liters/mol-s)$^{1/2}$ at 60°C	$\left(\dfrac{A_p^2}{2A_t}\right)^{1/2}$ (liters/mol-s)$^{1/2}$	$E_p - \tfrac{1}{2}E_t$ (kcal/mol)
Methacrylonitrile	0.053	4.3 × 10⁴	9.0
Methyl acrylate	0.99	1.03 × 10³	4.6
Methyl methacrylate	0.12	2.0 × 10²	4.9
Styrene	0.029	6.2 × 10²	6.6
Vinyl acetate	0.13	1.6 × 10²	4.7

The value of $(k_p^2/2k_t)^{1/2}$, determined from a plot such as the one in Figure 12.4, refers to a single temperature. If such values are obtained at several temperatures, a combination of these data with the Arrhenius formulation of a rate constant yields more information. Thus, writing the individual rate constants as

$$k_p = A_p e^{-E_p/RT} \tag{36}$$

$$k_t = A_t e^{-E_t/RT} \tag{37}$$

where the A's are the preexponential factors and the E's are the activation energies, the rate constant ratio may be written as

$$\left(\frac{k_p^2}{2k_t}\right)^{1/2} = \left(\frac{A_p^2}{2A_t}\right)^{1/2} \exp\left[-\frac{(E_p - \frac{1}{2}E_t)}{RT}\right] \tag{38}$$

Taking logarithms of both sides of (38) yields

$$\ln\left(\frac{k_p^2}{2k_t}\right)^{1/2} = \ln\left(\frac{A_p^2}{2k_t}\right)^{1/2} - \left(\frac{E_p - \frac{1}{2}E_t}{R}\right)\frac{1}{T} \tag{39}$$

According to (39), the Arrhenius parameters $(A_p^2/2A_t)^{1/2}$ and $(E_p - \frac{1}{2}E_t)$ may be obtained respectively from the intercept and slope of a plot of $\ln(k_p^2/2k_t)^{1/2}$ versus $1/T$. Values of these parameters for several monomers are also shown in Table 12.2.

From the Arrhenius parameters in Table 12.2, it is possible to calculate values for the ratio $(k_p^2/2k_t)^{1/2}$ at any temperature. Such information is of great utility because values of $(k_p^2/2k_t)^{1/2}$ and the rate of initiation r_i, permit the prediction not only of rates of polymerization but also, as will be seen, of average kinetic chain lengths, average degrees of polymerization, and the distribution of the degree of polymerization.

AVERAGE KINETIC CHAIN LENGTH

The average kinetic chain length, v, is defined as the average number of monomer molecules polymerized per chain initiated, or, equivalently, as the rate of polymerization per unit rate of initiation. Thus from (18) and (34) equation (40) can be obtained.

$$v = \frac{r_p}{r_i} = \frac{k_p[M]}{2k_t[R]} = \left(\frac{k_p^2}{2k_t}\right)^{1/2}\frac{[M]}{r_i^{1/2}} \tag{40}$$

This expression shows that, given a rate of initiation, kinetic data of the type shown in Table 12.2 can be used to calculate average kinetic chain lengths of polymerization.

AVERAGE DEGREE OF POLYMERIZATION AND AVERAGE MOLECULAR WEIGHT

According to the mechanism shown on page 272, termination by *combination* produces a polymer molecule larger than the terminating free radicals. On the other hand, termination by *disproportionation* yields two polymer molecules of the same size as the terminating free radicals. Because the average degree of polymerization, $\overline{DP}$, is defined as the average number of monomer molecules per polymer molecule, a distinction must be made between these two types of termination. Thus the total rate of formation of polymer is defined by (41) (the factor of 2 appears because two polymer molecules are formed in termination by disproportionation,

$$\frac{d[P]}{dt} = (k_{tc} + 2k_{td})[R]^2 \tag{41}$$

whereas only one is formed by combination). The instantaneous average degree of polymerization is then given by

$$\overline{DP} = \frac{-d[M]/dt}{d[P]/dt} = \frac{k_p[M]}{(k_{tc} + 2k_{td})[R]} \tag{42}$$

From (40), (42), and the relationship $k_t = k_{tc} + k_{td}$, we may express $\overline{DP}$ in terms of the average kinetic chain length:

$$\overline{DP} = 2v\left(\frac{k_{tc} + k_{td}}{k_{tc} + 2k_{td}}\right) \tag{43}$$

Often, one type of termination will predominate. If so, the following simplifications of (43) can be assumed.

1. If termination by combination predominates, $k_{tc} \gg k_{td}$, and

$$\overline{DP} = 2v \tag{44}$$

2. If termination by disproportionation predominates, $k_{td} \gg k_{tc}$, and

$$\overline{DP} = v \tag{45}$$

Thus, for any free-radical chain polymerization in which chain transfer does not occur, $\overline{DP}$ must be between v and $2v$.

Because the average degree of polymerization represents the average number of monomer molecules contained in a polymer molecule, the average molecular weight may at once be written as

$$\overline{M} = M_0 \overline{DP} \tag{46}$$

where M_0 is the molecular weight of the monomer. The average molecular weight obtained in this way is the *number*-average molecular weight, as will be discussed further in Chapters 14 and 15. A measurement of the number average molecular weight is the most straightforward way to determine $\overline{DP}$.

DISTRIBUTION OF THE DEGREE OF POLYMERIZATION AND OF MOLECULAR WEIGHT[1]

According to the mechanism given on page 272 and the steady-state approximation, the rate equation for propagating radicals *of degree of polymerization, n,* may be written

$$\frac{d[R_n]}{dt} = k_p[M][R_{n-1}] - k_p[M][R_n] - 2k_t[R_n][R] = 0 \tag{47}$$

which after division by $[R_{n-1}]$, followed by rearrangement, yields

$$\frac{[R_n]}{[R_{n-1}]} = \frac{k_p[M]}{k_p[M] + 2k_t[R]} = \left(1 + \frac{2k_t[R]}{k_p[M]}\right)^{-1} \tag{48}$$

In view of the definition of v, (48) may be written conveniently as

$$\frac{[R_n]}{[R_{n-1}]} = \left(1 + \frac{1}{v}\right)^{-1} \tag{49}$$

We may obtain an expression for the ratio $[R_n]/[R_1]$ by successive multiplication of the ratios in (49). Thus,

$$\frac{[R_n]}{[R_1]} = \left(\frac{[R_n]}{[R_{n-1}]}\right)\left(\frac{[R_{n-1}]}{[R_{n-2}]}\right)\left(\frac{[R_{n-2}]}{[R_{n-3}]}\right) \cdots \frac{[R_2]}{[R_1]} \tag{50}$$

or

$$[R_n] = [R_1]\left(1 + \frac{1}{v}\right)^{-(n-1)} \tag{51}$$

[1] A. M. North, *The Kinetics of Free Radical Polymerization* (New York: Pergamon Press, 1966), pp. 14–16.

If f_n is defined as the fraction of propagating radicals that have a degree of polymerization n, then, in view of (51), we have the relationship

$$f_n = \frac{[R_n]}{[R]} = \frac{[R_1]}{[R]}\left(1 + \frac{1}{v}\right)^{1-n}$$

(52)

Consider now the steady-state expressions for the total radical concentration, [R], and for the concentration of the smallest propagating radical, $[R_1]$, shown in (53) and (54).

$$\frac{d[R]}{dt} = r_i - 2k_t[R]^2 = 0$$

(53)

$$\frac{d[R_1]}{dt} = r_i - k_p[M][R_1] - 2k_t[R_1][R] = 0$$

(54)

Equating (53) and (54) and rearranging gives

$$\frac{[R_1]}{[R]} = \frac{1}{v}\left(1 + \frac{1}{v}\right)^{-1}$$

(55)

Substitution of (55) into (52) then yields the distribution function shown in (56) for the degree of polymerization of the propagating radicals.

$$f_n = \frac{1}{v}\left(1 + \frac{1}{v}\right)^{-n}$$

(56)

The distribution of the degree of polymerization (or of the molecular weight) in the polymer will be determined by f_n and by the *mechanism* of formation of the polymer. In the simple mechanism described on page 272, polymer molecules are formed only by combination and disproportionation of propagating radicals. It is convenient to consider these cases separately.

Polymer Formation Solely by Combination ($k_{td} = 0$)

The mole fraction of polymer having a DP of n (denoted by X_n) will be given by the rate ratio

$$X_{n,c} = \frac{d[P_n]/dt}{\sum_n d[P_n]/dt} = \frac{d[P_n]/dt}{d[P]/dt}$$

(57)

where the subscript c denotes explicitly that X_n in (57) refers to termination by combination. Because polymer is formed only in the radical termination reactions, the denominator of (57) is given by (58).

$$\frac{d[P]}{dt} = k_{t_c}[R]^2 \tag{58}$$

To obtain an expression for $d[P_n]/dt$ we need only to write down all the ways in which P_n can be formed and then sum the rates of these processes. The possible reactions that form P_n are shown in (59) to (62).

$$R_{n-1} + R_1 \longrightarrow P_n \tag{59}$$

$$R_{n-2} + R_2 \longrightarrow P_n \tag{60}$$

$$\cdots\cdots\cdots\cdots\cdots\cdots$$

$$R_{n-m} + R_m \longrightarrow P_n \tag{61}$$

$$\cdots\cdots\cdots\cdots\cdots\cdots$$

$$R_1 + R_{n-1} \longrightarrow P_n \tag{62}$$

Note that, in summing the rates of (59) to (62) [i.e., in allowing the summation index, m, to go from the value 1 in (59) to $n - 1$ in (62)], every reaction is counted twice, with one exception. The one exception occurs for that reaction for which $m = n - m = n/2$. Such a reaction (between like radicals) will occur only if the integer n is even. If n is even, then, we have

$$\frac{d[P_n]}{dt} = k_{tnnc}[R_{n/2}][R_{n/2}] + \frac{1}{2}\sum_{\substack{m=1 \\ (m \neq n/2)}}^{n-1} k_{tmnc}[R_{n-m}][R_m] \tag{63}$$

where the $\frac{1}{2}$ term corrects for the addition of each reaction twice. Recall now that the rate constant for the termination of unlike radicals is twice that for like radicals (i.e., $k_{tmn} = 2k_{tnn} = 2k_t$). So equation (63) may be written as (64).

$$\frac{d[P_n]}{dt} = k_{tc}\sum_{m=1}^{n-1}[R_{n-m}][R_m](1)^m \tag{64}$$

There can be no combination of like radicals to produce a polymer having an odd number for the degree of polymerization, and (64) follows directly.

The mole fraction of polymer with $DP = n$ is now obtained simply by dividing (64) by (58) to give

$$X_{n,c} = \frac{\sum_{m=1}^{n-1}[R_{n-m}][R_m](1)^m}{[R]^2} \tag{65}$$

The radical concentrations $[R_m]$ and $[R_{n-m}]$ may be written in terms of the total radical concentration, $[R]$, according to (56). When these are substituted into (65), equations (66) and (67) are obtained.

$$X_{n,c} = \frac{1}{v^2}\left(1 + \frac{1}{v}\right)^{-n} \sum_{m=1}^{n-1} (1)^m \tag{66}$$

or

$$X_{n,c} = \frac{n-1}{v^2}\left(1 + \frac{1}{v}\right)^{-n} \tag{67}$$

X_n is the fraction of polymer molecules having molecular weight nM_0, where M_0 is the molecular weight of the monomer. Therefore, (67) yields the numerical distribution function for molecular weight of the polymeric product if termination occurs solely by combination.

Polymer Formation Exclusively by Disproportionation ($k_{tc} = 0$)

If radical termination and polymer formation occur exclusively by disproportionation, the general step that produces polymer of $DP = n$ is

$$R_n + R_s \longrightarrow P_n + P_s \tag{68}$$

By a procedure completely analogous to that just described (but which is left as an exercise for the reader) it can be shown that the fraction of polymer molecules having $DP = n$ is given by

$$X_{n,d} = \frac{1}{v}\left(1 + \frac{1}{v}\right)^{-n} \tag{69}$$

The degree-of-polymerization distribution functions $X_{n,c}$ and $X_{n,d}$ given by (67) and (69) are shown plotted in Figure 12.5 for two values of the average kinetic chain length v. It may be shown from the distribution functions $X_{n,c}$ and $X_{n,d}$ that the average degrees of polymerization in the case of termination by combination or disproportionation are $2v$ and v, respectively. In any actual polymerization, the number-average degree of polymerization obtained from these distributions must lie *between* v and $2v$. Therefore, the number-average molecular weight must lie between vM_0 and $2vM_0$, where M_0 is the molecular weight of monomer.

An experimentally determined distribution for polystyrene produced in benzene solution at $50°C$ and having an average degree of polymerization of 626 is shown by the points in Figure 12.6. Also shown in Figure 12.6, for comparison, is the distribution calculated from (67) for $v = \frac{1}{2}(\overline{DP})_{\exp} = 313$. The agreement between the theoretical and experimental distributions attests to the validity and utility of the derived distribution functions. It also confirms that the termination in styrene polymerization is predominantly by combination of free radicals.

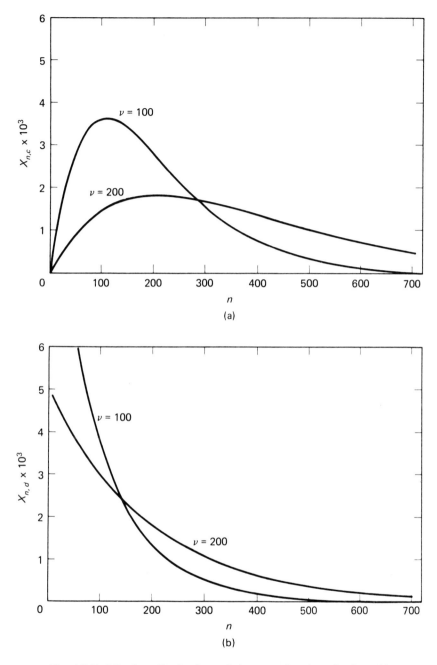

Fig. 12.5 Number distribution of degrees of polymerization. (a) Termination by combination. (b) Termination by disproportionation.

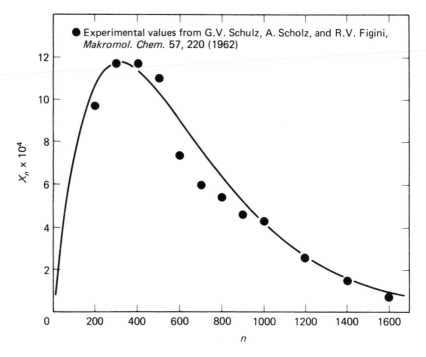

Fig. 12.6 Experimental and theoretical number distributions of the degree of polymerization (calculated using equation 67).

The reader should be aware that the distributions in (67) and (69) and in Figures 12.5 and 12.6 are *number* distributions or *mole-fraction* distributions. Often the distributions are expressed as a *weight-fraction* distribution. It is left as an exercise for the reader to show that the weight-fraction distributions $W_{n,c}$ and $W_{n,d}$ may be obtained from (67) and (69), respectively, by multiplication by the factor $n/\overline{DP}$.

CHAIN TRANSFER

In practice it is often observed that the average degree of polymerization is less than v, and sometimes considerably less than v, a fact that cannot be reconciled with the simple mechanism described on page 272. Moreover, it has been found that the addition of certain reagents to a free-radical polymerization reduces the average degree of polymerization from values greater than v to values considerably less than v *without affecting the rate of polymerization*.

The simplest explanation of these facts is that a reaction is occurring in which a growing free radical reacts with a stable molecule to form a polymer molecule *with the simultaneous generation of a new (and generally small) free radical*. Such a process, which is shown by (70), has not been included in the

$$R_n + X \xrightarrow{\ k_x\ } P_n + X\cdot \tag{70}$$

mechanism on page 272. In (70), $X \cdot$ is a free radical derived from the molecule X. If $X \cdot$ has a reactivity toward the monomer that is similar to that of R_n, so that (70) is nearly always followed by

$$X \cdot + M \longrightarrow R_1 \tag{71}$$

there will be no change in the rate of polymerization, r_p, or in the average kinetic chain length, v, because the concentration of propagating radicals, $[R]$, is unchanged. However, there will be a definite change in the average degree of polymerization and in the distribution of the degree of polymerization, and therefore also in the average molecular weight and distribution of molecular weights. The reaction (70) is termed chain transfer when it is followed nearly always by (71), and X is then called a chain transfer agent.

If $X \cdot$ [formed in (70)] has a sufficiently lower reactivity toward the monomer than does R_n, so that other reactions of $X \cdot$ can compete effectively with (71), then (70) represents inhibition, and X is called an inhibitor or retarder. If the free radical X does not react with monomer (M), it is not likely to react with the other stable molecules present in the system, such as polymer (P_n), initiator (I), or added reagent (X). The ultimate fate of an unreactive $X \cdot$ in such a severe case of inhibition is generally to react with other free radicals to terminate the kinetic chain. We shall consider here only the process that we have called chain transfer, namely, the case in which $X \cdot$ is reactive toward M.

If (70) is always followed by (71), the occurrence of (70) results in no change in the total radical concentration $[R]$. Therefore, the rate of polymerization and the kinetic chain length will be unchanged from the expressions given in (35) and (40), respectively. However, the average degree of polymerization and the distribution of the degrees of polymerization are altered significantly.

Effect of Chain Transfer on Average Degree of Polymerization

We may incorporate the occurrence of chain transfer into the expression for the average degree of polymerization simply by recognizing that reaction (70) produces a stable polymer molecule. Because chain transfer can occur for radicals that have any degree of polymerization, the total rate of polymer formation becomes

$$\frac{d[P]}{dt} = \sum_n \frac{d[P_n]}{dt} = (k_x[X] + k_{tc}[R] + 2k_{td}[R])[R] \tag{72}$$

By analogy to (42), which refers to $k_x = 0$, the average degree of polymerization becomes

$$\overline{DP} = \frac{k_p[M]}{k_x[X] + (k_{tc} + 2k_{td})[R]} \tag{73}$$

It is more convenient to consider the *reciprocal* of the average degree of polymerization. Utilizing (34) or (40) to eliminate [R], we obtain the expressions

$$(\overline{DP})^{-1} = \frac{k_x[X]}{k_p[M]} + \frac{(k_{tc} + 2k_{td})r_p}{k_p^2[M]^2} = \frac{k_x[X]}{k_p[M]} + \frac{k_{tc} + 2k_{td}}{2v(k_{tc} + k_{td})} \tag{74}$$

It must be recognized that the chain transfer reaction is not restricted to substances that are added specifically for that purpose. Chain transfer may also occur with monomer (M), polymer (P), solvent (S), or initiator (I), as well as with impurities that may be present. Thus, the term $k_x[X]$ is a composite that should actually be written as

$$k_X[X] = k_M[M] + k_S[S] + k_I[I] + k_{polymer}[P] + k_Y[Y] \tag{75}$$

where Y represents a chain transfer agent added specifically for this purpose and k_i represents chain transfer to the species i. Substitution of (75) into (74) and definition of the *chain transfer constant*, C_i, as $C_i \equiv k_i/k_p$, leads to

$$(\overline{DP})^{-1} = (\overline{DP})_0^{-1} + C_I \frac{[I]}{[M]} + C_S \frac{[S]}{[M]} + C_Y \frac{[Y]}{[M]} + C_{polymer} \frac{[P]}{[M]} \tag{76}$$

where

$$(\overline{DP})_0^{-1} = C_M + \frac{(k_{tc} + 2k_{td})r_p}{k_p^2[M]^2} \tag{77}$$

Very often, conditions can be chosen that eliminate or at least minimize chain transfer to the initiator or to the polymer, and further considerations here assume that this is the case. Therefore, the corresponding terms in (75) may be neglected.

According to (76), the chain transfer constants C_S and C_Y may be evaluated from experimental measurements of the average degree of polymerization as a function of the concentrations of the solvent and chain transfer agent at fixed values of the ratio $r_p/[M]^2$. Thus, a plot of $(\overline{DP})^{-1}$ versus [S]/[M] or [Y]/[M] should yield a straight line with an intercept of $(\overline{DP})_0^{-1}$ and a slope of C_S or C_Y, respectively. Typical plots for chain transfer to various solvents in vinyl acetate polymerization at 60°C are shown in Figure 12.7.

Because the chain transfer constants are actually *ratios of rate constants* of elementary reactions, they will be temperature-dependent, except in the improbable case that the activation energies of propagation and of chain transfer are the same. Thus, the temperature dependence of the chain transfer constant to the solvent is given by

$$C_s = \frac{A_s e^{-E_s/RT}}{A_p e^{-E_p/RT}} = \left(\frac{A_s}{A_p}\right) e^{(E_p - E_s)/RT} \tag{78}$$

where A_s and A_p are the frequency factors of the transfer and propagation reactions, respectively, and E_s and E_p are the respective activation energies. According to (78)

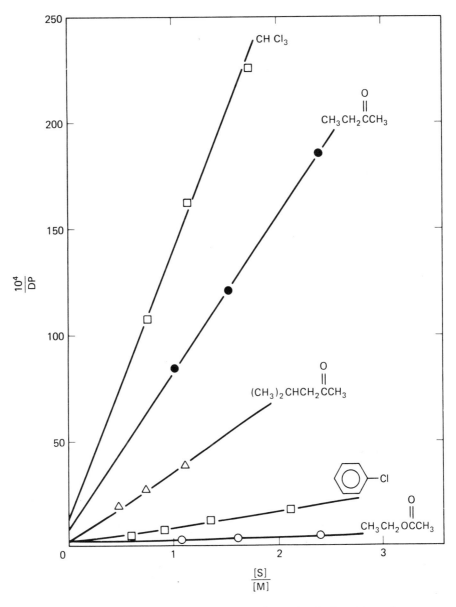

Fig. 12.7 Chain transfer to various solvents in vinyl acetate polymerization at 60°C. [From S. R. Palit and S. K. Das, *Proc. Roy. Soc.*
(*London*), **A226**, 82 (1954).]

the kinetic parameters A_s/A_p and $E_s - E_p$ may be determined from the intercept and
slope, respectively, of plots of the logarithm of C_s as a function of $1/T$. The C_s
values are determined at each temperature as described above.

Chain transfer constants have been determined for the reactions of polymerizing
monomers with a large number of reagents. A small collection of such constants at

TABLE 12.3   Chain Transfer Constants in Free-Radical Polymerization

POLYMERIZING MONOMER	CHAIN TRANSFER AGENT (Y)	$C \times 10^4$ (60°C)	E_y-E_p (kcal/mol)
Styrene*	Styrene	0.6	7.9
	Benzene	0.018	15
	Cyclohexane	0.024	13
	Triphenylmethane	35	5.1
	Carbon tetrachloride	92	5.0
	Carbon tetrabromide	13,600	3.0
Vinyl acetate†	Vinyl acetate	2.5	3.1
	Benzene	2.2	15
	Cyclohexane	6.6	
	Triphenylmethane	850	
	Carbon tetrachloride	960	1.4
	Carbon tetrabromide	390,000	−7.6
Methyl methacrylate*	Methyl methacrylate	0.07	
	Benzene	0.075	
	Cyclohexane	0.1	
	Carbon tetrachloride	2.4	
	Carbon tetrabromide	2700	

* G. E. Ham, *Vinyl Polymerization*, G. E. Ham, ed. (New York: Dekker, 1967), Chap. 1.

† M. K. Lindemann, *Vinyl Polymerization*, G. E. Ham, ed. (New York: Dekker, 1967), Chap. 4.

60°C, and the corresponding activation energy differences, $E_s - E_p$ (or $E_y - E_p$), are shown in Table 12.3.

Effect of Chain Transfer on the Distribution of the Degree of Polymerization

Chain transfer does not affect the rate of polymerization, r_p, or the average kinetic chain length, v. However, as shown in the preceding section, it can have a marked effect on the average degree of polymerization and hence on the average molecular weight. Obviously, the distribution of the degree of polymerization and the molecular weight will also be affected. Chain transfer may be introduced into the distribution function for the degrees of polymerization with the use of the same procedure employed in the development of (67) and (69). Although these derivations are left as exercises for the reader, we may note that the distribution function for termination by combination is given by (79),

$$X_{n,c} = \left[\frac{(n-1)(1+\gamma v)}{v^2} + 2\gamma \right] \left(\frac{1+\gamma v}{1+2\gamma v} \right) \left(1 + \gamma + \frac{1}{v} \right)^{-n} \tag{79}$$

where γ is the chain transfer term defined by (80),

$$\gamma = C_M + C_S \frac{[S]}{[M]} + C_Y \frac{[Y]}{[M]} \tag{80}$$

where S represents solvent and Y represents a reagent added for the purpose of chain transfer.

Numerical Example of the Effect of Chain Transfer on $\overline{DP}$ and Distribution of DP

Chain transfer can have a drastic effect on the average degree of polymerization and on the molecular weight distribution. As an example, consider the polymerization of 1 M styrene at 60°C in benzene and carbon tetrachloride solutions. If termination is by combination, then, for an initiation rate of $r = 1.9 \times 10^{-8}$

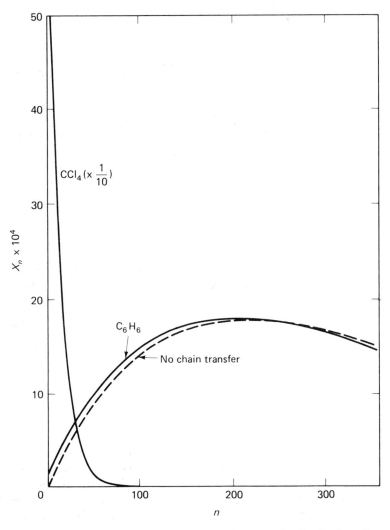

Fig. 12.8   Effect of chain transfer to solvent on the distribution of DP for polymerization of 1 M. Styrene at 60°C ($r_I = 10^{-8}$ mol/liter-s).

mol/liter-s (appropriate for 0.019M azobisisobutyronitrile at 60°C), the data in Tables 12.2 and 12.3 and equations (40) and (76) yield the following values:

$$v = 210$$
$$\overline{DP}(\text{benzene}) = 407 = 1.94v$$
$$\overline{DP}(CCl_4) = 12 = 0.057v$$

The distributions of the degrees of polymerization, as calculated from (79), are shown in Figure 12.8.

DEPENDENCE OF THE DEGREE
OF POLYMERIZATION ON TEMPERATURE

The manner in which the average degree of polymerization varies with temperature depends on the degree to which the initiation is temperature-dependent (e.g., thermal dissociation of an initiator) or temperature-independent (e.g., photolytic or radiolytic initiation) and on the extent of chain transfer.

Consider first the situation where chain transfer may be neglected. Then according to (44) and (45), the average degree of polymerization will lie between v and $2v$, depending on the relative contributions of combination and disproportionation to the chain termination step. Suppose, for simplicity, that $k_{td} = 0$, so that (44) applies. Then expression (81) is valid.

$$\overline{DP} = 2v = \frac{2(k_p^2/2k_t)^{1/2}[M]}{(2fk_i[I])^{1/2}} \tag{81}$$

If the rate constants are written according to the Arrhenius formulation and f is taken to be independent of temperature,[1] equation (82) or (83) follows:

$$\overline{DP} = \frac{A_p[M]}{(A_t A_i f[I])^{1/2}} \exp\left[\frac{-(E_p - \frac{1}{2}E_t - \frac{1}{2}E_i)}{RT}\right] \tag{82}$$

or

$$\ln \overline{DP} = \ln K - \frac{E_p - \frac{1}{2}E_t - \frac{1}{2}E_i}{RT} \tag{83}$$

where the temperature-independent term of (82) has been written as K. Typical values for $(E_p - \frac{1}{2}E_t)$ and E_i for several monomers and thermal initiators are given in Table 12.2 and Table 3.1, respectively. From the magnitudes of these activation energies it is apparent that when thermal initiation occurs the $\overline{DP}$ decreases as the temperature is increased. On the other hand, for photolytic, radiolytic (or any

[1] K. C. Berger, *Makromol. Chem.*, **176**, 3575 (1975).

temperature-independent method of initiation), $E_i = 0$. Then, according to (83), the $\overline{DP}$ will *increase* with increasing temperature.

As a specific example, consider a styrene polymerization, in which $(E_p - \frac{1}{2}E_t)$ = 6600 cal/mol, initiated as follows: (1) by the thermal dissociation of azobisiso-butyronitrile for which $(E_i = 30{,}800 \text{ cal/mol})$; and (2) by the photolytic dissociation of azopropane, for which $E_i = 0$. For case (1), expression (84) applies,

$$\frac{d \ln (\overline{DP})_1}{dT} = -\frac{8800}{RT^2} < 0 \tag{84}$$

while for case (2), expression (85) is appropriate.

$$\frac{d \ln (\overline{DP})_2}{dT} = +\frac{6600}{RT^2} > 0 \tag{85}$$

When chain transfer cannot be neglected, equation (74) applies. If the two terms in this equation are comparable, the temperature dependence is complicated, and an expression that is linear in $1/T$ cannot be written. In the very extreme case, where chain transfer to an added substance Y is the dominant mode of polymer formation, equation (86) is applicable:

$$\ln \overline{DP} = \ln \frac{[M]}{[Y]} + \ln \frac{A_P}{A_Y} + \frac{E_Y - E_P}{RT} \tag{86}$$

where the notation is the same as in Table 12.3. Since in most cases $E_y - E_p$ for chain transfer agents ·or solvents is positive, $\overline{DP}$ would *decrease* with increasing temperature.

ABSOLUTE PROPAGATION
AND TERMINATION RATE CONSTANTS

All the kinetic quantities derived in the preceding treatment depend on the rate of initiation, the monomer concentration, and the rate constant ratios $(k_p^2/k_t)^{1/2}$, C_M, and C_Y. As we have seen, if the initiation rate is known, the rate constant *ratios* given in Tables 12.2 and 12.3 can be determined quite easily from simple experimental measurements of the rates of polymerization and the average degrees of polymerization for varying concentrations of monomer and chain transfer substances. However, it is much more difficult to obtain *absolute* values for k_p, k_t, and k_y, since to do so requires a departure from the steady-state situation that led to the simple kinetic expressions for r_p, v, $\overline{DP}$, and so on.

A method that is used frequently to obtain the individual rate constants involves the intermittent or pulsed illumination of a photolytically initiated polymerization. The intermittent illumination produces a periodic departure of the propagating radical concentration from its steady-state value in such a way that a study of the polymerization rate (and therefore the mean propagating radical

TABLE 12.4   Rate Constants for Propagation and Termination at 60°C

Monomer	k_p (liters-mol^{-1}-s^{-1}) × 10^{-3}	k_t (liters-mol^{-1}-s^{-1}) × 10^{-7}
Methacrylonitrile	0.36	2.3
Methyl acrylate	1.6	0.13
Methyl methacrylate	0.71	1.8
Styrene	0.074	0.33
Vinyl acetate	1.0	3.2

concentration) as a function of the time of interruption of the illumination yields the steady-state *lifetime* of the radical chain, $\bar{\tau}_s$. Because it can be shown that, the steady-state lifetime is given by

$$\bar{\tau}_s = \frac{[R]}{2k_t[R]^2} = \frac{k_p}{2k_t}\frac{[M]}{r_p} \tag{87}$$

A measurement of $\bar{\tau}_s$, $[M]$, and r_p permits computation of the ratio k_p/k_t. A combination of this latter ratio with the corresponding $(k_p^2/k_t)^{1/2}$ of Table 12.2 yields k_p and k_t separately. A combination of k_p with C_Y of Table 12.3 then yields k_y.

For further details of this method, the reader is referred to more specialized treatments.[1] Individual rate constants at 60°C for propagation and termination are shown in Table 12.4. As mentioned, absolute rate constants for chain transfer at 60°C may be obtained by combining k_p from Table 12.4 with the appropriate chain transfer constant from Table 12.3.

COPOLYMERIZATION

The Copolymer Composition Equation.
Reactivity Ratios

The kinetic considerations of the preceding sections have been restricted to the polymerization of a single monomer. Suppose now that two monomers are present in the reaction mixture and that both are susceptible to free-radical polymerization. Four types of propagation reactions will now exist, as shown in (88) to (91),

$$\text{W—M·}_1 + \text{M}_1 \xrightarrow{k_{p11}} \text{W—M·}_1 \tag{88}$$

$$\text{W—M·}_1 + \text{M}_2 \xrightarrow{k_{p12}} \text{W—M·}_2 \tag{89}$$

$$\text{W—M·}_2 + \text{M}_2 \xrightarrow{k_{p22}} \text{W—M·}_2 \tag{90}$$

$$\text{W—M·}_2 + \text{M}_1 \xrightarrow{k_{p21}} \text{W—M·}_1 \tag{91}$$

[1] J. G. Calvert and J. N. Pitts, Jr., *Photochemistry* (New York: Wiley, 1966), pp. 651 ff.

where it is assumed that the reactivity of a particular radical is independent of its size and also independent of the nature of the polymeric chain bound to the radical sites $M\cdot_1$ and $M\dot{_2}$. Generally, the polymer obtained will be one in which both monomer units are incorporated together in the polymer molecules, namely into a *copolymer*. We can recognize from (88) to (91) two extreme types of kinetic behavior for a copolymerization. In the first, $k_{p12}/k_{p11} = k_{p21}/k_{p22} = 0$, and no copolymerization will occur. Instead, parallel polymerizations of the two monomers M_1 and M_2 will lead to the formation of two homopolymers $(M_1)_n$ and $(M_2)_m$. In the second extreme case, $k_{p11}/k_{p12} = k_{p22}/k_{p21} = 0$ and a copolymerization will take place to produce an alternating copolymer $(M_1M_2)_n$. Most copolymerizations fall between these two extremes.

Using the "long-chain assumption," that monomer molecules are consumed solely by the propagation reactions, the rates of monomer depletion in (88) to (91) are

$$-\frac{d[M_1]}{dt} = k_{p11}[M\cdot_1][M_1] + k_{p21}[M\dot{_2}][M_1] \tag{92}$$

$$-\frac{d[M_2]}{dt} = k_{p12}[M\cdot_1][M_2] + k_{p22}[M\dot{_2}][M_2] \tag{93}$$

Let us now define $\gamma = n_{M_1}/n_{M_2}$ as the ratio of the number of molecules of M_1 to the number of molecules of M_2 in the copolymer. Then from (92) and (93) we have

$$\gamma = \frac{-d[M_1]/dt}{-d[M_2]/dt} = \left(\frac{[M_1]}{[M_2]}\right)\left(\frac{k_{p11}[M\cdot_1] + k_{p21}[M\dot{_2}]}{k_{p12}[M\cdot_1] + k_{p22}[M\dot{_2}]}\right) \tag{94}$$

where $[M_1]/[M_2]$ is the mole ratio of monomers in the reactant mixture.

The steady-state approximation applied to the radicals $M\cdot_1$ and $M\dot{_2}$ leads to the expression

$$-\frac{d[M\cdot_1]}{dt} = \frac{d[M\dot{_2}]}{dt} = k_{p12}[M\cdot_1][M_2] - k_{p21}[M\dot{_2}][M_1] = 0 \tag{95}$$

from which rearrangement yields

$$\frac{[M\dot{_2}]}{[M\cdot_1]} = \frac{k_{p12}[M_2]}{k_{p21}[M_1]} \tag{96}$$

Substitution of (96) into (94) leads finally to the expression

$$\gamma = \frac{1 + r_1([M_1]/[M_2])}{1 + r_2([M_2]/[M_1])} \tag{97}$$

in which the *reactivity ratios*, r_1 and r_2, are defined by $r_1 = k_{p11}/k_{p12}$ and $r_2 = k_{p22}/k_{p21}$. Equation (97) is known as the *copolymer composition equation*, since it gives the composition of the copolymer being formed, at a given instant, in terms of the composition of the monomer mixture and the reactivity ratios of the two monomers.

The copolymer composition equation (97) has been well tested, and the following predicted properties have been verified.

1. The equation is independent of dilution because r_1 and r_2 and the concentration ratios are dimensionless.

2. It is independent of the initiation rate because no rate constants for initiation or termination appear.

3. It is independent of the medium in which the reaction is carried out because no changes in r_1 and r_2 are found when reagents are added that change significantly the nature of the medium.

In order to show the dependence of the polymer composition on the composition of the reaction mixture, it is convenient to write the copolymer composition equation in terms of the mole fractions in the polymer, f_i, and in the reactant mixture, F_i. Thus,

$$\frac{f_1}{f_2} = \frac{1 + r_1(F_1/F_2)}{1 + r_2(F_2/F_1)} = \frac{F_1(F_2 + r_1 F_1)}{F_2(F_1 + r_2 F_2)} \tag{98}$$

is obtained which, following the use of the relationships $f_2 = 1 - f_1$ and $F_2 = 1 - F_1$, yields

$$f_1 = \frac{F_1(1 + [r_1 - 1]F_1)}{(r_1 + r_2 - 2)F_1^2 + 2(1 - r_2)F_1 + r_2} \tag{99}$$

The functional dependence of f_1 on F_1 is shown graphically in Figure 12.9 for several sets of values of r_1 and r_2.

For the special case of $r_1 = r_2$, a family of curves is obtained that pass through the single point (0.5, 0.5). The shape depends on whether the value of r is greater or less than unity. In such a system all reactant mixtures show azeotropic properties. By this we mean that the mole fractions of the reactants in any mixture will approach 0.5 as the polymerization proceeds. When the mole fraction of each reactant becomes 0.5, no further change will occur in monomer composition, and the monomer ratio in the polymer is the same as the concentration ratio of the monomers in the reaction mixture.

It is also possible to have azeotropic reaction mixtures for systems in which $r_1 \neq r_2$. Thus, an application of the azeotropic condition, namely, $\gamma = [M_1]/[M_2]$, leads to the result

$$\left(\frac{[M_1]}{[M_2]}\right)_{\text{Azeotropic}} = \frac{1 - r_2}{1 - r_1} \tag{100}$$

Because the ratio of monomer concentrations must be a positive number, equation (100) shows that azeotropic reaction mixtures are possible when r_1 and r_2 are both

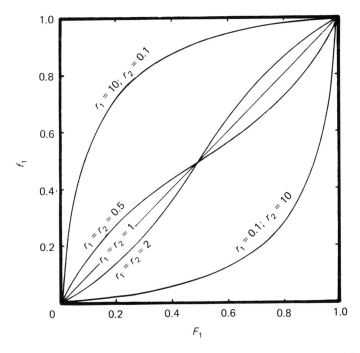

Fig. 12.9   Dependence of copolymer composition on composition of reactant mixtures.

less than unity or both greater than unity. In practice, many azeotropic systems are known for which both r_1 and r_2 are less than unity, but cases in which both reactivity ratios are greater than unity are very rare.

Experimental Determination of Reactivity Ratios

Actual numerical values of the reactivity ratios are determined from experimental measurements of the molar ratios of the monomers in the copolymer that is formed from reactant mixtures of known initial monomer concentration ratios. Thus, if (97) is solved explicitly for r_2, equation (101) is obtained.

$$r_2 = \frac{1}{\gamma}\left(\frac{[M_1]}{[M_2]}\right)^2 r_1 + \frac{[M_1]}{[M_2]}\frac{1}{\gamma} - 1 \tag{101}$$

For a chosen value of $[M_1]/[M_2]$, it is possible to determine *by chemical analysis* a corresponding value of γ. Substitution of these values into (101) yields a linear equation of positive slope for r_2 as a function of r_1. Each separate experiment at different $[M_1]/[M_2]$ values yields a straight-line equation of positive slope. The coordinates of the intersection of these straight lines on a graph of r_2 versus r_1 is the reactivity ratio for the particular monomer pair under investigation. Of course,

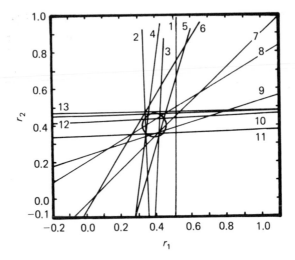

Fig. 12.10 Determination of reactivity ratios by the slope-intersection method. The circle represents the most probable area for r_1, r_2 value. [Reproduced from T. Alfrey, Jr., J. J. Bohrer, and H. Mark, *Copolymerization* (New York: John Wiley & Sons, Inc. 1952), p. 20.]

TABLE 12.5 Free-Radical Copolymerization Reactivity Ratios

M_1	M_1	TEMPERATURE (°C)	r_1	r_2
Methyl methacrylate	Acrylonitrile	80	1.22	0.15
Methyl methacrylate	Butadiene	90	0.25	0.75
Methyl methacrylate	p-Chlorostyrene	60	0.42	0.89
Styrene	Butadiene	60	0.78	1.39
Styrene	Methyl methacrylate	60	0.52	0.46
Styrene	Methyl methacrylate	131	0.59	0.54
Styrene	Vinyl acetate	60	55	0.01
Vinyl acetate	Acrylonitrile	70	0.07	6.0
Vinyl acetate	Methyl methacrylate	60	0.015	20
Vinyl acetate	Vinyl chloride	60	0.23	1.68

Source: H. Mark, B. Immergut, E. H. Immergut, L. J. Young, K. I. Beynon, *Copolymerization*, G. E. Ham, ed., (New York: Wiley–Interscience, 1964), pp. 695ff.

because r_1 and r_2 are ratios of reaction rate constants, their values will depend on the temperature at which the copolymerization is carried out. Figure 12.10 shows a typical plot of this type with the "intersection" for an actual case being an area instead of a point. This is due mainly to experimental error. Table 12.5 shows some representative values of r_1 and r_2.

Individual Monomer Reactivity in Copolymerization

Each copolymerization reactivity ratio r_j describes the relative tendency of two monomers to add to a particular growing chain. The reactive end of the growing chain is a free radical derived from one of the two monomers. Obviously, two types of reactive ends can exist and, for this reason, reactivity ratios must be determined in pairs. Moreover, because the values obtained experimentally are *relative* values, they pertain only to *one particular pair* of monomers. Thus, in the absence of any

Thermodynamics and Kinetics of Polymerization / Part II

correlation procedure, reactivity ratios must be determined experimentally for each pair of monomers. On the other hand, a correlation procedure that permits the assignment to each monomer of a reactivity parameter that is applicable to its copolymerization with all other monomers would represent a great economy in data accumulation and tabulation.[1]

Several such correlations having varying degrees of complexity and theoretical foundation have been proposed, the best known and most widely used being the Alfrey–Price $Q - e$ scheme. In this correlation procedure, each propagation rate constant is written as

$$k_{ij} = P_i Q_j \exp\left(-e_i e_j\right) \tag{102}$$

where P_i describes the reactivity of radical i, Q_j describes the reactivity of monomer j, and e_i and e_j describe the polarity interactions of radical and molecule, respectively. From (102) the reactivity ratios become

$$r_1 = \frac{Q_1}{Q_2} \exp\left(-e_1[e_1 - e_2]\right) \tag{103}$$

$$r_2 = \frac{Q_2}{Q_1} \exp\left(-e_2[e_2 - e_1]\right) \tag{104}$$

and the product of the reactivity ratios is

$$r_1 r_2 = \exp\left(-[e_1 - e_2]^2\right) \tag{105}$$

If one monomer is chosen as a reference, and arbitrary values of Q and e are assigned to it, it is possible to construct a table of Q, e values from experimental r_1, r_2 values. This is analogous to the construction of tables of standard electrode potentials in electrochemical cells. With this procedure we have two parameters for each monomer. These may be used as input to equations (103) and (104) to calculate reactivity ratios for the copolymerization of any monomer pair for which Q, e values are available. Of course, the Q, e values determined in the manner described depend on the temperature, since r_1 and r_2 are temperature-dependent. Table 12.6 shows some representative values at 60°C, in which the reference monomer, styrene, has been assigned the values $Q_{\text{styrene}} = 1.0$ and $e_{\text{styrene}} = -0.8$. The data in this table may be used to calculate reactivity ratios and also to construct copolymer–monomer composition diagrams for 45 copolymerization pairs.

Although the theoretical significance of the $Q - e$ scheme is still not clear (despite numerous attempts to provide it with a theoretical basis), it remains, as it was originally proposed, an extremely useful correlation framework for relative reactivities in copolymerization.

[1] For a collection of n monomers, there are $n(n - 1)/2$ copolymerization pairs. Thus, taking a number as conservative as $n = 100$ indicates that with a correlation procedure employing two parameters, we need to tabulate only 200 numbers, whereas, without such a procedure, we require 9900 numbers.

TABLE 12.6 Alfrey–Price Q-e Values at 60°C

MONOMER	Q	e	MONOMER	Q	e
Acrylonitrile	0.60	1.20	Methyl methacrylate	0.74	0.40
Butadiene	2.39	−1.05	Methyl vinyl ketone	1.0	0.7
p-Methoxystyrene	1.36	−1.11	Styrene	(1.0)	(−0.8)
p-Cyanostyrene	1.61	0.30	Vinyl acetate	0.026	−0.22
Methylacrylate	0.42	0.60	Vinyl chloride	0.044	0.20

Distributions of the Monomers in a Copolymer[1]

Although chemical analysis of the copolymeric product indicates the mole ratio of monomers present, it does not reveal the manner in which the monomer units are distributed in the copolymer. Thus, for two monomers M_1 and M_2, the mole ratio $\gamma = (N_{M_1}/N_{M_2})_{\text{polymer}}$ yields no information concerning the average lengths of the $-(M_1)_n-$ and $-(M_2)_n-$ sequences in a typical copolymer as illustrated by

$$-M_1-M_1-M_1-M_1-M_2-M_2-M_1-M_2-M_2-M_2-M_1-M_1-M_1$$

where the sequences are underlined. It is now necessary to consider how kinetic reasoning may be used to calculate the mean sequence lengths of M_1 and M_2 units from reactivity ratios.

Let P_{11} be the probability that a growing radical chain $M\cdot_1$ will add to monomer M_1. To a good approximation the only two possible fates of the growing chain $M\cdot_1$ are addition of M_1 or addition of M_2. Hence, it is possible to write this probability as

$$P_{11} = \frac{k_{p11}[M\cdot_1][M_1]}{k_{p11}[M\cdot_1][M_1] + k_{p12}[M\cdot_1][M_2]} \tag{106}$$

or

$$P_{11} = \frac{r_1[M_1]}{r_1[M_1] + [M_2]} = 1 - P_{12} \tag{107}$$

where P_{12} is the probability that $M\cdot_1$ will react with M_2. Given a radical site of type $M\cdot_1$ in the copolymer, consider now the probability of forming a sequence of *exactly m* units of monomer M_1 and denote this probability by $P_{M_1}(m)$. It is instructive for this purpose to construct Table 12.7. To form a sequence of *exactly m* units the sequence must end with the last entry in the table. This means that the last growing radical in the table must react with M_2. Thus,

$$P_{M_1}(m) = P_{11}^{m-1}P_{12} \tag{108}$$

[1] A. M. North, *The Kinetics of Free Radical Polymerization* (New York: Pergamon Press, 1966), pp. 90–92.

TABLE 12.7 Buildup of a Sequence of M_1 Units

REACTION	SEQUENCE LENGTH	PROBABILITY
$M\cdot_1 + M_1 \rightarrow M_1M\cdot_1$	2	P_{11}
$M_1M\cdot_1 + M_1 \rightarrow (M_1)_2 M\cdot_1$	3	P_{11}^2
. . . .		
$(M_1)_{\overline{m-2}}M\cdot_1 + M_1 \rightarrow (M_1)_{\overline{m-1}}M\cdot_1$	m	P_{11}^{m-1}

Similarly, the probability that a sequence of m units of M_2 will be formed, given a radical site derived from M_2, is given by

$$P_{M_2}(m) = P_{22}^{m-1}P_{21} \tag{109}$$

The average sequence lengths, $\overline{m}_{M_1}$ and $\overline{m}_{M_2}$, may now be determined from (108) and (109) using the definition of an arithmetic mean. Thus,

$$\overline{m}_{M_1} = \frac{\sum_{m=1}^{\infty} mP_{M_1}(m)}{\sum_{m=1}^{\infty} P_{M_1}(m)} = \frac{\sum_{m=1}^{\infty} mP_{11}^{m-1}}{\sum_{m=1}^{\infty} P_{11}^{m-1}} \tag{110}$$

The reader may easily verify (by writing out the first few terms of the sums in (110) and comparing with known algebraic series) that the numerator is given by $(1/1 - P_{11})^2$ and the denominator by $(1/1 - P_{11})$. Therefore, expression (111) holds:

$$\overline{m}_{M_1} = \frac{1}{1 - P_{11}} = \frac{1}{P_{12}} \tag{111}$$

In view of the relationship between P_{11} and r_1, (107), we finally obtain (112).

$$\overline{m}_{M_1} = 1 + r_1 \frac{[M_1]}{[M_2]} \tag{112}$$

Similarly, the average sequence length of M_2 units is given by

$$\overline{m}_{M_2} = 1 + r_2 \frac{[M_2]}{[M_1]} \tag{113}$$

The *run number*, R, of the copolymer is defined as the average number of sequences of either type per 100 monomer units. To illustrate the meaning of this term, consider the segment of a hypothetical copolymer shown below in which the sequences are underlined. The number of sequences here are nine and twenty monomer units are present. Hence, $R = 45$.

$$\underline{M_1-M_1}-\underline{M_2-M_2-M_2}-\underline{M_1-M_1-M_1-M_1}-\underline{M_2}-\underline{M_1-M_1}$$

$$-\underline{M_2-M_2-M_2-M_2}-\underline{M_1-M_1}-\underline{M_2}-\underline{M_1}$$

The rate of sequence formation, dS/dt, regardless of length, is simply the rate at which sequences are ended. Neglecting chain termination, this is given by

$$\frac{dS}{dt} = k_{12}[M\cdot_1][M_2] + k_{21}[M\cdot_2][M_1] \tag{114}$$

where the subscript p on the rate constants has been dropped for convenience. The total rate of polymerization is given by

$$-\frac{d([M_1] + [M_2])}{dt} = k_{11}[M\cdot_1][M_1] + k_{12}[M\cdot_1][M_2] + k_{21}[M\cdot_2][M_1] + k_{22}[M\cdot_2][M_2]$$

$$\tag{115}$$

Elimination of the time by combination of (114) and (115) and application of the steady-state approximation (95) yields

$$-\frac{d([M_1] + [M_2])}{dS} = \frac{k_{11}[M_1] + k_{12}[M_2]}{2k_{12}[M_2]} + \frac{k_{22}[M_2] + k_{21}[M_1]}{2k_{21}[M_1]} \tag{116}$$

or

$$-\frac{d([M_1] + [M_2])}{dS} = 1 + \frac{r_1}{2}\frac{[M_1]}{[M_2]} + \frac{r_2}{2}\frac{[M_2]}{[M_1]} \tag{117}$$

As mentioned above in the definition, the *run number* is the average number of sequences per 100 monomer units. This may be written as

$$R = 100\left(-\frac{dS}{d([M_1] + [M_2])}\right) \tag{118}$$

which, after substitution of (117), yields, finally,

$$R = \frac{200}{2 + r_1([M_1]/[M_2]) + r_2([M_2]/[M_1])} = \frac{200}{\bar{m}_{M_1} + \bar{m}_{M_2}} \tag{119}$$

Equations (112), (113), and (119) indicate that a knowledge of $[M_1]$, $[M_2]$, r_1, and r_2 enables a prediction to be made, not only of the average mole ratio γ, but also of the average number of sequences of monomer units per unit length of polymer and the average sequence length of each monomer. As many significant properties of the copolymers depend on the distribution of monomer units, the ability to make such predictions from a relatively small amount of experimental data can be very useful indeed.

STUDY QUESTIONS

1. For the free-radical chain polymerization of styrene at 60°C, $k_p = 74$ and $k_t = 3.3 \times 10^6$ liters-mol^{-1}-s^{-1}. A typical rate of initiation of propagating radicals may be taken as 10^{-6} mole-liter^{-1}-s. Assuming that the rate of initiation is constant:
   (a) Calculate the steady-state concentration of free radicals.
   (b) Derive an expression for the time dependence of the free-radical concentration and calculate the time required to reach 95% of the steady-state value.
   (c) If the initial monomer concentration is 5 mol/liter, calculate the time required to polymerize 20% of the monomer.
   (d) Use your results to discuss the validity of the steady-state hypothesis. For example, can you say how much error is involved in using the steady-state hypothesis in answering (c).

2. Initial rates of polymerization of methyl methacrylate (MMA) initiated by the decomposition of azobisisobutyronitrile in benzene at 77°C have been reported [L. M. Arnett, *J. Am. Chem. Soc.*, **74**, 2027 (1952)] to be as shown in the accompanying table.

[MMA] (mol/liter)	[ABIN] × 10⁴ (mol/liter)	Rate × 10³ (mol/liter-min)	[MMA] (mol/liter)	[ABIN] × 10⁴ (mol/liter)	Rate × 10³ (mol/liter-min)
9.04	2.35	11.61	4.75	1.92	5.62
8.63	2.06	10.20	4.22	2.30	5.20
7.19	2.55	9.92	4.17	5.81	7.81
6.13	2.28	7.75	3.26	2.45	4.29
4.96	3.13	7.13	2.07	2.11	2.49

   (a) Is the rate law expressed by (35) in accord with these data? Why or why not?
   (b) Given that the rate constant for dissociation of azobisisobutyronitrile is described by $k_I = 10^{15.2} e^{-30,800\,cal/RT}$ s^{-1} and that the efficiency of initiation is 0.7, derive from these data a value of the rate constant ratio $k_p/k_t^{1/2}$.

3. G. V. Schulz and G. Harborth [*Makromol. Chem.*, **1**, 106 (1947), published by Hüthig and Wepf Verlag, Basel] reported the following data for polymerization of methyl methacrylate in benzene at 50°C and 70°C using the thermal decomposition of 0.0413 *M* benzoyl peroxide for initiation.

[MMA] (M)	$r_p \times 10^5$ at 50°C (M/s)	$\overline{DP}$ (50°C)	$r_p \times 10^5$ at 70°C (M/s)	$\overline{DP}$ (70°C)
0.944	1.53	630	8.2	210
1.89	3.34	1200	18.6	450
3.78	6.74	2120	38.4	840
5.66	9.72	2900	56.6	1190

   (a) Show that these data are in accord with (35).
   (b) Assuming that the efficiency of initiation is independent of temperature, estimate $E_p - \frac{1}{2}E_t + \frac{1}{2}E_i$ from these data.
   (c) Using data given in this chapter, evaluate the activation energy for the decomposition of benzoyl peroxide.
   (d) Explain the effects of monomer concentration and temperature on $\overline{DP}$.

4. Derive equation (69).

5. Calculate and plot the weight-fraction distributions of DP for (a) $k_{tc} = 0$ and (b) $k_{td} = 0$ when chain transfer is absent.

6. Derive the number distribution function for DP when chain transfer is operating and $k_{td} = 0$, namely, equation (79).

7. Derive the number distribution function for DP analogous to that in Problem 4 when chain transfer is operating and $k_{tc} = 0$.

8. Show that the weight-fraction distribution of DP, W_n, is equal to the number-fraction distribution, X_n, multiplied by $n/\overline{DP}$.

9. Vinyl acetate at a concentration of 4 M in benzene is polymerized at 60°C using benzoyl peroxide (0.05 M) as an initiator. The rate constant for benzoyl peroxide decomposition is given by $k_d = 3.0 \times 10^{13} e^{-29,600 \text{cal}/RT}$ s^{-1}, and you may assume an initiation efficiency of 0.75. You may also assume that $k_{tc} = 0$ (see Table 3.5). Using data from this chapter, calculate (a) the rate of polymerization; (b) the kinetic chain length; (c) the average degree of polymerization; (d) the distribution function for DP, including an appropriate plot; (e) the lifetime of the kinetic chain; (f) the lifetime of the propagating radical of $DP = 15$. The densities of final acetate and benzene are 0.93 and 0.87 g/cm^3, respectively.

10. A reaction mixture containing 8.6 mol/liter of styrene and 0.1 mol/liter carbon tetrachloride is polymerized at 0°C using photosensitization. The average kinetic chain length is observed to be 1000. Calculate, using data given in this chapter: (a) the rate of initiation; (b) the average degree of polymerization; (c) the average degree of polymerization at 100°C if the initiation rate remains the same.

11. A sample of polystyrene prepared by a free-radical polymerization was separated into 21 fractions of different molecular weight. [G. V. Schulz, A. Scholz, and R. V. Figini, *Makromol. Chem.*, **57**, 220 (1962), published by Hüthig and Wepf Verlag, Basel]. The data obtained were as follows:

WEIGHT OF FRACTION (mg)	(DP) OF FRACTION	WEIGHT OF FRACTION (mg)	(DP) OF FRACTION
25.60	138	23.05	885
51.65	274	47.55	960
55.95	365	51.15	1050
37.40	428	52.65	1160
45.05	480	50.10	1260
52.85	535	47.70	1420
31.30	605	44.75	1600
33.75	673	36.25	1890
40.10	740	27.00	2100
38.75	795	28.90	2030
42.60	835		

(a) From these data construct the weight distribution and number distribution curves of the molecular weight. (*Hint*: First construct the integral distribution curves by plotting the weight fraction of molecules with M less than a given M versus the given M. Then differentiate the curve obtained. A similar procedure is used to determine the number distribution curve.)

(b) Compare the experimental distribution points obtained in (a) with the theoretical curve calculated by equation (67).

12. A benzene solution contains vinyl acetate at 3.5 M and vinyl chloride at 1.5 M. A free-radical polymerization is initiated by adding 0.1 M azo-bis-isobutyronitrile and heating the solution to 60°C. Calculate:

    (a)   Composition of the copolymer first formed.

    (b)   Average sequence lengths of vinyl acetate and vinyl chloride in the copolymer first formed.

    (c)   The run number of the copolymer first formed.

    (d)   The probability of forming a vinyl acetate sequence that is 8 units long.

13. Using the Alfrey–Price Q-e values, calculate the reactivity ratios for the following copolymerization systems: (a) acrylonitrile–butadiene; (b) methyl acrylate–vinyl chloride; (c) methyl vinyl ketone–p-methoxystyrene; (d) styrene–p-methoxystyrene.

SUGGESTIONS FOR FURTHER READING

ALLEN, P. E. M. AND C. R. PATRICK, *Kinetics and Mechanisms of Polymerization Reactions*, Chaps. 2, 3, 7. New York: Wiley, 1974.

BEVINGTON, J. C., *Radical Polymerization*. New York: Academic Press, 1961.

DAINTON, F. S., *Chain Reactions*, Chap. 7. London: Methuen, 1956.

FLORY, P. J., *Principles of Polymer Chemistry*, Chap. 4. Ithaca, N.Y.: Cornell University Press, 1953.

NORTH, A. M., *The Kinetics of Free Radical Polymerization*. New York: Pergamon Press, 1966.

SCOTT, G. E., AND E. SENOGLES, "Kinetic Relationships in Radical Polymerization," *J. Macromol. Sci.—Rev. Macromol. Chem.*, **C9**, 49 (1973).

Kinetics
of Ionic Polymerization

DIFFERENCES BETWEEN IONIC
AND FREE-RADICAL KINETICS

As discussed earlier in Chapter 4, ionic polymerizations take place by chain mechanisms in which many monomer molecules add to a single chain center. Thus, ionic polymerization resembles free-radical polymerization in terms of the initiation, propagation, transfer, and termination reactions. The only difference between the two modes is the nature of the active chain end, that is, whether it is a free radical, a positive ion, or a negative ion.

However, the kinetics of ionic polymerizations are significantly different from free-radical polymerizations. The initiation reactions of ionic polymerization have only very low activation energies and, thus, they more closely resemble photoinitiated free-radical polymerizations. In free-radical polymerizations, chain termination occurs by the mutual destruction of two polymeric radicals, but in ionic polymerization such a process is impossible because the charge is not neutralized in the reaction between two positive or two negative ions. Solvent effects are much more pronounced in ionic polymerizations because of the role that the solvent plays in assisting the separation of electric charge. Thus, in a solvent of high dielectric constant, a polymerization may proceed via free ions. In a solvent of low dielectric constant the chain centers may be ion pairs, or both ion pairs and free ions. No such solvent role is encountered in free-radical polymerization.

The overall result of the foregoing features is to make the kinetics of ionic polymerization much more complex than the kinetics of free-radical polymerization. Most of the complications in ionic polymerization arise from the initiation reactions. The variety of initiation possibilities often gives the appearance that each ionic polymerization is unique and that no general kinetic treatment is possible. However, some cases do exist, particularly in anionic polymerization, in which the initiator dissociates *completely* into the active ionic form and does so *before* any significant amount of polymerization has occurred. In such cases, the polymerization kinetics are so simple that it is useful to classify ionic polymerization initiators according to whether they are quantitatively and instantaneously dissociated or not.

ANIONIC POLYMERIZATION

Quantitative and Instantaneous Dissociation of Initiator: Living Polymers

As mentioned above, particularly simple kinetics are found when an ionic polymerization initiator is completely dissociated before the polymerization begins. Examples of such initiators (cf. Table 4.1) are (1) alkali metal suspensions in liquids that are Lewis bases: (2) organolithium compounds (also used in solvents that act as Lewis bases toward the lithium ion); and (3) sodium naphthalenide, which is prepared by the electron exchange reaction between sodium and naphthalene.

If we represent the undissociated initiator by GA, then we can assume, for the present, that in the polymerization medium the dissociation reaction (1a) is instantaneous and complete.

$$GA \longrightarrow G^+ + A^- \tag{1a}$$

A monomer molecule must add to A^-, to complete the initiation of a polymerization, namely,

$$G^+ + A^- + M \longrightarrow G^+ + AM^- \tag{1b}$$

Depending on the solvent, the propagating anion may behave as a free ion, AM^-, or as an ion pair, AM^-G^+, or as both. For simplicity we will consider only one type (i.e., free ions) for the present kinetic treatment. The propagation reactions may then be written

$$A-M^- + M \longrightarrow A-M-M^- \tag{2}$$

$$A-M-M^- + M \longrightarrow A-M_2-M^- \tag{3}$$

$$A-M_{n-1}-M^- + M \longrightarrow A-M_n-M^- \tag{4}$$

It is assumed that the positive ion G^+ is always in the vicinity of the negative chain center at each step. The extent of charge separation between G^+ and the negative ions depends on the dielectric constant of the solvent. In the absence of impurities or of substances added deliberately for the purpose of chain termination, no termination reactions exist in this special case. The polymerization ceases only when the monomer is consumed. It begins again if more monomer is added. For this reason, the polymers produced in this special case are the so-called "living polymers."

In such a situation the kinetics are especially simple because no initiation reaction takes place *during* the polymerization. The number of chain centers to which the monomer molecules may add reaches its maximum value before polymerization begins. Moreover, the number of chain centers does not change during the polymerization because there is no termination step.

Rate of polymerization

The rate of polymerization is defined in a similar manner to that described earlier for free-radical polymerization,

$$r_p = -\frac{d[M]}{dt} = k_p[A^-][M] \tag{5}$$

where $[A^-]$ is the total concentration of anions of all degrees of polymerization. The total concentration $[A^-]$ is constant and is given by the concentration of the initiator before dissociation, namely $[GA]_0$. Hence,

$$r_p = -\frac{d[M]}{dt} = k_p[GA]_0[M] \tag{6}$$

Integration of this first-order rate equation gives the time dependence of the monomer concentration as

$$[M] = [M]_0 e^{-k_p[GA]_0 t} \tag{7}$$

Rates of anionic polymerizations are measured experimentally in the same way as described for free-radical polymerizations in Chapter 12, namely, by the measurement of $[M]$ at various times after the polymerization is started. The value of $[M]$ may be measured directly or by a secondary method once the relationship between the two has been established. From the experimental measurements and a knowledge of $[GA]_0$, k_p may be determined either by (6) or (7). The significance of the values obtained will be discussed later in this chapter.

Average kinetic chain length

Because no termination step exists in a true "living" polymerization, the kinetic chain growth is ended only when the monomer is completely consumed. The average kinetic chain length by definition is

$$v = \frac{\text{Monomer consumed}}{\text{Number of chain centers}} = \frac{[M]_0 - [M]}{[GA]_0} \tag{8a}$$

Thermodynamics and Kinetics of Polymerization / *Part II*

which on substitution of [M] from (7) becomes a function of time as given by (8b)

$$v = \frac{[M]_0}{[GA]_0} (1 - e^{-k_p[GA]_0 t}) \tag{8b}$$

According to (8b), in the limit of $t \to \infty$, or, in other words, at the completion of reaction,

$$v_\infty = \frac{[M]_0}{[GA]_0} \tag{8c}$$

Average degree of polymerization

The average degree of polymerization is given by the number of monomer molecules polymerized per polymer molecule formed. The number of polymer molecules formed is equal to the number of chain centers, or initiators, so that (9) is obtained.

$$\overline{DP} = \frac{[M]_0 - [M]}{[GA]_0} = \frac{[M]_0}{[GA]_0} (1 - e^{-k_p[GA]_0 t}) = v \tag{9}$$

In some cases it is necessary to make a simple modification of (9) to account for the fact that dianions are involved (see Chapter 4). For example, in the sodium naphthalenide initiated polymerization of styrene, the mechanism shown in (10) and (11) operates.[1]

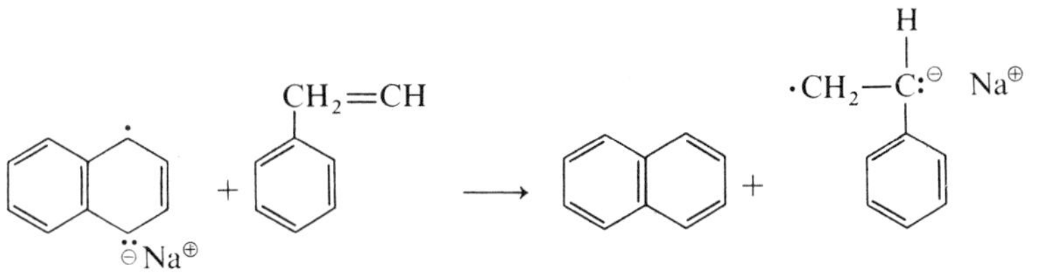

$$\tag{10}$$

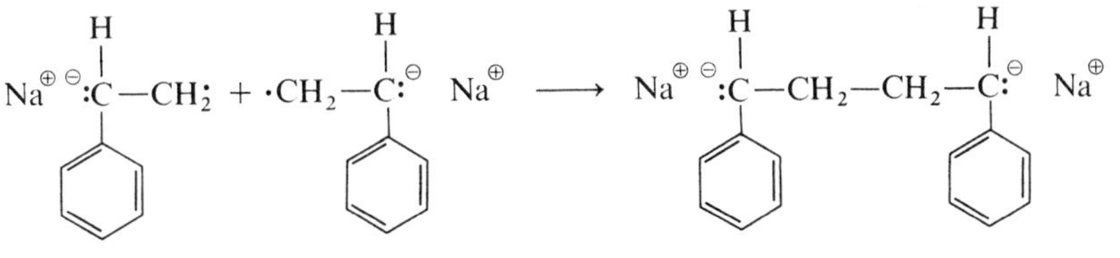

$$\tag{11}$$

The product of (11) is the true initiating species so that, for this particular case, the number of polymer molecules formed is $\frac{1}{2}$ of the number of chain centers or initiators. The degree of polymerization is then given by

$$\overline{DP} = \frac{[M]_0 - [M]}{\frac{1}{2}[GA]_0} = 2 \frac{[M]_0}{[GA]_0} (1 - e^{-k_p[GA]_0 t}) = 2v \tag{12}$$

[1] M. Szwarc, M. Levy, and R. Milkovich, *J. Am. Chem. Soc.*, **78**, 2656 (1956).

Distribution of the degree of polymerization

The most important difference between an ionic polymerization which has no termination or transfer mechanism and free-radical or ionic processes that do have termination or chain transfer steps is that the distributions of the degrees of polymerization are quite different. The distribution function may be derived by a kinetic treatment due to Flory[1], which is analogous to that discussed earlier for free-radical reactions. However, the steady-state approximation with its many simplifications cannot be used in this present case.

Consider the mechanism, (1) to (4), for this polymerization. If the usual assumption is made that k_p is independent of size and also that the initiation steps (1a and 1b) are instantaneous, application of the mass action law gives the set of rate equations shown in (13) to (16).

$$\frac{d[AM^-]}{dt} = -k_p[AM^-][M] \tag{13}$$

$$\frac{d[AMM^-]}{dt} = k_p[M]\{[AM^-] - [AMM^-]\} \tag{14}$$

$$\frac{d[AMMM^-]}{dt} = k_p[M]\{[AMM^-] - [AMMM^-]\} \tag{15}$$

$$\vdots$$

$$\frac{d[AM_nM^-]}{dt} = k_p[M]\{[AM_{n-1}M^-] - [AM_nM^-]\} \tag{16}$$

Substitution of the expression for $[M]$, namely (7), into the first of these rate equations yields

$$\int \frac{d[AM^-]}{[AM^-]} = -k_p[M]_0 \int e^{-k_p[GA]_0 t}\, dt \tag{17}$$

On integration, under the condition that at $t = 0$, $[AM^-] = [GA]_0$,

$$[AM^-] = [GA]_0 \exp\left[-\frac{[M]_0}{[GA]_0}(1 - e^{-k_p[GA]_0 t})\right] \tag{18}$$

In view of the expression for average kinetic chain length, namely (8), this last result may be written in the compact form

$$[AM^-] = [GA]_0 e^{-\nu} \tag{19}$$

[1] P. J. Flory, *J. Am. Chem. Soc.*, **62**, 1561 (1940).

It is useful at this point to eliminate the time, t, from the remainder of the rate equations (14) to (16). This may be done conveniently by the use of the expression for the average kinetic chain length, (8b). Differentiation of (8b) with respect to time gives

$$dv = k_p[M]_0 e^{-k_p[GA]_0 t}\, dt \tag{20}$$

Substitution of (7), (19), and (20) into (14), with elimination of t, transforms this rate equation for $[AMM^-]$ into the differential equation

$$\frac{d[AMM^-]}{dv} + [AMM^-] = [GA]_0 e^{-v} \tag{21}$$

The rate equation (21) is in the standard form of a linear, first-order differential equation whose solution is given by

$$e^v[AMM^-] = \int e^v[GA]_0 e^{-v}\, dv + C \tag{22}$$

where e^v is the integrating factor and C is a constant of integration. After integration of (22) and evaluation of C by the condition that at $v = 0$ (i.e., $t = 0$), $[AMM^-] = 0$, we obtain

$$[AMM^-] = [GA]_0 v e^{-v} \tag{23}$$

The process by which the rate equation (14) was solved to yield (23) may now be repeated for the rate equation (15). Thus, elimination of t by substitution of (7), (20), and (23) transforms (15) into the differential equation (24),

$$\frac{d[AMMM^-]}{dv} + [AMMM^-] = [GA]_0 v e^{-v} \tag{24}$$

which may be solved in the same manner as was (21) using the integrating factor e^v. The result is

$$[AMMM^-] = \tfrac{1}{2}[GA]_0 v^2 e^{-v} \tag{25}$$

Repetition of this process soon makes it evident that the concentration of the anion containing n monomer molecules is given by

$$[AM_{n-1}M^-] = [GA]_0 \frac{v^{n-1} e^{-v}}{(n-1)!} \tag{26}$$

When polymerization is complete, the kinetic chain length as given by (8c) is $[M]_0/[GA]_0$. Furthermore, each initial anion has produced one polymer species, since there is no termination. Therefore, the fraction of polymer of degree of polymerization n at the end of the reaction is

$$X_n = \frac{\text{Number of anions containing } n \text{ monomers}}{\text{Number of anions}} = \frac{[AM_{n-1}M^-]}{[GA]_0} \tag{27}$$

or

$$X_n = \frac{v_\infty^{n-1}e^{-v_\infty}}{(n-1)!} = \frac{1}{(n-1)!}\left(\frac{[M]_0}{[GA]_0}\right)^{n-1}e^{-[M]_0/[GA]_0} \tag{28}$$

The distribution of v (or DP) given by (28) is shown graphically for $v_\infty = 50$ in Figure 13.1. Also shown for comparison is the distribution of DP for a polymer produced by free-radical polymerization (with termination by combination) with the same value of v. The narrowness of the distribution of DP for the living polymer is very striking. Actually, the observed distributions of DP are usually somewhat broader than predicted by (28). This fact is attributed to the existence of a propagation–depropagation equilibrium,

$$A-M_n-M^- + M \; \rightleftharpoons \; A-M_{n+1}-M^- \tag{29}$$

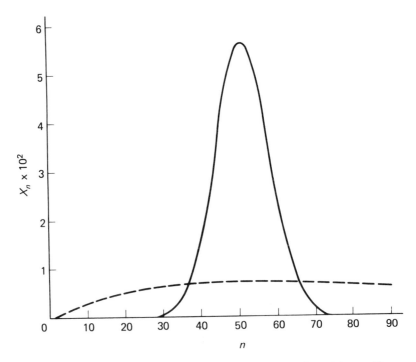

Fig. 13.1 Distribution of the degree of polymerization with $v = 50$ for "Living" Polymers (solid line) (equation 28) and for free-radical polymerization (dashed line) (Chapter 12, equation 67).

which was not considered in the derivation of (28). Nonetheless, the observed distributions are often very close to those predicted.

It should be kept in mind that the simple kinetics and the narrow distributions of *DP* are not a general characteristic of anionic polymerization, nor are they unique to it. They arise because the entire initiation process occurs before propagation and because no termination process is present to destroy the chain centers.

Rate Constants for Propagation

The use of initiators that dissociate quantitatively before propagation occurs and the absence of termination reactions permits the straightforward determination of k_p from measurements of the extent or rate of polymerization and equations (6) and (7). Some values obtained in this way at 25°C in tetrahydrofuran solvent with the counterion (G^+) being Na^+ are shown in Table 13.1.

It was mentioned earlier that the propagation reaction could involve free anions, ion pairs, or both. Experimentally, it is found that k_p depends on the nature of the counterion and on the solvent. Neither of these effects would exist if the *sole* chain centers in the propagation steps were free ions. It must be concluded that ion pairs are important chain carriers, and may perhaps be the sole chain centers in anionic polymerizations. Therefore, the k_p values of Table 13.1 may represent composite values for propagation by ion pairs and free ions.

In principle, the degree of dissociation in any equilibrium of the type

$$A^-G^+ \; \rightleftharpoons \; A^- + G^+ \tag{30}$$

will be shifted to the right as the system is diluted with solvent, or, in other words, as the concentration $[GA]_0$ is reduced. Extrapolations of experimental values of k_p to infinite dilution should yield the k_p for propagation that occurs solely by free ions. Where such studies have been possible they have indicated that the value of k_p for a free-ion propagation is much greater than that for propagation by ion pairs. Therefore, the contribution of free ions to the overall propagation may be significant even in solvents of low dielectric constant. Hence, it must be concluded that the experimental values of k_p determined from polymerization rates represent a composite value for mechanisms that involve ion pairs *and* free ions. These

TABLE 13.1 Propagation Rate Constants for Anionic Polymerization at 25°C; Solvent = Tetrahydrofuran, Counterion = Na^+

MONOMER	k_p (liters-mol^{-1}-s^{-1})	MONOMER	k_p (liters-mol^{-1}-s^{-1})
α-Methylstyrene	2.5	Styrene	950
p-Methoxystyrene	52	1-Vinylnaphthalene	850
o-Methylstyrene	170	2-Vinylpyridine	7300
p-t-Butylstyrene	220	4-Vinylpyridine	3500

Source: M. Szwarc and S. Smid, *Progress in Reaction Kinetics*, Vol. 2, G. Porter, ed. (New York: Pergamon Press, Ltd., 1964), p. 249.

complicating effects are typical of ionic reactions. Unfortunately, they reduce the utility of the k_p values because such rate constants can be used to make predictions only for polymerizations carried out under exactly the same set of conditions (i.e., exactly the same solvent, initiator, concentration of initiator, and temperature). The k_p values measured in free-radical polymerizations are independent of such conditions, except for the influence of temperature.

Incomplete Dissociation of Initiator

The simple kinetics of the anionic polymerization discussed earlier exist because the initiator is converted *completely* from the inactive form, GA, to the active form, G^+A^-, (or $G^+ + A^-$) before any propagation reactions take place. However, some initiators (i.e., lithium alkyls and aryls) maintain an equilibrium between the active form and the inactive form. Moreover, this equilibrium may extend to the growing anionic chains also. In such a situation we must write the initiation steps as

$$GA \; \rightleftharpoons \; G^+A^- \; \rightleftharpoons \; G^+ + A^- \tag{1a'}$$

$$G^+A^- + M \; \longrightarrow \; AM^-G^+ \; \rightleftharpoons \; AMG \tag{1b'}$$

and the propagation steps as

$$AM^-G^+ + M \; \longrightarrow \; AMM^-G^+ \; \rightleftharpoons \; AM_2G \tag{2'}$$

$$\cdots\cdots\cdots\cdots\cdots\cdots\cdots\cdots\cdots\cdots\cdots\cdots\cdots\cdots\cdots$$

$$AM_{n-1}-M^-G^+ + M \; \longrightarrow \; AM_n-M^-G^+ \; \rightleftharpoons \; AM_{n+1}G \tag{4'}$$

We have included here propagation by ion pairs only. Propagation by free anions in mechanisms that correspond to (2') and (4') must also be considered in the complete scheme.

The mechanism described here is much more complex than that of "living" polymerization because of existence of the equilibrium

$$AM_nG \; \rightleftharpoons \; AM_n^-G^+ \tag{31}$$

As discussed, such equilibria exist not only in the initiation but in the propagation reactions as well. Furthermore, the equilibria may involve solvation contributions by the solvent, although this is not shown explicitly in the mechanism. Finally, as a further complicating feature, the monomer may also affect the equilibria. In such a case, the initiation rate would depend on the nature of the monomer, even though all other factors, such as solvent, temperature, concentrations, initiator, and so on, were fixed.

The complexities described above make it virtually impossible to write explicit general equations for the rate of polymerization, kinetic chain length, average degree of polymerization, and distribution of degree of polymerization as was done in Chapter 12 for free-radical polymerization, and earlier in this chapter for a completely dissociated ionic initiator. Hence, with the exception of those cases discussed above, in anionic polymerization (and in cationic reactions as well) each system represents a kinetically unique problem that must be solved separately.

Anionic Copolymerization

For the polymerization of two monomers by an anionic mechanism, we may write a set of elementary propagation reactions analogous to those described in Chapter 12:

$$\text{WW}-M_1^- + M_1 \xrightarrow{k_{11}} M_1^- \tag{32}$$

$$\text{WW}-M_1^- + M_2 \xrightarrow{k_{12}} M_2^- \tag{33}$$

$$\text{WW}-M_2^- + M_2 \xrightarrow{k_{22}} M_2^- \tag{34}$$

$$\text{WW}-M_2^- + M_1 \xrightarrow{k_{21}} M_1^- \tag{35}$$

Similarly, we may define reactivity ratios: $r_1 = k_{11}/k_{12}$ and $r_2 = k_{22}/k_{21}$, and determine such ratios from the composition of the copolymer product. However, a serious complication exists. The propagation rate constants, k_{ij}, are *composite* rate constants, being composed of free-ion contributions and ion-pair contributions. Therefore, the reactivity ratios will also be composite quantities, having contributions from both ion pairs and free ions. Because the relative abundances of free ions and ion pairs are strongly dependent on the reaction conditions, the reactivity ratios will also depend on these conditions. Therefore, the utility of such ratios is much more limited in anionic than in free-radical polymerization, because they can be applied only to systems identical to those for which they were determined.

Typical reactivity ratios for the anionic copolymerization of styrene with several monomers are shown in Table 13.2. Most of the values shown in this table were determined from measurements of the copolymer compositions as a function of monomer concentrations, as discussed in Chapter 12. The use of "living copolymerizations" (in which one monomer is first polymerized and the second monomer then added) permits reactions (33) and (35) to be studied independently and allows the respective rate constants k_{12} and k_{21} to be measured directly. The rate constants k_{11} and k_{22} are known or can be determined, in principle, from the polymerization of the pure monomers. Thus, in some anionic systems, the simplicity afforded by complete initiator dissociation (i.e., living polymers) permits not only a determination of the rate-constant ratios but also the measurement of

TABLE 13.2 Reactivity Ratios in the Anionic Copolymerization of Styrene (M_1)

M_2	INITIATOR	SOLVENT	TEMPERATURE (°C)	r_1	r_2
Acrylonitrile*	C_6H_5MgBr in toluene	Cyclohexane	-45	0.05	15.0
Methyl methacrylate*	C_6H_5MgBr in ether	Toluene	-30	0.01	25.0
Methyl methacrylate*	C_6H_5MgBr in ether	Ether	-30	0.05	14.0
Methyl methacrylate*	C_6H_5MgBr in ether	Ether	-78	0.02	20.0
Methyl methacrylate*	C_6H_5MgBr in ether	Ether	$+20$	0.30	2.0
p-Methoxystyrene†	C_4H_9Li	Toluene	0	10.9	0.05
p-Methoxystyrene†	Li	Tetrahydrofuran	0	2.9	0.23
α-Methylstyrene‡	Na–K alloy	Tetrahydrofuran	$+25$	35	0.003
p-Methylstyrene§	Na–K alloy	Tetrahydrofuran	$+25$	5.3	0.18
p-Methylstyrene†	Na	Tetrahydrofuran	0	1.97	0.38

* F. Dawans and G. Smets, *Makromol. Chem.*, **59**, 163 (1963).
† A. V. Tobolsky and R. J. Boudreau, *J. Polymer Sci.*, **51**, S53 (1961).
‡ C. L. Lee, J. Smid, and M. Szwarc, *J. Am. Chem. Soc.*, **83**, 2961 (1961).
§ M. Shima, D. N. Bhattacharyya, J. Smid, and M. Szwarc, *J. Am. Chem. Soc.*, **85**, 1306 (1963).

individual rate constants. The reactivity ratios for the copolymerization of styrene with α-methylstyrene and p-methylstyrene in Table 13.2 were determined in this manner.

The drastic effect that the reaction medium and the reaction conditions have on the anionic copolymerization reactivity ratios may be appreciated by inspection of the data for styrene–methyl methacrylate in Table 13.2.

CATIONIC POLYMERIZATION

Rate of Polymerization

The elementary reactions involved in addition polymerization via positive ions are formally identical to those found in anionic polymerization. The only difference is that the charges of the propagating chain center and the counterion are reversed. As in anionic polymerization, propagation may proceed by both ion pairs and free ions. Despite this similarity, the kinetics of cationic polymerizations are less well understood than those of anionic polymerizations. This is probably due to the relatively small number of cationic polymerization systems that are known to show the phenomena of complete and prior dissociation of the initiator (i.e., "living" polymerization). Essentially all the known cationic polymerizations involve initiation reactions that occur simultaneously with the propagation steps. As we have seen in anionic polymerization, the variety of initiation mechanisms and the accompanying complex association–dissociation equilibria preclude the writing of a general kinetic scheme. Moreover, the initiation of a cationic polymerization is often complicated by the need for cocatalysts. Cocatalysis occurs when traces of certain substances are required along with the initiator before polymerization can proceed (see page 88).

For the purpose of discussion, we may depict a cationic polymerization by the following set of elementary reactions, in which C is the catalyst or initiator, RX is the cocatalyst, and M is the monomer:

$$C + RX \underset{k_{-1}}{\overset{k_1}{\rightleftharpoons}} R^+CX^- \underset{k_{-2}}{\overset{k_2}{\rightleftharpoons}} R^+ + CX^- \quad \left.\begin{array}{l} \\ \\ \end{array}\right\} \quad \text{(36)}$$

$$R^+CX^- + M \xrightarrow{k_i} RM^+CX^- \qquad \text{Initiation} \quad \text{(37)}$$

$$RM^+CX^- + M \xrightarrow{k_p} RMM^+CX^- \qquad \text{(38)}$$

$$\cdots\cdots\cdots\cdots\cdots\cdots\cdots\cdots\cdots\cdots\cdots \quad \left.\begin{array}{l} \\ \\ \\ \end{array}\right\} \quad \text{Propagation}$$

$$RM_{n-1}-M^+CX^- + M \xrightarrow{k_p} RM_n-M^+CX^- \qquad \text{(39)}$$

$$R-M_n-M^+CX^- + M \longrightarrow R'-M^+CX^- + R-M_nM' \quad \begin{array}{l}\text{Chain}\\\text{transfer}\end{array} \quad \text{(40)}$$

$$RM_n-M^+CX^- \longrightarrow RM_{n+1}-X + C \quad \begin{array}{l}\text{Chain}\\\text{termination}\end{array} \quad \text{(41)}$$

In (37) to (41), only the polymerization that involves ion pairs is shown explicitly. It must be kept in mind that a similar polymerization chain carried by the free ions, RM_n-M^+, may be occurring simultaneously and that an equilibrium involving ion pairs and free ions, such as shown in (36), exists for ion pairs of all sizes. An example of such a mechanism is the one in which isobutylene is polymerized in methyl chloride solution in the presence of aluminum chloride, in which case

$$M = CH_2\!=\!\overset{\displaystyle CH_3}{\underset{\displaystyle CH_3}{\overset{|}{\underset{|}{C}}}} \qquad RM_n-M^+ = CH_3\!\!\left(\!CH_2\overset{CH_3}{\underset{CH_3}{\overset{|}{\underset{|}{C}}}}\!\!\right)_{\!\!n}\!\!CH_2\overset{CH_3}{\underset{CH_3}{\overset{|}{\underset{|}{C^+}}}}$$

$$C = AlCl_3$$

$$R'M^+ = CH_3\overset{CH_3}{\underset{CH_3}{\overset{|}{\underset{|}{C^+}}}}$$

$$RX = CH_3Cl$$

$$R^+ = CH_3^+$$

$$RM_{n+1}X = CH_3\!\!\left(\!CH_2\overset{CH_3}{\underset{CH_3}{\overset{|}{\underset{|}{C}}}}\!\!\right)_{\!\!n}\!\!CH_2\overset{CH_3}{\underset{CH_3}{\overset{|}{\underset{|}{C}}}}\!\!-Cl$$

$$CX^- = AlCl_4^-$$

The rate of polymerization by ion pairs may be written

$$-\frac{d[M]}{dt} = r_p = k_p[M] \sum_{n=0}^{\infty} [RM_n\!-\!M^+CX^-] \tag{42}$$

The rate of termination in a catalyzed ionic polymerization will be first-order with respect to the growing chain concentration. This is understandable on the basis of electrostatic considerations. In the media usually employed, the counterion must remain so close to the positive chain center that the ion pair (or "free ions") behaves in a kinetic sense as a single entity. Thus,

$$r_t = k_t \sum_n [M_n\!-\!M^+CX^-] \tag{43}$$

If we assume a steady state for the concentration of growing chains (the validity of which is not nearly as certain as for free-radical polymerizations), then $r_t = r_i$ and we have, from (43),

$$\sum_{n=0}^{\infty} [RM_n\!-\!M^+CX^-] = \frac{r_i}{k_t} \tag{44}$$

Therefore, the rate of polymerization by ion pairs is given by

$$r_p = \left(\frac{k_p}{k_t}\right)[M]r_i \tag{45}$$

with an analogous expression holding for the rate of polymerization by free ions.

It is the complexity and variety of the kinetic expressions for r_i, the rate of initiation, that defeat attempts to write general rate laws for cationic polymerizations. The rate of initiation at a fixed temperature and in a given solvent will be determined by the equilibria or steady-state existing among ion pairs, free ions, catalyst, cocatalyst, and monomer. Thus, it is possible to write

$$r_i = f([C], [RX], [M]) \tag{46}$$

but it must be recognized that the function f can take on such a variety of forms that it can be determined with confidence only by actual experiment for any given system.

Degree of Polymerization

In the discussion of free-radical kinetics in Chapter 12, the average degree of polymerization was given simply by the ratio of the rate of depletion of the monomer to the rate of formation of the polymer. The same relationship holds in cationic polymerization. Thus, because polymer is formed in both the termination and the transfer reactions, namely (40) and (41), we have

$$\overline{DP} = \frac{r_p}{r_t + r_{Tr}} \tag{47}$$

In terms of the cationic mechanism shown in (36) to (41), in which transfer occurs only to monomer, equation (47) becomes

$$\overline{DP} = \frac{k_p[M]}{k_t + k_{Tr}[M]} \tag{48}$$

or, in the usual inverted form,

$$\frac{1}{\overline{DP}} = \frac{k_{Tr}}{k_p} + \frac{k_t}{k_p} \frac{1}{[M]} \tag{49}$$

Additional substances are often present to which transfer can occur. These materials are sometimes inadvertently introduced as impurities. In such a case (49) must be modified to take into account the additional opportunities for chain transfer as shown in (50).

$$\frac{1}{\overline{DP}} = \frac{k_{Tr, M}}{k_p} + \frac{k_t}{k_p} \frac{1}{[M]} + \frac{\sum_x k_{Tr, X}}{k_p} \frac{[X]}{[M]} \tag{50}$$

In cationic polymerization, the formation of the final polymeric product takes place mainly by transfer rather than by termination. In free-radical polymerization, the opposite is usually the case. Therefore, if cationic systems could be rigorously freed from all impurities (i.e., [X] = 0), equation (50) predicts that when $k_t \ll k_{Tr}[M]$, $\overline{DP}$ should be independent of [M] and should be determined only by the value $k_p/k_{Tr, M}$. In practice, this is rarely true because impurities need to be present only at trace levels to decrease $\overline{DP}$. For example, in the polymerization of isobutylene, trace amounts of water may prevent $\overline{DP}$ from attaining the level $k_p/k_{Tr, M}$ because the reaction shown in (51) is strongly preferred.

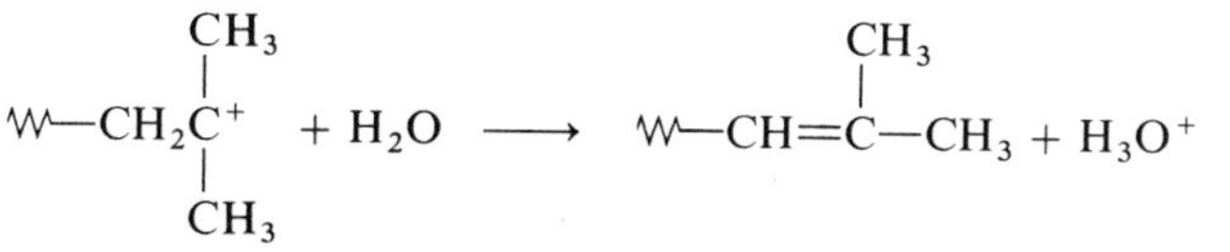

$$\underset{\substack{| \\ CH_3}}{\overset{\substack{CH_3 \\ |}}{\text{W—CH}_2\text{C}^+}} + H_2O \longrightarrow \underset{}{\overset{\substack{CH_3 \\ |}}{\text{W—CH=C—CH}_3}} + H_3O^+ \tag{51}$$

Figure 13.2 illustrates an example of this effect, and of the applicability of equation (50), by showing the effect of small amounts of water on the stannic chloride-catalyzed polymerization of isobutylene in ethyl chloride solution at −78°C. From the slope of this plot it is found that $k_{Tr, H_2O}/k_p = 27.7 \times 10^{-4}$.

The linear plot of $(DP)^{-1}$ in Figure 13.2 is in accord with equation (50), and such linear plots are obtained in many cationic systems. However, some systems

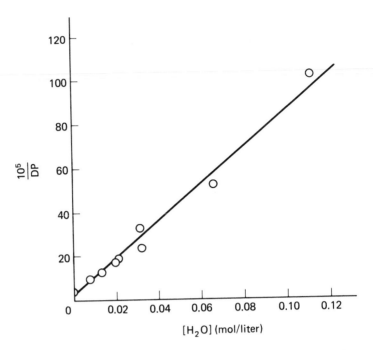

Fig. 13.2   Chain transfer to H_2O in the $SnCl_4$-catalyzed polymerization of iso-C_4H_8 in C_2H_5Cl solution at $-78°C$. $[C_4H_8] = 3.22\ M$; $[SnCl_4] = 0.185\ M$. [From R. G. W. Norrish and K. E. Russell, *Trans. Faraday Soc.*, **48**, 91 (1952).]

show a complex dependency of $\overline{DP}$ on the concentrations of monomer, catalyst, and added substances. These completely disobey equation (50). Thus, again no useful generalization is possible.

Effect of Temperature

The rate of the initiation of a cationic polymerization via the formation of ion pairs and free ions is only very slightly dependent on the temperature. This means that the effect of temperature on the rate of polymerization will be determined by the temperature dependence of k_p/k_t or k'_p/k'_t, where the primes denote rate constants for free ions and the unprimed units signify rate constants for ion pairs. If the rate constants are written in the Arrhenius formulation, then when one form of propagation dominates (i.e., free ions or ion pairs),

$$r_p \propto \frac{A_p}{A_t}\ e^{(E_t - E_p)/RT} \tag{52}$$

In cationic polymerization, the activation energy for termination, (41), is usually greater than that of propagation (39). As a result, the rate of cationic polymerization will generally *increase* as the temperature is *lowered*. This is in direct contrast to free-radical polymerization. The effect is due primarily to the much lower activation energies for cationic initiation and propagation than is found for free-radical initiation and propagation.

The effect of temperature on the degree of polymerization (and therefore on the molecular weight of the polymer) in impurity-free systems can be seen from a consideration of (48). If termination is more important than chain transfer to monomer, then $k_t \gg k_{Tr}[M]$. The degree of polymerization will then show the same dependence on temperature as the rate of polymerization, namely that given by (52). Because E_t is usually greater than E_p, as already mentioned, $\overline{DP}$ will increase as the temperature is lowered.

On the other hand, if chain transfer is more important than chain termination, $k_{Tr, M}[M] \gg k_t$ and (48) reduces to

$$\overline{DP} = \frac{k_p}{k_{Tr, M}} = \frac{A_p}{A_{Tr, M}} e^{(E_{Tr} - E_p)/RT} \tag{53}$$

Usually, the activation energy for chain transfer to the monomer is greater than that for propagation, so $\overline{DP}$ will again increase as the temperature is lowered.

Rate Constants for Propagation

In catalyzed cationic polymerizations, the observed rate constants are composite quantities that contain contributions from free ions and from ion pairs. The relative importance of the two types of contribution depends on the experimental conditions. It has been found possible to obtain rate constants for cationic propagation only for systems in which it is thought that the propagation is solely by free ions. Such systems are (1) polymerizations initiated by ionizing radiation, where no counterions exist, and (2) polymerizations initiated by stable carbonium ion salts, such as $(C_6H_5)_3C^+SbCl_6^-$ and $C_7H_7^+SbCl_6^-$, in which dissociation to the free ions is instantaneous and may be complete. Typical rate constants for free ions, k_p', determined from such systems are shown in Table 13.3. It will be noted that these rate constants are very large. In fact, even at the low temperatures used, all the rate

TABLE 13.3 Rate Constants for Propagation by Free Cations

MONOMER	SOLVENT	TEMPERATURE (°C)	INITIATOR	k_p' (liters/mol-s)
Styrene	None	15	Radiation	3.5×10^6
α-Methylstyrene	None	0	Radiation	4×10^6
Isobutyl vinyl ether	None	30	Radiation	3×10^5
Isobutyl vinyl ether	CH_2Cl_2	0	$C_7H_7^+SbCl_6^-$	5×10^3
t-Butyl vinyl ether	CH_2Cl_2	0	$C_7H_7^+SbCl_6^-$	3.5×10^3
Methyl vinyl ether	CH_2Cl_2	0	$C_7H_7^+SbCl_6^-$	1.4×10^2

Source: A. Ledwith and D. C. Sherrington, *Reactivity, Mechanism and Structure in Polymer Chemistry*. A. D. Jenkins and A. Ledwith, eds. (New York: © John Wiley and Sons, Ltd., 1974), p. 278.

constants are comparable to or greater than the rate constants for free-radical polymerizations at 60°C.

While it has been stated that the rate constants in Table 13.3 refer to free-ion propagation, it should be noted that much larger values are obtained by radiation initiation than by initiation using stable carbonium salts. This suggests that ion-pair propagation may be playing a significant role in the initiation of those cationic polymerizations by the salt $C_7H_7^+SbCl_6^-$ that are shown in Table 13.3.

Cationic Copolymerizations

Cationic copolymerization can be treated in an identical manner to anionic copolymerization. The mechanistic scheme for propagation is obtained simply by replacing the negative signs in (32) to (35) by positive signs. The limitations discussed previously for reactivity ratios in anionic copolymerization apply also to cationic reactivity ratios. However, in cationic systems, the living-polymer technique of measuring k_{11}, k_{12}, k_{21}, and k_{22} independently has not yet been exploited. Thus, in cationic systems even less is known, in an absolute sense, than in anionic systems.

Some typical reactivity ratios for the cationic copolymerization of styrene are shown in Table 13.4.

TABLE 13.4 Reactivity Ratios in the Cationic Copolymerization of Styrene (M_1)

M_2	INITIATOR	SOLVENT	TEMPERATURE (°C)	r_1	r_2
α-Methylstyrene	$BF_3O(C_2H_5)_2$	SO_2	−40	<0.1	>20
	$BF_3O(C_2H_5)_2$	CH_2Cl_2	−20	0.2–0.5	12 ± 2
	$TiCl_4$—CCl_3COOH	CH_2Cl_2	−78	0.24 ± 0.05	1.12 ± 0.09
Isobutylene	$TiCl_4$	Toluene	−78	1.20 ± 0.10	1.78 ± 0.10
	$TiCl_4$	n-Hexane	−20	1.20 ± 0.11	0.54 ± 0.24
	$SnCl_4$	SO_2	−78	1.1	3.1
Isoprene*	$SnCl_4$	C_2H_5Cl	−30 to 0	0.8	0.1

* T. E. Lipatova, A. R. Gantmakher, and S. S. Medvedev, *Dokl. Akad. Nauk S.S.S.R.*, **100**, 925 (1955).

Source: A. Tsukamoto and O. Vogl, *Progr. Polymer Sci.* (A. D. Jenkins, ed.), **3**, 199 (1971).

STUDY QUESTIONS

1. Styrene is added to a solution of sodium naphthalenide in tetrahydrofuran so that the initial concentrations of styrene and sodium napthalenide in the reaction mixture are 0.2 M and 1×10^{-3} M, respectively. After 5 s of reaction at 25°C, the styrene concentration is determined to be 1.73×10^{-3} M. Calculate:
   (a) The rate constant for propagation of the polymerization.
   (b) The initial rate of polymerization.
   (c) The rate of polymerization after 10 s.

(d) The number-average molecular weight of the polymer after 10 s.

(e) The width at half-height of the maximum in the mole fraction distribution of the molecular weight at the completion of reaction.

2. After completion of the reaction in Problem 1, p-methylstyrene is added to the mixture so that the initial concentration is 0.15 M. Using data from this chapter, calculate:

(a) The initial rate of polymerization.

(b) The rate of polymerization after 10 s.

(c) The average degree of polymerization of the copolymer after 100 s.

3. A copolymerization of p-methoxystyrene (1 M) with styrene (0.5 M) in tetrahydrofuran solution initiated by C_4H_9Li is carried out at 25°C. Calculate:

(a) The composition of the copolymer first formed.

(b) The mean sequence lengths of p-methoxystyrene and styrene in the copolymer.

(c) The probability of forming a styrene sequence that has a length of 5 units.

4. For the copolymerization of Problem 3, calculate and plot the monomer concentrations versus time, making use of the following assumptions: (1) the concentrations of the growing anionic centers derived from the two monomers does not change; and (2) the relative probabilities of addition of $C_4H_9^- Li^+$ to styrene and p-methoxystyrene are equal.

5. 1-Vinylnaphthalene is polymerized anionically at 25°C in a tetrahydrofuran solution containing initially 5×10^{-3} M C_4H_9Li and 0.75 M 1-vinylnaphthalene. Calculate:

(a) The average degree of polymerization.

(b) The number-fraction and weight-fraction distributions of the degree of polymerization.

6. In studies of the low-temperature polymerization of isobutylene using $TiCl_4$ as catalyst and H_2O as cocatalyst [R. H. Biddulph, P. H. Plesch, and P. P. Rutherford, J. Chem. Soc., 275 (1965)], the following results have been obtained at $-35°C$ for the effect of monomer concentration on the average degree of polymerization:

$[C_4H_8]$ (mol/liter)	0.667	0.333	0.278	0.145	0.059
$\overline{DP}$	6940	4130	2860	2350	1030

From these data, evaluate the rate constant ratios k_{Tr}/k_p and k_t/k_p.

7. In similar studies over a range of temperatures, Biddulph, Plesch, and Rutherford (see Problem 6) found the following values for the intercepts of plots of $(\overline{DP})^{-1}$ versus $[C_4H_8]^{-1}$:

t (°C)	$+18$	-14	-35	-48
$10^3/\overline{DP}$	4.37	0.50	0.098	0.027

(a) Evaluate from these data the difference in activation energy between chain propagation and chain transfer to monomer.

(b) Evaluate the ratio of the preexponential factor for transfer to that for propagation.

(c) Assuming chain transfer to be much more important in producing polymer than termination, calculate and plot a curve showing the dependence of $\overline{DP}$ on temperature.

SUGGESTIONS FOR FURTHER READING

Anionic polymerization

ALLEN, P. E. M., AND C. R. PATRICK, "The Kinetics of Addition Polymerization," *Kinetics and Mechanisms of Polymerization Reactions*, Chap. 7. New York: Wiley, 1974.

BYWATER, S., "Anionic Polymerization," *Progress in Polymer Science* (A. D. Jenkins, ed.), Vol. 4, Chap. 2. New York: Pergamon Press, 1975.

CUBBON, R. C. P., AND D. MARGERISON, "The Kinetics of Polymerization of Vinyl Monomers by Lithium Alkyls," *Progress in Reaction Kinetics*, Vol. 3 (G. Porter, ed.), Chap. 9. New York: Pergamon Press, 1965.

MORTON, M., "The Mechanism of Stereospecific Polymerization of Propylene," *Vinyl Polymerization*, Vol. 1, Part 2 (G. E. Ham, ed.), Chap. 5. New York: Dekker, 1969.

MULVANEY, J. E., C. G. OVERBERGER, AND A. M. SCHILLER, "Anionic Polymerization," *Fortschr. Hochpolym.-Forsch.*, **3**, 106 (1961).

PARRY, A., "Anionic Polymerization," *Reactivity, Mechanism and Structure in Polymer Chemistry* (A. D. Jenkins and A. Ledwith, eds.), Chap. 11. New York: Wiley, 1974.

SZWARC, M., AND J. SMID, "The Kinetics of Propagation of Anionic Polymerization and Copolymerization," *Progress in Reaction Kinetics*, Vol. 2 (G. Porter, ed.), Chap. 5. New York: Macmillan, 1964.

Cationic polymerization

ALLEN, P. E. M., AND C. R. PATRICK, "The Kinetics of Addition Polymerization," *Kinetics and Mechanisms of Polymerization Reactions*, Chap. 7. New York: Wiley, 1974.

ALLEN, P. E. M., AND P. H. PLESCH, "A Comparison of the Radical, Cationic and Anionic Mechanisms of Addition Polymerization," *The Chemistry of Cationic Polymerization* (P. H. Plesch, ed.), Chap. 3. New York: Macmillan, 1963.

LEDWITH, A., AND D. C. SHERRINGTON, "Reactivity and Mechanism in Cationic Polymerization," *Reactivity, Mechanism and Structure in Polymer Chemistry* (A. D. Jenkins and A. Ledwith, eds.), Chap. 9. New York: Wiley, 1974.

PEPPER, D. C., "Ionic Polymerisation," *Quart. Rev.*, **8**, 88 (1954).

PLESCH, P. H., "The Propagation Rate-Constants in Cationic Polymerisations," *Adv. Polymer Sci.*, **8**, 137 (1971).

ZLAMAL, Z., "Mechanisms of Cationic Polymerization," *Vinyl Polymerization*, Vol. 1, Part 2 (G. E. Ham, ed.), Chap. 6. New York: Dekker, 1969.

PHYSICAL CHARACTERIZATION OF POLYMERS

14

Determination
of Absolute Molecular Weights

A knowledge of the molecular weight of a polymer is vital for even a preliminary understanding of the relationship between structure and properties. Moreover, as discussed in the preceding chapters, the molecular weight *distribution* provides valuable clues to the polymerization reaction mechanism. Two fundamentally different approaches are used for the measurement of polymer molecular weights—absolute and secondary methods. Absolute methods give values that provide a *direct* estimate of the molecular weight. Secondary methods yield *comparisons* between the molecular weights of different polymers, and must be *calibrated* by reference to a system that has been studied by one of the absolute approaches. Absolute methods are considered in this chapter. Secondary methods are discussed in Chapter 15.

The measurement of absolute molecular weights of polymers is not an easy task. The experimenter is faced with the problem of studying materials which are nonvolatile, of very high molecular weight, and sometimes poorly soluble in organic media. Moreover, the sample is not homogeneous in molecular weight, but contains molecules whose molecular weights may span a broad range. The physical methods commonly used to determine the average molecular weights of high-polymer samples on an absolute scale require that the polymer sample should first be dissolved in a solvent. It is appropriate, therefore, to consider briefly some characteristics of polymer solubility.

SOLUBILITY OF HIGH POLYMERS

The dissolution of a polymer sample in a solvent takes place in two distinct stages. In the first stage the polymer sample "takes up" or imbibes solvent and expands to a swollen gel. This first stage is exhibited by all amorphous linear, branched, or lightly crosslinked polymer samples, regardless of whether or not they will ultimately form a true solution. The second stage of dissolution consists of a breakdown of the swollen gel to give an actual solution of polymer molecules in the given solvent. This second stage will not be shown by polymers that are crosslinked and may not be shown by polymers that contain microcrystalline domains. Crosslinked polymers do not dissolve in any solvent (unless the crosslinks are broken). Hence, crosslinked network polymers will be excluded from our consideration of molecular-weight determination.

A liquid will be a "good" solvent for a polymer if the molecules of the liquid chemically and physically resemble the structural units of the polymer. If this situation exists, the adhesive forces between the solvent and the polymer are similar to the cohesive forces that exist between solvent molecules or between polymer molecules. An exchange of a solvent molecule by a polymer structural unit can then occur with little or no change in the interaction forces that exist between solvent or polymer molecules. In a thermodynamic sense, we can say that the dissolution of a polymer in a "good" solvent occurs with a negligibly small heat of mixing. This is a more elegant way in which to state the well-known chemist's rule that "like dissolves like." For example, this rule suggests that cumene and ethylbenzene (**1** and **2**) should be good solvents for polystyrene (**3**) because of the similarity between the solvent molecules and the structural units of the polymer.

$$CH_3CHCH_3 \qquad\qquad CH_2CH_3$$

and

1   **2**

$$\text{W}—CH_2CHCH_2CHCH_2CH—\text{W}$$

3

The criterion of "like dissolves like" can be interpreted more quantitatively. When a substance vaporizes, energy is needed to overcome the cohesive forces between molecules. The magnitude of these cohesive forces may be described simply by the amount of energy needed to vaporize a certain volume of the substance. Such a quantity is called the *cohesive energy density*[1] and is defined by (1),

$$\delta^2 = \frac{E_0}{V_0} \tag{1}$$

---

[1] J. H. Hildebrand and R. L. Scott, *Solubility of Non-electrolytes*, 3rd ed. (New York: Reinhold, 1950), p. 124.

in which E_0 is the latent energy of vaporization for a volume V_0 of the substance in question. If the adhesive forces between the solvent and the polymer are to be similar to the solvent–solvent and polymer–polymer cohesive forces, then it is obvious that the cohesive energy densities of the solvent and the polymer should be nearly the same. In general, for endothermic dissolution (the usual case with polymers), the enthalpy change per unit volume is given by (2), in which ϕ_1 is

$$\Delta H_{\text{dissolution}} = \phi_1 \phi_2 (\delta_1 - \delta_2)^2 \qquad (2)$$

the volume fraction of solvent and ϕ_2 is the volume fraction of polymer in the solution. A good solvent for a given polymer is one in which ΔH approaches as closely as possible to zero. In other words, δ_1^2 and δ_2^2—*the cohesive energy densities*—should be nearly equal.

Values for the cohesive energy densities of liquids that could be used as solvents are readily available from density data and from heats of vaporization. However, the heat of vaporization cannot be obtained for polymers because all high polymers decompose before they vaporize. Hence, the δ^2 for polymers cannot be obtained directly. The method most often used to determine the cohesive energy densities of polymers is based on the idea that the degree of solubility of a polymer (or degree of swelling of a crosslinked polymer) is at a maximum when the cohesive

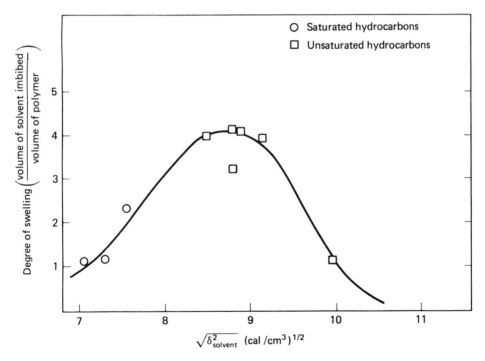

Fig. 14.1   Degree of swelling of natural rubber as a function of the cohesive energy density of the solvent. [Data from R. L. Scott and M. Magat, *J. Polymer Sci.*, **4**, 555 (1949).]

TABLE 14.1 Cohesive Energy Densities of Solvents and Polymers

POLYMER	δ^2 (cal/cm^3)	SOLVENT	δ^2 (cal/cm^3)
Polyethylene	62.4	n-C_6H_{14}	52.4
Polystyrene	74.0	CCl_4	73.6
Poly(methyl methacrylate)	82.8	C_6H_6	83.7
Poly(vinyl chloride)	90.3	$CHCl_3$	85.4
Nylon 66	185	$(CH_3)_2C{=}O$	94.3
Polyacrylonitrile	237	CH_3OH	210

energy densities of the solvent and the polymer are equal. The degree of solubility (or degree of swelling) of a given polymer is measured in solvents with varying δ^2 values. A plot is then made of the solubility of the polymer as a function of δ^2. A maximum in this plot should be observed, and the value of δ^2 at this maximum is taken to be the cohesive energy density of the polymer. An actual experimental curve of this type is shown in Figure 14.1 and typical values for some solvents and polymers are shown in Table 14.1.

The data in Table 14.1 indicate that benzene and carbon tetrachloride should be good solvents for polystyrene and poly(methyl methacrylate), but poor solvents for nylon 66 and polyacrylonitrile. Methanol should be a good solvent for polyacrylonitrile.

COLLIGATIVE PROPERTIES AND MOLECULAR WEIGHTS

In principle, any of the colligative properties of solutions such as freezing-point depression, osmotic pressure, and so on, may be used to determine the average molecular weight of a dissolved polymer sample. A colligative property of a solution is any property that depends on the lowering of the chemical potential of a solvent by the introduction of a solute.

According to classical thermodynamics, equation (3) gives the chemical potential of a liquid solvent in a solution if the vapor of the solvent behaves as an ideal gas.

$$\mu_s = \mu_s^\circ + RT \ln \frac{P_s}{P_s^\circ} \tag{3}$$

In this equation μ_s° is the chemical potential of the pure solvent at temperature T, P_s° is the vapor pressure of the pure solvent at temperature T, and P_s is the pressure of the solvent vapor above the solution at temperature T.

According to Raoult's law, which is valid for all solutions provided they are sufficiently dilute, equation (4) holds:

$$\frac{P_s}{P_s^\circ} = X_s \tag{4}$$

Here X_s is the mole fraction of solvent in the solution. In view of equations (3) and (4), we may write the expression shown in (5) for the chemical potential of the solvent in solution

$$\mu_s = \mu_s^\circ + RT \ln X_s \tag{5}$$

For a solution, X_s must be less than unity and, according to equation (5), μ_s must therefore be less than μ_s°. Hence, the presence of the solute lowers the chemical potential of the solvent. The observable quantities that depend on this effect are called *colligative properties*. These are:

1. Vapor-pressure lowering (see equation 4).
2. Boiling-point elevation.
3. Freezing-point depression.
4. Osmotic pressure.

All four of these properties can be transposed into the form

$$\Delta y_i = K_i \frac{c}{\overline{M}_n} \tag{6}$$

in which c is the concentration of the solute (in weight per unit volume); $\overline{M}_n$ is the number average molecular weight of solute;[1] K_i is a constant which depends on the solvent and on the effect being investigated; and Δy_i is the experimental factor actually being observed [e.g., the depression of the freezing point $(T_0 - T)$]. From equation (6) it is clear that Δy_i becomes smaller as $\overline{M}$, the average molecular weight, becomes larger. This leads to serious inadequacies in the use of the first three colligative properties mentioned above for the determination of the molecular weights of high polymers.

For example, consider the freezing-point depression of benzene brought about by addition of a high polymer. For a concentration of 10^{-3} g/cm^3 (a concentration at which the dilute solution laws begin to be obeyed approximately) a freezing-point depression of only about 0.0002°C would be produced if the average molecular weight of the polymer was 25,000. This temperature change is below the limits of accurate measurement. The boiling-point elevation or vapor-pressure lowering by the same polymer would be even more difficult to measure.

OSMOTIC-PRESSURE MEASUREMENT OF ABSOLUTE MOLECULAR WEIGHTS

Theory

Osmotic pressure is the only one of the four colligative properties that provides a practical method for the measurement of the average molecular weights of polymers. In fact, osmotic pressure measurements can be used to determine molecular weights in the range 3×10^4 to 1×10^6.

---

[1] The number-average molecular weight is the total mass of the solute divided by the number of solute particles present.

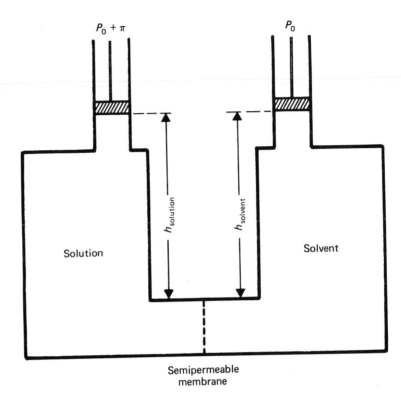

Fig. 14.2 Schematic diagram of an osmometer.

The principle of osmotic-pressure measurements may be illustrated[1] by consideration of the schematic apparatus shown in Figure 14.2. The dashed line represents a semipermeable membrane through which the solvent, but not the polymer, may pass. If both sides of the apparatus contained pure solvent, the liquid levels at equilibrium would be at the same height, and the external pressure on both pistons would be P_0. In terms of the chemical potential of the solvent, expression (5) for this trivial case would simplify to

$$\mu_s(\text{left}) = \mu_s(\text{right}) = \mu_s^\circ \tag{7}$$

If the polymer solute is now added to the left-hand side of the apparatus shown in Figure 14.2, it must remain on the left-hand side, since it cannot pass through the membrane. A system such as this, which contains a solution on the left side and pure solvent on the right side, can be described in terms of the chemical potential of the solvent, which, from equation (5), may be stated $\mu_s(\text{left}) < \mu_s(\text{right})$. If the external pressure on the solution side is maintained equal to that on the

[1] I. M. Klotz, *Chemical Thermodynamics* (Englewood Cliffs, N.J.: Prentice-Hall, 1950), pp. 261–263.

Physical Characterization of Polymers / Part III

solvent side, namely P_0, the level of the liquid on the left-hand side will rise. This rise is due to the flow of solvent through the membrane into the solution, a region in which its chemical potential is lower. This flow of solvent could be prevented (in order to keep the heights of the two liquid levels fixed) by increasing the external pressure on the solution. The amount of this increase in external pressure is the *osmotic pressure*, π, of the solution.

The driving force for the flow of solvent through the membrane into the solution is the reduction of the chemical potential of a solvent that occurs when a polymer dissolves in it (equation 5). On the other hand, it is possible to prevent solvent flow and, therefore, to restore μ_s on the solution side to μ_s° by increasing the external pressure on the solution. Obviously, then, the chemical potential of the solvent is a function of both the concentration and the external pressure (at constant temperature). Thus, the general expression shown in (8) can be employed,

$$\mu_s = f(P, X_p) \tag{8}$$

where X_p is the mole fraction of polymer in the solution. If equation (8) is differentiated (assuming constant temperature), expression (9) is obtained:

$$d\mu_s = \left(\frac{\partial \mu_s}{\partial P}\right)_{T, X_p} dP + \left(\frac{\partial \mu_s}{\partial X_p}\right)_{T, P} dX_p \tag{9}$$

By increasing the pressure on the solution to prevent flow of solvent, the value of μ_s in solution is maintained constant at μ_s°. Hence, under this condition $d\mu_s = 0$ and equation (10) is obtained:

$$\left(\frac{\partial \mu_s}{\partial P}\right)_{T, X_p} dP = -\left(\frac{\partial \mu_s}{\partial X_p}\right)_{T, P} dX_p \tag{10}$$

The definition of the chemical potential[1] of the solvent is given by

$$\mu_s = \left(\frac{\partial G}{\partial n_s}\right)_{T, P, n_p} \tag{11}$$

where G is the Gibbs free energy, n_s is the number of moles of solvent, and n_p is the number of moles of the polymer. From elementary thermodynamics the derivative of the Gibbs free energy with respect to pressure at constant temperature and composition is given by the volume of the solution or by

$$\left(\frac{\partial G}{\partial P}\right)_{T, X_p} = V \tag{12}$$

[1] W. J. Moore, *Physical Chemistry*, 4th ed. (Englewood Cliffs, N.J.: Prentice-Hall, 1972), p. 205.

By differentiation of (11) with respect to P at constant composition, and differentiation of (12) with respect to n_s, all other variables being kept constant, expressions (13) and (14), respectively, are obtained. In (14), $\overline{V}_s$ is the *partial molar volume* of the

$$\left(\frac{\partial \mu_s}{\partial P}\right)_{T, X_p} = \frac{\partial^2 G}{\partial P\, \partial n_s} \tag{13}$$

$$\frac{\partial^2 G}{\partial n_s\, \partial P} = \left(\frac{\partial V}{\partial n_s}\right)_{T, P, n_p} = \overline{V}_s \tag{14}$$

solvent. Since the order of differentiation is of no consequence, (13) and (14) may be equated so that (15) results:

$$\left(\frac{\partial \mu_s}{\partial P}\right)_{T, X_p} = \overline{V}_s \tag{15}$$

Differentiation of (5) with respect to X_p at constant T and P leads directly to

$$\left(\frac{\partial \mu_s}{\partial X_p}\right)_{T, P} = \left(\frac{dX_s}{dX_p}\right)\left(\frac{\partial \mu_s}{\partial X_s}\right)_{T, P} = -\frac{RT}{1 - X_p} \tag{16}$$

Substitution of (15) and (16) into (10) then yields

$$\int_{P_0}^{P_0 + \pi} \overline{V}_s\, dP = RT \int_0^{X_p} \frac{dX_p}{1 - X_p} \tag{17}$$

The partial molar volume of solvent may be assumed to be independent of the external pressure over the pressure range P_0 to $P_0 + \pi$, and integration of (17) then yields

$$\pi = -\frac{RT}{\overline{V}_s} \ln (1 - X_p) \tag{18}$$

for the osmotic pressure.

In the very dilute solutions to which (4) is applicable, X_p is so small that the following approximations are generally assumed to be valid:

$$\ln (1 - X_p) \approx -X_p \approx -\frac{n_p}{n_s}$$

$$n_s \overline{V}_s = V_s \approx V_{\text{solution}}$$

Use of these approximations with equation (18) then leads to the well-known approximate expression of Van't Hoff:

$$\frac{\pi}{c} \approx \frac{RT}{\overline{M}_n} \tag{19}$$

where c is the concentration in weight of polymer per unit volume of solution, and π and R are in the corresponding units. A measurement of the colligative properties effectively is a count of the *number of particles* in solution. Hence, the average molecular weight in (19) is *the number-average molecular weight*.

As mentioned, expression (19) applies only to ideal solutions for which Raoult's law is obeyed. Generally, ideality can only be approached in the limit of infinite dilution, and this is particularly true for solutions of high polymers. Therefore, the correct expression relating osmotic pressure to the molecular weight must be written

$$\lim_{c \to 0} \left(\frac{\pi}{c} \right) = \frac{RT}{\overline{M}_n} \tag{20}$$

The form of (20) indicates that, in order to determine molecular weights by the osmotic pressure method, it is necessary to extrapolate a plot of π/c as a function of c to the value at $c = 0$.

The dependence of π/c on the concentration is often written in the form of a virial equation:

$$\left(\frac{\pi}{c} \right) = \frac{RT}{\overline{M}_n} (1 + \Gamma c + g\Gamma^2 c^2 + \cdots) \tag{21}$$

As will be discussed further in Chapter 16, the second virial coefficient Γ is given by the Flory–Huggins theory of polymer solutions as

$$\Gamma = \frac{\overline{M}_n \rho_s}{M_s \rho_p^2} \left(\frac{1}{2} - \chi_1 \right) \tag{22}$$

where ρ_s and ρ_p are the densities of the solvent and the polymer, respectively; M_s is the molecular weight of the solvent; and χ_1 is a polymer–solvent interaction constant. The value of the constant g in (21) is often assumed to be $\sim\frac{1}{4}$ in good solvents. With $g = \frac{1}{4}$, equation (21) reduces to (23). This latter form is sometimes convenient to use for the extrapolation of π/c data to zero concentration because, for good solvents, plots given by the expression (21) are not linear.

$$\left(\frac{\pi}{c} \right)^{1/2} = \left(\frac{RT}{\overline{M}_n} \right)^{1/2} (1 + \tfrac{1}{2}\Gamma c) \tag{23}$$

In poor solvents $g \approx 0$, and hence plots given by (21) are found to be linear over the usually measured ranges of concentration.

Practical Osmometry

A simple, much-used osmometer is shown in Figure 14.3.[1] In this apparatus two semipermeable membranes, M, are held against the finely ground glass walls of an open-ended cylindrical cell, A, by perforated metal plates, P. Two capillary tubes, B and C, are sealed onto the side of the cylindrical cell. C is the filler tube. This has an inside diameter of 2 mm. B is the measuring capillary tube, which has an inside diameter of 0.5 mm. The reference capillary, D, also has an inside diameter of 0.5 mm.

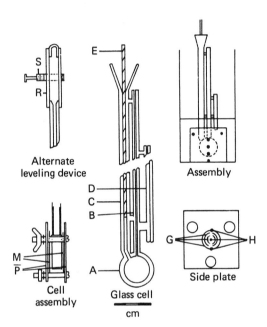

Alternate leveling device

Assembly

Cell assembly

Glass cell

cm

Side plate

Fig. 14.3 Simple osmometer. [Reproduced from B. H. Zimm and I. Myerson, *J. Am. Chem. Soc.*, **68**, 911 (1946); © The American Chemical Society.]

The cell is filled with the polymer solution through the tube C, and a snugly fitting metal rod, E, is then inserted into C to close the filling tube. The apparatus is then immersed in the solvent and the osmotic pressure of the solution is measured as the difference in the levels of the liquid in capillaries B and D. Generally, the difference in the levels of liquid in the two capillaries is measured by a cathetometer.

The difference in heights of the liquid levels, Δh, in an osmometer such as this is related to the osmotic pressure of the solution by the expression

$$\pi = \rho_{\text{solution}} g' \Delta h \tag{24}$$

in which ρ_{solution} is the density of the solution and g' is the acceleration due to gravity. If the weight is measured in grams, volume in cm^3, and height in centimeters, π

[1] More elaborate commercial instruments are available.

will be given in units of dyn/cm². It is customary to report osmotic pressures as π/g', which has the units grams of solution/cm². These units must be kept in mind clearly when converting extrapolated values of π/c to number-average molecular weights.

Some typical results of osmotic pressure measurements for a polymer solution are shown in Figures 14.4 and 14.5 for the specific case of polyisobutylene in chlorobenzene. The plot of π/c versus c in Figure 14.4 (see equation 21) shows some

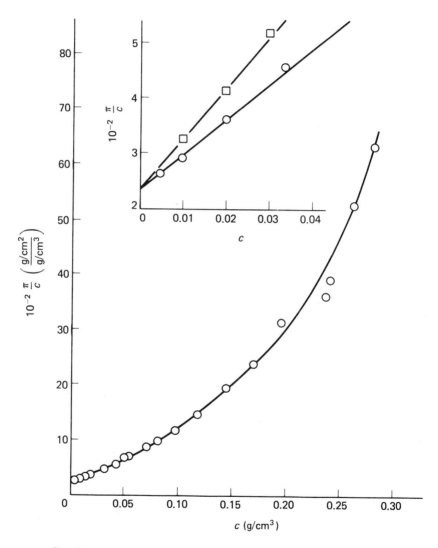

Fig. 14.4 Concentration dependence of the osmotic pressure of polyisobutylene–chlorobenzene solutions. [Reproduced from J. Leonard and H. Daoust, *J. Polymer Sci.*, **57**, 53 (1962); with permission of John Wiley & Sons, Inc., New York.] ○, 25°C; □, 40°C.

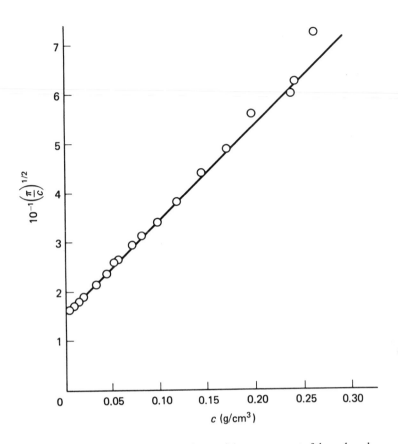

Fig. 14.5 Concentration dependence of the square root of the reduced
osmotic pressure of polyisobutylene–chlorobenzene solutions. [Re-
produced from J. Leonard and H. Daoust, *J. Polymer Sci.*, **57**, 53
(1962); with permission of John Wiley & Sons, Inc., New York.]

curvature. This indicates that chlorobenzene is a good solvent for polyisobutylene.
However, as shown by the inset section of Figure 14.4, an extrapolation of the data
for the lower concentrations to $c = 0$ can be made easily to obtain $\lim_{c \to 0} (\pi/c)$.
Figure 14.5 shows that a plot of $(\pi/c)^{1/2}$ versus c for the same data (according to
equation 23) yields a straight line over the entire range of concentrations measured.
This indicates that g in equation (21) is $\sim \frac{1}{4}$.

It is instructive to use the data obtained from Figures 14.4 (or 14.5) to illustrate
the calculation of a number-average molecular weight and the second virial coefficient. From Figure 14.4, or the inset to this figure, it is possible to obtain the
following result for the intercept at 25°C:

$$\lim_{c \to 0} \left(\frac{\pi}{c}\right) = 235 \text{ g/cm}^2\text{-concn.}$$

According to equation (20), the number-average molecular weight is

$$\overline{M}_n = \frac{RT}{\lim_{c \to 0} (\pi/c)} = \frac{(82.06 \text{ cm}^3\text{-atm/mol-deg})(298 \text{ deg})(1.013 \times 10^6 \text{ g/cm-s}^2\text{-atm})}{(235 \text{ g/cm}^2\text{-concn.})(980 \text{ cm/s}^2)}$$

or

$$\overline{M}_n = 1.07 \times 10^5 \text{ cm}^3\text{-concn./mol} = 1.07 \times 10^5 \text{ g/mol}$$

From the inset of Figure 14.4, the initial slope at 25°C is found to be 6350 g/cm²-concn.². From equation (21), the following result is then obtained for the second virial coefficient:

$$\Gamma = (6350 \text{ g/cm}^2\text{-concn.}^2)\left(\frac{\overline{M}_n}{RT}\right)$$

$$\times \frac{(6350 \text{ g/cm}^2\text{-concn.}^2)(1.075 \times 10^5 \text{ g/mole})(980 \text{ cm/s}^2)}{(1.013 \times 10^6 \text{ g/cm-s}^2\text{-atm})(82.06 \text{ cm}^3\text{-atm/mol-deg})(298 \text{ deg})}$$

or

$$\Gamma = 27.0 \text{ cm}^3/\text{g}$$

Accuracy of Molecular Weights Determined from Osmotic Pressure

Despite the excellent *precision* of the data shown in Figures 14.4 and 14.5, the absolute *accuracy* of the molecular weights of polymers obtained from osmotic pressure measurements is not nearly so good. This may be seen from inspection of Figure 14.6. This figure shows plots of π/c versus c obtained independently by eight laboratories in the United States, Canada, and Europe for samples of a single polystyrene fraction. Measurements were made in two solvents, toluene and methyl ethyl ketone (which are representative of good and poor solvents for polystyrene, respectively). Obviously, a great deal of scatter in the data exists. This can be attributed mainly to the different semipermeable membranes used by the various laboratories. The average molecular weights and second virial coefficients obtained from Figure 14.6 are:

(a) Toluene: $\overline{M}_n = 4.50 \pm 1.08 \times 10^5$
$\Gamma = 214 \pm 27 \text{ cm}^3/\text{g}$

(b) Methyl ethyl ketone: $\overline{M}_n = 4.60 \pm 0.82 \times 10^5 \text{ g/mol}$
$\Gamma = 60 \pm 32 \text{ cm}^3/\text{g}$

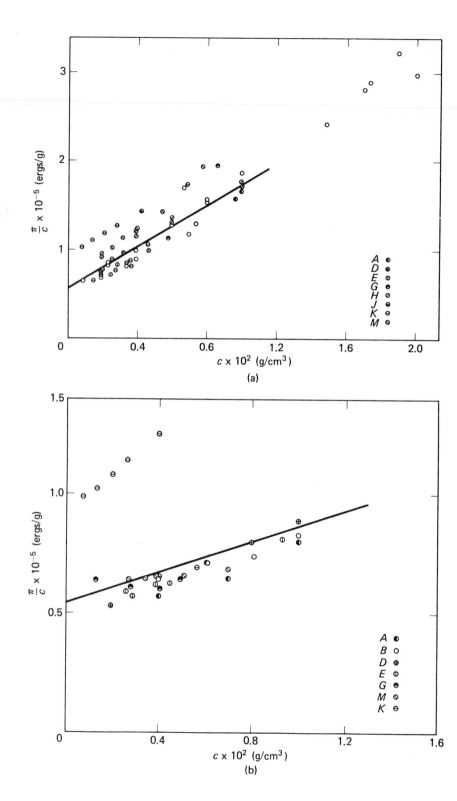

(a)

(b)

Fig. 14.6 (*Opposite*) Concentration dependence of the osmotic pressure of polystyrene solutions in two solvents at 25°C. (a) Toluene; (b) methyl ethyl ketone. [Reproduced from H. P. Frank and H. Mark, *J. Polymer Sci.*, **17**, 1 (1955); with permission of John Wiley & Sons, Inc., New York.] The letters in the figure represent different laboratories where the measurements were made.

It is somewhat reassuring that essentially the same molecular weight was obtained in the two solvents. However, the accuracy is clearly not as good as might be assumed from an inspection of the data obtained from a single osmometer and membrane (see Figure 14.4). It is also noteworthy (as illustrated by the foregoing results) that the more accurate extrapolation to $c = 0$ that is possible in a poor solvent (i.e., a shallower slope) leads to considerably less uncertainty in the average molecular weight. Of course, this greater accuracy is obtained at the expense of a much greater percentage error in the second virial coefficient. This is also demonstrated by the results given above.

Semipermeable Membranes

The semipermeable membrane is the most important experimental component in the determination of molecular weights by osmometry. In fact, it is generally accepted that variations in the degree of semipermeability of the various membranes employed is the major factor for the scatter of π/c data shown in Figure 14.6.

Pretreated gel cellophane is probably the most widely used organic membrane material. Generally, such membranes are treated first with aqueous sodium hydroxide, then washed with water, and immersed successively for an hour or two first in ethanol, next in a mixture of ethanol and the solvent to be used, and finally in pure solvent. Such membranes have been found to be satisfactory for the measurement of molecular weights above 10^4 g/mol.

Other membranes commonly used with organic solvents include rubber, poly(vinyl alcohol), polyurethane, and poly(trifluorochloroethylene). Nitrocellulose membranes are often used for aqueous solutions. They are of little use with organic solvents because they are soluble in or are swelled by most organic media.

LIGHT SCATTERING FOR MEASUREMENT OF ABSOLUTE MOLECULAR WEIGHT AND SIZE

Intensity of Scattered Light: Rayleigh Ratio

When a beam of monochromatic light traverses a system and no absorption of light occurs, the transmission is still not complete. A fraction of the incident light is scattered and a resultant attenuation of the intensity of the incident light occurs. The light is scattered in all directions relative to the incident beam and the major portion of the scattered light has the same wavelength as the incident beam.

Over 100 years ago Lord Rayleigh[1] considered the scattering of light in terms of the optical properties of individual molecules. He showed that, for a dilute gas, the intensity of scattered light is given by

$$\frac{i(\theta)r^2}{I_0} = R(\theta) = \frac{2\pi^2}{\lambda^4 N_0} \frac{(n-1)^2 M}{c} (1 + \cos^2\theta) \qquad (25)$$

In this expression I_0 is the intensity of light of wavelength λ incident on the system; $i(\theta)$ is the intensity of scattered light per unit volume of the system that is detected at angle θ to the incident beam direction and at distance r from the center of the system; M is the molecular weight of the gas; n is the refractive index; c is the density or concentration of the gas in mass per unit volume; and N_0 is Avogadro's number. The term $R(\theta)$ is called the *Rayleigh ratio*. Note that the units of Rayleigh ratio are length^{-1}. It therefore represents the fraction of light scattered at angle θ per unit path length through the system.

Serious difficulties are encountered in attempts to extend the Rayleigh treatment of light scattering in dilute gases to liquids. The main reason for these difficulties lies in the existence of strong intermolecular forces in liquids that are absent in dilute gases. A completely different approach to light scattering in liquids, worked out by Einstein, showed how these difficulties could be avoided.[2] He considered the scattering to arise from local fluctuations in the density due to the thermal motions of the molecules. The density fluctuations lead directly to local fluctuations in the refractive index and hence to scattering of the incident light. Einstein's expression for the Rayleigh ratio of a pure liquid is given by (26). In this equation p is the hydrostatic pressure on the liquid

$$R(\theta) = \frac{i(\theta)r^2}{I_0} = \frac{2\pi^2}{\lambda^4 N_0} \frac{RT}{\beta} \left(n\frac{dn}{dp}\right)^2 (1 + \cos^2\theta) \qquad (26)$$

and β is the compressibility, that is, $\beta = -(1/V)(\partial V/\partial p)_T$. The other terms are as described earlier.

High polymers do not usually exist as gases or pure liquids but they can often be dissolved in liquid solvents. However, the use of light scattering measurements to determine the molecular weights and sizes of high polymers was held up by the lack of a theoretical treatment of light scattering in liquid solutions. Such a treatment was made by Peter Debye in 1944.[3] Debye pictured the additional scattering of light by a solution (over and above that of the pure solvent) to result from local fluctuations in the concentration of the solute. In a treatment analogous to that of Einstein for the density fluctuations in pure liquids, Debye considered the local fluctuations in solute concentration, due to random thermal motion, to

[1] J. W. Strutt (Lord Rayleigh), *Phil. Mag.*, **41**, 107, 274, 447 (1871).
[2] A. Einstein, *Ann. Physik.*, **33**, 1275 (1910).
[3] P. Debye, *J. Appl. Phys.*, **15**, 338 (1944).

be opposed by the osmotic pressure of the solution. With this model he derived the expression shown in (27) for the Rayleigh ratio of the scattering due to solute.

$$R'(\theta) = \frac{i'(\theta)r^2}{I_0} = \frac{2\pi^2}{\lambda^4 N_0} n_0^2 (n - n_0)^2 \frac{RT}{c(\partial\pi/\partial c)_T} (1 + \cos^2 \theta) \qquad (27)$$

Here the prime denotes the excess scattering from the liquid due to the solute, n_0 is the refractive index of the solvent, c is the concentration of solute in mass per unit volume, π is the osmotic pressure of the solution, and the other terms are as described previously.

Turbidity

The decrease in the intensity of a beam of light because of scattering is used to define the *turbidity* of a solution. The decrease or attenuation depends on the length of the light path through the system and, by analogy to the Lambert law, it is possible to write

$$\frac{I}{I_0} = e^{-\tau l} \qquad (28)$$

where I_0 is the incident light intensity, I the transmitted light intensity, l the length of the light path in the solution, and τ the turbidity. We may write (28) in the form

$$e^{-\tau l} = \frac{I_0 - I_s}{I_0} = \frac{I_0 - I_s' l}{I_0} = 1 - \frac{I_s' l}{I_0} \qquad (29)$$

where I_s is the total intensity of light that is scattered by the solution and I_s' the total intensity scattered per unit path length. The fraction of light scattered is generally very small and it is a good approximation to express the exponential in (29) as

$$e^{-\tau l} = 1 - \tau l + \tfrac{1}{2}(\tau l)^2 - \tfrac{1}{6}(\tau l)^3 + \cdots \approx 1 - \tau l \qquad (30)$$

A combination of (29) and (30), with neglect of higher powers of τl, leads to

$$\tau = \frac{I_s'}{I_0} \qquad (31)$$

as the relationship between the total intensity of scattered light per unit path and the turbidity.

The total light intensity scattered per unit path length through the solution is the intensity scattered through all angles of polar coordinates, or

$$I_s' = \int_0^\pi \int_0^{2\pi} r^2 i'(\theta) \sin\theta \, d\theta \, d\phi \qquad (32)$$

Since $r^2 i'(\theta) = I_0 R'(\theta)$ (cf. 27), (31) becomes

$$\frac{I'_s}{I_0} = \tau = \int_0^\pi \int_0^{2\pi} R'(\theta) \sin \theta \, d\theta \, d\phi \tag{33}$$

or

$$\tau = \frac{2\pi^2}{\lambda^4 N_0} n_0^2 (n - n_0)^2 \frac{RT}{c(\partial\pi/\partial c)_T} \int_0^\pi (1 + \cos^2 \theta) \sin \theta \, d\theta \int_0^{2\pi} d\phi \tag{34}$$

The value of the product of the definite integrals is $16\pi/3$, so that we obtain the relationship between the turbidity and the total scattered light intensity as

$$\tau = \frac{I'_s}{I_0} = \left(\frac{32\pi^3}{3\lambda^4 N_0}\right) \frac{n_0^2 (n - n_0)^2 RT}{c(\partial\pi/\partial c)_T} \tag{35}$$

Note that the relationship between the turbidity and the Rayleigh ratio is

$$\tau = \frac{16\pi}{3} R_{(90°)} = \frac{16\pi i(90°) r^2}{3 I_0} \tag{36}$$

or, more generally,

$$\tau = \left(\frac{16\pi}{3}\right) \frac{R(\theta)}{1 + \cos^2 \theta} = \left(\frac{16\pi}{3}\right) \frac{i(\theta) r^2}{I_0 (1 + \cos^2 \theta)} \tag{37}$$

In (36) and (37) the prime on $i(\theta)$ (that referred to excess scattering by solution over that of solvent) has been dropped. These expressions now provide us with a way to determine the turbidity by measurement of the intensity of light scattered at given angles, that is, by measurement of $i(\theta)$.

Turbidity and Molecular Weight of Polymer Solutions

The connection between turbidity and the molecular weight of a single molecular-weight solute may be obtained by substitution of $(\partial\pi/\partial c)_T$ into (35). However, before making this substitution it is convenient to rearrange (35) to (38):

$$\frac{Hc}{\tau} = \frac{1}{RT} \left(\frac{\partial\pi}{\partial c}\right)_T \tag{38}$$

where the function H is given by

$$H = \frac{32\pi^3}{3\lambda^4 N_0} n_0^2 \left(\frac{n - n_0}{c}\right)^2 \tag{39}$$

TABLE 14.2 Light-scattering Determination of Molecular Weight of Polystyrene in Toluene ($\lambda = 4360$ Å; $n_0 = 1.4976$)

CONCN. (g/cm³)	$(n - n_0)$	τ (cm^{-1})	$Hc/\tau \times 10^6$ (mol/g)
0.066	0.0075	0.0061	2.79
0.128	0.0145	0.0100	3.34
0.255	0.0288	0.0159	4.17
0.510	0.0576	0.0202	6.55

Source: H. P. Frank and H. P. Mark, *J. Polymer Sci.*, **17**, 1 (1955).

Then substitution of $(\partial \pi / \partial c)_T$ from (21) yields

$$\frac{Hc}{\tau} = \frac{1}{M}(1 + 2\Gamma c + 3g\Gamma^2 c^2 + \cdots) \tag{40}$$

According to (39) and (40), the determination of molecular weight by light scattering requires (a) a measurement of the refractive index differences between solutions of varying concentration and pure solvent, (b) a measurement of the turbidities of the solutions, and (c) an extrapolation of Hc/τ to $c = 0$, the intercept of the plot being $1/M$. By analogy to the treatment of osmotic pressure data, the initial slope yields the second virial coefficient Γ. Typical data for solutions of polystyrene in toluene are shown in Table 14.2, and a plot according to (40) is shown in Figure 14.7.

Weight-Average Molecular Weight

The molecular weight obtained by application of (40) to a polymer solution will be some *average* over the molecular-weight distribution characteristic of the polymer. It is instructive to inquire as to what type of average this is.

Let us define τ° as the turbidity of a polymer solution in the limit of infinite dilution. In other words, τ° represents the turbidity of a solution in which inter-particle interference of the scattered light intensity has been eliminated. For a solute such as a polymer, which contains a distribution of molecular weights, we may write (in the limit of infinite dilution)

$$\tau^\circ = \sum_i \tau_i^\circ \tag{41}$$

where τ_i° is the turbidity contributed by species having molecular weight M_i. According to (40), τ approaches HMc as $c \to 0$. Therefore, (41) may be written as shown in (42), namely

$$\tau^\circ = H^\circ \sum_i M_i c_i \tag{42}$$

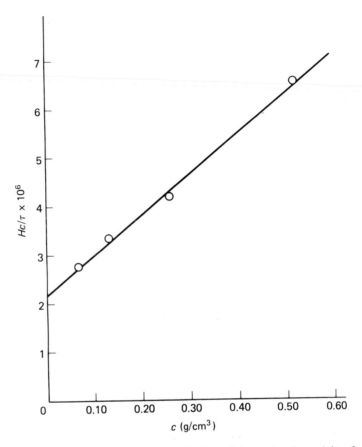

Fig. 14.7 Light-scattering determination of the molecular weight of polystyrene in toluene. (Data from Table 14.2.)

where $H°$ is the value of the constant defined by (39) in the limit $c \to 0$. The concentration c_i is the weight per unit volume of solution of molecules having molecular weight M_i. Therefore, we may write

$$c_i = \frac{w_i}{V} = \frac{(N_i/N_0)M_i}{V} \tag{43}$$

in which w_i is the weight of molecules of molecular weight M_i, N_i is the number of molecules of molecular weight M_i, N_0 is Avogadro's number, and V is the volume of the solution. Substitution of (43) into (42) leads to

$$\tau° = \frac{H°}{N_0 V} \sum_i N_i M_i^2 \tag{44}$$

It will be obvious from (44) that the turbidity is proportional to the square of the polymer molecular weight. This means that the heavier molecules will contribute more than the lighter ones to the turbidity of the solution. *Hence, any average molecular weight derived from the turbidity will be influenced more by heavier molecules than by lighter ones.*

The weight-average molecular weight of a polymer is defined by

$$\overline{M}_w = \frac{\sum_i w_i M_i}{\sum_i w_i} = \frac{\sum_i N_i M_i^2}{\sum_i N_i M_i} \qquad (45)$$

where the symbols are as defined previously. From (45) it is seen that

$$\sum_i N_i M_i^2 = \overline{M}_w \sum_i N_i M_i \qquad (46)$$

After substitution of (46) into (44), equation (47) is obtained:

$$\tau^\circ = \frac{H^\circ}{N_0 V} \overline{M}_w \sum_i N_i M_i = H^\circ \overline{M}_w \sum_i \frac{(N_i/N_0) M_i}{V} \qquad (47)$$

which, by reference to (43) may be written

$$\tau^\circ = H^\circ \overline{M}_w \sum_i c_i = H^\circ \overline{M}_w c \qquad (48)$$

where c is the total concentration of polymer. Recalling now that τ° and H° refer to the limit of infinite dilution, we may write (48) as

$$\lim_{c \to 0} \frac{Hc}{\tau} = \frac{1}{\overline{M}_w} \qquad (49)$$

Thus, light-scattering measurements yield the weight-average molecular weight. For any solute having a distribution of molecular weights, the weight-average molecular weight is greater than the number-average molecular weight. The ratio $\overline{M}_w/\overline{M}_n$, determined from light-scattering measurements and from osmotic pressure measurements on the same solutions, is a measure of the width of the molecular-weight distribution.

Polymer Dimensions and Corrections for Dissymmetry of Scattering

The relationships between light scattering and molecular weights of polymers, discussed in the previous sections, are satisfactory provided that the scattering is isotropic (i.e., symmetric about 90°). This will be true for polymer solutions if the average size of the largest dimension of the polymer is less than about $\frac{1}{20}$ of the wavelength of the incident light. If the average largest dimension of the polymer becomes greater than $\lambda/20$, a dissymmetry of the scattered light about 90° is observed. The dissymmetry arises because of destructive interference of light scattered from different parts *of the same molecule.* This *intra*particle interference cannot be eliminated by an extrapolation to infinite dilution.

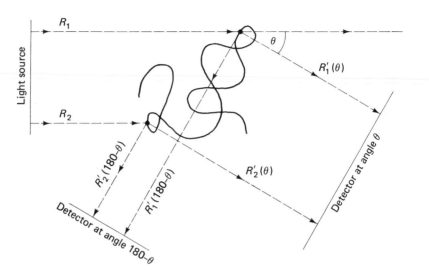

Fig. 14.8 Intraparticle destructive interference for large polymer molecules.

The origin of intraparticle interference is shown schematically in Figure 14.8. In this figure two incident light rays, R_1 and R_2, of a parallel beam are scattered at an angle θ to the incident direction from different parts of the same polymer molecule. Initially, the light waves associated with the two rays are in phase but, because they travel different distances before being detected, the corresponding scattered rays R_1' and R_2' are, in general, out of phase. The mutal canceling at the detector of the out-of-phase scattered rays is termed *destructive interference*. The extent of destructive interference depends on the path-length difference of the two rays reaching the detector. Geometric considerations, which the reader may easily verify, establish the following two facts:

1. For a given value of the angle θ, the path-length difference of the two scattered rays detected at θ is less than that of the two scattered rays detected at $180° - \theta$; that is, $[R_1 + R_1'(\theta) - R_2 - R_2'(\theta)] < [R_1 + R_1'(180 - \theta) - R_2 - R_2'(180° - \theta)]$. As a result, destructive interference is greater at $180 - \theta$ (backward direction) than at θ (forward direction). In turn, this results in a greater observed scattered intensity in the forward direction than in the backward direction. That is, $I_s(\theta) > I_s(180 - \theta)$.

2. As θ approaches $0°$, the path-length differences between forward scattered rays from different parts of the molecule also approach zero. Thus, in the limit $\theta = 0$, no intraparticle destructive interference of the scattered light occurs.

To take into account the reduction of scattered intensity by intraparticle interference, it is convenient to replace the turbidity, τ, by the Rayleigh ratio, $R(\theta)$ (e.g., equation 37) and further modify (40) to read

$$\frac{Kc}{R(\theta)}(1 + \cos^2 \theta) = \frac{1}{\overline{M}_w P(\theta)} + \frac{2\Gamma}{\overline{M}_w} c + \frac{3g\Gamma^2}{\overline{M}_w} c^2 + \cdots \qquad (50)$$

where $K = (3/16\pi)H$ and $P(\theta)$ is a scattering function that has the property $P(\theta) \to 1$ as $\theta \to 0$. Thus, the determination of the molecular weights of large polymer molecules requires an extrapolation, not only to the limit $c \to 0$ to eliminate interparticle interference, but also to the limit of $\theta \to 0$ to eliminate intraparticle interference. That is,

$$\frac{1}{\overline{M}} = \lim_{\substack{c \to 0 \\ \theta \to 0}} \frac{Kc}{R(\theta)}(1 + \cos^2 \theta) \tag{51}$$

Although the extrapolation of experimental scattering data to $\theta = 0$ is independent of any model of the polymer in solution, the scattering functions, $P(\theta)$, depend on the model chosen. Moreover, the determination of average polymer dimensions also depends on the model chosen. The following scattering functions have been derived for three models in terms of a polymer dimension, the wavelength of the light, and the refractive index of the solvent.

1. *Random coil polymer*

$$P(\theta) = \frac{2}{v^2}[e^{-v} - (1 - v)] \tag{52}$$

where

$$v = \tfrac{8}{3}\pi^2 n_0^2 \left(\frac{\overline{S^2}}{\lambda^2}\right) \sin^2 \frac{\theta}{2}$$

$\overline{S^2}$ = Mean-square separation of polymer ends

2. *Spherical polymer molecules*

$$P(\theta) = \left[\left(\frac{3}{u^3}\right)(\sin u - u \cos u)\right]^2 \tag{53}$$

where

$$u = 2\pi\left(\frac{n_0 d}{\lambda}\right)\sin \frac{\theta}{2}$$

d = Diameter of sphere

3. *Rigid-rod polymer*

$$P(\theta) = \frac{1}{x}S(2x) - \left(\frac{1}{x}\sin x\right)^2 \tag{54}$$

where

$$S(x) = \int_0^x \frac{\sin y}{y}\,dy$$

$$x = 2\pi\left(\frac{Ln_0}{\lambda}\right)\sin \frac{\theta}{2}$$

L = Length of rod

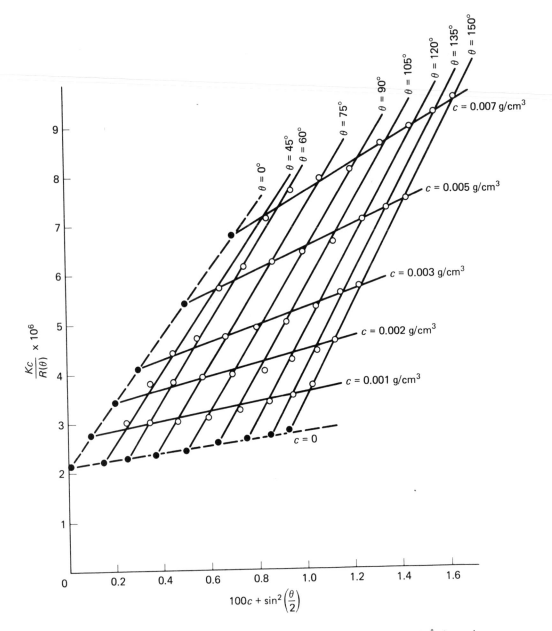

Fig. 14.9 Light-scattering measurements of $\lambda = 5461$ Å for solutions of poly(L-lactic acid) in bromobenzene at 85°C. [Reproduced from A. E. Tonelli and P. J. Flory, *Macromolecules*, **2**, 225 (1969); © the American Chemical Society.]

Note that all the scattering functions depend on the angle θ through $\sin(\theta/2)$ or (in the important case of the random coil) on $\sin^2(\theta/2)$. This suggests that the double extrapolation to obtain molecular weights (cf. equation 51) can be made conveniently by plotting $Kc/R(\theta)$ versus $kc + \sin^2(\theta/2)$, where k is an arbitrary constant. When this is done, a grid of points is obtained as shown in Figure 14.9 for poly(L-lactic acid) in bromobenzene. The quantity $1/\overline{M}_w$ is evaluated from the plot as the intercept on the vertical axis, for which $\theta = 0$ and $c = 0$. The molecular weights obtained by this extrapolation procedure are independent of the model chosen, because, for all models, $P(\theta) \to 1$ as $\theta \to 0$. Grid plots such as shown in Figure 14.9 are often called Zimm plots after their originator.[1]

An average size of the polymer molecule may be obtained from the dissymmetry coefficient,[2] defined as

$$Z_\theta = \frac{I(\theta)}{I(180° - \theta)} = \frac{P(\theta)}{P(180° - \theta)} \tag{55}$$

It must be realized that the polymer sizes so obtained depend on the extent to which the model chosen represents the actual polymer. For three models of a polymer in solution, the theoretical values of Z_{45}° as functions of polymer dimensions [i.e., S, d, and L of (52) to (54)] are shown in Figure 14.10. A determination of the dimensions appropriate to the particular model chosen requires an experimental measurement of the intensity ratio, I_{45}/I_{135}. The value obtained experimentally may be located on the plot in Figure 14.10 to give immediately the ratios $(\overline{S^2})^{1/2}n_0/\lambda$, $n_0 d/\lambda$, or $L n_0/\lambda$. Since n_0, the refractive index of solvent, and λ, the wavelength of light, are known, $(\overline{S^2})$, d, or L is determined.

Experimental Apparatus and Technique

According to (39) and (40), the quantities that must be determined to calculate the molecular weight are the turbidity τ and the refractive index increment $(n - n_0/c)$. It is not practical to measure the turbidity from the attenuation of the incident light intensity (cf. equation 28) because this attenuation is generally too small. The turbidity τ of a solution must be determined, then, from the Rayleigh ratio (cf. equations 36 and 37). This means that the intensity of light scattered per unit volume of solution at some angle θ (usually 90°) to the incident beam must be measured. Furthermore, if the polymer molecules have, on the average, a dimension that is greater than about 1/20 of the wavelength of the light used (typically 4000 to 5000 Å), it is necessary to measure the scattered intensity at several angles.

Therefore, a light-scattering determination of the molecular weight and of the molecular size requires the measurement of (1) the intensity of light scattered at several angles by solutions of various concentrations, and (2) the difference in refractive index between solution and solvent for various concentrations.

[1] B. H. Zimm, *J. Chem. Phys.*, **16**, 1099 (1948).
[2] P. Debye, *J. Phys. Coll. Chem.*, **51**, 18 (1947).

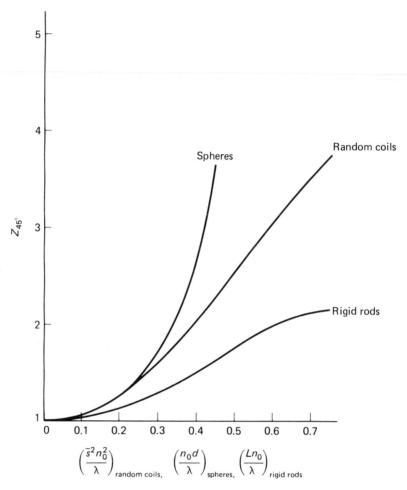

Fig. 14.10   Dissymmetry coefficient at 45°C as a function of polymer size.

Light-scattering photometry

A commercially available light-scattering photometer[1] is shown schematically in Figure 14.11. Light is generated by a mercury-arc source,[2] L, to yield predominately the principal emission lines of the mercury spectrum. The wavelength desired is isolated by the filter F_1, and this monochromatic light is collimated by lenses L_1 and L_2. If desired, neutral filters F_2 may be used to reduce the intensity of the primary beam by known amounts. The beam then impinges on a cell containing the polymer solution C. Most of the light is transmitted and is either absorbed by the light trap T or collected and measured by the photomultiplier, FT,

[1] B. A. Brice, M. Halwer, and R. Speiser, *J. Opt. Soc. Am.*, **40**, 768 (1950).
[2] More recent instruments utilize a laser light source, which permits measurements to much lower angles. However, many older instruments which use a mercury arc source are still in use.

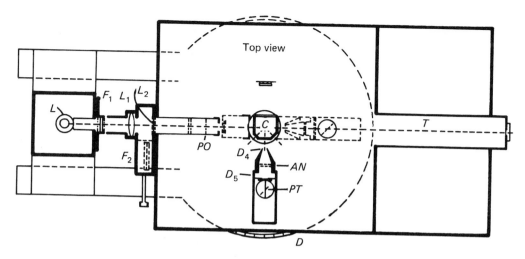

Fig. 14.11 Commercial light-scattering photometer. (Reproduced with permission of the Phoenix Precision Instrument Company, Gardiner, New York 12525.)

when this is located, as shown by the dashed lines, in the 0° position. The light scattered by the solution at various angles passes through the slit system D_4 and D_5 and then impinges on the photomultiplier tube, PT. The photomultiplier and slit system can be rotated by known angles about the scattering cell. As mentioned above, measurement of the incident beam intensity I_0 is accomplished by placing the detection system at 0° to the incident beam with solvent in the cell. Given a measurement of $i(\theta)$, I_0, θ, and r^2, the turbidity τ may be calculated from (36) or (37). For large particles which show anisotropic scattering, a correction for the depolarization of the light is required. For this purpose, plane polarizers PO and AN may be included in the optical train.

Scattering cells of a variety of cross-sectional shapes may be used. Top views of typical cells are shown in Figure 14.12. Cell (a) is used for polymer molecules that are small (i.e., $(\overline{S^2})^{1/2} < \lambda/20$), because a measurement of the scattered intensity at 90° only is necessary. For larger molecules, in which $i(\theta)$ must be measured at various values of θ, cell (b) would be preferred. Cell (c) is a dissymmetry cell which may be used to measure Z_{45} as well as $R(90°)$.

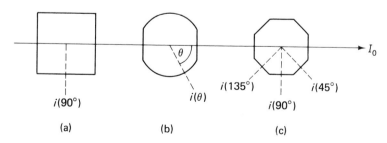

Fig. 14.12 Top view of typical light-scattering cells.

Refractive index increment

A differential refractometer for measurement of the difference in refractive index between solution and solvent is shown schematically in Figure 14.13. The particular instrument shown has been reported to be capable of measurements of $n - n_0$ to an accuracy of about ± 0.000003.

Monochromatic light produced by the mercury arc H4 and filter F illuminates the adjustable slit S. An image I of the slit S is viewed by a micrometer which is capable of measuring displacements of this image of 0.001 cm. A rectangular outer cell C, made of glass and containing the solvent, surrounds a hollow prism cell P that contains the solution. The lenses L_1 and L_2 are collimating lenses to produce a parallel beam and to focus it at the position I. Any difference in refractive index between solution and solvent, will cause the image I to be deflected. Moreover, if the refractive index difference is not too large, the deflection of the image is proportional to $n - n_0$. For example, if A is the apex angle of the prism cell and f is the focal length of the lenses, the deflection of the image can be shown to be given by (56). Measurement of Δd for known values of f and A thus yield $n - n_0$.

$$\Delta d = 2f(n - n_0) \tan \frac{A}{2} \tag{56}$$

For the concentration ranges generally employed, the refractive index difference, $n - n_0$, is a linear function of the concentration. In other words, $(n - n_0)/c$, which is called the *specific refractive index increment*, is a constant for a given polymer–solvent system and a given wavelength of light. This means that $n - n_0$ need be measured for only one or two different concentrations. Typical specific refractive index increments are shown in Table 14.3. As can be seen in the table, $(n - n_0)/c$ depends on the wavelength of light used, being slightly larger at the shorter wavelength.

Preparation of solutions

It is very important to remove all suspended dust particles from a solution before making measurements of the intensity of scattered light. Dust particles are generally larger than polymer molecules and may produce greater scattering. Solvents and solutions are generally clarified by filtration (usually under pressure)

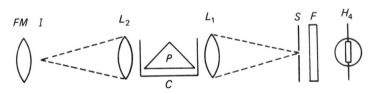

Fig. 14.13 Schematic arrangement of a differential refractometer. [Reproduced from P. P. Debye, *J. Appl. Phys.*, **17**, 392 (1946); with permission of the American Institute of Physics, New York.]

TABLE 14.3 Specific Refractive Index Increments of Typical Polymer Solutions

POLYMER	SOLVENT	$\lambda = 4360\,\text{Å}$	$\lambda = 5460\,\text{Å}$
Polystyrene	Methyl ethyl ketone	0.231	0.220
Poly(ethyl acrylate)	Acetone	0.109	0.106
Poly(vinyl palmitate)	Neohexane	0.122	0.120
Poly(vinyl laurate)	Neohexane	0.118	0.114
β-Lactoglobulin	0.1 M NaCl, pH 5.2	0.189	0.182
Lysozyme	0.1 M NaCl, pH 6.2	0.196	0.189

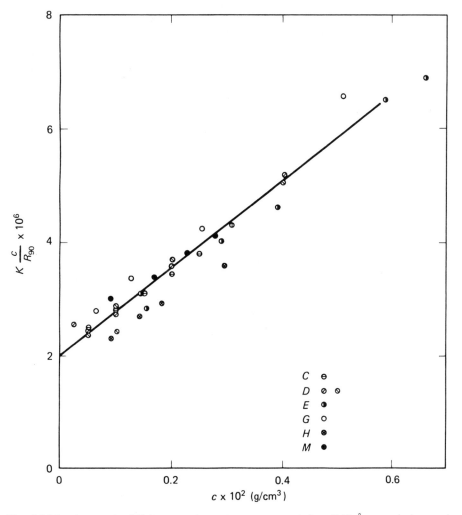

Fig. 14.14 Accuracy of light-scattering measurements at $\lambda = 4360$ Å on polystyrene in toluene at 25°C. [Reproduced from H. P. Frank and H. F. Mark, *J. Polymer Sci.*, **17**, 1 (1955); with permission of John Wiley & Sons, Inc., New York.] The letters refer to results from different laboratories.

or by ultracentrifugation. In these procedures, care must be taken not to remove very large polymer molecules that have molecular weights at the heavy end of the distribution.

Accuracy of Light-Scattering Measurements of Molecular Weight

The accuracy of molecular weights determinated by light scattering is not as good as might be expected on the basis of the precision shown in Figures 14.7 and 14.9. This may be seen from an inspection of Figures 14.14 and 14.15. These figures show plots of Hc/τ (or equivalently $K_c/R_{90°}$) at 4360 Å versus c, as determined independently by six laboratories in the United States, Canada, and Europe for samples of the same polystyrene fraction. Figure 14.14 refers to solutions of the polymer in toluene and Figure 14.15 to solutions of the polymer in methyl ethyl ketone. Fewer discrepancies are noted than for the corresponding osmotic pressure measurements. The still-appreciable scatter of the data has been attributed to difficulties in the calibration of the light-scattering photometer.

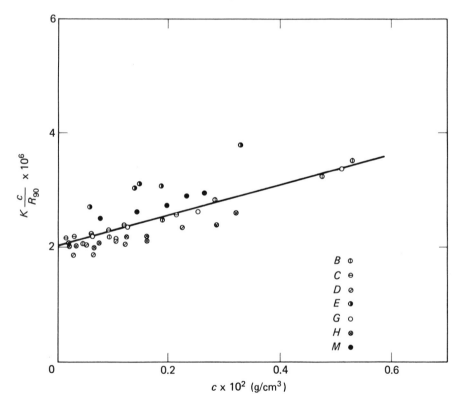

Fig. 14.15 Accuracy of light-scattering measurements at $\lambda = 4360$ Å on polystyrene in methyl ethyl ketone at 25°C. [Reproduced from H. P. Frank and H. F. Mark, *J. Polymer Sci.*, **17**, 1 (1955); with permission of John Wiley & Sons, Inc., New York.] The letters refer to results from different laboratories.

The average molecular weights and second virial coefficients found for this polystyrene sample by light scattering are:

(a) Toluene: $\overline{M}_w = 5.90 \pm 0.58 \times 10^5$ g/mol
$$\Gamma = 224 \pm 30 \text{ cm}^3/\text{g}$$

(b) Methyl ethyl ketone: $\overline{M}_w = 5.92 \pm 0.75 \times 10^5$ g/mol
$$\Gamma = 79.3 \pm 19 \text{ cm}^3/\text{g}$$

The agreement of the weight-average molecular weights obtained from the measurements in the two solvents is well within the experimental errors. Moreover, from a theoretical viewpoint it is reassuring that, despite the difference in weight-average and number-average molecular weights of this polymer sample, the second virial coefficients determined by the two very different methods are in agreement, within experimental error.

It should be pointed out that the polystyrene molecules in this test sample were sufficiently large to cause an asymmetry in the light scattering. The weight-average molecular weights given above were determined by the double extrapolation procedure (Zimm plot), in which values of K_c/R_θ were extrapolated to $c \to 0$ and $\theta \to 0$ (cf. Figure 14.9).

ULTRACENTRIFUGATION AS A METHOD FOR MEASUREMENT OF ABSOLUTE MOLECULAR WEIGHTS

Principle of the Method

Sedimentation velocity

A particle in the earth's gravitational field is subjected to a downward force, mg, where m is the mass of the particle and g is the acceleration due to gravity. If the particle is in a vacuum, it will fall with a velocity u that continuously increases with time. According to Newton's second law of motion, this can be expressed by

$$m\frac{du}{dt} = mg \tag{57}$$

or

$$u = gt \tag{58}$$

where t is the time. If the particle is not in a vacuum, but rather is immersed in a fluid of density ρ_s, a buoyancy force equal to the weight of the displaced fluid will oppose the gravitational force. If the volume displaced by the particle is V_p, the mass of displaced fluid will be $V_p \rho_s$ and the weight or buoyancy force will be given by $V_p \rho_s g$. In addition, a frictional force proportional to the velocity (i.e., Fu, where the proportionality constant F is called the frictional coefficient) will oppose the gravitational force. Hence, we may write for the net downward force on the particle in a solvent

$$m \frac{du}{dt} = (m - V_p \rho_s)g - Fu \tag{59}$$

Eventually, the velocity of fall will increase to such a value that the net force on the particle becomes zero and, from this time on, the particle falls with a constant velocity u_s, called the *terminal velocity* or *sedimentation velocity*. Thus, from (59) we have

$$(m - V_p \rho_s)g = Fu_s \tag{60}$$

In the special case of a spherical particle of radius r, the value of F is given by Stokes' law, which, on substitution into (60), yields

$$m(1 - \bar{v}_p \rho_s)g = \frac{M}{N_0}(1 - \bar{v}_p \rho_s)g = 6\pi \eta r u_s \tag{61}$$

in which η is the viscosity of the fluid, $\bar{v}_p$ the partial specific volume of the particle in the fluid, M the mass of the particle per mole, and N_0 is Avogadro's number. Eliminating r from (61), by the use of the relationship $\bar{v}_p = 4/3\pi r^3 N_0/M$, and solving for M yields

$$M = 9\pi \sqrt{2\bar{v}_p} N_0 \left(\frac{\eta u_s}{(1 - \bar{v}_p \rho_s)g} \right)^{3/2} \tag{62}$$

Of course, if the particles are polymer molecules, the molecular weight M in (62) is an average molecular weight. Hence, (62) provides a means to determine an average molecular weight for a spherical polymer simply by measurement of the rate of sedimentation and the partial specific volume of the polymer.

In practice the matter is not quite so simple. First, as is obvious from the fact that polymer solutions are stable, the sedimentation rate under the influence of gravity is vanishingly small. Elaborate ultracentrifugation techniques must be used to supply the much larger accelerations needed. Second, (62) applies only to spherical polymer molecules, whereas most polymer molecules exist in solution as random coils. Therefore, it is usually necessary to use (60) in which g, the acceleration due to gravity, is replaced by the centrifugal acceleration, $\omega^2 x$, where ω is the angular velocity of rotation and x is the distance of the sample from the center of rotation. The proportionality constant, F, is not known explicitly and must be eliminated by experimental measurement of some other quantity that depends on it. Usually, this other quantity is the diffusion coefficient D, since many years ago Nernst[1] showed that the coefficient for free diffusion in an infinitely dilute solution is given by

$$D = \frac{RT}{N_0 F} \tag{63}$$

[1] W. Nernst, *Z. Phys. Chem.*, **2**, 613 (1888).

It is expected that the proportionality constant, F, for diffusion should be the same as that for sedimentation. Combination of (60) and (63) so as to eliminate F, and replacement of g by $\omega^2 x$ leads to

$$(m - V_p \rho_s)\omega^2 x = \left(\frac{N_0 D}{RT}\right)^{-1} u_s \qquad (64)$$

The *sedimentation coefficient* is defined by

$$s = \frac{u_s}{\omega^2 x} \qquad (65)$$

and if we also introduce into (64) the partial specific volume of polymer in the solution, $\bar{v}_p$, the expression

$$M = N_0 m = \frac{sRT}{D(1 - \bar{v}_p \rho_s)} \qquad (66)$$

is finally obtained. This is known as the Svedberg equation. According to this equation, the molecular weight of polymer particles in solution can be determined by measurement of the sedimentation coefficient, s, the diffusion coefficient, D, and the partial specific volume of the polymer in solution, $\bar{v}_p$.

Actually, it is found that the experimental values of s and D depend significantly on the concentration of polymer in the solution. Moreover, the Nernst equation (63) and therefore (66) are strictly valid only in the limit of infinite dilution. Therefore, in order to eliminate intermolecular effects and obtain reliable molecular weights, the values of s and D inserted into (66) must be values that have been extrapolated to zero concentration. Strictly speaking, the partial specific volume of the polymer $\bar{v}_p$ also depends on the concentration, but the dependence is sufficiently weak that extrapolation of measured values of $\bar{v}_p$ to zero concentration is generally not necessary. Finally, the derivation of (66) considered solute particles that have a single molecular weight M. A generalization[1] to polymer samples having a distribution of molecular weights shows that the molecular weight obtained is *approximately* equal to the weight-average molecular weight, $\overline{M}_w$.

In view of the foregoing discussion, the Svedberg equation (66) is more properly written as

$$\overline{M}_w = \frac{s_0 RT}{D_0(1 - \bar{v}_p \rho_s)} \qquad (67)$$

where s_0 and D_0 represent values that have been extrapolated to zero concentration.

As in light-scattering and osmotic-pressure measurements on polymer solutions, it is useful to describe the concentration dependence of s and D in terms of a

[1] G. Meyerhoff, *Angew. Chem.*, **72**, 699 (1960).

virial equation [cf. equation 21 and equation 40]. Thus, it can be shown[1] that the concentration dependence of (66) is well described by

$$\frac{D(1 - \bar{v}_p \rho_s)}{sRT} = \frac{1}{M_w}(1 + 2\Gamma c + 3g\Gamma^2 c^2 + \cdots) \tag{68}$$

where Γ, a fundamental parameter of the Flory–Huggins theory of polymer solutions is given by equation (22). The weight-average molecular weight $\overline{M}_w$ is obtained either from (67) or from the zero-concentration limit of the left-hand side of (68). The second virial coefficient, Γ, is determined from the initial slope of a plot of the left-hand side of (68) as a function of concentration.

The sedimentation equilibrium method

In a homogenous polymer solution the concentration gradient is zero in all directions. During centrifugation the sedimentation process produces a migration of polymer particles toward the bottom of the container and, in so doing, this creates a nonzero concentration gradient, dc/dx, in the direction from the center of rotation to the bottom of the centrifugation cell. The process of diffusion, on the other hand, produces a migration of polymer molecules from a region of high concentration to a region of lower concentration. Thus, sedimentation and diffusion operate in opposite directions during centrifugation and, after a sufficient time, they must produce a pseudo equilibrium or steady state that is characterized by a nonzero but constant value of dc/dx. The faster the rotation, the larger the gradient dc/dx. It can be shown that at equilibrium the ratio of concentrations at two positions in the cell is given by

$$\ln \frac{c_2}{c_1} = \frac{\omega^2(1 - \bar{v}_p \rho_s)(x_2^2 - x_1^2)\overline{M}}{2RT} \tag{69}$$

where c_2 and c_1 are the concentrations of polymer at positions x_2 and x_1, respectively. The concentration ratio is generally determined by optical methods that depend on the change of refractive index with concentration.

Molecular weights obtained by this method will deviate appreciably from the weight-average molecular weights. Another disadvantage in the use of this method, compared with the sedimentation velocity method, is that, because of the small diffusion coefficients of polymer molecules, very long times are often required to attain equilibrium.

Experimental measurements

Calculation of the weight-average molecular weight $\overline{M}_w$ by (67) requires the experimental measurement of the sedimentation constant s, the diffusion coefficient D, and the partial specific volume of the polymer $\bar{v}_p$. Moreover, s and D must be

[1] G. V. Schulz, Z. Phys. Chem., **193**, 168 (1944).

measured in solutions of varying concentrations so that the respective values at infinite dilution, s_0 and D_0, may be obtained by extrapolation.

Partial specific volume. The partial specific volume of a polymer in solution is generally calculated from the densities of the solutions and solvent. The respective densities are determined by a pycnometric method[1] in which an accurately known volume containing, in turn, solvent and solution is weighed. The relationship between the volume of the solution (pycnometer volume) and the partial specific volumes of polymer and solvent is given by

$$V = m_p \bar{v}_p + m_s \bar{v}_s \tag{70}$$

The partial specific volumes are defined thermodynamically by (71) and (72).

$$\bar{v}_p = \left(\frac{\partial V}{\partial m_p} \right)_{m_s, T, p} \tag{71}$$

$$\bar{v}_s = \left(\frac{\partial V}{\partial m_s} \right)_{m_p, T, p} \tag{72}$$

In actual solutions of finite concentration of polymer, $\bar{v}_s$ is not equal to the reciprocal of the solvent density (i.e., to ρ_s^{-1}). However, in dilute solutions little error is introduced by assuming that $\bar{v}_s \cong \rho_s^{-1}$. The partial specific volume of the polymer is then given by a rearrangement of (70) as

$$\bar{v}_p = \frac{V - m_s/\rho_s}{m_p} \tag{73}$$

Introduction of the density of the solution, $\rho = (m_p + m_s)/V$, and the concentration of polymer, $c = m_p/V$, into (73) transforms the expression into (74), a more useful form in terms of the solution and solvent densities.

$$\bar{v}_p = \frac{1}{\rho_s} \left(1 - \frac{\rho - \rho_s}{c} \right) \tag{74}$$

In (74) ρ_s and ρ are the densities of solvent and solution of concentration c, respectively.

Strictly speaking, the experimental values of $\bar{v}_p$ obtained from (74) should be extrapolated to zero concentration and the extrapolated value used in (67). In practice, the values of $\bar{v}_p$ in the dilute solutions used do not differ appreciably from the value at infinite dilution, and extrapolation is generally not necessary.

The reader should note carefully that $\bar{v}_p$ is a property of the *polymer solution* and is not the same as the reciprocal of the density of the solid polymer. Hence, in general, $\bar{v}_p \neq \rho_p^{-1}$.

[1] D. P. Shoemaker, C. W. Garland, and J. I. Steinfeld, *Experiments in Physical Chemistry*, 3rd ed. (New York: McGraw-Hill, 1974), p. 171.

Sedimentation constant. According to the definition in (65), the sedimentation constant may be calculated from measurements of the terminal sedimentation velocity, u_s, that is reached when the solution is subjected to a centrifugal force of $\omega^2 x$. Such measurements are carried out in an ultracentrifuge, an instrument that is capable of rotational speeds up to 80,000 revolutions per minute. The dimensions are such that these angular velocities correspond to accelerations that are in the range 7000 to 420,000 times that of gravity.

A schematic diagram of an ultracentrifuge is shown in Figure 14.16. A dilute solution of the polymer, at a concentration c, is placed in the centrifugation cell, such as shown in Figure 14.17, and this cell, as well as a balancing cell, are placed in the rotor of the ultracentrifuge. The rotor is operated in a vacuum to reduce friction and the temperature is controlled to $\pm 0.1°C$.

Initially, the solution is homogenous throughout the cell (i.e., $dc/dx = 0$), but when the centrifugal force is applied by spinning the rotor, the polymer molecules migrate, ultimately with a constant velocity, u_s, toward the bottom of the cell. This migration of the polymer molecules results in the formation of a *boundary* between solution and solvent. The sharpness of this boundary depends upon the nature of the molecular-weight distribution of the polymer and is characterized by the value

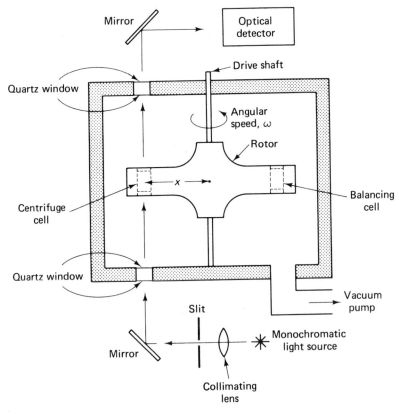

Fig. 14.16 Schematic diagram of an ultracentrifuge.

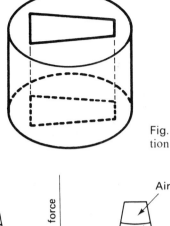

Fig. 14.17 Typical centrifugation cell of the sector type.

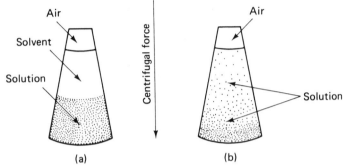

Fig. 14.18 Light-beam view of a cell during centrifugation. (a) Mono-
dispense polymer. (b) Polydisperse polymer.

of the concentration gradient, dc/dx. In Figure 14.18, the result of polymer migration in the cell due to centrifugation is shown for two extreme cases of polymer. In
the left-hand picture is seen the sharp boundary that results if all polymer molecules have the same molecular weight. The right-hand diagram shows the diffuse
type of boundary resulting from a sample with a broad distribution of molecular
weights.

A quantitative measurement of dc/dx is generally accomplished by the use of
schlieren or optical interference techniques. Both techniques measure the change
in refractive index of the system as a function of the position in the cell. Because the
refractive index depends on the concentration, both techniques ultimately yield
the concentration gradient, dc/dx, as a function of distance x. For details of these
fairly complex optical techniques, the reader should consult more specialized
accounts.[1]

At each revolution the centrifugation cell is in the position at which the light
beam passes through the cell (cf. Figure 14.16) and, therefore, a distribution curve
of dc/dx as a function of x is obtained for each revolution (or equivalently for each
time of measurement). For a polymer having a broad but continuous distribution

---

[1] F. Daniels, J. W. Williams, P. Bender, R. H. Alberty, C. D. Cornwell, and J. E. Harriman,
Experimental Physical Chemistry, 7th ed. (New York: McGraw-Hill, 1970), pp. 460–465.

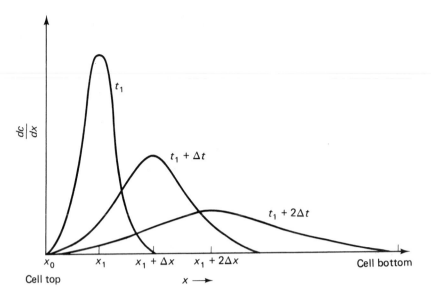

Fig. 14.19   Variation of the distribution of dc/dx in the cell as a function of centrifugation time.

of molecular weights, data such as those depicted in Figure 14.19 would be obtained. From such curves it is possible to determine the sedimentation velocity u_s from the relationship

$$u_s = \frac{x_1 + \Delta x - x_1}{t_1 + \Delta t - t_1} = \frac{\Delta x}{\Delta t} \tag{75}$$

The sedimentation coefficient, s, is then calculated by the expression

$$s = \frac{u_s}{\omega^2 \bar{x}} = \frac{1}{\omega^2} \left(\frac{\Delta x}{\Delta t}\right)\left(\frac{2}{2x_1 + \Delta x}\right) \tag{76}$$

where $\bar{x} = (2x_1 + \Delta x)/2$ is the average distance from the center of rotation.

As has already been mentioned, the values of s calculated from the data by (76) depend on the concentration, and an extrapolation to zero concentration must be made to eliminate interparticle effects. It is found empirically that the extrapolation is most readily made according to

$$\frac{1}{s} = \frac{1}{s_0} + k_s c \tag{77}$$

Thus, the reciprocals of the experimental s values are plotted versus the concentration, and s_0 is determined as the reciprocal of the intercept. A typical plot of $1/s$ versus c for polyisobutylene is shown in Figure 14.20.

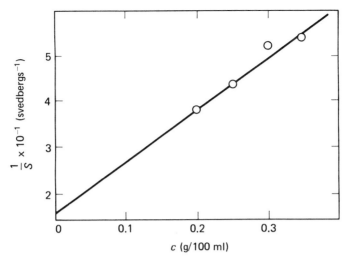

Fig. 14.20 Extrapolation of the reciprocal sedimentation constant to infinite dilution. [Reproduced from W. J. Closs, B. R. Jennings, and H. G. Jerrard, "Concentration Dependence of the Sedimentation Coefficient," *Eur. Polymer J.*, **4**, 639 (1968); with permission of Pergamon Press, New York.]

Diffusion coefficient. The process of diffusion acts in opposition to sedimentation. Thus, sedimentation begins with a uniform solution (i.e., $dc/dx = 0$) and produces concentration gradients. Diffusion begins with a finite value for dc/dx and the process drives the system toward the equilibrium value $dc/dx = 0$. Accordingly, the same optical techniques mentioned previously may be used to measure the distribution of dc/dx as a function of time and, from these measurements, the diffusion coefficient D may be calculated.

Consider the schematic diffusion cell in Figure 14.21(a). Initially, a partition divides the polymer solution of concentration c from the solvent. The initial values of c and dc/dx are then described by

$$c_0 = c \quad \text{for } x < 0$$

$$c_0 = 0 \quad \text{for } \quad x > 0$$

$$\left(\frac{dc}{dx}\right)_0 = \infty \quad \text{at } x = 0$$

$$\left(\frac{dc}{dx}\right)_0 = 0 \quad \text{anywhere else}$$

When the partition is removed, polymer diffuses from left to right, that is, from the region of high concentration to that of low concentration. The variation of concentration and concentration gradient with position x at two subsequent times after removal of the partition are shown in Figures 14.21b and 14.21c, respectively.

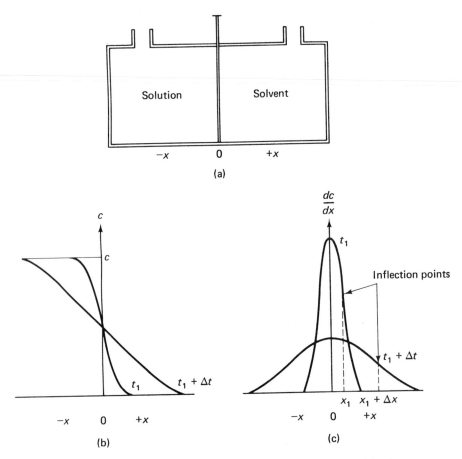

Fig. 14.21 Principle of a diffusion-constant measurement. (a) schematic diagram of a diffusion cell. (b) Variation of concentration with position and time. (c) Variation of concentration gradient with position and time.

The position of the inflection points x_{in} of the dc/dx distribution curves (Figure 14.21c) are taken as a measure of the progress of the diffusion. The mean rate of diffusion is then given by x_{in}/t, where t is the time at which the curve with inflection point x_{in} was measured. It is found that this mean rate of diffusion is inversely proportional to the progress of the diffusion (i.e., inversely proportional to x_{in}). Therefore, it is possible to write

$$\frac{x_{in}^2}{t} = \text{Constant} \tag{78}$$

It may be shown[1] that the constant in (78) can be identified with $2D$, where D is the

[1] W. J. Moore, *Physical Chemistry*, 4th ed. (Englewood Cliffs, N.J.: Prentice-Hall, 1972), pp. 159–163.

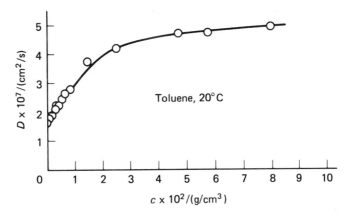

Fig. 14.22 Concentration dependence of the diffusion coefficient of polystyrene ($\overline{M}_w$ = 670,000) in toluene at 20°C. [Reproduced from G. Büldt and G. Meyerhoff, *Makromol. Chem.* **176**, Suppl. 1, 359 (1975); with permission of Hüthig and Wepf Verlag, Basel, Switzerland.]

diffusion coefficient. Thus, the diffusion coefficient may be determined by a measurement of x_{in} and t by the equation

$$D = \frac{x_{in}^2}{2t} \qquad (79)$$

In cases where the inflection points are nonexistent or difficult to determine, an alternative definition of the progress can be used. One convenient method defines the average value of x_{in}^2 as in (80).

$$\overline{x_{in}^2} = \frac{\int_{-\infty}^{\infty} x^2 (dc/dx)\,dx}{\int_{-\infty}^{\infty} (dc/dx)\,dx} \qquad (80)$$

The necessary extrapolation of D-values, obtained from solutions of various concentration, to infinite dilution may be made simply by plotting D versus c. Straight lines of small slope are generally obtained *at low concentrations* which may be described by

$$D = D_0 + D_0 k_D c \qquad (81)$$

An example of such a plot is shown in Figure 14.22.

Molecular Weights

After the determination of $\bar{v}_p$, s_0, and D_0, as described above, these values are inserted into (67) and the weight-average molecular weight is calculated. The versatility of the method, and its comparative freedom from extraneous disturbing effects, has led to the extensive use of ultracentrifugation for the determination of

TABLE 14.4 Partial Specific Volumes, Sedimentation Constants, Diffusion Coefficients, and Molecular Weights of Some Polymers

PROTEIN	$\bar{v}_p$ (cm³/g)	$S_0 \times 10^{13}$ (s)	$D_0 \times 10^7$ (cm²/s)	$\bar{M}_w \times 10^{-3}$ (g/mol)
Hemoglobin*	0.749	4.48	6.9	63
Serum albumin*	0.736	4.67	5.9	72
Urease*	0.73	18.6	3.46	480
Tobacco mosaic virus*	0.73	185	0.53	31,400
Polystyrene†	0.99	17.9	3.6	592

* In aqueous solution.
† In toluene solution.

Source: W. J. Moore, *Physical Chemistry*, 4th ed. (Englewood Cliffs, N.J.: Prentice-Hall, 1972), p. 939.

molecular weights of biological polymers. Typical values of $\bar{v}_p$, s_0, D_0, and M at 20°C for a number of such biological polymers and for a synthetic polystyrene are shown in Table 14.4.

The reliability of sedimentation velocity measurements can be judged by an examination of the results obtained independently by five laboratories for methyl ethyl ketone solutions of the same fractionated sample of polystyrene. The results are shown in Figure 14.23, from which an extrapolated value of $s_0 = 17.9 \pm$

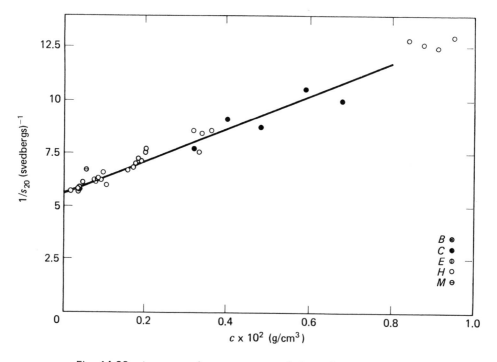

Fig. 14.23 Accuracy of measurements of the sedimentation constant of polystyrene in methyl ethyl ketone at 20°C. [Reproduced from H. P. Frank and H. F. Mark, *J. Polymer Sci.*, **17**, 1 (1955); with permission of John Wiley & Sons, Inc., New York.] The letters *B*, *C*, *E*, *H*, and *M* refer to results from different laboratories.

0.4×10^{-13} s at 20°C is obtained. The diffusion coefficient was determined by only one of the participating laboratories and a value of $3.6 \pm 0.2 \times 10^{-7}$ cm^2/s was reported.

Although ultracentrifugation is the most elaborate, expensive, and time-consuming technique for the determination of polymer molecular weights, it is perhaps the most versatile of all the absolute methods. Unlike osmotic pressure, it is not limited in the range of molecular weights that can be measured. Also, unlike light scattering, it is not influenced by the presence of minute particulate matter in the solutions. In the hands of skilled experimenters, sedimentation constants are very reproducible (cf. Figure 14.23), and it must be concluded that the method is very reliable. However, probably because of the expense, ultracentrifugation has not been used as widely for the determination of molecular weights of synthetic polymers as have the other absolute methods. On the other hand, as mentioned earlier, ultracentrifugation has been used very extensively,[1] for molecular-weight determinations of biological polymers such as proteins.

STUDY QUESTIONS

1. The densities and enthalpies of vaporization of three common solvents at 20°C are as follows:

	TOLUENE	CARBON DISULFIDE	WATER
ρ (g/cm^3)	0.867	1.263	0.998
ΔH_{vap} (cal/mol)	9016	6620	10,540

Calculate the cohesive energy density of each solvent.

2. The density of a sample of poly(methyl methacrylate) is 1.18 g/cm^3. Using data available in this chapter, estimate the enthalpy change involved in dissolving 1 g of poly(methyl methacrylate) in 50 cm^3 of benzene at 20°C. What assumptions have you made in obtaining your estimate?

3. A sample of polyisobutylene has a number-average molecular weight of 428,000 g/mol. The second virial coefficient in chlorobenzene solution at 25°C is $\Gamma = 94.5$ cm^3/g. Calculate the osmotic pressure in g/cm^2 of a 7.0×10^{-6} m solution of this polymer in chlorobenzene at 25°C. (The density of chlorobenzene at 25°C is 1.11 g/cm^3.) Compare this with the value calculated for an ideal solution.

4. Solutions of a sample of polychloroprene ($\rho = 1.25$ g/cm^3) in toluene at 30°C have been reported to have the following osmotic pressures [K. Sato, F. R. Eirich, and J. E. Mark, *J. Polymer Sci., Polymer Phys.*, **14**, 619 (1976)]:

Concn. (g/cm^3) $\times 10^3$	1.33	2.10	4.52	7.18	9.87
Osmotic pressure (dyn/cm^2) $\times 10^{-3}$	0.30	0.51	1.32	2.46	3.90

[1] D. Freifelder, *Physical Biochemistry* (San Francisco: W. H. Freeman, 1976), Chaps. 11 and 12.

From these data determine (a) the number-average molecular weight of the sample; (b) the second virial coefficient, Γ; and (c) the polymer–solvent interaction constant X_1. Give appropriate units for all calculated quantities.

5. Calculate values of the polystyrene–solvent interaction constants, X_1, for the samples discussed in Figure 14.6 ($\rho_p = 1.050$ g/cm³).

6. An osmotic pressure measurement of a solution of poly(vinyl chloride) ($c = 1.5 \times 10^{-3}$ g/cm³) in toluene at 25°C ($\rho = 0.867$ g/cm³) in the apparatus shown in Figure 14.3 indicated a difference of 4.67 mm in the heights of the solution and solvent levels. (a) What is the osmotic pressure of the solution? (b) If the second virial coefficient for poly(vinyl chloride) in toluene is $\Gamma = 200$ cm³/g, calculate the number-average molecular weight of the polymer.

7. Calculate the fraction of light of $\lambda = 546$ nm that would be scattered in a 1-cm path through a solution of poly(vinyl palmitate) in neohexane ($c = 2.0 \times 10^{-4}$ g/cm³). The weight-average molecular weight of the polymer is 5.20×10^4 g/mol and the refractive index of the solvent is 1.3688.

8. Derive an expression for the turbidity of a gas analogous to that given in equation (35) for a solution. Use your expression to calculate the fraction of light of $\lambda = 546$ nm that is scattered by 1 cm of SO_2 at 1 atm pressure and 25°C. The refractive index of SO_2 is 1.000665. Compare your result with the fraction of light scattered by the polymer solution in Problem 7.

9. In a study of light scattering ($\lambda = 436$ nm) from toluene solutions of a polystyrene sample [D. Rahlwes and R. G. Kirste, *Makromol. Chem.*, **178**, 1793 (1977)], the following results were obtained for the Rayleigh ratio, $R(\theta)$, at various concentrations and scattering angles:

| | $\dfrac{R(\theta)}{1 + \cos^2\theta} \times 10^4$ (cm⁻¹) | | | | | |
$c \times 10^3$ (g/cm³)	15°	45°	75°	105°	135°	150°
0.20	2.47	1.53	0.860	0.560	0.428	0.395
0.40	4.40	2.84	1.65	1.09	0.839	0.775
0.60	5.91	3.96	2.37	1.59	1.23	1.14
0.80	7.07	4.91	3.02	2.05	1.60	1.49
1.00	7.98	5.71	3.61	2.49	1.96	1.82

Construct a Zimm plot of the data and evaluate the weight-average molecular weight of the sample. Determine also the second virial coefficient Γ. (The refractive index of toluene is 1.4976 and the specific refractive index increment of polystyrene–toluene solutions is 0.1121.)

10. A sample of poly(α-methylstyrene) having a weight-average molecular weight of 4.10×10^6 g/mol behaves as a random coil in toluene solutions with a root-mean-square end-to-end separation of 2609 Å. The second virial coefficient of the solution is $\Gamma = 709$ cm³/g, the refractive index of toluene is 1.4976, and the specific refractive index increment of the solution is 0.1370 cm³/g. Use the appropriate scattering function and any other expressions in this chapter to calculate the Rayleigh ratio, $R(\theta)$, as a function of scattering angle in the range 15° to 150° for a solution with $c = 1.0 \times 10^{-4}$ g/cm³. Plot the Rayleigh ratios

on polar coordinates to show the dissymmetry of the scattering. What is the dissymmetry coefficient Z_{45}?

11. Using anionic polymerization techniques, D. Rahlwes and R. G. Kirste [*Makromol. Chem.*, **178**, 1793 (1977)] produced highly uniform styrene/α-methyl styrene block copolymers. The results of light-scattering studies on one such sample in toluene ($n_0 = 1.4976$) were as follows:

	$\dfrac{R(\theta)}{1 + \cos^2\theta} \times 10^4 \ (\text{cm}^{-1})$					
$c \times 10^3 \ (\text{g/cm}^3)$	15°	45°	75°	105°	135°	150°
0.20	1.91	1.47	1.01	0.725	0.577	0.537
0.40	3.55	2.78	1.95	1.41	1.13	1.05
0.60	4.95	3.94	2.80	2.05	1.65	1.54
0.80	6.16	4.96	3.59	2.66	2.15	2.01
1.00	7.19	5.87	4.31	3.22	2.62	2.46

The specific refractive index increment of the block copolymer in toluene was 0.1263 cm³/g.
   (a) Construct a Zimm plot and determine the weight-average molecular weight of the polymer.
   (b) Evaluate the dissymmetry coefficient Z_{45} and determine the root-mean-square end-to-end length of the polymer in toluene solution.
   (c) Evaluate the second virial coefficient.
   (d) Calculate the turbidity of the solution of $c = 0.2$ g/cm³ and the fraction of light scattered by this solution.

12. A certain monodisperse polymer has a molecular weight of 10^6 g/mol. Assuming that the polymer may be considered to be a sphere of diameter 70 Å in methyl ethyl ketone, calculate the sedimentation velocity under the influence of gravity at 20°C. The viscosity of methyl ethyl ketone at this temperature is 4.284 millipoises. The density of MEK = 0.805 g/cm³.

13. An ultracentrifuge in which the centrifugation cell is located at a distance of 5.0 cm from the center of rotation operates at a maximum speed of 65,000 rpm. (a) Calculate the ratio of the centrifugal force generated to the force due to gravity. (b) Calculate the sedimentation velocity of the spherical polymer in Problem 12 in methyl ethyl ketone at 20°C that would be observed with this ultracentrifuge.

14. A solution of polystyrene in toluene with $c = 5 \times 10^{-2}$ g/cm³ is found by a pycnometric measurement to have a density of 0.8855 g/cm³ at 25°C as compared with a density of 0.8788 found by the same technique with pure toluene. The sedimentation coefficient and diffusion coefficient at infinite dilution were found to be 17.9×10^{-13} s and 3.6×10^{-7} cm²/s, respectively. (a) Calculate the average molecular weight of the polystyrene. (b) If the ultracentrifuge is that described in Problem 13 and is operating at its maximum speed, calculate the sedimentation velocity of the polymer sample.

15. The diffusion coefficient of hemoglobin in aqueous solution at 20°C is 6.9×10^{-7} cm²/s and the partial specific volume of the polymer is 0.749 cm³/g. If the protein molecules are assumed to be spherical, calculate the molecular weight. (The viscosity of water at 20°C is 10.05 millipoises.)

16. G. Meyerhoff [*Z. Physik. Chem.*, **4**, 336 (1955)] used ultracentrifugation techniques to measure the average molecular weight of a sample of polystyrene in toluene at 20°C. The partial specific volume of the polymer was found to be 0.91 cm^3/g and the sedimentation and diffusion coefficients measured were as follows:

$c \times 10^3$ (g/cm^3)	4.3	2.2	1.1	0.6	0.4
$S \times 10^{13}$ (s)	4.31	5.69	6.72	7.60	7.88

$c \times 10^3$ (g/cm^3)	7.5	5.0	2.5
$D \times 10^7$ (cm^2/s)	2.36	2.17	2.07

From these data determine: (a) the average molecular weight of the polymer, and (b) the second virial coefficient of the polymer in toluene solution.

17. The following results have been reported for the sedimentation coefficients of a polystyrene sample in cyclohexane at 35°C [W. J. Closs, B. R. Jennings, and H. G. Jerrard, *Eur. Polymer J.*, **4**, 639 (1968)]:

$c \times 10^3$ (g/cm^3)	2.0	3.0	4.0	5.0	6.0	7.0
$S \times 10^{13}$ (s)	14.8	13.9	13.1	12.4	11.8	11.2

The density of cyclohexane at 35°C is 0.765 g/cm^3 and the partial specific volume of polystyrene is 0.93 cm^3/g. The dependence of the diffusion constant of polystyrene in cyclohexane at 35°C on weight-average molecular weight has been shown to be $D_0 = 1.3 \times 10^{-4} M_w^{-0.497}$ cm^2/s. [T. A. King, A. Knox, W. I. Lee, and J. D. G. McAdam, *Polymer*, **14**, 151 (1973)]. From these data determine the average molecular weight of the polystyrene sample.

SUGGESTIONS FOR FURTHER READING

BILLINGHAM, N. C., *Molar Mass Measurements in Polymer Science*. New York: Wiley, 1977.

CHIEN JEN YUAN, *Determination of Molecular Weights of High Polymers*. Jerusalem: Israel Program for Scientific Translations, 1963.

FLORY, P. J., *Principles of Polymer Chemistry*, Chap. 7. Ithaca, N.Y.: Cornell University Press, 1953.

HUGLIN, M. B. (ed.), *Light Scattering from Polymer Solutions*. New York: Academic Press 1972.

McINTYRE D., AND F. GORNICK (eds.), *Light Scattering from Dilute Polymer Solutions*. New York: Gordon and Breach, 1964.

RAFIHOV, S. R., S. A. PAVLOVA, AND I. I. TVERDOKHLEBOVA, *Determination of Molecular Weights and Polydispersity of High Polymers*. Jerusalem: Israel Program for Scientific Translations, 1964.

STACEY, K. A., *Light Scattering in Physical Chemistry*, New York: Academic Press, 1956.

SVEDBERG, T., AND K. O. PEDERSEN, *The Ultracentrifuge*. London: Oxford University Press, 1940.

TANFORD, C., *Physical Chemistry of Macromolecules*, Chaps. 4–6. New York: Wiley, 1961.

TOMPA, H., *Polymer Solutions*, Chaps. 6 and 10. New York: Academic Press, 1956.

VOLLMERT, B., *Grundriss der Macromolekularen Chemie*, Chap. 3. Springer-Verlag, Berlin, 1962.

WILLIAMS, J. W., *Ultracentrifugation of Macromolecules*. New York: Academic Press, 1972.

15

Secondary Methods for Molecular-Weight Determination

The absolute methods for the determination of molecular weights that were discussed in Chapter 14 are well established both theoretically and experimentally. Unfortunately, the absolute measurements are difficult to carry out, are time-consuming, and often require expensive apparatus. For these reasons, most molecular-weight determinations are routinely carried out by the much faster methods of solution viscosity and gel permeation chromatography. However, these techniques are not absolute methods, and their use requires a prior determination of empirical relationships that relate the molecular weight to the viscosity of a polymer solution or to the retention volume of a polymer solution being eluted from a gel permeation column. Once the calibration has been accomplished, the secondary methods provide the polymer chemist with a fast, simple, and accurate way to obtain molecular weights.

SOLUTION VISCOSITY

Solution Viscosity and Molecular Size

In the early days of polymer chemistry, Staudinger[1] observed that even a low concentration of a dissolved polymer markedly increased the viscosity of a solution relative to that of the pure solvent. This increase in viscosity is caused principally by

[1] H. Staudinger, *Kolloid Z.*, **51**, 71 (1930).

the unusual size and shape of the dissolved polymer and by the nature of solutions of high polymers.

Most polymer molecules are best described not as long thin rods but as random statistical coils. In dilute solution these coils are free from entanglement with other coils but are completely solvated, which means they have taken up as much solvent as they can hold. Thus, the smallest entities of solute in a polymer solution are not the actual polymer molecules but rather the large, irregularly shaped "particles" made up of polymer coils and large numbers of absorbed solvent molecules. As far as motion through the solution is concerned, the polymer coil and absorbed solvent form a single entity which is actually much heavier than the polymer molecule itself. In many respects, each polymer "particle" resembles a completely saturated sponge. On the basis of the size of these solute "particles," polymer solutions are correctly classified as *colloidal dispersions*. Each colloidal particle is a solvent-filled polymer coil; hence, they are sometimes called *molecular colloids*.

For a long time it has been known that the large particles in colloidal solutions or dispersions tend to impede the flow of adjacent layers of liquid when the liquid is subjected to a shearing force. In other words, the viscosity of the liquid is increased relative to that of the pure solvent by the presence of a colloidal or polymeric solute. As long ago as 1906, it was shown by Einstein[1] that, in the case of spherical colloid particles, the relative viscosity is given by the expression

$$\eta_r = \frac{\eta}{\eta_0} = 1 + 2.5\phi_2 \tag{1}$$

where η_r is the relative viscosity, η the viscosity of the solution, η_0 the viscosity of the pure solvent, and ϕ_2 the volume fraction of the colloidal particle. According to (1), as the overall size or volume of the colloidal particle (i.e., polymer molecule plus imbibed solvent) increases, so do the volume fraction ϕ_2 and the relative viscosity. Because the molecular weight of a polymer molecule also increases with size, it is possible to relate the increase in solution viscosity to the molecular weight.

Measurement of Viscosity

Principles

According to Newton's law of viscous flow, the frictional force, F, that resists the flow of any two adjacent layers of liquid is given by

$$F = \eta A \frac{dv}{dx} \tag{2}$$

where A is the area of contact of the layers, dv/dx the velocity gradient between them, and the proportionality constant, η, is called the coefficient of viscosity or, simply, the viscosity. The unit of viscosity is the poise, i.e., 1 poise $= 1$ g-cm^{-1}-s^{-1}.

[1] A. Einstein, *Ann. Physik*, **19**, 289 (1906).

When an external driving force is applied to overcome the frictional resistance and cause the liquid to flow uniformly through a tube, the rate of flow is given by Poiseuille's law,[1]

$$\frac{dV}{dt} = \frac{\pi R^4 \Delta P}{8\eta L} \tag{3}$$

In (3), dV/dt is the volume of liquid that flows through the tube per unit time; R and L are the radius and length of the tube, respectively; and ΔP is the difference in external pressure between the ends of the tube. In practice, measurements are usually carried out in viscometer tubes in which the capillary is in a vertical position and the driving force is simply the weight of the liquid itself. Therefore, the pressure difference, ΔP, which is the driving force per unit area, is given by

$$\Delta P = h\rho g \tag{4}$$

where h is the average height of the liquid during measurement, ρ the density, and g the acceleration due to gravity. Substitution of (4) into (3), along with the assumption of a constant flow rate, yields equation (5) for the viscosity.

$$\eta = \frac{\pi R^4 h g \rho t}{8 L V} \tag{5}$$

The applicability of (5) demands that the flow be "Newtonian" or "viscous." This will be true provided that a dimensionless quantity, called the Reynolds number, is less than 1000. In terms of the variables of (5) this condition is given by

$$\frac{2V\rho}{\pi R \eta t} < 1000 \tag{6}$$

and is readily satisfied for the apparatus and liquids usually used for measurements of the viscosities of polymer solutions. However, in addition to the requirements of viscous flow, the derivation of (5) relies on the following assumptions:

1. All of the potential energy of the driving force is expended in overcoming the frictional resistance. This is not strictly true, since some energy must be expended to accelerate the liquid in the tube. When this "kinetic-energy correction" is made, equation (5) becomes

$$\eta = \frac{\pi R^4 h g \rho t}{8 L V} - \frac{\rho V}{8 \pi L t} \tag{7}$$

For measurements of the *absolute* viscosity, this correction term can amount to 10 to 15% but, for measurements of the *relative* viscosities of interest in polymer chemistry (cf. equation 1), the error introduced by the use of (5) rather than (7) is usually less than 2%.

---

[1] W. J. Moore, *Physical Chemistry*, 4th ed. (Englewood Cliffs, N.J.: Prentice-Hall, 1972), pp. 153ff.

2. The second assumption is that the velocity of the liquid at the walls of the capillary is zero (i.e., there is no "slippage" of the liquid along the walls). This assumption is usually valid for liquids that "wet" the capillary walls. In any event, viscosities of polymer solutions are measured relative to the pure solvent and, unless the presence of small concentrations of polymer markedly affect the surface tension of the solvent, such capillary effects may be neglected.

Although expressions (5) and (7) are usually used, in practice it is not necessary to make precise measurements of the viscometer tube dimensions. Thus, if (7) is written in terms of the *kinematic viscosity*, η/ρ, we have

$$\frac{\eta}{\rho} = \alpha t - \frac{\beta}{t} \qquad (8)$$

where the constants, $\alpha = \pi R^4 h g/8LV$ and $\beta = V/8\pi L$, depend only on the geometry of the viscometer tube. Measurement of the times required for the fixed volume V of two liquids of known viscosity and density to flow through *the same* tube is sufficient to define the viscometer constants, α and β. The viscosity of a liquid depends markedly on the temperature. Hence, the calibration measurements and the measurement of the viscosities of the polymer solutions must be made at the same carefully controlled temperature ($\pm 0.1°C$).

The viscosities and densities of some typical solvents used in polymer chemistry are given in Table 15.1. It should be noted from (8) that the viscosities obtained from viscometer tube flow times are *kinematic* viscosities. The conversion of relative kinematic viscosity to relative viscosity can usually be made by assuming that the densities of solution and solvent are equal.

Experimental apparatus

A number of methods exist for the determination of the viscosity of a liquid. The most useful method from the viewpoint of simplicity, accuracy, and cost is based on a measurement of the flow rate of the liquid through a capillary tube. In practice, the capillary tube forms part of the "viscometer."

TABLE 15.1 Densities and Viscosities of Common Solvents

Solvent	ρ (g/cm^3) at:		100η (poises) at:	
	20°C	30°C	20°C	30°C
Benzene	0.8737	0.8684	0.652	0.564
Toluene	0.8669	0.8577	0.590	0.526
p-Xylene	0.8610	0.8523	0.644	0.568
Cyclohexane	0.7786	0.7693	0.935	0.820
n-Hexane	0.6594	0.6505	0.326	0.293
Ethanol	0.7893	0.7808	1.200	1.003
Acetone	0.7908	0.7793	0.326	0.295
Methyl ethyl ketone	0.8047	0.7945	0.400	0.365
Carbon tetrachloride	1.5940	1.5748	0.969	0.843
Chloroform	1.4892	1.4706	0.568	0.514

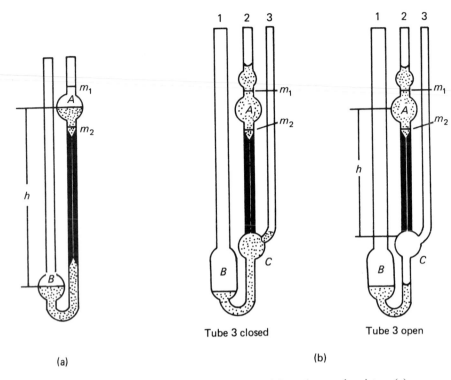

Fig. 15.1   Viscometers commonly used in polymer chemistry. (a)
Ostwald viscometer. (b) Ubbelohde viscometer. [Reproduced from
L. Ubbelohde, *Ind. Eng. Chem., Anal. Ed.*, **9**, 85 (1937); with permission
of the American Chemical Society, Washington, D.C.]

The most commonly used viscometers are of the Ostwald and Ubbelohde
types[1] shown in Figure 15.1. In the Ostwald viscometer, a given volume of liquid is
introduced into B and is drawn up by suction into A until the liquid level is above
the mark m_1. The suction is released and the time required for the liquid level to fall
from m_1 to m_2 is measured. The average driving force during the flow of this volume
of liquid through the capillary tube is proportional to the average difference in
heights of the liquids in tubes B and A (i.e., proportional to h, as shown in Figure
15.1a). In order that this driving force is the same in all cases, it is clearly essential
that the same amount of liquid should always be introduced into tube B.

This requirement that the same amount of liquid should always be used does not
apply in the case of the Ubbelohde viscometer shown in Figure 15.1b. Here, the
liquid is introduced into B. With tube 3 closed, the liquid is drawn up by suction into
A so that the liquid level is above mark m_1. The suction is released and, before the
liquid level in tube 2 reaches the mark m_1, tube 3 is opened to the air. Bulb C fills
with air and the liquid flowing out of bulb A must do so along the walls of bulb C.
In this case, the driving force for flow through the capillary is independent of the
level of the liquid in B, since the average height, h, is always the same.

[1] L. Ubbelohde, *Ind. Eng. Chem., Anal. Ed.*, **9**, 85 (1937).

TABLE 15.2 Viscosity of Solutions of Poly(methyl methacrylate) in Chloroform at 20°C

FRACTION	CONCENTRATION (g/cm³) × 10²	FLOW TIME* (s)	η_r (Eq. 1)	$\dfrac{\eta_{sp}}{c}$ (Eq. 2)
1	0.0000	170.1	1.000	—
	0.03535	178.1	1.047	133
	0.05152	182.0	1.070	136
	0.06484	185.2	1.089	137
	0.100	194.3	1.142	142
	0.200	219.8	1.292	146
	0.400	275.6	1.620	155
2	0.02242	180.8	1.063	281
	0.03520	187.3	1.101	287
	0.04620	192.7	1.133	288
	0.08682	214.2	1.259	298
	0.18806	273.0	1.605	322

* Ostwald viscometer, $R = 1.5 \times 10^{-2}$ cm, $L = 11$ cm.

Source: G. V. Schulz and F. Blaschke, *J. prakt. Chem.*, **158**, 130 (1941).

Typical flow times and relative viscosities for chloroform solutions of two fractions of poly(methyl methacrylate), as determined in an Ostwald viscometer, are shown in Table 15.2. The kinetic-energy correction was not made in this determination. The following sections illustrate how such data may be related to the average molecular weight of the polymer.

Definition of Solution–Viscosity Terms

The *relative viscosity*, which has already been defined in (1), may be written very simply in terms of the viscometer flow times if the kinetic energy correction is neglected:

$$\eta_r = \frac{\eta}{\eta_0} = \frac{t}{t_0} \tag{9}$$

where t and t_0 are the flow time of solution and solvent, respectively. Obviously, η and η_0 (i.e., t and t_0) must be measured under the same conditions. The relative viscosity is always greater than unity because the presence of the polymeric solute always increases the viscosity. It is appropriate, then, to define the *specific viscosity*, η_{sp}, as the fractional increase in viscosity caused by the presence of the dissolved polymer in the solvent, as shown in equation (10).

$$\eta_{sp} = \frac{\eta - \eta_0}{\eta_0} = \eta_r - 1 \tag{10}$$

The specific viscosity and the relative viscosity clearly depend on the concentration of the polymer in solution; they increase in magnitude with increasing concentration. This may be seen in Table 15.2 for solutions of poly(methyl methacrylate) in chloroform. The quantity η_{sp}/c, where c is the concentration of polymer in g/cm^3, is sometimes called the *reduced viscosity* or *reduced specific viscosity* and is a measure of the specific capacity of the polymer to increase the relative viscosity. Finally, the *intrinsic viscosity*, $[\eta]$, is defined as the limit of the reduced viscosity as the concentration approaches zero, and is given by

$$[\eta] = \lim_{c \to 0} \left(\frac{\eta_{sp}}{c} \right) \tag{11}$$

Note that none of the terms defined here actually has the dimensions of viscosity. The relative viscosity and the specific viscosity are dimensionless, but the reduced viscosity and the intrinsic viscosity have the dimensions of a specific volume (i.e., cm^3/g).

A linear dependence of the reduced viscosity on polymer concentration is usually found when $\eta_r < 2$. This linear dependence is described well by the expression

$$\frac{\eta_{sp}}{c} = [\eta] + k'[\eta]^2 c \tag{12}$$

where k' is a constant, usually in the range 0.35 to 0.40; it is sometimes called the Huggins constant.[1] In view of equations (11) and (12), it is evident that the intrinsic viscosity $[\eta]$ can be found by an extrapolation of the experimental values of the reduced viscosity (η_{sp}/c) to zero concentration.

Examples of these viscosity terms are illustrated by the typical data for two fractions of poly(methyl methacrylate) in chloroform that are shown in Table 15.2. In addition, plots of η_{sp}/c as a function of concentration are shown in Figure 15.2. The intrinsic viscosities of the polymer samples are given by the intercepts of these plots, in accordance with (11).

Intrinsic Viscosity and Molecular Weight

Suppose that the intrinsic viscosities $[\eta]_i$ (defined by equation 11) are determined for different molecular-weight fractions of a given polymer, each fraction having a very narrow range of molecular weights. Assume that the molecular weights, M_i, of the various fractions are known from the use of an absolute method such as ultracentrifugation or light scattering. It has been found that a straight line is obtained if the logarithms of the intrinsic viscosities are plotted versus the logarithms of the molecular weights of the different fractions. Such plots for fractions of

[1] M. L. Huggins, *J. Am. Chem. Soc.*, **64**, 2716 (1942).

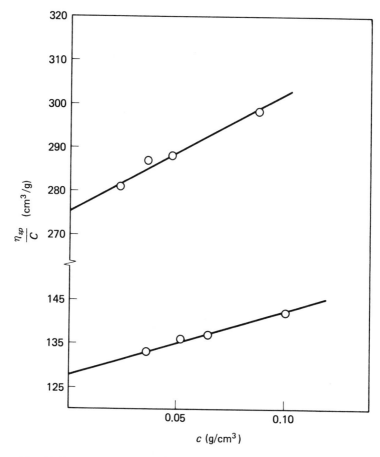

Fig. 15.2 Reduced viscosities of two samples of poly(methyl methacrylate) in chloroform at 20°C as a function of concentration. [From G. V. Schulz and F. Blaschke, *J. prakt. Chem.*, **158**, 130 (1941).]

polyisobutylene in cyclohexane[1] at 30°C and for polystyrene in cyclohexane[2] at 35°C and in methyl ethyl ketone[3] at 25°C are shown in Figure 15.3. It may be seen from this figure that a linear relationship does exist between $\log [\eta]_i$ and $\log M_i$ over the useful range of $10^4 < M < 10^6$. The slopes of the lines in Figure 15.3 are: polyisobutylene in cyclohexane, 0.69; polystyrene in cyclohexane, 0.50; polystyrene in methyl ethyl ketone, 0.58. Such nonintegral values of the slopes are typical of $\log [\eta]$ versus $\log [M]$ plots, with the values for all polymer–solvent combinations falling in the range 0.5 to 1.0. It is clear from the data in Figure 15.3 that, for a given polymer, the slope depends on the *solvent*. It is also found that, for a given polymer and solvent, the slope depends on the *temperature*.

[1] W. R. Krigbaum and P. J. Flory, *J. Polymer Sci.*, **11**, 37 (1953).
[2] P. Outer, C. I. Carr, and B. H. Zimm, *J. Chem. Phys.*, **18**, 830 (1950).
[3] H. J. Cantow, *Makromol. Chem.*, **30**, 169 (1959).

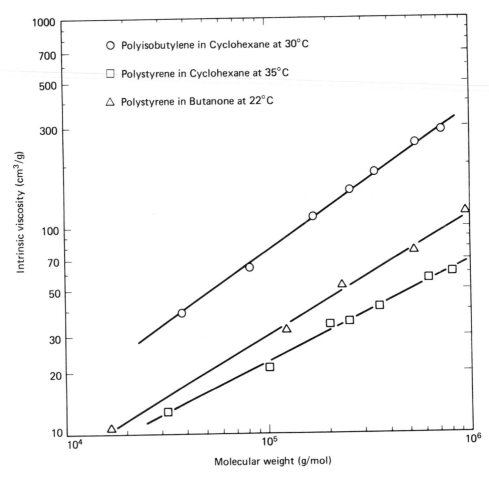

Fig. 15.3 Dependence of intrinsic viscosity of polymer solutions on molecular weight.

Since a plot of log $[\eta]_i$ versus log $[M]_i$ is linear for narrow molecular-weight fractions of a given polymer, we may write

$$\log [\eta]_i = \log K + a \log M_i \tag{13}$$

or

$$[\eta]_i = K M_i^a \tag{14}$$

where K and a are constants that are easily determined from calibration plots such as in Figure 15.3. The relationship given in (14) is usually known as the Mark–Houwink[1,2] equation.

[1] H. Mark, Z. Elektrochem., **40**, 499 (1934).
[2] R. Houwink, J. prakt. Chem., **157**, 15 (1940).

It must be kept in mind that the data in Figure 15.3 and, hence, equations (13) and (14), refer to *fractionated* samples of a given polymer, in which molecular-weight ranges of the fractions are very small. Carefully fractionated samples must be used in order to obtain numerical values for K and a for a given polymer–solvent pair at a given temperature. However, in practice, solution viscosity measurements are used to obtain, quickly and easily, a measure of the molecular weight of an unfractionated or crudely fractionated polymer. An *average* molecular weight of the sample is obtained by this procedure and it is necessary to inquire about the type of average that is involved.

Polymer molecules of a given molecular weight, M_i, will contribute an amount $(\eta_{sp})_i$ to the total observed specific viscosity (cf. equation 10). The observed specific viscosity of the solution of unfractionated polymer is then obtained as a sum over all the molecular weights (i.e., all the polymer molecules) present in the solution. Thus, equation (15) applies.

$$\eta_{sp} = \sum_i (\eta_{sp})_i \tag{15}$$

According to the Huggins equation, (12), for those molecules in each narrow molecular-weight range, i, whose concentration is c_i, we may write

$$(\eta_{sp})_i = [\eta]_i c_i + k_i'[\eta]_i^2 c_i^2 \tag{16}$$

In the limit of infinite dilution, we can neglect the second term on the right-hand side of (16). Combining (14), (15), and (16) results in the expressions shown in (17) and (18).

$$(\eta_{sp})_i = [\eta]_i c_i = K M_i^a c_i \tag{17}$$

$$\eta_{sp} = K \sum_i M_i^a c_i \tag{18}$$

Equation (18) describes the specific viscosity of an infinitely dilute solution of an unfractionated sample.

Because the extrapolation to infinite dilution has, in effect, already been carried out by the neglect of the second term in (16), the observed intrinsic viscosity is obtained simply by dividing through (18) by c, the total concentration of polymer in the solution. Thus,

$$[\eta] = \lim_{c \to 0} \frac{\eta_{sp}}{c} = K \sum_i M_i^a \left(\frac{c_i}{c}\right) \tag{19}$$

The concentration ratio $c_i/c = c_i/\sum_i c_i$ is equal to the weight ratio $w_i/w = w_i/\sum_i w_i$. Hence, finally we may write expression (20) for the intrinsic viscosity of an unfractionated polymer,

$$[\eta] = K \sum_i w_i M_i^a \bigg/ \sum_i w_i = K \sum_i W_i M_i^a \tag{20}$$

where the term $W_i = w_i/\sum_i w_i$ is the *weight fraction* of polymer of molecular weight M_i. Now, if we define the *viscosity-average* molecular weight by the relationship

$$\overline{M}_v = \left(\sum_i W_i M_i^a \right)^{1/a} = \left(\sum_i w_i M_i^a \middle/ \sum_i w_i \right)^{1/a} \tag{21}$$

then, for an unfractionated polymer, the intrinsic viscosity/molecular weight relationship (cf. equation 14) may be written

$$[\eta] = K(\overline{M}_v)^a \tag{22}$$

Therefore, use of the solution viscosity technique yields the *viscosity-average* molecular weight defined by (21). To see how this particular molecular weight relates to the number-average and weight-average molecular weights, consider the general expression for an average molecular weight, defined in terms of the index β, and shown in (23).

$$\overline{M}_\beta = \sum_i w_i M_i^\beta \middle/ \sum_i w_i M_i^{\beta-1} \tag{23}$$

It is easy to show from (23) that the number-average molecular weight corresponds to $\beta = 0$, and the weight-average molecular weight corresponds to $\beta = 1$. On the other hand, in terms of the Mark–Houwink equation, (21), the weight-average molecular weight corresponds to $a = 1$, and the number-average molecular weight to $a = -1$. Thus, in terms of the definitions (21) and (23), we have for the number-average and weight-average molecular weights,

$$\overline{M}_n = \overline{M}_v(a = -1) = \overline{M}_\beta(\beta = 0) \tag{24}$$

$$\overline{M}_w = \overline{M}_v(a = 1) = \overline{M}_\beta(\beta = 1) \tag{25}$$

The range of the empirical constant, a, is found experimentally to lie in the range 0.5 to 1.0. Because this constant is never as low as -1, the viscosity-average molecular weight *cannot* correspond to the number-average molecular weight. The relationship between β and a depends on the molecular weight distribution in a complicated manner,[1] but it has been shown to be such that $0.5 < a < 1.0$ corresponds to $0.75 < \beta < 1.0$.

In view of the discussion above, we may conclude that *the viscosity-average molecular weight lies between $\overline{M}_n$ and $\overline{M}_w$ but closer to $\overline{M}_w$*. Furthermore, in the special case of $a = 1$, $\overline{M}_v = \overline{M}_w$, but because $a \neq 1$, $\overline{M}_v$ can never become identical to $\overline{M}_n$.

In principle, any of the absolute methods of molecular-weight determination may be used to establish the calibration plot and determine K and a. However, for

[1] G. Meyerhoff, *Fortschr. Hochpolym.-Forsch.*, **3**, 59 (1961).

TABLE 15.3 Intrinsic Viscosity/Molecular-Weight Constants

$$[\eta] = KM^a$$

POLYMER	SOLVENT	TEMPERATURE (°C)	MOLECULAR-WEIGHT RANGE	$K \times 10^2$	a
Amylose	Dimethylsulfoxide	25	$1.5 \times 10^3 - 1.2 \times 10^6$	0.850	0.76
Gelatin	Water	35	$3 \times 10^4 - 2.1 \times 10^5$	0.166	0.885
Natural rubber	Toluene	25	$4 \times 10^4 - 1.5 \times 10^6$	5.0	0.67
Polyacrylontrile	Dimethylformamide	25	$3 \times 10^4 - 3.7 \times 10^5$	2.33	0.75
Poly(p-bromostyrene)	Benzene	20	$3 \times 10^4 - 3 \times 10^5$	9.4	0.53
Polybutadiene	Cyclohexane	20	$2.3 \times 10^5 - 1.3 \times 10^6$	3.6	0.70
Poly(dimethylsiloxane)	Toluene	25	$3.6 \times 10^4 - 1.1 \times 10^6$	0.738	0.72
Polyisobutylene	Cyclohexane	30	$5 \times 10^2 - 3.2 \times 10^6$	2.88	0.69
	Diisobutylene	20	$5 \times 10^2 - 3.2 \times 10^6$	3.63	0.64
	Toluene	25	$1.4 \times 10^5 - 3.4 \times 10^5$	8.70	0.56
Poly(methyl methacrylate)	Acetone	25	$8 \times 10^4 - 1.4 \times 10^6$	0.75	0.70
	Chloroform	25	$8 \times 10^4 - 1.4 \times 10^6$	0.48	0.80
	Methyl ethyl ketone	25	$8 \times 10^4 - 1.4 \times 10^6$	0.68	0.72
Polypropylene	Benzene	25	$1 \times 10^3 - 7 \times 10^4$	9.64	0.73
	Cyclohexane	25	$1 \times 10^3 - 7 \times 10^4$	7.93	0.81
Polystyrene	Benzene	20	$1.2 \times 10^3 - 1.4 \times 10^5$	1.23	0.72
	Methyl ethyl ketone	20–40	$8 \times 10^3 - 4 \times 10^6$	3.82	0.58
	Toluene	20–30	$2 \times 10^4 - 2 \times 10^6$	1.05	0.72
Poly(vinyl acetate)	Acetone	30	$2.7 \times 10^4 - 1.3 \times 10^6$	1.02	0.72
	Methanol	30	$2.7 \times 10^4 - 1.3 \times 10^6$	3.14	0.60
Poly(vinyl alcohol)	Water	25	$8.5 \times 10^3 - 1.7 \times 10^5$	30.0	0.50
Poly(vinyl bromide)	Cyclohexanone	20	$1.9 \times 10^4 - 1.0 \times 10^5$	3.28	0.55

the calibration to be strictly valid, the molecular-weight ranges of the various fractions must be very small.

In practice, incompletely fractionated polymers must normally be studied, so the molecular-weight ranges of the various fractions may not be as small as one would wish. Therefore, to obtain a clear relationship between $[\eta]$ and M, absolute methods in which a and β correspond as closely as possible should be used. The number-average molecular weight corresponds to $\beta = 0$, while the weight-average molecular weight corresponds to $\beta = 1$. Because the *physically real* values of the empirical constant a correspond to $0.75 < \beta < 1$, it is preferable to use an absolute method that yields the weight-average molecular weight for calibration. The reader should bear in mind that even this is usually only an approximation, because only in the rare case of $a = \beta = 1$ is $\overline{M}_v = \overline{M}_w$. However, the approximation is better when absolute methods that yield $\overline{M}_w$ rather than those that yield $\overline{M}_n$ are used. For this reason most determinations of K and a are carried out with the use of light-scattering and the various ultracentrifugation techniques, rather than with osmotic pressure or other colligative property measurements.

Values of K and a for a number of polymer–solvent systems are shown in Table 15.3. For more extensive tables, the reader should consult the literature.[1]

[1] Meyerhoff, op. cit.

Molecular Size from Intrinsic Viscosity

It has been mentioned previously that the intrinsic viscosity has the dimensions of a specific volume—namely, volume per unit mass. This agrees with a viscosity relationship derived by Einstein in 1906,[1] for the case of a solution of rigid spherical particles. This relationship, (1), may be expressed in terms of intrinsic viscosity as

$$[\eta] = 2.5 \frac{N_0 V_e}{M} = 2.5 \frac{N_0(\frac{4}{3}\pi R_e^3)}{M} \tag{26}$$

where N_0 is Avogadro's number, M is the molecular weight, and V_e and R_e are the volume and the radius, respectively, of the effective hydrodynamic sphere. However, polymer molecules usually are not spherical. Later theories have treated the polymer in solution more realistically as a random coil and have led to the expression

$$[\eta] = \Phi \frac{(\overline{r^2})^{3/2}}{M} \tag{27}$$

in which $\overline{r^2}$ is the mean-square end-to-end distance of the random coil solute and Φ may be regarded approximately as a universal constant.

Values for Φ may be determined by a comparison of light-scattering measurements of $\overline{r^2}$ and M with intrinsic viscosity measurements made in the same solvent and at the same temperature. If r is expressed in centimeters, $[\eta]$ in cm^3/g, and M in atomic mass units, Φ is a dimensionless quantity, the value of which (within $\pm 20\%$) may be taken as

$$\Phi = 2.0 \times 10^{23} \tag{28}$$

Using the foregoing value of Φ, values of $(\overline{r^2})^{1/2}$, the root-mean-square end-to-end distance of the polymer in solution, may be determined from a knowledge of the intrinsic viscosity and the molecular weight of the polymer. Thus, a combination of (14) with (27) yields

$$(\overline{r^2})^{1/2} = \left(\frac{KM^{1+a}}{\Phi}\right)^{1/3} \tag{29}$$

where K and a are the constants contained in Table 15.3.

In general, the end-to-end distance of a given polymer chain depends on the polymer–solvent interaction. In "poor" solvents the polymer would be expected to coil up so as to maximize the polymer–polymer interactions. In a "good" solvent the polymer chain would tend to stretch out in order to maximize the polymer–solvent interactions. A solvent in which the free energies of solvent–solvent, solvent–polymer, and polymer–polymer interactions are all the same is called a *theta* solvent. In principle, any solvent will become a theta solvent for a given polymer at the theta temperature [see Chapter 16 equations (43), (51), and (52)].

[1] Einstein op. cit.

In the special case of a theta solvent, we may define $\overline{r_0^2}$ as the mean-square end-to-end distance of the *unperturbed* polymer, and for *any* solvent we may write

$$\overline{r^2} = \alpha^2 \overline{r_0^2} \tag{30}$$

where α is known as the expansion coefficient. The usefulness of the unperturbed mean-square end-to-end distance, $\overline{r_0^2}$, may be appreciated when the polymer is viewed as a freely jointed chain of bound monomer units or chain segments. An application[1] of classical random walk theory to such a freely jointed chain leads to the relationship

$$(\overline{r_0^2}) = L^2 \cdot x \tag{31}$$

where x is the number of chain segments, or monomer units, and L is the length of each segment. Because the molecular weight is given by

$$M = M_0 x \tag{32}$$

where M_0 is the molecular weight of a monomer unit, equation (31) yields the result

$$\frac{\overline{r_0^2}}{M} = \frac{L^2}{M_0} \tag{33}$$

Thus, according to (33), $\overline{r_0^2}/M$ is a constant that is independent of the molecular weight and the solvent. Substitution of (30) into (27) leads to

$$[\eta] = \Phi(\overline{r_0^2})^{3/2}\frac{\alpha^3}{M} = \Phi\left(\frac{\overline{r_0^2}}{M}\right)^{3/2}\alpha^3 M^{1/2} \tag{34}$$

In view of the constancy of $\overline{r_0^2}/M$, (34) may be written as

$$[\eta] = K\alpha^3 M^{1/2} \tag{35}$$

where the viscosity constant K is given by

$$K = \Phi\left(\frac{\overline{r_0^2}}{M}\right)^{3/2} \tag{36}$$

As expressed by (35), the intrinsic viscosity increases with molecular weight according to the product $\alpha^3 M^{1/2}$. Because it is found experimentally that $[\eta]$ usually increases with M by a power somewhat larger than $\frac{1}{2}$, α must depend, at least weakly, on the molecular weight. Thus, by comparison of (35) and (14), this dependence must be

$$\alpha^3 = M^{a-1/2} \tag{37}$$

---

[1] P. J. Flory, *Principles of Polymer Chemistry* (Ithaca, N.Y.: Cornell University Press, 1953), Chap. 14.

However, by definition, in a theta solvent $\alpha = 1$, and in such a case the intrinsic viscosity is given simply as the product $KM^{1/2}$. This square-root dependence of $[\eta]$ on M in theta solvents (i.e., $a = \frac{1}{2}$) has been verified experimentally.

Accuracy of the Determination of Intrinsic Viscosity

Figures 15.4 and 15.5 indicate the accuracy to be expected in the determination of the intrinsic viscosity of a given polymer–solvent pair. These figures show plots of η_{sp}/c versus c as determined independently by six laboratories in Europe and four

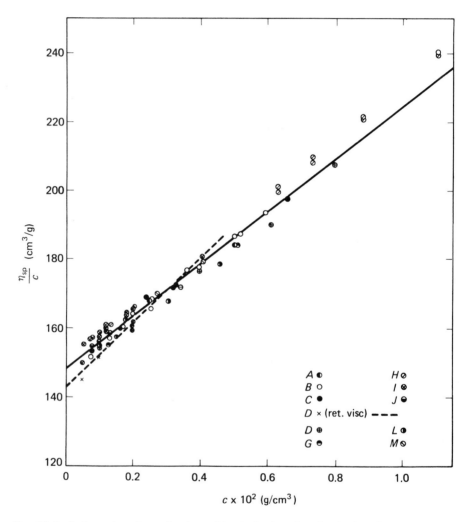

Fig. 15.4 Independent determinations of intrinsic viscosity of a single polystyrene fraction in toluene at 25°C by several laboratories. [Reproduced from H. P. Frank and H. F. Mark, *J. Polymer Sci.*, **17**, 1 (1955); with permission of John Wiley & Sons, Inc., New York.] The different symbols indicate results from different laboratories.

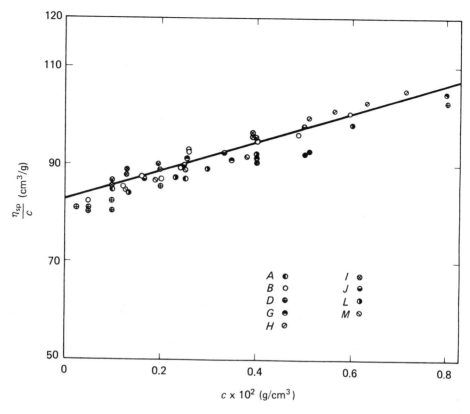

Fig. 15.5 Independent determinations of intrinsic viscosity of a single polystyrene fraction in methyl ethyl ketone at 25°C by several laboratories. [Reproduced from H. P. Frank and H. F. Mark, *J. Polymer Sci.*, **17**, 1 (1955); with permission of John Wiley & Sons, Inc., New York.] The different symbols indicate results from different laboratories.

laboratories in the United States and Canada on samples of the same polystyrene fraction.[1] Figure 15.4 refers to solutions of the polymer in toluene and Figure 15.5 to solutions of the polymer in methyl ethyl ketone.

The average intrinsic viscosities obtained from these plots and the viscosity-average molecular weights calculated from the data in Table 15.3 are as follows:

(a) Toluene: $[\eta] = 148.3 \pm 2.3$ cm^3/g
$$\overline{M}_v = 5.60 \pm 0.08 \times 10^5 \text{ g/mol}$$

(b) Methyl ethyl ketone: $[\eta] = 82.9 \pm 2.8$ cm^3/g
$$\overline{M}_v = 5.66 \pm 0.19 \times 10^5 \text{ g/mol}$$

In the above values of $\overline{M}_v$, the error limits shown represent only those due to the indicated error in $[\eta]$.

[1] H. P. Frank and H. F. Mark, *J. Poly. Sci.*, **17**, 1 (1955).

It will be seen that the average molecular weights determined from these data agree within experimental error despite a difference of nearly a factor of 2 in the intrinsic viscosities of the polymer in the two different solvents. Moreover, in view of the discussion earlier in this chapter, it is possible to confirm by this specific example that $\overline{M}_v$ for a given polymer sample lies between $\overline{M}_n$ and $\overline{M}_w$, but much closer to $\overline{M}_w$. To confirm this the reader should compare the foregoing results for $\overline{M}_v$ with $\overline{M}_n$ and $\overline{M}_w$ determined for this same polystyrene sample by osmotic pressure and light scattering (cf. p. 343 and p. 361).

It is apparent that the solution viscosity method for the determination of polymer molecular weights can provide a satisfactory accuracy. Moreover, provided that the constants K and a are known for the solvent and temperature used, the method is by far the fastest, simplest, and most inexpensive way to determine a reliable average molecular weight of a high polymer.

GEL PERMEATION CHROMATOGRAPHY

The Underlying Principle

The gel permeation chromatography (GPC) method is essentially a process for the *separation* of macromolecules according to their *size*. The method has been used extensively in biochemistry to separate biological macromolecules from small-molecule contaminants (with the use of Sephadex columns). Its general application to synthetic polymer chemistry in the 1970s has revolutionized the procedures for polymer characterization and molecular-weight determination.

The principle that underlies the method is as follows. Imagine that a dilute solution is available that contains a broad molecular-weight distribution of polymer chains, oligomers, and perhaps even the monomer from which the other species were derived. Assume that this solution is allowed to flow through a column that is packed with finely divided solid particles, each particle being permeated by *pores* (tunnels) that have a diameter of say 1000 Å. As the dissolved solute passes each particle, the smaller molecules (those with dimensions smaller than 1000 Å) will enter the openings of the pores and will "explore" the pore space under the influence of the usual thermal motions. Thus, the smaller molecules will be "delayed" in their elution through the column. On the other hand, the larger polymer molecules (those with a random coil radius of larger than 1000 Å) will be unable to penetrate the pores and will be swept along with the solvent front to be eluted before the smaller molecules (Figure 15.6).

In practice, even if the substrate particles have only one uniform pore size, the process can separate molecules that form part of a continuous molecular-weight distribution. This is because those molecules that have "diameters" below 1000 Å will be differentially delayed according to their molecular size. The very smallest molecules can presumably penetrate far into the tunnel system, whereas the medium-sized molecules may merely "sample" the openings into the tunnels. Thus, in principle, a single-pore-size column would be expected to *separate* high-molecular-weight molecules (in this example, those with diameters over 1000 Å) from the rest, and *fractionate* the smaller molecules according to their size. Clearly, the smallest

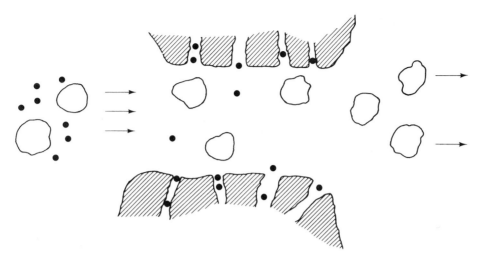

Fig. 15.6 Gel permeation chromatography (GPC). The mixture of different-sized polymer and oligmer molecules is eluted in a solvent through a column of porous particles. The smaller molcules (black circles) can enter the pores and be retarded, whereas the larger molecules are swept through relatively unhindered. If a *distribution* of different molecular sizes enters the column at the same time, the molecules will emerge from the column in sequence, distributed according to molecular size.

molecules (the monomers, dimers, trimers, etc.) will differ very little in their retention by a 1000-Å column, and hence they will be eluted together. On the other hand, those molecules that have a coil diameter close to that of the pore diameter will be fractionated the most effectively.

The effectiveness of the fractionation process for the whole molecular-weight distribution can be improved by the use of several different columns in series, each of which contains particles with a different pore size. For example, a series of columns that contain 1×10^6, 1×10^5, 1×10^3, and 500-Å pore-diameter particles should, in principle, be capable of fractionating a molecular-weight distribution that encompasses this entire range.

Thus, the gel permeation method is essentially a process for the *fractionation* of polymers according to their size and, therefore, according to their molecular weight. The molecular weight as such cannot be determined directly, but only after calibration of the system in terms of the elution time (or volume of solution eluted) expected for a particular polymer molecular-weight fraction with the use of that particular piece of equipment.

Equipment

It will be clear that the use of this method to measure molecular weights depends critically on being able to ensure that the elution time along the column is reproducible for two different specimens of the same polymer that have the same molecular weight and molecular-weight distribution. This requirement can only be met if (1) the flow rate of the eluting solvent through the column remains the same, and (2) the size of the tunnels within the stationary particles remains the same in different experiments. Both of these factors may change if the column temperature varies or if the solvent composition changes. The equipment used for gel permeation chromatography is usually designed to avoid such problems.

A schematic layout of a typical gel permeation chromatography unit is shown in Figure 15.7. Although crude *separations* of macromolecules from small molecules can be achieved by gravity elution through a vertical column (as in conventional chromatography), the elution rates under such conditions will be slow and non-reproducible. Hence, a mechanical pump is usually employed to force the sample and the elution solvent through the columns at pressures of up to 1000 to 4000 psi and at a rate of 2 to 3 ml/min. When a reciprocating pump is employed, the individual pressure pulses must be smoothed out by some form of constrictor coil. Injection of the sample into the line is usually accomplished from a graduated hypodermic syringe (typically 0.5 to 3 ml of a 0.05 to 0.1 % solution of the polymer) by means of a mechanical inlet device.

The columns are usually $\frac{3}{8}$-in. ($\sim$1-cm)-diameter stainless steel tubes with a combined length of from 3 ft to 10 ft or more, dependent on the type of packing material used. Two principal types of column packing materials have been employed—microporous glass beads and powdered, swelled, crosslinked polystyrene. The latter material is in more common use. With polystyrene particles, the pore size is determined by the amount of crosslinking, since the tunnels are formed by the solvent-swelled cavities that exist between the crosslinks (hence the name *gel* permeation). The particle size of the stationary phase has a profound effect on the *resolution* of the separation. Gel permeation is a diffusion-controlled phenomenon and, clearly, the speed and efficiency of the differential imbibition will be increased as the stationary particle size is decreased and the relative surface area is increased.

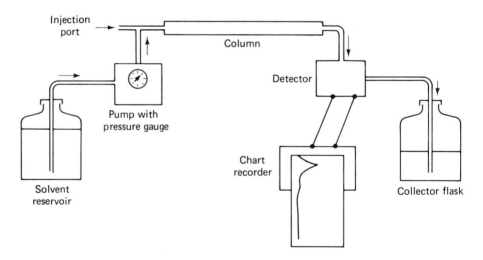

Fig. 15.7   Schematic diagram of a gel permeation chromatography apparatus.

Thus, small particles (≈ 10 μm diameter, called microstyragel particles) allow faster separations with smaller samples and shorter columns than do larger particles (37 to 74 μm).

After passage through the column system, the eluent passes through a detector. Two alternative detection methods are commonly employed—differential refractive index measurements or ultraviolet absorption. A differential refractometer measures the difference in refractive index between the eluted solution and the pure solvent. A plot of the refractive index difference as a function of time (on a chart recorder) can yield directly a plot of the molecular-weight distribution. Of course, the accuracy of this procedure will depend on the requirement that the refractive index difference between the polymer and solvent depends only on concentration and is independent of molecular weight.

When ultraviolet detection is used, the spectrometer is usually set to a particular wavelength (e.g., to the aromatic absorption region of a polymer that contains phenyl rings) and the absorbance is monitored as a function of elution time.

Problems in Gel Permeation Chromatography

Two main types of problem are often encountered with this technique—practical problems and problems of data interpretation. Perhaps the most serious practical problem is an overloading of the stationary phase by the polymer. It is fairly obvious that if all the pores in the column are occupied by polymer molecules, an effective separation will be impossible. Thus, only very small samples of very dilute polymer solutions are used—the smaller the better. A second practical problem may be encountered in conjunction with a polar stationary substrate such as porous glass. Under such circumstances, the polymer may become chemically *adsorbed* onto the surface of the substrate. In extreme cases it may be impossible to remove it by simple solvent elution. In other cases, the gel permeation elution pattern may be severely distorted by fractionation effects that depend on differential adsorption. Third, with swelled, crosslinked polystyrene used as a stationary phase, the degree of swelling of the polystyrene (and hence the pore size) will depend on the nature of the elution solvent. Fluids in which polystyrene is "insoluble" may close the pores completely. Thus, only certain specific elution solvents can be employed (e.g., tetrahydrofuran, benzene, xylene, chloroform, dimethylformamide, or fluorinated alcohols). This constitutes one of the main reasons for the continued use of porous glass as a substrate, since water, alcohols, and a wide variety of other solvents can be used with columns of this type.

The interpretation of gel permeation data can be complicated by two important factors. First, the ease with which a polymer molecule will penetrate a pore depends on whether it assumes a random coil or an extended rodlike conformation in the solvent being used. Thus, the chromatographic behavior of a polymer might be quite different in two different solvent systems. Second, one of the most valuable features of the gel permeation technique is the *speed* with which the average molecular weight and the molecular-weight distribution can be measured. Strictly speaking, an accurate assessment of the molecular weight can only be made if the average molecular weights of specific fractionated samples are first measured by another method and then used to calibrate the apparatus. This can be a time-consuming process that largely negates the speed advantage of the GPC technique. Of course, once the system is calibrated for a given polymer, other samples of the same polymer can be examined easily and rapidly. Often, in order to obtain preliminary indications of molecular weight, the equipment is calibrated with well-characterized fractions of a different but chemically or structurally related polymer and it is *assumed* that the calibration applies to the polymer of interest. Well-characterized "standard" samples of polystyrene and some other polymers are available commercially. A more satisfactory "universal" calibration technique has been developed which permits the calibration of gel permeation columns for a wide range of polymers using a single set of standard samples (i.e., polystyrene fractions having narrow ranges of molecular weights).

Universal Calibration in Gel Permeation Chromatography

The calibration of a gel permeation column for a given polymer–solvent system requires the establishment of a relationship between the volume of solution eluted (or, equivalently, the elution time for a given flow rate of solution) and the molecular weight of *monodisperse* fractions of the polymer. The main problem encountered is that monodisperse samples of most polymers are not generally available. However, such samples are available for a few specific polymers. A notable example is polystyrene for which samples having a molecular-weight range of 10^3 to 10^6 and a ratio of the weight-average molecular weight to number-average molecular weight of less than 1.15 can be obtained commercially.

If a set of monodisperse samples of a single polymer can be obtained, the remaining problem is to establish a relationship for a particular GPC column (or columns) between the volume of solution eluted and the molecular weight of some chemically different polymer. Clearly, in order to be able to do this, a calibration parameter is required which is independent of the chemical nature of the polymer, that is, a *universal calibration parameter*. Such a parameter has been found[1] experimentally to be the product of the intrinsic viscosity and the molecular weight (i.e., $[\eta]M$). Thus, as shown in Figure 15.8, with tetrahydrofuran used as a solvent, the logarithm of the product $[\eta]M$ plotted against the volume of solution eluted from the column provides a *single curve* from all the points determined for a wide variety of polymers. This is not possible on a plot of log M versus elution volume, which would be the simplest way to display the data.

The experimental finding that $[\eta]M$ is the same function of elution volume, V_e, for many different polymers suggests the possibility that a universal calibration procedure may be possible. First, the functional relationship between the molecular weight of the monodisperse standard samples and the elution volume of the solution must be determined for a given solvent and column under fixed conditions. Such a relationship is shown graphically in Figure 15.9 for a set of commercial polystyrene samples at 25°C using tetrahydrofuran as solvent and with spherical

[1] Z. Grubisic, P. Rempp, and H. Benoit, *Poly. Lett.*, **5**, 753 (1967).

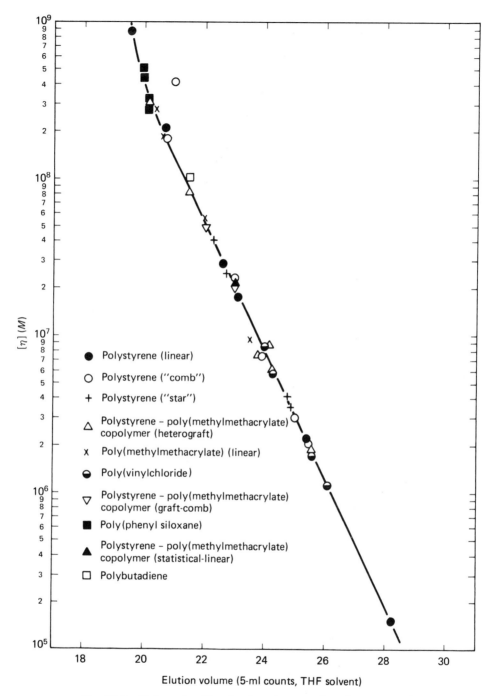

Fig. 15.8 Universal calibration in gel-permeation chromatography for a variety of polymers in tetrahydrofuran. [Reproduced from Z. Grubisic, P. Rempp and H. Benoit, *Polymer Lett.*, **5**, 753 (1967); with permission of John Wiley & Sons, Inc., New York.]

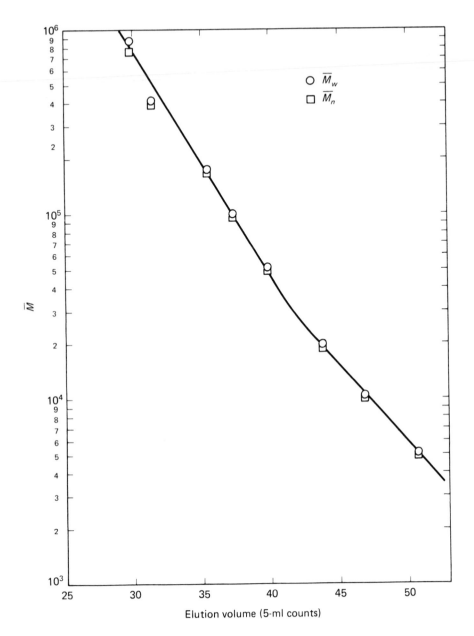

Fig. 15.9   Molecular weight of monodisperse polystyrene standards as a function of elution volume in tetrahydrofuran. [From M. Kolinsky and J. Janca, *J. Polymer Sci., Chem. Ed.*, **12**, 1181 (1974).]

porous silica beads used as the column packing.[1] If it is assumed (as shown[2] in Figure 15.8 to be true for a wide variety of polymers) that $\log [\eta]M$ is a constant for all polymers in a given solvent at a given temperature, *at the same elution volume*, it is possible to write (38), in which the

$$\log [\eta]_x M_x = \log [\eta]_s M_s \tag{38}$$

subscripts x and s indicate the unknown polymer and the standard polymer, respectively. If each intrinsic viscosity term in (38) is replaced by its Mark–Houwink expression,

$$[\eta]_j = K_j M_j^{a_j} \tag{14}$$

and the resulting expression is solved for $\log M_x$, (39) is obtained, which describes the elution volume calibration curve for M_x.

$$\log M_x = \left(\frac{1}{1 + a_x}\right) \log \frac{K_s}{K_x} + \frac{1 + a_s}{1 + a_x} \log M_s \tag{39}$$

The value of V_e that corresponds to a GPC peak in the unknown polymer is used to obtain a value of $\log M_s$ from Figure 15.9, and M_x is then calculated from (39). An alternative procedure is simply to choose values of V_e and construct a new calibration curve for the unknown polymer from a curve such as Figure 15.9 and equation (39). All this presupposes that the Mark–Houwink constants, $K_s, a_s, K_x,$ and a_x, are known. K_s and a_s are available in the literature for a variety of solvents (i.e., Table 15.3), and so, in many cases are values of K_x and a_x. On the other hand, if the desired Mark–Houwink constants are not available for the polymer under study or for the standard in the solvent to be used, they can easily be determined by measurement of the intrinsic viscosity. Unfractionated or crudely fractionated samples of known average molecular weight can be used for a determination of K_x and a_x from intrinsic viscosity measurements and a reasonable accuracy can be obtained.

As an example of the universal calibration technique using monodisperse polystyrene samples as standards, suppose that it is necessary to determine the molecular weight of polyisobutylene in toluene solution by gel permeation chromatography. First, it would be necessary to construct a calibration curve, such as the one shown in Figure 15.9 from the polystyrene standards in toluene. The Mark–Houwink constants for the two polymers in toluene are given in Table 15.3 and, when these are substituted in (39), expression (40) is obtained.

$$\log M_x = -0.589 + 1.10 \log M_s \tag{40}$$

Suppose now that the elution volume of a peak in the gel permeation chromatogram of polyisobutylene was such that a value of 10^5 was deduced for M_s from the calibration curve. According to (40), this peak corresponds to polyisobutylene with $\overline{M}_x = 8.15 \times 10^4$.

[1] M. Kolinsky and J. Janca, *J. Polymer Sci., Polymer Chem. Ed.*, **12**, 1181 (1974).
[2] Grubisic et al., op. cit.

Some disagreement exists in the literature about the type of molecular-weight average that is given by the positions of the peak maxima in a gel permeation chromatography measurement. Some investigators assume the peak maxima represent number average ($\overline{M}_n$) molecular weights but others consider that the method determines the weight-average molecular weight $\overline{M}_w$. It has been proposed[1] recently that the root-mean-square molecular weight, $\overline{M}_{rms} = (\overline{M}_w \cdot \overline{M}_n)^{1/2}$, correlates better than $\overline{M}_w$ with the peak maxima in a gel permeation chromatogram when the calibration technique just outlined is used. On the other hand, an elaboration[2] on the calibration technique that replaces (39) by

$$\log M_x = \left(\frac{1}{1 + a_x}\right) \log \frac{K_s}{K_x} \frac{f(\varepsilon_x)}{f(\varepsilon_s)} + \left(\frac{1 + a_s}{1 + a_x}\right) \log M_s \tag{41}$$

where

$$f(\varepsilon_j) = 1 - 2.63\varepsilon_j + 2.86\varepsilon_j^2 \tag{42}$$

$$\varepsilon_j = \tfrac{1}{3}(2a_j - 1) \tag{43}$$

leads to just as good a correlation of $\overline{M}_w$ with the peak maxima.

Although it was mentioned that the universal calibration parameter in gel permeation chromatography is the product $[\eta]M$, the more fundamental universal parameter is the partial molar volume of the polymer at infinite dilution. This may be seen as follows: according to the Einstein relationship for the relative viscosity of spherical colloidal solutes,[3]

$$\eta_r - 1 = \eta_{sp} = 2.5\phi_2 \tag{1}$$

where, as mentioned before, ϕ_2 is the volume fraction of solute (i.e., polymer) in the solution. In view of the definition of volume fraction, ϕ_2 is given by

$$\phi_2 = \frac{n_2 \overline{V}_2}{n_1 \overline{V}_2 + n_2 \overline{V}_2} \tag{44}$$

where n_1 and n_2 are the numbers of moles of solvent and dissolved polymer, respectively, and $\overline{V}_1$ and $\overline{V}_2$ are the partial molar volumes of solvent and dissolved polymer. Since the denominator of (44) is simply the volume of solution, (44) may be written as (45)

$$\phi_2 = \frac{(w_2/M_2)\overline{V}_2}{V_{soln}} = \frac{\overline{V}_2}{M_2} c \tag{45}$$

[1] Kolinsky and Janca, op. cit.
[2] H. Coll and D. K. Gilding, *J. Polymer Sci.* (*A2*), **8**, 89 (1970).
[3] Einstein, op. cit.

where c is the concentration of the solution in mass per unit volume and M_2 is the molecular weight of polymer. Substitution of (45) into (1) with rearrangement yields (46)

$$\frac{\eta_{sp}}{c} = 2.5 \frac{\overline{V}_2}{M_2} \tag{46}$$

which in the limit of infinite dilution becomes

$$\lim_{c \to 0} \frac{\eta_{sp}}{c} = [\eta] = 2.5 \frac{\overline{V}_2^{\circ}}{M_2} \tag{47}$$

$\overline{V}_2^{\circ}$ is the partial molar volume of polymer at infinite dilution,

$$\overline{V}_2^{\circ} = \lim_{n_2 \to 0} \left(\frac{\delta V_{soln}}{\delta n_2} \right)_{n_1, T, P} \tag{48}$$

and is equivalent to the hydrodynamic volume of the polymer. An expression similar to (47), but with a different proportionality constant, is obtained when the polymer solute is not spherical but rather appears as a random coil (i.e., equations 26 and 27).

A simple rearrangement of (47) shows that the product $[\eta]M$ is simply proportional to $\overline{V}_2^{\circ}$. Therefore, the statement that $[\eta]M$ is the same function of elution volume for all polymers in a given solvent is equivalent to the observation that the partial molar volumes (or hydrodynamic volumes) of all polymers in a given solvent at infinite dilution are given by the same function of the GPC elution volume. The experimental results discussed in this section demonstrate the validity of this statement for a wide variety of polymers in a single solvent. However, the statement does not hold true for cases in which a specific chemical interaction exists between polymer and the stationary gel. Moreover, the same $[\eta]M$ versus V_e curve would not normally apply to polymers of the rigid-rod type in the same way as it does to those of the random coil variety.[1]

STUDY QUESTIONS

1. Show from equation (15) that $\overline{M}_{\beta=0} = \overline{M}_n$ and $\overline{M}_{\beta=1} = \overline{M}_w$.

2. The following data have been obtained for the intrinsic viscosity of polystyrene fractions in dichloroethane at $22°C$ using light scattering as the absolute measurement of molecular weight.

$[\eta]$ (cm^3/g)	260	278	142	138	12.2	4.05
$\overline{M}_w \times 10^4$	178	157	56.2	48.0	1.55	0.308

Evaluate the constants in the intrinsic viscosity/molecular weight relationship.

[1] H. Coll and L. R. Prusinowski, *Polymer Lett.*, 5, 1153 (1967).

3. Suppose that you have an Ostwald viscometer with $R = 2.00 \times 10^{-2}$ cm, $L = 11.0$ cm, $V = 4.00$ cm^3, and $h = 16.0$ cm. What percentage error will be introduced by neglect of the kinetic-energy correction in equation (21) in determining:
   (a) The absolute viscosity of chloroform which has a flow time of 170 s at 20°C.
   (b) The relative viscosity of a solution of poly(methyl methacrylate) in chloroform whose flow time through the viscometer is 230 s.

4. With the viscometer described in Problem 3, a solution of poly(methyl methacrylate) in acetone, having a concentration of 0.0865 g/cm^3, is found to have a flow time of 232 s. Assuming a value of $k' = 0.40$ for the Huggins constant in equation (12), calculate the viscosity-average molecular weight of the polymer.

5. Using data given in this chapter, calculate the viscosity-average molecular weights of the poly(methyl methacrylate) fractions to which Table 15.2 and Figure 15.2 refer.

6. The following data on intrinsic viscosities and elution volumes from a gel permeation chromatograph at 25°C for standard polystyrene samples dissolved in tetrahydrofuran have been reported [M. Kolinsky and J. Janca, *J. Polymer Sci., Chem. Ed.*, **12**, 1181 (1974)]:

$\bar{M}_w \times 10^{-3}$ (g/mol)	867	411	173	98.2	51	19.85	10.3	5.0
$[\eta]$ (cm^3/g)	206.7	125.0	67.0	43.6	27.6	14.0	8.8	5.2
V_e (cm$^3 \times \frac{1}{5}$)	29.8	31.4	35.4	37.3	39.9	43.8	46.8	50.7

The Mark–Houwink constants for poly(vinyl bromide) in tetrahydrofuran at 25°C may be taken as $K = 1.59 \times 10^{-2}$ cm^3/g and $a = 0.64$ [A. Ciferri, M. Kryezewski, and G. Weil, *J. Polymer Sci.*, **27**, 167 (1958)]. Construct an appropriate GPC calibration curve (i.e., log $\bar{M}_w$ as a function of V_e) for poly(vinyl bromide) in tetrahydrofuran.

7. A sample of poly(vinyl bromide) is dissolved in tetrahydrofuran and introduced at a liquid flow rate of 2 cm^3/min into the GPC column of Problem 6. When the refractive index difference between the eluted solution and pure solvent was plotted versus the elution time, the result was a broad peak, the maximum of which occurred at an elution time of 90 min. Calculate the average molecular weight, corresponding to the peak maximum, of the poly(vinyl bromide) sample.

8. The following elution volumes were obtained in a gel permeation chromatograph at 35°C for a set of monodisperse polystyrene standards dissolved in chloroform [J. V. Dawkins and M. Hemming, *Makromol. Chem.*, **176**, 1777 (1975)]:

M (g/mol) $\times 10^{-3}$	1900	867	670	411	160	98.2	51	19.8	10.3	3.7
V_e (cm^3) $\times \frac{1}{5}$	23.75	24.55	25.20	25.80	27.30	28.20	29.40	31.30	32.50	34.00

The Mark–Houwink constants for polystyrene in chloroform at 35°C may be taken as $K = 4.9 \times 10^{-3}$ cm^3/g and $a = 0.79$. Assuming that universal calibration is valid, construct a calibration curve for the molecular weight-elution volume of poly(dimethylsiloxane) in chloroform at 35°C; the Mark–Houwink constants for this polymer in chloroform at 35°C are $K = 5.4 \times 10^{-3}$ cm^3/g and $a = 0.77$.

9. A sample of poly(dimethylsiloxane) in chloroform is injected into the same gel permeation chromatograph used in Problem 8. Using a differential refractometer as detector, the following primary data of refractive index difference, Δn, and elution volume, V_e, were obtained:

$\Delta \eta \times 10^5$	0.6	3.4	12.4	15.0	11.7	4.1	1.0
V_e (cm^3) $\times \frac{1}{5}$	32.00	31.20	30.41	29.72	29.02	28.19	27.40

(a) Using your results from Problem 8, calculate and show graphically the molecular-weight distribution of the poly(dimethylsiloxane). You may assume that Δn is proportional to concentration and that the proportionality factor is independent of molecular weight.

SUGGESTIONS FOR FURTHER READING

FLORY, P. J., *Principles of Polymer Chemistry*, Chaps. 7 and 14. Ithaca, N.Y.: Cornell University Press, 1953.

MEYERHOFF, G., "Die Viscosimetrische Molekulare Wichtsbestimmung von Polymeren," *Fortschr. Hochpolym.-Forsch.*, **3**, 59 (1961).

MOORE, W. R., "Viscosities of Dilute Polymer Solutions," *Progr. Polymer Sci.* (A. D. Jenkins, ed.), **1**, 1 (1967).

OUANO, A. C., "Quantitative Data Interpretation Techniques in Gel Permeation Chromatography," *J. Macromol. Sci.—Rev. Macromol. Chem.*, **C9**, 123 (1973).

PORTER, R. S., AND J. F. JOHNSON, "Gel Permeation Chromatography," *Progr. Polymer Sci.* (A. D. Jenkins, ed.), **2**, 201 (1970).

16

Thermodynamics
of Solutions
# of High Polymers

Probably the most important single physical property of a high polymer is its molecular weight and, as seen in Chapter 14, the absolute measurement of this property is based on the properties of *solutions* of high polymers. It is therefore important that the polymer chemist should have a general understanding of the thermodynamics of polymer solutions and an appreciation of how the thermodynamic properties of such solutions differ from those formed by small molecules. Because of the very large size of the polymeric solute molecules compared to solvent molecules, many of the traditional concepts of solutions must be modified. For example, even the concept of an ideal solution requires modification.

The theoretical basis for the understanding of polymer solutions was developed independently by Flory[1] and Huggins[2] some 35 years ago in essentially equivalent treatments. In this chapter the treatment and notation of the former will be followed.

DEFINITION OF AN IDEAL SOLUTION

A traditional definition of an ideal solution is that it is a system in which Raoult's law (1) is obeyed.

$$a_1 = X_1 = (1 - X_2) = \frac{P_1}{P_1^\circ} \tag{1}$$

[1] P. J. Flory, *J. Chem. Phys.*, **10**, 51 (1942).
[2] M. L. Huggins, *J. Phys. Chem.*, **46**, 151 (1942).

In this equation a_1 is the thermodynamic activity of the solvent, X_1 the mole fraction of the solvent, X_2 the mole fraction of the solute, P_1 the pressure of solvent vapor above the solution, and P_1° the vapor pressure of the pure solvent. A thermodynamic consequence of this definition is that the chemical potential of the solvent in an ideal solution is given by (2), where μ_1° is the chemical potential of the pure solvent, or, in other words, the Gibbs free energy per mole.

$$\mu_1 = \mu_1^\circ + RT \ln X_1 \tag{2}$$

It will be recalled that (1) and (2) were used in Chapter 14 in the derivation of the osmotic pressure of ideal solutions.

All solutions, including polymer solutions, obey (1) and (2) in the limit of infinite dilution where they become ideal. For solutions of small solute molecules, deviations from ideality become negligible when both the mole fractions and weight fractions of the solute are small. However, the molecular weights of high-polymer solutes are so drastically different from those of typical solvents that vanishingly small *mole* fractions of solutes (i.e., $X_2 \rightarrow 0$) are obtained even though the *weight* fraction of the polymer is very large. Under such conditions, the mole fraction and the adherence of the system to Raoult's law are not useful indicators of ideality. As a numerical example, consider a polymer of molecular weight $M_2 = 10^6$, a solvent of molecular weight $M_1 = 10^2$ and a solution that is 91% by weight of polymer. The mole fraction of the solvent is

$$X_1 = \frac{n_1}{n_1 + n_2} = \frac{w_1/M_1}{w_2/M_2 + w_1/M_1} = \frac{M_2/M_1}{w_2/w_1 + M_2/M_1} \tag{3}$$

where n_i is the number of moles and w_i is the weight of component i. Inserting the numbers $w_2/w_1 = 10$ and $M_2/M_1 = 100$ into (3), we find that $X_1 = 0.999$. Thus, while the solvent makes up only 9% of the solution by weight (and the solution must be expected to behave very nonideally), the mole fraction of the solvent is sufficiently close to unity to *suggest* ideal behavior. This contradiction indicates that the thermodynamic *activity* of a solvent in an ideal polymeric solution is not equal to the mole fraction, whereas the two are equal for solutions of small molecule solutes. Therefore, Raoult's law is of little use for polymer solutions. As will be shown, the ideal polymer solution is better described as one in which the activity of the solvent is equal to the *volume fraction* of the solvent. This definition can be extended to ordinary solutions, since volume fraction and mole fraction for such solutions are very nearly the same, and this definition is, therefore, of more general validity than the traditional one.

The traditional definition of an ideal solution [i.e., (1) and (2)] is based on the *interchangeability* of solvent and solute particles. This means that the replacement of a solvent molecule by a solute molecule results in no change in the net molecular attractions and repulsions. As a consequence, an equivalent traditional definition of an ideal solution is one in which the formation of the solution from n_1 moles of pure solvent and n_2 moles of pure solute meets the following thermodynamic requirements:

$$\Delta H_{\text{mix}} = 0 \tag{4}$$

$$\Delta S_{\text{mix}} = -R(n_1 \ln X_1 + n_2 \ln X_2) \tag{5}$$

Early experimental work on polymer solutions indicated that deviations from ideality depend only weakly on the temperature. In view of the thermodynamic relationship describing the temperature dependence of the free energy of mixing,

$$\frac{\partial}{\partial T}\left(\frac{\Delta G_{\text{mix}}}{T}\right)_P = -\frac{\Delta H_{\text{mix}}}{T^2} \tag{6}$$

this observation suggests that ΔH_{mix} is not generally large. Therefore, the major cause for deviations from ideality lies in the failure of (5) to describe the *entropy* of mixing in the preparation of polymer solutions. Accordingly, we shall first devote our attention to a theoretical treatment of the entropy of mixing of solvent and solute, beginning with a simple treatment applicable to small molecules. An extension will then be made to macromolecular solutes. Finally, we shall consider the enthalpy and free energy of mixing that accompany the formation of a polymer solution.

ENTROPY OF MIXING OF SOLVENT AND SOLUTE

Small-Molecule Solutes Dissolved in Small-Molecule Solvents

Let us approach the entropy of mixing of solute and solvent from the point of view of a statistical theory in which the solvent and solute particles are assigned to positions in an imaginary lattice. For the present, consider both the solute and solvent molecules to be spherical particles of the same size. Assume also that the replacement of a solvent molecule by a solute molecule results in no change in the interactions of neighboring particles. Under these conditions the entropy of mixing of the solvent and solute arises solely from the greater number of lattice arrangements (i.e., configurations) possible for the solution, as compared to the solvent.

In Figure 16.1 a finite two-dimensional representation of the imaginary lattice is shown, with open circles representing solvent molecules and closed circles denoting solute molecules. In this situation there are no restrictions on the placing of particles in the lattice positions.

Let N_0 be the number of lattice positions, N_1 be the number of solvent molecules, and N_2 the number of solute particles. The assumption is made that all the lattice positions are occupied, and this may be described by

$$N_0 = N_1 + N_2 \tag{7}$$

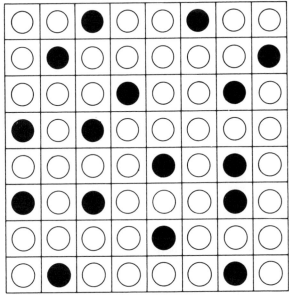

 Solute ◯ Solvent

Fig. 16.1   Two-dimensional lattice representation of a solution.

The problem is to calculate the number of ways that the N_0 molecules may be assigned to the N_0 positions in the lattice. If we imagine for the moment that all the N_0 molecules are distinguishable, then there are N_0 ways to choose the first molecule to drop randomly into the lattice. For each of these N_0 ways of choosing the first molecule there are $N_0 - 1$ ways to choose the second one, and for each of the $N_0(N_0 - 1)$ ways of choosing the first two molecules there are $N_0 - 2$ ways to choose the third, and so on. Therefore, for N_0 *distinguishable* particles, the number of arrangements in the lattice, Ω', is given by

$$\Omega' = N_0(N_0 - 1)(N_0 - 2)(N_0 - 3)\cdots(1) = N_0! \tag{8}$$

However, although a solvent molecule may be distinguished from a solute molecule, we cannot distinguish solvent molecules from each other nor solute molecules from each other. Since (8) assumes that we can, we must correct Ω' by the number of ways of permuting N_1 solvent molecules and N_2 solute molecules among themselves. Thus, the number of *distinguishable* arrangements in the lattice is

$$\Omega = \frac{N_0!}{N_1! N_2!} \tag{9}$$

For the starting materials (i.e., pure solvent and solute), the number of distinguishable arrangements is

$$\Omega_1 = \Omega_2 = \frac{N_1!}{N_1!} = \frac{N_2!}{N_2!} = 1 \tag{10}$$

According to Boltzmann, the entropy of a system is given by

$$S = k \ln \Omega \tag{11}$$

where k is Boltzmann's constant (i.e., $k = 1.38 \times 10^{-23}$ J/deg-molecule) and Ω is the number of distinguishable configurations or arrangements of the system as calculated above. The entropy of mixing of the solvent and solute, in the simple case at hand, is due solely to changes in the possible number of configurations of the mixed and unmixed systems and may be written as

$$S_c = \Delta S_{mix} = S - S_1 - S_2 \tag{12}$$

or

$$S_c = \Delta S_{mix} = k \ln \Omega - k \ln \Omega_1 - k \ln \Omega_2 \tag{13}$$

where the symbol S_c denotes this configurational entropy. Following substitution of (9) and (10) into (13), equation (14) is obtained.

$$S_c = \Delta S_{mix} = k[\ln N_0! - \ln N_1! - \ln N_2!] \tag{14}$$

To proceed further, use is made of the Stirling approximation for the factorials of large numbers. This states that

$$N! = \left(\frac{N}{e}\right)^N \tag{15}$$

or

$$\ln N! = N \ln N - N \tag{16}$$

Substitution of (7) and (16) into (14) leads directly to the expression

$$S_c = \Delta S_{mix} = -k\left[N_1 \ln \frac{N_1}{N_1 + N_2} + N_2 \ln \frac{N_2}{N_1 + N_2}\right] \tag{17}$$

Finally, from the relationships $R = N_A k$ and $N_i = N_A n_i$, where N_A is Avogadro's number and n_i represents the number of moles of the ith component, (17) may be transformed to the form shown in (18), in which X_i represents the mole fraction.

$$S_c = \Delta S_{mix} = -R[n_1 \ln X_1 + n_2 \ln X_2] \tag{18}$$

Polymeric Solutes Dissolved in Small-Molecule Solvents

The simplicity of the treatment described above depends on the interchangeability of solute and solvent molecules. Despite its simplicity, the expression shown in (18) describes quite well the entropy of mixing of solvent and solute molecules whose ratio of sizes (i.e., molar volumes) range from unity to about 3 or 4. However, when the solute is a polymer molecule whose molar volume may be thousands of times greater than that of a solvent molecule, the concept of interchangeability of a solvent and a solute particle is absurd and must be abandoned. Yet, this simple general approach to the entropy of mixing is so attractive that it is worthwhile to retain it and modify the model to take into account the vast difference in size of solvent and solute molecules.

The model chosen[1,2] for a polymer solute is that of a long-chain molecule consisting of x chain segments, each *segment* being of the same size (i.e., volume) as a solvent molecule. Solvent *molecules* and polymer chain *segments* may now be considered interchangeable in the lattice model of the solution. A simple analogy is to regard each solvent molecule as a white pearl and the polymer molecule as a string of x black pearls. The sizes of the black and white pearls are the same and hence are interchangeable in the lattice positions. Thus, according to this model, the number of chain segments (i.e., the number of pearls in the string) is related to the size ratio by

$$x = \frac{\bar{V}_2}{\bar{V}_1} \qquad (19)$$

where $\bar{V}_1$ and $\bar{V}_2$ are the molar volumes of solvent and solute, respectively.

The assumption that solvent molecules and chain segments are interchangeable permits the derivation to proceed in an analogous manner to the simple case just described for small-molecule solutes. The only difference is that the x chain segments of the polymer solute must be connected. This means that chain segments cannot be assigned to lattice positions in a completely random manner because each segment must have at least one other polymer segment adjacent to it. The lattice model of the polymer solution may be illustrated as in Figure 16.2. The relationship between the number of lattice positions and the number of solvent and solute molecules now becomes

$$N_0 = N_1 + xN_2 \qquad (20)$$

where, as before, N_0, N_1, and N_2 are the number of lattice positions, solvent molecules, and solute molecules, respectively.

To calculate the number of configurations of the mixture, first consider the number of ways in which a polymer molecule of x chain segments may be added to the lattice when i polymer molecules are already present. The number of vacant positions into which the first segment of this $(i + 1)$st molecule may be placed, and hence the number of ways in which this may be done is $(N_0 - xi)$. Having chosen one of these vacant sites in which to place the first segment of the $(i + 1)$st polymer molecule, we must now consider how many ways there are to place the second segment of the polymer. Letting Z be the coordination number of a lattice site (i.e., the number of nearest neighbor sites to any given site), the second segment must go

[1] Flory, op. cit.
[2] Huggins, op. cit.

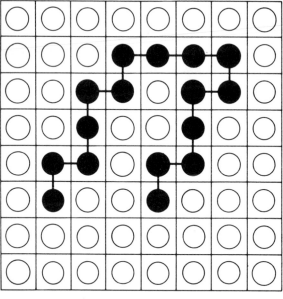

◯ Solvent   ⬤ Chain segments of the polymer

Fig. 16.2   Two-dimensional lattice representation of a polymer molecule in solution.

into one of the Z sites that are nearest neighbors to the one in which the first segment was placed. However, not all of these Z sites may be available. Some may already be occupied by segments from the first i polymer molecules present in the lattice. Let the symbol f_i be the probability that a site adjacent to the one occupied by a segment of the $(i + 1)$st molecule is already occupied by a segment from one of the first i molecules. Then the number of ways in which the second segment may be added is $Z(1 - f_i)$. For the addition of the third segment, one of the sites adjacent to the second segment is already occupied by the first segment. Hence the number of ways to add the third segment, and succeeding segments, is $(Z - 1)(1 - f_i)$. The number of configurations of the $(i + 1)$st molecule in the lattice, v_{i+1}, is the product of these numbers for the individual segments, namely,

$$v_{i+1} = \underbrace{(N_0 - xi)}_{\substack{\text{1st} \\ \text{segment}}} \cdot \underbrace{Z(1 - f_i)}_{\substack{\text{2nd} \\ \text{segment}}} \cdot \underbrace{(Z - 1)(1 - f_i)}_{\substack{\text{3rd} \\ \text{segment}}} \cdot \underbrace{(Z - 1)(1 - f_i)}_{\substack{\text{4th} \\ \text{segment}}} \cdots \qquad (21)$$

or

$$v_{i+1} = (N_0 - xi)Z(Z - 1)^{x-2}(1 - f_i)^{x-1} \qquad (22)$$

As an approximation to f_i, it may be assumed (with a reasonably small error) that the average probability that a given site is not occupied by segments of the first i molecules is equal to the fraction of sites remaining empty after the first i molecules have been added. Thus,

$$(1 - f_i) \approx \frac{N_0 - xi}{N_0} \tag{23}$$

The use of (23) and the simplifying approximation, $Z(Z - 1)^{x-2} \approx (Z - 1)^{x-1}$, enables (22) to be reduced to the more compact form shown in (24).

$$v_{i+1} = (N_0 - xi)^x \left(\frac{Z - 1}{N_0}\right)^{x-1} \tag{24}$$

Finally, as a third and convenient approximation, it can be shown by Stirling's formula (15) that the first term of (24) can be written, with little error, in the factorial form which yields

$$v_{i+1} = \frac{(N_0 - xi)!}{[N_0 - x(i + 1)]!} \left(\frac{Z - 1}{N_0}\right)^{x-1} \tag{25}$$

Expression (25) describes the number of configurations of just one polymer molecule in the lattice. The number of ways to place the N_2 indistinguishable polymer molecules is the product of these individual numbers of configurations divided by the number of ways of permuting the N_2 molecules among themselves. Thus,

$$\Omega = \frac{1}{N_2!} \left(\prod_{i=1}^{N_2} v_i\right) = \frac{1}{N_2!} \left(\prod_{i=0}^{N_2-1} v_{i+1}\right) \tag{26}$$

Substitution of (25) into (26) and writing out the terms in the product yields

$$\Omega = \frac{1}{N_2!} \left[\frac{N_0!}{(N_0 - x)!} \cdot \frac{(N_0 - x)!}{(N_0 - 2x)!} \cdot \frac{(N_0 - 2x)!}{(N_0 - 3x)!} \cdots \frac{[N_0 - (N_2 - 1)x]!}{(N_0 - N_2 x)!}\right] \left(\frac{Z - 1}{N_0}\right)^{N_2(x-1)} \tag{27}$$

which on cancellation of terms simplifies to

$$\Omega = \frac{N_0!}{N_2!(N_0 - xN_2)!} \left(\frac{Z - 1}{N_0}\right)^{N_2(x-1)} = \frac{N_0!}{N_1!N_2!} \left(\frac{Z - 1}{N_0}\right)^{N_2(x-1)} \tag{28}$$

Because the solvent molecules can occupy the remaining lattice sites in only one way, (28) is the total number of arrangements or configurations of the solution. The reader should note that the expression in (28) is the same as that for the ordinary solution, i.e. (9), except for the factor $[(Z - 1)/N_0]^{N_2(x-1)}$. Substitution of typical numbers into this factor (i.e., $Z \sim 10$, $x = 10^3$, $N_0 \sim 10^{23}$, $N_2 \sim 10^{18}$) shows that $[(Z - 1)/N_0]^{N_2(x-1)} \ll 1$. This means that there are many fewer configurations possible for the polymer solutions compared to small-molecule solutions.

The total configurational entropy is given by (11). Substitution of (28) into (11), and the use of Stirling's approximation for the factorials, leads in a straightforward way to

$$S_c = -k\left[N_1 \ln \frac{N_1}{N_1 + xN_2} + N_2 \ln \frac{N_2}{N_1 + xN_2} - N_2(x - 1) \ln \frac{Z - 1}{e}\right] \quad (29)$$

where e is the base of natural logarithms. The configurational entropy in (29) represents the entropy of mixing of the perfectly ordered pure solid polymer, for which $S = 0$, with pure solvent. This mixing process can be broken down into two reversible steps. The first step is conversion of the perfectly ordered polymer to a randomly oriented polymer and this process corresponds, in our model, to the random placement of polymer molecules into the lattice without a solvent. The second process consists of adding solvent molecules to the empty sites in the lattice and represents the entropy of mixing of the randomly oriented polymer with the solvent. If the entropy change of the first process is designated as ΔS_{dis} and that of the second process ΔS_{mix}, expression (30) holds.

$$\Delta S_{mix} = S_c - \Delta S_{dis} \quad (30)$$

In order to use (30) to evaluate the entropy of mixing of a randomly oriented polymer with the solvent, it is important to note that S_c is given by (29) and ΔS_{dis} is given by (29) *under the special condition* that $N_1 \to 0$ (i.e., no solvent has been added to the lattice). Thus,

$$\Delta S_{dis} = \lim_{N_1 \to 0} S_c = k\left[N_2 \ln x + N_2(x - 1) \ln \frac{Z - 1}{e}\right] \quad (31)$$

and so, subtracting (31) from (29), we obtain

$$\Delta S_{mix} = -k\left(N_1 \ln \frac{N_1}{N_1 + xN_2} + N_2 \ln \frac{xN_2}{N_1 + xN_2}\right) \quad (32)$$

If the approximation is made that x can be replaced by the ratio of the partial molar volumes (i.e., $x = \bar{V}_2/\bar{V}_1$), the expression can be changed to a molar basis (i.e., $k = R/N_A$) and this last result may be written as

$$\Delta S_{mix} = -R(n_1 \ln \phi_1 + n_2 \ln \phi_2) \quad (33)$$

where n_i is the number of moles of ith component and ϕ_i is the volume fraction:

$$\phi_i = \frac{n_i \bar{V}_i}{\sum_i n_i \bar{V}_i} \quad (34)$$

A comparison of (33) with (18) shows that the ideal entropy of mixing of a polymeric solute with a solvent is given by an expression that is similar to the classical ideal entropy of mixing of small-molecule solute and solvent molecules. The only difference is that, for polymer solutions, the volume fraction rather than the mole fraction is the dimensionless measure of concentration. The mole fractions and volume fractions of small molecule solutes in solution are essentially the same, and it would appear that (33) is the more general expression which reduces to (18) as the molecular sizes become equal.

The expression in (33) refers to a monodisperse polymer solute in which all the molecules are the same size. For a polydisperse polymer with a distribution of molecular weights, the term $n_2 \ln \phi_2$ must be replaced by $\sum_i n_i \ln \phi_i$, where the summation goes over the solute particles only.

ENTHALPY OF MIXING OF SOLVENT AND POLYMERIC SOLUTE

When a polymeric solute is added to a solvent, an enthalpy change occurs because solvent–solvent and solute–solute interactions are replaced by solvent–solute interactions. According to the lattice theory, such interactions may be represented by the numbers and types of nearest neighbors in the lattice. A nearest-neighbor interaction may be defined as a lattice contact, so there will be three types of such contacts (i.e., [1, 1], [2, 2], and [1, 2], respectively). The process of dissolution may then be written in terms of the change in these contacts

$$\tfrac{1}{2}[1, 1] + \tfrac{1}{2}[2, 2] \rightarrow [1, 2] \tag{35}$$

The energy change associated with the formation of one solvent–solute contact, $\Delta w_{1, 2}$, is given by

$$\Delta w_{1, 2} = w_{1, 2} - \tfrac{1}{2}(w_{11} + w_{22}) \tag{36}$$

Now if $P_{1, 2}$ is the average number of solvent–solute contacts (i.e., 1,2 contacts) over all the lattice configurations, then the enthalpy of mixing of the solvent and solute is

$$\Delta H_{\mathrm{mix}} = \Delta w_{1, 2} P_{1, 2} \tag{37}$$

per solute particle. The fraction of the lattice sites that are adjacent to those which contain a polymer segment and are at the same time occupied by solvent molecules (i.e., the probability of a 1,2 contact) should be given approximately by ϕ_1, the volume fraction of solvent. The total number of all the different types of contacts of each of the $x - 2$ internal polymer segments (not counting segments to which each is chemically bound) is $Z - 2$, while the two terminal segments will each have $Z - 1$ such contacts. The total number of 1,2 contacts for each polymer molecule is then

$$P_{1, 2} = [(x - 2)(Z - 2) + 2(Z - 1)]\phi_1 \tag{38}$$

For large values of Z, $P_{1,2} \approx Zx\phi_1$ and the enthalpy of mixing of N_2 polymer molecules with N_1 solvent molecules is given by

$$\Delta H_{\text{mix}} = Zx\phi_1\Delta w_{1,2}N_2 \tag{39}$$

From the definition of volume fractions, ϕ_1 and ϕ_2, it is easily shown that $xN_2\phi_1 = N_1\phi_2$. Then, on a molar basis, the enthalpy of mixing is given by

$$\Delta H_{\text{mix}} = Z\Delta w_{1,2}n_1\phi_2 N_A = Z\Delta W_{1,2}n_1\phi_2 \tag{40}$$

where $\Delta W_{1,2} = N_A\Delta w_{1,2}$. It is convenient to describe the interaction energy per mole of solvent, $Z\Delta W_{12}$, in terms of a dimensionless interaction parameter multiplied by RT. Thus, defining $Z\,\Delta W_{1,2} = \chi_1 RT$, the enthalpy of mixing (40) becomes

$$\Delta H_{\text{mix}} = RT\chi_1 n_1\phi_2 \tag{41}$$

The interaction parameter χ_1, given by $Z\Delta W_{1,2}/RT$, is the energy change (in units of RT) that occurs when a mole of solvent molecules is removed from the pure solvent (where $\phi_2 = 0$) and is immersed in an infinite amount of pure polymer (where $\phi_2 = 1$). Because of the approximate nature of the lattice theory, χ_1 is found to depend on the concentration of the solution. According to its definition, χ_1 depends inversely on the temperature. χ_1 is generally positive, with values at 25°C and at infinite dilution being near 0.5. According to (41) the fact that χ_1 is positive means that the dissolution of a polymeric solute in a solvent is generally an endothermic process.

FREE ENERGY OF MIXING OF POLYMERIC SOLUTE WITH SOLVENT

The Gibbs free energy change for the dissolution of a polymeric solute is easily obtained from the well-known thermodynamic expression

$$\Delta G = \Delta H - T\Delta S \tag{42}$$

because substitution of (33) and (41) into (42) leads immediately to the result

$$\Delta G_{\text{mix}} = RT(\chi_1 n_1\phi_2 + n_1 \ln \phi_1 + n_2 \ln \phi_2) \tag{43}$$

It is now possible to answer the question of whether dissolution of a polymer in a solvent occurs with positive or negative free energy. The answer, (43), clearly depends on the concentration of the solution and on the sign and magnitude of χ_1. As the temperature is increased, χ_1 decreases and dissolution becomes thermodynamically more favorable.

CHEMICAL POTENTIAL AND ACTIVITY
OF SOLVENT

It was shown in Chapter 14 that the presence of a solute lowers the chemical potential of a solvent from its value in the pure solvent. This is of fundamental importance for the derivation of osmotic pressure changes. A theoretical expression for the reduction of the chemical potential of the solvent is readily obtained from the free energy of mixing since, by definition, the chemical potential of a solvent in a solution relative to that in the pure solvent is given by

$$\mu_1 - \mu_1^\circ = \left(\frac{\partial[G_{\text{soln}} - G_1^\circ]}{\partial n_1}\right)_{T, P, n_2} = \left(\frac{\partial \Delta G_{\text{mix}}}{\partial n_1}\right)_{T, P, n_2} \tag{44}$$

Partial differentiation of ΔG_{mix}, (43), with respect to n_1 at constant T gives

$$\mu_1 - \mu_1^\circ = RT\left[\frac{n_1}{\phi_1}\left(\frac{\partial \phi_1}{\partial n_1}\right)_{n_2} + \ln \phi_1 + \frac{n_2}{\phi_2}\left(\frac{\partial \phi_2}{\partial n_1}\right)_{n_2} + \chi_1\phi_2 + \chi_1 n_1\left(\frac{\partial \phi_2}{\partial n_1}\right)_{n_2}\right] \tag{45}$$

The partial derivatives of the expression above may be evaluated from the definition of volume fraction. Volume fractions may be written in terms of the molar volume ratio, $x = \bar{V}_2/\bar{V}_1$, as

$$\phi_1 = \frac{n_1}{n_1 + xn_2} \tag{46}$$

and

$$\phi_2 = \frac{xn_2}{n_1 + xn_2} \tag{47}$$

The result, after some manipulation, is

$$\mu_1 - \mu_1^\circ = RT\left[\ln(1 - \phi_2) + \left(1 - \frac{1}{x}\right)\phi_2 + \chi_1\phi_2^2\right] \tag{48}$$

For an ideal solution, in which the solvent and solute molecules are identical in size and shape (i.e., $x = 1$), in which $\Delta H_{\text{mix}} = 0$ (i.e., $\chi_1 = 0$), and in which volume fraction and mole fraction are equal, equation (48) reduces to the classical expression shown in (49), (see also Chapter 14, equation 5)

$$\mu_1 - \mu_1^\circ = RT \ln X_1 \tag{49}$$

where X_1 is the mole fraction of solvent. In the case of a heterogenous polymer, x in equation (48) is replaced by $\bar{x}$ (i.e., by the *average* degree of polymerization).

In classical solution theory, equation (49) is valid only for ideal solutions. However, to retain the simple form of this equation for nonideal solutions, the activity of the solvent in a solution is defined by

$$\mu_1 - \mu_1^\circ = RT \ln a_1 \tag{50}$$

Hence, the activity of the solvent in a solution of polymer is given by

$$\ln a_1 = \ln (1 - \phi_2) + \left(1 - \frac{1}{x}\right)\phi_2 + \chi_1\phi_2^2 \tag{51}$$

and an analogous treatment beginning with (43) yields equation (52) for the activity of the solute:

$$\ln a_2 = \ln \phi_2 + (1 - x)(1 - \phi_2) + \chi_1 x\phi_2^2 \tag{52}$$

THE OSMOTIC PRESSURE OF POLYMERIC SOLUTIONS

It will be recalled from Chapter 14 that the fundamental precept underlying our understanding of the osmotic pressure of a solution is the reduction in the chemical potential of a solvent that occurs when a solute is added to it and the compensating increase in solvent chemical potential that accompanies an increase in the external pressure applied to the solution. A repeat of the osmotic pressure derivation of Chapter 14, but with (48) instead of (49) being used to describe the dependence of chemical potential on concentration, leads to the expression

$$\Pi = -\frac{RT}{\overline{V}_1}\left[\ln (1 - \phi_2) + \left(1 - \frac{1}{x}\right)\phi_2 + \chi_1\phi_2^2\right] \tag{53}$$

where Π is the osmotic pressure and $\overline{V}_1$ is the molar volume of solvent. If the logarithmic term in (53) is expanded in the well-known series (54),

$$\ln (1 - \phi_2) = -\phi_2 - \tfrac{1}{2}\phi_2^2 - \tfrac{1}{3}\phi_2^3 - \cdots \tag{54}$$

the result shown in (55) is obtained.

$$\Pi = \frac{RT}{\overline{V}_1}\left[\frac{\phi_2}{x} + (\tfrac{1}{2} - \chi_1)\phi_2^2 + \tfrac{1}{3}\phi_2^3 + \cdots\right] \tag{55}$$

It is sometimes more convenient to express the concentration of the solution in terms of weight per unit volume (i.e., $c = w_2/V_{\text{soln}}$) than in terms of volume fraction.

This can be introduced into (55) by recognizing that the volume fraction of the polymer in the solution may be written as

$$\phi_2 = \frac{n_2 \bar{V}_2}{n_1 \bar{V}_1 + n_2 \bar{V}_2} = \frac{w_2 \bar{V}_2}{M_2 V_{\text{soln}}} = \frac{c \bar{V}_2}{M_2} \tag{56}$$

Strictly speaking, the symbols $\bar{V}_i$ in (56) are partial molar volumes. However, we may assume them to be equal to the respective molar volumes, an assumption that is compatible with the other approximations of the theory. Substitution of (56) into (55) and elimination of x by its definition, $x = \bar{V}_2/\bar{V}_1$, gives after some rearrangement

$$\frac{\Pi}{c} = \frac{RT}{M_2} \left[1 + \frac{\bar{v}_2^2 M_2}{\bar{V}_1} \left(\frac{1}{2} - \chi_1 \right) c + \frac{\bar{v}_2^3 M_2}{3 \bar{V}_1} c^2 + \cdots \right] \tag{57}$$

In the expression in (57), $\bar{v}_2$ is the specific volume of polymer (actually the partial specific volume) or, in other words, $\bar{v}_2 = \bar{V}_2/M_2$. In the case of a heterogenous polymer, M_2 should be replaced by the number-average molecular weight of the polymer.

It is seen that (57) is in the form of a virial equation which was written in Chapter 14 in the form

$$\frac{\Pi}{c} = \frac{RT}{M_2} (1 + \Gamma c + g\Gamma^2 c^2 + \cdots) \tag{58}$$

and in which the second and third virial coefficients Γ and $g\Gamma^2$, respectively, are given by

$$\Gamma = \frac{\bar{v}_2^2 M_2}{\bar{V}_1} \left(\frac{1}{2} - \chi_1 \right) \tag{59}$$

and

$$g\Gamma^2 = \frac{\bar{v}_2^3 M_2}{3 \bar{V}_1} \tag{60}$$

From experimental studies, such as an examination of the dependence of Π/c on concentration, it is possible to derive values of χ_1 provided, of course, that the densities or specific volumes of the polymer and the solvent are known. Some values of χ_1 for several polymers in a variety of solvents at 25°C and in the limit of zero concentration are shown in Table 16.1. Also included in Table 16.1 are the interaction energies $RT\chi_1$ which represent the energy change that occurs when 1 mol of solvent is transferred from the pure solvent and is immersed in an infinite amount of polymer.

All the polymer–solvent systems in Table 16.1 show positive values of χ_1. These positive values indicate that replacement of a solvent molecule by a polymer molecule occurs with a positive enthalpy change (i.e., is an endothermic process). Negative values of χ_1 would indicate exothermic dissolution, with $\Delta H_{\text{mix}} < 0$. Such

TABLE 16.1 Polymer–Solvent Interaction Energies at 25°C and at Infinite Dilution

POLYMER	SOLVENT	χ_1	$RT\chi_1$ (cal/mol)
Natural rubber	Benzene	0.42	249
Poly(dimethyl-siloxane)	Chlorobenzene	0.47	278
Polyisobutylene	Benzene	0.50	296
Polyisobutylene	Cyclohexane	0.43	254
Polyisobutylene	n-Pentane	0.49	290
Polystyrene	Methyl ethyl ketone	0.47	278
Polystyrene	Ethylbenzene	0.40	237
Polystyrene	Cyclohexane	0.505	299
Poly(methyl methacrylate)	Chloroform	0.377	223
Poly(methyl methacrylate)	4-Heptanone	0.509	301
Poly(methyl methacrylate)	Tetrahydrofuran	0.447	265

negative values of χ_1 are observed only very rarely, even though they would be more likely in systems in which either the polymer or the solvent is polar (thereby increasing the attractive interactions on mixing). It must be concluded, then, that at 25°C dissolution is an endothermic process. Dissolution will only be favored thermodynamically (i.e., $\Delta G < 0$) at those temperatures and compositions for which the negative terms in the free-energy expression (43) are numerically greater than the enthalpy of mixing. Thus, for thermodynamically favored dissolution, expression (61) must hold.

$$- (n_1 \ln \phi_1 + n_2 \ln \phi_2) > \chi_1 n_1 \phi_2 \qquad (61)$$

LIMITATIONS OF THE THEORY

According to (57), a single value of χ_1 should be sufficient to describe the osmotic pressure, as well as other thermodynamic properties, over a wide range of polymer concentrations. However, experimental tests show that χ_1 depends on the *concentration* of the solution with the values usually increasing as ϕ_2 is increased. Some typical results designed to test the theory are shown in Figure 16.3 for polystyrene in methyl ethyl ketone and for polyisobutylene in cyclohexane.

The failure of the theory to account for the dependence of χ_1 on the composition of the solution is due to the approximations inherent in the theory. However, despite these shortcomings, the simple lattice theory gives us, in a relatively simple and instructive way, a semiquantitative appreciation of the factors involved in the thermodynamics of polymer solutions. Further developments of the theory do account crudely for the dependence of χ_1 on composition, but these treatments are quite complex and are beyond the scope of this text.

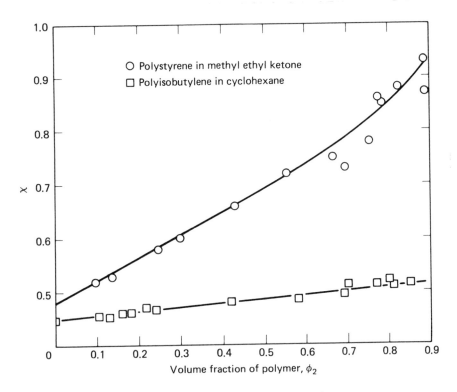

Fig. 16.3 Dependence of polymer–solvent interaction parameter χ on concentration at 25°C. [Reproduced from P. J. Flory, *Disc. Faraday Soc.*, **49** (1970); with permission of the Faraday Division of the Chemical Society of London.]

STUDY QUESTIONS

1. Using the simple lattice theory, evaluate the number of distinguishable arrangements, Ω, for a solution of styrene in *m*-xylene in which the concentration of styrene is $10^{-2}\,M$ at $20°C$ ($\rho_{xylene} = 0.861$ g/cm³ at 20°C). Calculate the entropy of mixing in the preparation of 100 cm³ of this solution.

2. Suppose that the dissolved styrene monomer in Problem 1 is completely converted to a dissolved polymer of $\overline{DP} = 1000$. Assuming a coordination number of 12, calculate the number of distinguishable arrangements of solute and solvent in this solution and compare with the result of Problem 1. Calculate the entropy of mixing in the preparation of 100 cm³ of this solution and the entropy of polymerization of the styrene.

3. Using the lattice theory, calculate the entropy change for the conversion of 10^{-6} mol of polymer of $\overline{DP} = 500$ from a perfectly ordered state to a randomly ordered state. (Assume a coordination number of 12.)

4. Calculate the enthalpy of mixing when 10^{-5} mol of poly(methyl methacrylate) of $\overline{M}_n = 10^5$ g/mol and of $\rho = 1.20$ g/cm³ are dissolved in 150 g of chloroform ($\rho = 1.49$ g/cm³) at 20°C. Assume that the volumes are additive.

5. Show that $\chi N_2 \phi_1 = N_1 \phi_2$ when the symbols are as defined in the text.

6. Calculate the Gibbs free energy change in the preparation of the solution of Problem 4.

7. Derive equation (48).

8. Derive equation (57).

9. The second virial coefficient in toluene solution of a test sample of polystyrene was found to be 219 cm^3/g at 25°C. Evaluate the interaction parameter χ_1.

10. A solution of poly(methyl methacrylate) ($\rho = 1.20$ g/cm^3, $\overline{M}_n = 3.52 \times 10^5$) in chloroform ($\rho = 1.49$ g/cm^3) has been prepared by adding 50.0 mg of polymer to 150 g of the solvent. Estimate the osmotic pressure of the resulting solution.

SUGGESTIONS FOR FURTHER READING

FLORY, P. J., "Thermodynamics of Polymer Solutions," *Disc. Faraday Soc.*, **49**, 7 (1970).

FLORY, P. J., *Principles of Polymer Chemistry*, Chap. 12. Ithaca, N.Y.: Cornell University Press, 1953.

ISIHARA, A., AND E. GUTH, "Theory of Dilute Macromolecular Solutions," *Adv. Polymer Sci.*, **5**, 233 (1967).

MORAWETZ, H., *Macromolecules in Solution* (2nd ed.), Chap. 2. New York: Wiley, 1975.

TOMPA, H., *Polymer Solutions*, Chap. 4. New York: Academic Press, 1956.

WOLF, B. A., "Zur Thermodynamik der enthalpisch und der entropisch bedingten Entmischung von Polymerlösungen," *Adv. Polymer Sci.*, **10**, 109 (1972).

17

Morphology, Glass Transitions, and Polymer Crystallinity

MORPHOLOGICAL CHANGES IN POLYMERS

Most long-chain synthetic polymers show a characteristic sequence of changes as they are heated. All linear polymers are glasses at low temperatures. As the temperature is raised, a certain point is reached at which the polymer changes from a glass to a rubber. This change is known as the glass transition temperature (T_g). When heated above T_g, amorphous polymers pass successively through rubbery, gumlike, and finally liquid states with no clear demarcation between the different phases. On the other hand, crystalline polymers remain flexible and thermoplastic above T_g until the temperature is raised to the crystalline melting temperature, T_m. At this point the polymer melts to a viscous liquid at a sharply defined temperature. The crystalline melting phenomenon occurs when sections of adjacent chains are packed together in a regular array, and the melting point represents the temperature at which these microcrystallites are thermally disrupted. The different characteristics of amorphous and crystalline polymers are illustrated in Figure 17.1. Extensive crosslinking may distort this picture and mask the transitions.

When a crystalline and a noncrystalline modification of the same polymer are compared, it is sometimes found that the glass transition temperature is higher or lower in the crystalline form, and the temperature span of the rubbery phase may be truncated. In fact, in many crystalline polymers the rubbery phase of the amorphous state is replaced by a flexible, thermoplastic phase which is less extensible than that of a conventional elastomer, but much tougher.

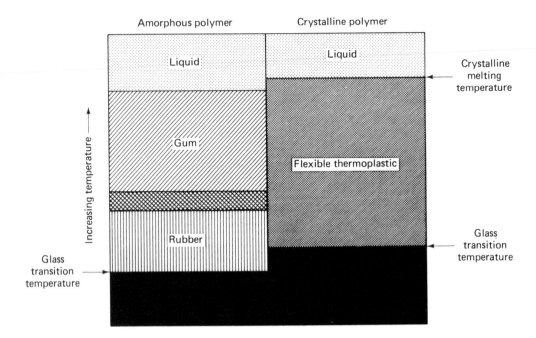

Fig. 17.1   Different transition behavior exhibited by amorphous and crystalline polymers.

CHARACTERISTICS OF THE VISCOELASTIC STATE

Elementary physics divides matter into three quite distinct categories: solids, liquids, and gases. *Solids* are substances that occupy a fixed shape and volume (i.e., they do not flow). *Liquids* flow readily but occupy a fixed volume. *Gases* flow and change their volume easily. This simple view of the universe does not account for the properties of open-chain or moderately crosslinked polymers. Most polymers are neither classical solids nor liquids. They are *viscoelastic* materials. The viscoelastic state has the characteristics of *both* the solid and liquid states.

Consider a piece of lightly crosslinked natural rubber. At rest on a laboratory bench, it has all the characteristics of a solid (definite shape, fixed volume, no evidence of liquid flow). But if we stretch the material or apply pressure to one part of it, it will change shape like a liquid. Of course, if we release the tension or pressure, it will revert to its original shape. These are some of the characteristics of the viscoelastic state, and these unusual properties account for the valuable properties of macromolecules. By possessing these properties, polymers can be used for many applications where conventional solids (like sodium chloride or benzoic acid) or ordinary liquids would be unsuitable.

In this section we examine some of the peculiarities of the viscoelastic state; later we will consider the two critical phenomena: the glass transition (T_g) and polymer crystallinity.

Terminology

It is relatively easy for an experimenter to describe a particular substance as a glass, elastomer, gum, or liquid. However, more precise terminology is needed if, for example, we wish to say that one elastomer is "stiffer" than another.

The following terms are commonly used. *Tensioning* is the act of attempting to stretch a material. Quantitatively, the amount of force applied is known as the tensile stress, σ. The amount of stretch induced in the sample is the tensile *strain*, ε. These two factors are related by the equation, $\sigma = E\varepsilon$, where the constant, E, is known as the *modulus* or Young's modulus. This constant is a characteristic of a particular polymer. It is a measure of stiffness or rigidity. A substance that has a high modulus has a high rigidity and can be deformed only by the application of appreciable stress. It is important to recognize that resistance to stretching (modulus or rigidity) is not the same as resistance to breaking. A material may be extremely rigid but may break under low tension (although this combination of properties is quite rare for polymers). Thus, another term—*tensile strength*—is needed to define the ultimate load that a material can bear without breaking. Piano wire has a tensile strength of roughly 2×10^{10} dyn/cm^2. Poly(methyl methacrylate) in the glassy state has a value of only 8×10^8 dyn/cm^2. *Tenacity* is the stress at the breaking point of the material. *Impact strength* is the resistance of a material to breakage when subjected to a sharp blow. *Toughness* is another measure of resistance to breakage, and is defined as the total energy input to the breaking point.

Viscoelastic behavior can also be defined in terms of *shear*. Shear occurs when, for example, a piece of polymer is deformed without a volume change by the application of force to the top of the sample in a direction parallel to the surface on which the sample is resting. The resistance to deformation is described as the *shear modulus*, G. The application of hydrostatic pressure to a material will bring about a decrease in its volume. The resistance to contraction is given by the *bulk modulus*, B.

Creep is the process by which a polymer undergoes a slow change of shape or a flowing action when subjected to a constant force such as gravity. It is the most obvious "liquidlike" feature of the viscoelastic state, and one that is obviously detrimental to most polymer applications.

Some of these terms are employed in the discussion of the testing of polymers (Chapter 21).

The Glassy State

Two typical glassy polymers are poly(methyl methacrylate) and polystyrene. They are dimensionally stable at room temperature and do not creep. Although they have some flexibility as thin samples, they are not elastomeric. They have moderate strength and some impact resistance, but they can be shattered by a sharp blow with a hammer. At room temperature both are well below their glass transition temperatures ($\simeq 100°C$). Because most glassy polymers are used as structural materials, rigidity and resistance to creep, high impact strength, and a high glass transition temperature are the most desirable properties. Unfortunately, high impact strength is more a characteristic of elastomers than of glasses. Hence, a compromise may be needed between high rigidity and high impact strength.

The phenomenon of brittleness or susceptibility to shattering is connected with the ease with which a small fracture can be propagated throughout the polymer. The application of impact force to a glassy polymer will result in a separation of chains at the point of impact and in a cleavage of skeletal bonds. The separation of individual chains takes place by overcoming the weak van der Waals attractions. More energy is required to break covalent bonds. The problem is that even a small impact-induced re-entrant discontinuity at the surface of a glassy polymer will cause an increase in the local stress at that point. This focuses most of the impact stress on relatively few chemical bonds. Hence, the crack propagates rapidly as both covalent bonds and the van der Waals attractions give way. The fundamental problem with a glass is that the conversion of the impact energy into the breakage of bonds is one of the few mechanisms available for dissipation of that energy. By contrast, an elastomer *absorbs* the energy into harmless molecular motions. In the glassy state, the polymer molecules are rigidly fixed in place, and such impact-absorbing molecular mobility is not present. Hence, the material shatters.

Three ways have been developed to overcome this problem. First, the formation of covalent crosslinks between chains has the effect of increasing the amount of energy needed to propagate a crack. There are now more covalent bonds to be broken. Second, finely divided materials known as "fillers" may be added to the polymer. These serve to interrupt the propagation of cracks. Third, an impact-absorbing second phase may be incorporated into the polymer matrix. This last approach is discussed in a later section.

The Rubbery State

Elastomers are polymers that are well above their glass transition temperatures. Although a few elastomers contain microcrystallites, most are amorphous materials. The elastomeric characteristics become lost if the temperature is high enough to induce gumlike behavior. However, as discussed earlier, crosslinks between the chains maintain the elastomeric character of the material at high temperatures.

The phenomenon of elasticity has been mentioned earlier. Elastomeric materials have the properties of liquids—they change shape or flow readily when subjected to weak forces. But they differ from liquids in their capacity to reassume their original shape once the distorting force has been removed. This behavior is a consequence of the high mobility of the backbone bonds. The chains can readily undergo conformational changes to yield contracted (coiled) or extended conformations in response to an external force. The ability of an elastomer to reassume its original shape after deformation can be attributed to two features. First, the highly coiled molecular conformations are preferred for entropic reasons, and a stretched polymer will contract when the tension is released to allow a maximization of the entropy. Second, crosslinks prevent the chains from slipping past each other when the applied tension is excessive or prolonged. As a consequence, many crosslinked elastomers have a high modulus and high strength when stretched. The crosslinks also reduce the tendency for the polymer to creep.

Elastomers have two other characteristics: they are "resilient" and they absorb "solvents" and swell to a surprising degree. *Resilience* is the ability of a material to bounce back. A rubber ball dropped from a height onto the floor rebounds. The degree to which the rebound height compares with the initial height is a measure of resilience. Resilience reflects the ability of an elastomer to absorb energy, store it, and then *rapidly* return that energy following the impact. In molecular terms, it reflects the ease with which the polymer chains can undergo a rapid conformational distortion from their preferred state, and the ease and rapidity with which they reassume their original preferred condition and release the stored free energy. In turn, this depends on the ease of torsion of the backbone bonds (Chapter 18) and on a low cohesive energy between adjacent chains (individual segments of chains must be able to slip past each other readily). The same molecular characteristics explain why elastomers have a high impact resistance. They absorb energy in conformational changes on impact and then release it. They do not absorb the energy by cleavage of chemical bonds.

The ability of an elastomer to absorb solvents and undergo a marked expansion of volume is a well-known phenomenon. The swelling of rubber in benzene or toluene is an example of this process (p. 333). The solvent absorption is facilitated because the polymer chains are in flexural motion at room temperature. Thus, the solvent molecules can readily penetrate the polymer lattice and bring about a separation of the large molecules. Eventually, the polymer will dissolve unless crosslinks are present. Crosslinks place a limit on the degree to which the chains can separate and, hence, on the extent of swelling. In fact, the degree of swelling in a solvent can be used as a measure of the crosslink density. Some elastomers are used in technology in a solvent-expanded form. The "solvents" are known as plasticizers or "oil extenders."

Elastomers are not generally used for structural applications because of their dimensional instability. However, they are widely employed for energy absorbing applications, tires, and as flexible, impact resistant coatings.

The Flexible, Nonelastomeric State

Many microcrystalline polymers, at temperatures above their glass transitions, exist in the form of flexible film- or fiber-forming materials. Such materials show some elastomeric-type properties (i.e., flexibility, impact resistance, some limited elasticity, ability to swell in solvents, etc.), but their main characteristic is a *combination* of flexibility and dimensional stability.

Such materials can be viewed as amorphous elastomers in which are embedded temporary crosslink sites in the form of microcrystalline domains. The nature of these crystalline domains is discussed in a later section. Here, it is sufficient to point out that the crystalline regions impart a certain stiffness to the material (compared to the purely amorphous material), and reduce the tendency of the polymer to creep. Very high degrees of crystallinity can induce brittleness, and compromises may be necessary between the advantages of crystallinity and the advantages of impact resistance. "Tough" polymers represent a result of this compromise.

Plasticizers are often used to extend the range of properties of flexible polymers. The effect of the plasticizer is to swell the amorphous regions, lower the cohesion between the chains, and allow these regions to function as flexible elastomeric domains. Plasticizers are also employed to modify polymers that have a normal T_g value above room temperature. In such cases they serve to lower the glass transition temperature until it is below room temperature.

The Liquid State

If crosslinks are absent, both amorphous and microcrystalline polymers melt at high temperatures. The melting process allows the chains to separate from each other and permits viscous flow to occur readily. Few high polymers are used as technological materials in the molten state. Decomposition reactions occur at the high temperatures required for melting. However, the molten state is used extensively for the fabrication of polymers (see Chapter 20).

Two-Phase Systems

As mentioned earlier, a major problem in the technological use of polymers is to design materials that have dimensional stability and yet have high impact strength. One way in which this can be accomplished is by the addition of "reinforcement" fillers such as glass fiber, asbestos, or powdered carbon to a polymer that has a high glass transition temperature. This method is discussed in Chapter 20. The second approach involves the preparation of a polymer system that possesses both glassy and elastomeric domains. The glassy regions provide the resistance to deformation, and the elastomeric regions function as impact-absorbing domains. For example, block copolymers can be prepared from styrene and butadiene by the use of anionic polymerization techniques. The polymers are known as SBR copolymers. The polystyrene block has a T_g at 100°C and the polybutadiene block at -63°C. At normal temperatures the polystyrene blocks function as "anchors" or glassy "filler" domains, preventing the polybutadiene regions from exhibiting their normal extensibility and potential for creep. However, the impact resistance is much higher than that of pure polystyrene. Acrylonitrile–butadiene–styrene (ABS) polymers have the same characteristics.

THE GLASS TRANSITION TEMPERATURE

The glass transition represents the rather sharp change that occurs from the glassy to the rubbery or flexible thermoplastic states in nearly all linear-type polymers. It is known as a "second-order transition." This transition is characteristic of a particular polymer in much the same way that a melting point is characteristic of ordinary low-molecular-weight compounds. In fact, the glass transition temperature varies with the types of skeletal atoms present, with the types of side groups, and even with the spatial disposition of the side groups. Table 17.1 lists values for a number of different polymers.

To a very large extent, the practical ultility of polymers and their different properties depend heavily on their glass transition temperatures. Thus, an important area of polymer research is the drive to understand how different molecular features affect the glass transition temperature. One approach to this problem is described in Chapter 18. Here it is sufficient to note that there appears to be a very close connection between the T_g value and the *flexibility* of the polymer chain. The chain flexibility depends more on the rotation or torsion of skeletal bonds than on changes in bond angles or lengths. When a randomly coiled chain is pulled out into an elongated conformation, the skeletal bonds "unwind" rather than undergo angular distortion (Figure 17.2). Thus, flexibility on a macroscopic scale depends on *torsional mobility* at the molecular level. If a highly flexible chain is present, the glass transition temperature will generally be low. If the chain is rigid, the T_g value will be high. For example, poly(dimethylsiloxane) has one of the lowest T_g values known ($-123°C$), presumably because the silicon–oxygen bonds have considerable torsional mobility.

However, the inherent rigidity or flexibility of the backbone structure is only one contributing factor. The torsional mobility of the skeletal bonds will also be

TABLE 17.1 Glass Transition Temperatures (T_g) and Crystalline Melting Temperatures (T_m) for Selected Polymers*

POLYMER	T_g (°C)	T_m (°C)
Polystyrene (isotactic)	100	240
Poly(*m*-methylstyrene) (isotactic)	70	215
Poly(methyl methacrylate) (atactic)	114	—
Poly(methyl methacrylate) (isotactic)	48	160
Poly(methyl methacrylate) (syndiotactic)	126	200
Poly(*cis*-1,4-isoprene)	−67	36
Poly(*trans*-1,4-isoprene)	−68	74
Poly(dimethylsiloxane)	−123	−29
Poly(dichlorophosphazene)	−63	−10
Poly[bis(trifluoroethoxy)phosphazene]	−66	242
Poly[bis(ethoxy)phosphazene]	−84	—
Poly(trifluoroethoxypentafluoroprop- oxyphosphazene) rubber	−77	—
Polyacrylontrile	85	317
Nylon 66	45	267
Poly(ethylene terephthalate)	17	—
Polyethylene	−20	141
Polyethylene	−107	95

* A compilation of T_g and T_m values can be found in O. G. Lewis, *Physical Constants of Linear Homopolymers* (New York: Springer-Verlag, 1968).

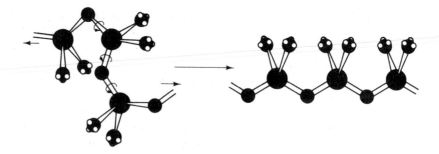

Fig. 17.2   Elasticity of a polymer such as silicone rubber depends on
the ease with which the chains can be stretched out from a random
coil. The chain elongation is a consequence of the *unwinding* of bonds
rather than a marked widening of band angles.

affected by the side groups: large side groups or charged structures on the same
chain will repel or attract each other and this could appreciably lower the chain
mobility. Examination of Table 17.1 will provide examples of this effect. Furthermore, polar interactions *between* neighboring chains could raise the T_g. Even the
chain length has an effect. For low polymers, the T_g generally rises with increasing
chain length until a limiting value is reached. Thus, an analysis of glass transitions
in terms of molecular structure is a complex problem that is only now in the process
of being unraveled.

DETECTION OF GLASS TRANSITIONS

It might be imagined that the most straightforward way to measure a glass transition temperature is to simply manipulate the polymer as it is cooled or heated in
order to find the temperature at which it changes from a hard glass to a rubber or a
flexible thermoplastic. In principle, this can be done, although the results are likely
to be far less accurate than if one of the following methods is employed.

''Indentation'' Techniques (Penetrometer)

Below the T_g a polymer is hard and glassy. Above T_g the material is soft and flexible.
Thus, the degree to which a sharp point can penetrate the surface of the polymer
at a given temperature can be used to detect the transition. In practice, the point
of a weighted needle is allowed to rest on the polymer surface as the temperature is
raised. As the polymer passes through its T_g the needle penetrates the surface and
the movement of the needle can be monitored by means of an amplification gauge.
This method is less accurate than those described in the following sections, and
requires relatively large samples of polymer. However, it is a useful method for
the preliminary, engineering-oriented examination of polymers.

Torsional Rigidity Methods

The resistance of a polymer sample to torsion depends on whether the polymer is in the glassy or flexible state and this principle can be used to measure T_g. Perhaps the most engineering-oriented technique based on this principle involves the application of a torsional vibration to the end of a bar of the polymer. The resistance to torsion and the energy loss of the polymer are then measured at different temperatures.

In the laboratory a more convenient technique makes use of a torsional pendulum (Figure 17.3). In one device, an inert matrix such as a braided glass fiber is impregnated with a solution of the polymer and then dried. The fiber is suspended in a torsional pendulum device (see Figure 17.4) and the period of the pendulum and its damping frequency are measured (Figure 17.5). As the sample is heated through its glass transition temperature, a drastic loss in rigidity is detected, accompanied by a sharp maximum in the damping curve. This provides a very

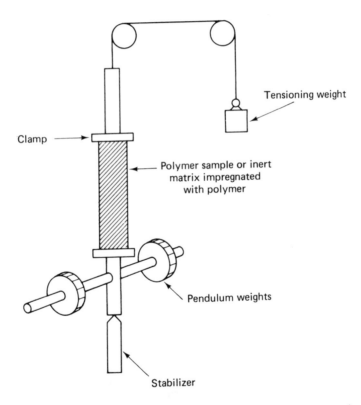

Fig. 17.3 Simple device in which a polymer sample or a porous matrix (paper) impregnated with polymer forms part of a torsional pendulum. The rigidity of the polymer and its capacity to absorb torsional energy are measured from the period of the pendulum and its damping characteristics.

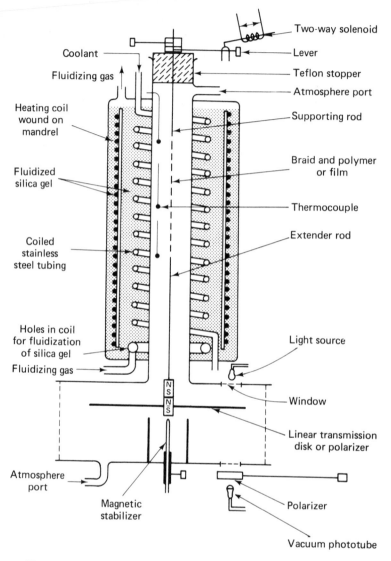

Fig. 17.4 Diagram showing the construction of a torsional pendulum and torsional braid analyzer. [Printed with permission from J. K. Gillham, *CRC Critical Rev. Macromol. Sci.*, **1**, 83 (1972–73). © The Chemical Rubber Co., CRC Press, Inc.]

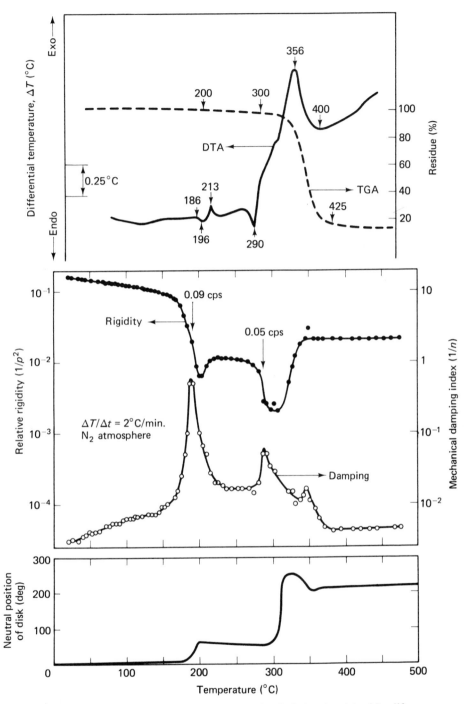

Fig. 17.5 Comparison of thermomechanical (torsional braid), differential thermal analysis, and thermogravimetric analysis data for cellulose triacetate. The bottom figure shows the twisting of the sample in the absence of oscillations as a result of expansion or contraction of the sample at T_g and T_m. [Reprinted with permission from J. K. Gillham, *CRC Critical Rev. Macromol. Sci.*, **1**, 83 (1972–73). © The Chemical Rubber Co., CRC Press, Inc.]

sensitive technique for the detection and measurement of the transition. Furthermore, the method requires the use of very small (0.25-g) samples of polymer and is thus ideal for exploratory work.

Broadline Nuclear Magnetic Resonance

The typical nmr signal seen for a low-molecular-weight compound in dilute solution or as a molten sample is a series of sharp peaks. Nmr spectra of solids, on the other hand, are broad, diffuse, and often difficult to detect. When a polymer is heated through the glass transition region it is, in a sense, converted from a solid to a pseudo liquid, and this transition can be detected by broadline nmr methods. In practice, several grams of the polymer are needed. Some care is required to distinguish between the glass transition and the onset of *side-group* torsional motions. Figure 17.6 illustrates the way in which nmr changes can be utilized to detect glass transitions.

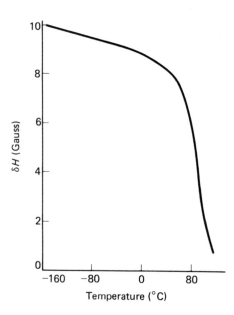

Fig. 17.6 Curve showing change in ^1H nmr line width as a sample of commercial poly-(vinyl chloride) is heated through the glass transition region. [From A. M. Hassan, *CRC Critical Rev. Macromol. Sci.*, **1**, 83 (1972–73). © The Chemical Rubber Co., CRC Press, Inc.]

Dilatometry

The rate of volume expansion of a polymer with temperature depends on whether the polymer occupies the glassy, rubbery, thermoplastic, or liquid states. Thus, the change of slope of a volume versus temperature plot can be used to identify the glass transition (Figure 17.7). In practice, a dilatometer (Figure 17.8) is used for these measurements, and the position of the capillary miniscus is plotted as a function of temperature.

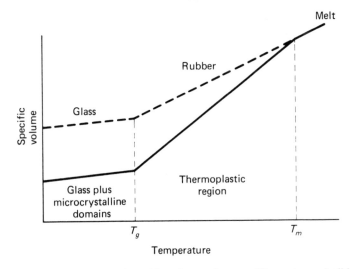

Fig. 17.7   Variation of specific volume of a crystalline polymer (solid line) and an amorphous modification of the same polymer (broken line) as a function of temperature. Transitions are indicated by abrupt changes in the slopes of the lines.

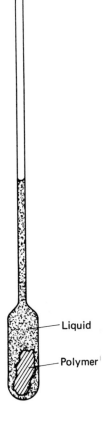

Fig. 17.8   Dilatometer, used to monitor the change in volume of a polymer plus the surrounding liquid as a function of temperature by changes in the height of the liquid level in the capillary tube. Because the liquid does not exhibit sharp transitions when heated, but the polymer does, any changes in the slope of volume versus temperature may be attributed to T_g or T_m transitions in the polymer.

Differential Thermal Analysis (DTA) and Differential Scanning Calorimetry (DSC)

These two methods are perhaps the most popular techniques for the measurement of glass transition temperatures. The DTA method requires the heating of a small polymer sample at a constant rate of temperature increase. The temperature of the polymer is compared continuously with the temperature of a control substance, such as alumina, which itself undergoes no transitions in the temperature range being scanned. The temperature difference between the polymer and the control material is then a function of the different specific heats of the two substances. The specific heat of a polymer changes rapidly as a transition region is approached with exothermic changes being characteristic of glass transitions (Figure 17.9). Endotherms indicate first-order transitions (melting temperatures). Two advantages of the DTA method are that only a small amount of the polymer is required, and the measurement is quite rapid. Also, the transition temperature can be identified to within 1 or 2°C. A disadvantage is that highly crystalline polymers may show only weak exotherms that may prove difficult to identify.

Differential scanning calorimetry operates by a slightly different principle. Two small metal containers, one containing the polymer sample and the other a

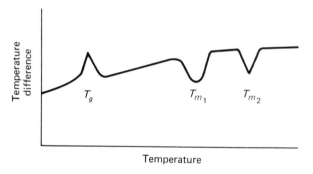

Fig. 17.9   Differential thermal analysis (DTA) scan of a polymer, showing an exotherm typical of a glass transition and the endotherms that are characteristic of T_m transitions.

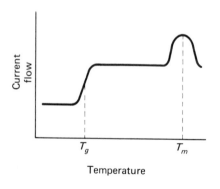

Fig. 17.10   Differential scanning calorimetry scan (DSC) showing sections of the curve characteristic of T_g and T_m transitions.

control substance, are heated by individual electric heaters. The temperature of each container is monitored by a heat sensor. If the sample suddenly absorbs heat during a transition, this change will be detected by the sensor, which will initiate a greater current flow through the heater to compensate for the loss. Thus, absorption of heat by the sample results in a greater current flow. Since the change in electric current can be monitored accurately, this provides a sensitive measure of transition temperatures. Figure 17.10 shows a DSC scan.

MICROCRYSTALLINITY

A distinction must be drawn between polymers that are commonly described as "crystalline" and the single crystals formed by low-molecular-weight substances. In the latter type of crystal, the crystalline order results from a regular packing of molecules or ions in a three-dimensional lattice (Figure 17.11). Crystals formed from low-molecular-weight compounds retain their integrity as the temperature is raised, until the point is reached at which the vibrational forces become more important than the intermolecular attractions. At this point the lattice breaks down over a very narrow temperature range, and the crystal melts sharply.

In microcrystalline polymers, on the other hand, the crystallinity results from the *regular packing of chains* (Figure 17.12). However, it is important to recognize that such regular packing arrangements usually exist only in small domains within the polymer. Hence, a microcrystalline polymer really consists of microcrystallites embedded in a matrix of amorphous polymer. In fact, a polymer chain may pass through several amorphous and crystalline regions as shown in Figure 17.13. It will be obvious from Figure 17.13 that a maximum degree of order can be obtained only when the crystallites themselves are lined up on the same axis, a process that is known as *orientation*. The microcrystalline regions are characterized by a more efficient use of the available space. Hence, polymers that possess microcrystalline domains have a higher density than forms of the same material that are totally amorphous. In some polymer systems, especially polyamides, the regular packing

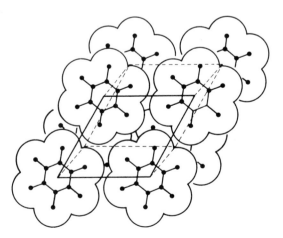

Fig. 17.11 Regular packing of small molecules of hexamethyl-benzene in the space lattice of a single crystal. [From C. W. Bunn, *Chemical Crystallography* (London: Oxford University Press, 1963).]

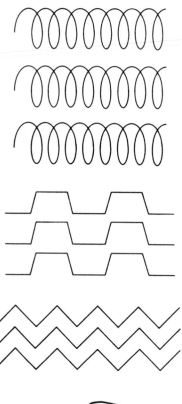

Fig. 17.12 Microcrystalline regions of a polymer are composed of individual chains (or different segments of the same chain) packed side by side in a regular manner.

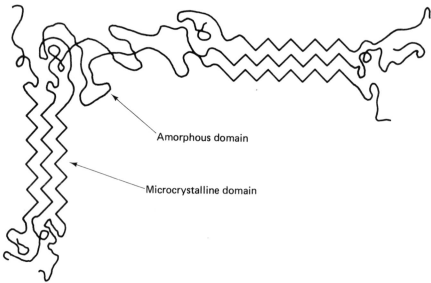

Amorphous domain

Microcrystalline domain

Fig. 17.13 Structure of a microcrystalline polymer is generally considered to consist of microcrystalline domain separated by amorphous, random coil regions. A single polymer chain may traverse several microcrystalline and amorphous regions.

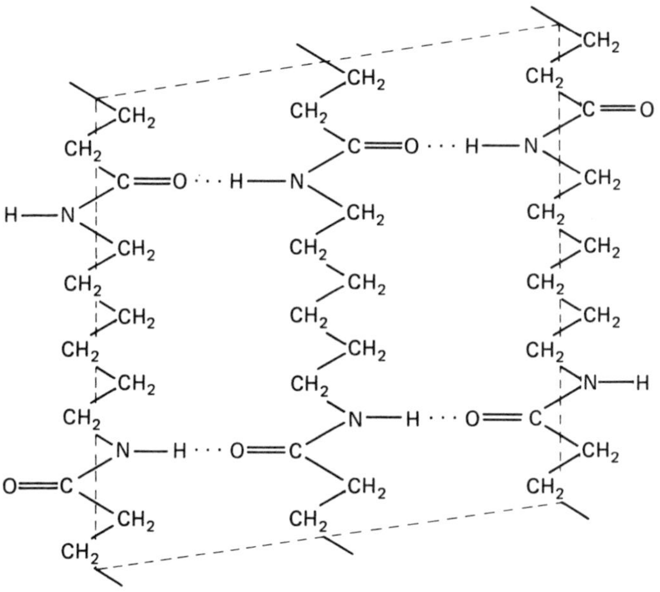

Fig. 17.14   Orientation of the chains in the microcrystalline region of a polyamide (nylon 66). Individual chains are held together by hydrogen bonds.

of adjacent chains is facilitated by intermolecular hydrogen bonding as well as by the normal van der Waals attractive forces (Figure 17.14). Such polymers often have especially high crystalline melting temperatures.

INFLUENCE OF CRYSTALLINITY ON PHYSICAL PROPERTIES

Polymers crystallize to attain a state of lower free energy. The regular packing of chains means *closer* packing and enhanced opportunities for intermolecular attractions. This is the driving force behind crystallization. However, an opposing force is the need for the system to maximize its entropy by increases in the conformational disorder. Hence, the observed structure represents a balance (sometimes unstable) between these opposing forces. Typically, 30 to 70% of the polymer may remain in the amorphous state, and the logical question must, therefore, be raised as to why the polymer does not continue to behave like an amorphous material.

In the microcrystalline regions the chains are essentially held together by dipolar, hydrogen bonding, or van der Waals forces. For this reason the crystalline domains function as *crosslinks* for the amorphous regions. The crosslinks are labile at the melting temperature or even during manipulation of the polymer, but crosslinks they are, with all the physical property influences that would be expected from such structures. The crystalline crosslinks stiffen and toughen the polymer and reduce the swelling in solvents. On a macroscopic level the introduction of microcrystallinity changes a rubbery elastomeric polymer into a tough, flexible material.

For example, at room temperature, polyethylene is roughly 100°C above its glass transition temperature, but it remains a tough plastic material. Without the microcrystallites it would be a soft elastomer, as indeed it is above the crystalline melting temperature of 115°C. This effect is strikingly demonstrated with organophosphazene high polymers. Poly[bis(trifluoroethoxy)phosphazene], $[NP(OCH_2CF_3)_2]_n$, and poly(diphenylphosphazene), $[NP(OPh)_2]_n$, are both highly crystalline, flexible, tough thermoplastics. However, if the symmetry of the structure is destroyed by the random introduction of a second substituent group, the polymers become rubbery and highly elastomeric.

Knowledge of effects such as this allows the polymer chemist to modify his materials to an even greater degree than is permitted by chemical means alone. In fact, the properties of many technologically important polymers are controlled to within fine limits by the degree of crystallinity introduced into the system.

ENHANCEMENT OF CRYSTALLINITY

Microcrystalline polymers are generally tougher than totally amorphous ones. They can be bent more without breaking, they resist impact better, and they are less affected by temperature changes or solvent penetration than are completely

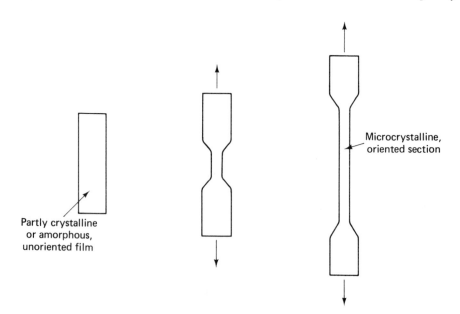

Fig. 17.15 Piece of polymer film can be oriented by stretching. Often, the central section of the film elongates first with the orientation spreading toward the ends of the film as continued tension is applied.

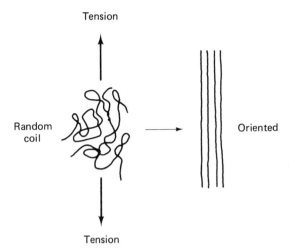

Tension

Random
coil

→

Oriented

Tension

Fig. 17.16 On a molecular
level the stretching of a polymer
film or fiber results in a roughly
parallel alignment of the macromolecular chains. This facilitates crystallization.

amorphous polymers. For these reasons, it is frequently advantageous, especially
in manufacturing processes, to attempt to increase the degree of crystallinity.

The simplest technique available for this purpose is to stretch a fiber or a film
of the polymer (Figure 17.15). When a polymer is cast as a film or extruded as a
fiber, some microcrystallinity is often introduced. However, the microcrystallites
tend to be few in number and randomly oriented relative to each other. The act of
orienting the sample by stretching serves to pull the individual chains into a roughly
parallel orientation (Figure 17.16). This enhances the chance that regular packing
of adjacent chains will take place. Additional crystallization can also be introduced
by heating and cooling the tensioned polymer in an "annealing" process. Alto-
gether, a stretching of the material to four or five times its original length is not
uncommon during crystallization and annealing. This process can be carried out

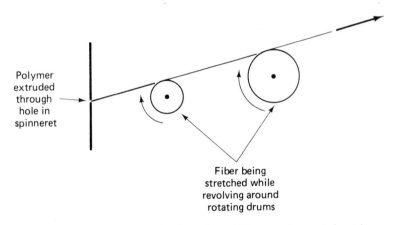

Polymer
extruded
through
hole in
spinneret

Fiber being
stretched while
revolving around
rotating drums

Fig. 17.17 Stretching and orientation of fibers can be carried out in
a continuous process by allowing the fiber to rotate around a revolving
drum. Each drum in the sequence is of larger diameter than the preceding one, or of the same diameter but revolving at a faster speed.

by hand or with the use of stretching devices (see Chapter 20), but in large scale fiber manufacture it is performed continuously by passing the fiber around heated rotating drums, as shown in Figure 17.17. The influence of the orientation process is especially evident with nylon, where unoriented fibers are brittle, but oriented fibers are strong, tough, and somewhat elastic.

DETECTION AND MEASUREMENT OF CRYSTALLINITY

Nearly all the methods discussed earlier for the measurement of glass transition temperatures can be used to measure crystalline melting temperatures. For example, endotherms in differential thermal analysis curves (Figure 17.9) or breaks in dilatometric curves (Figure 17.7) yield T_m values. Crystallization results in a volume contraction that can be followed dilatometrically. The torsional pendulum method (Figures 17.3 to 17.5) also identifies T_m values from points of abrupt decrease in rigidity and from damping maxima. Nuclear magnetic resonance techniques can also be used. However, convincing proof that microcrystallinity exists in a particular polymer requires the use of two additional techniques: optical birefringence and X-ray crystallography.

The optical birefringence technique makes use of a polarizing microscope, preferably fitted with a heating stage for raising the temperature of the sample. When viewed through crossed polarizers, a microcrystalline polymer will show a dark background broken by bright specks which originate from the crystalline domains. Some polymers show a more complex structure in which individual microcrystallites radiate in all directions from specific points to form spherulites (Figure 17.18). The structure of each spherulite can be visualized as made up of microcrystals oriented along the radii of individual spheres. In practice, the spherulites grow until they contact each other to form the unusual patterns shown in Figure 17.18. The bright specks extinguish or reappear as the sample is rotated relative to the polarizing filter. The specular pattern should persist until the temperature is raised to the crystalline melting point, at which temperature the bright specks or spherulites should be totally extinguished.

The X-ray diffraction approach makes use of the fact that a totally amorphous polymer will not give rise to a sharp diffraction pattern. Diffuse scattering rings only are seen. These resemble the X-ray patterns obtained from liquids. An unoriented microcrystalline polymer will yield an X-ray photograph that consists of sharp rings (see Chapter 19). Such photographs bear a striking resemblance to Debye–Scherrer "powder" photographs of powdered small-molecule crystalline materials, and for the same reason. The sharp concentric areas or rings result from diffraction by randomly oriented microcrystallites.

An *oriented* microcrystalline polymer fiber or film will yield a pattern of diffraction *spots* or arcs. The apparatus used to obtain photographs of this type is discussed in Chapter 19. Equipment is available that will allow X-ray patterns to be obtained as the temperature is raised. The appearance of amorphous-type patterns and the disappearance of a crystalline-type pattern provides a measure of

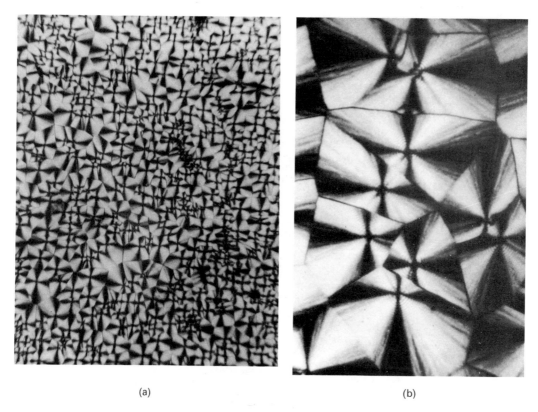

(a)   (b)

Fig. 17.18 Polarized-light microscope photograph of a sample of
low-density polyethylene showing the spherulitic structure at two
different magnifications. The light polarization directions are horizontal
and vertical. (By courtesy of R. S. Stein, University of Massachusetts.)

the crystalline melting temperature. However, it is not uncommon that, as the
temperature is raised, one crystalline pattern is replaced by another, and in this
way a sequence of T_m values can be measured.

POLYMER SINGLE CRYSTALS

The preceding comments have been concerned with *micro*crystalline polymers.
However, a few polymers are known to exist also as single crystals. Polyethylene is
the best known example. Two main techniques are available for the preparation of
polymer single crystals. There are (1) the crystallization of the polymer from an
inert solvent at very high dilutions: for example, the slow cooling of linear polyethylene from hydrocarbon media at dilutions near 1 : 2000, and (2) the simultaneous polymerization of a monomer and crystallization of the polymer. This
technique has been used to prepare single crystals of polyoxymethylene.

The crystals obtained from polyethylene are visible only under a microscope.
They have a flat, lozenge shape (Figure 17.19). X-ray diffraction photographs of

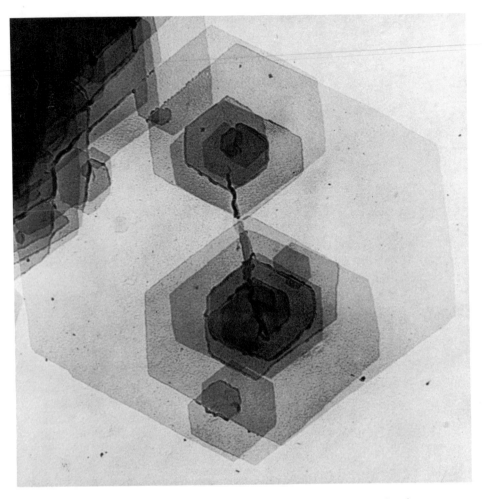

Fig. 17.19   Photograph of lozenge-shaped single crystals of polyethylene. (By courtesy of I. R. Harrison, The Pennsylvania State
University.)

such crystals resemble those of low-molecular-weight materials. However, the
molecular structure within the crystal consists of a folded arrangement of zigzag
chains (Figure 17.20). It has been found that the folding interval depends on the
temperature of crystallization rather than on the polymer chain length or the
nature of the solvent.

It should be noted that globular proteins and viruses crystallize to form single
crystals. However, these molecules are arrayed individually in the lattice in the
manner typified by crystals of low-molecular-weight compounds. This, in itself, is
an astonishing phenomenon. The Bushy stunt virus, for instance, has a molecular
weight of 13 million, with 4 molecules per cell and with a unit cell edge of 318 Å.
The X-ray analysis of systems of this kind will be considered briefly in Chapter 19.

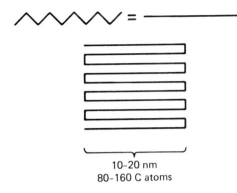

Fig. 17.20 Presumed packing arrangement of polyethylene chains in a single crystal. This represents the so-called adjacent reentrant model of the structure in which a single chain bends back and forth in order to provide a parallel alignment of the chains. Another model (the so-called "switchboard" model) allows re-entry of the chain at a place some distance from the point of emergence.

10-20 nm
80-160 C atoms

STUDY QUESTIONS

1. If the physical changes that occur at the glass transition represent a sudden onset in backbone torsional motions, why should the value of T_g vary with the absence or presence of microcrystallinity?

2. Of the alternative methods mentioned in this chapter for the detection of glass transitions and melting temperatures, which methods would be more suitable to use: (a) in an exploratory synthetic research laboratory, (b) in a physical testing laboratory, or (c) in a manufacturing plant. Why?

3. What problems might you anticipate if, in the torsional braid method, instead of the use of a glass braid you decided to use a braid made from (a) Nylon 66; (b) cellulose acetate; (c) paper; (d) copper.

4. After reference to Figure 17.14, suggest ways in which you might raise or lower the glass transition temperature of a polyamide and raise or lower the melting temperature.

5. Suggest reasons why polyethylene is one of the few synthetic polymers known that forms single crystals.

SUGGESTIONS FOR FURTHER READING

AKLONIS, J. J., W. J. MacKNIGHT, AND M. SHEN, *An Introduction to Polymer Viscoelasticity.* New York: Wiley–Interscience, 1972.

BAER, E., "Relaxation Processes at Cryogenic Temperatures." *CRC Critical Rev. Macromol. Sci.*, **1**, 215 (1972–73).

BLACKADDER, D. A., "Ten Years of Single Crystals," *J. Macromol. Sci.—Rev. Macromol. Chem.*, **C1**, 297 (1967).

CHANZY, H., "Nascent Morphology of Polyolefins," *CRC Critical Rev. Macromol. Sci.*, **1**, 315 (1972–73).

CHIU, J. (ed.), *Polymer Characterization by Thermal Methods of Analysis* (Symposium-Selected Papers). New York: Dekker, 1974.

DESPER, C. S., "Technique of Measuring Orientation in Polymers," *CRC Critical Rev. Macromol. Sci.*, **1**, 501 (1972–73).

DOLE, M., "Calorimetric Studies of States and Transitions in Solid High Polymers," *Fortschr. Hochpolym.-Forsch.*, **2**, 221 (1960).

DUSEK, K., AND W. PRINS, "Structure and Elasticity of Non-crystalline Polymer Networks," *Adv. Polymer Sci.*, **6**, 1 (1969).

EISENBERG, A., "Ionic Forces in Polymers," *Adv. Polymer Sci.*, **5**, 59 (1967).

FAVA, R. A., "Polyethylene Crystals," *J. Polymer Sci.* (D) (*Macromol. Rev.*), **5**, 1 (1971).

FERRY, D., *Viscoelastic Properties of Polymers*. New York: Wiley, 1970.

FISA, B., "Nascent Morphology of Polyolefins," *CRC Critical Rev. Macromol. Sci.*, **1**, 315 (1972–73).

GEIL, P. H., *Polymer Single Crystals* (*Polymer Reviews*, Vol. 5). New York: Wiley–Interscience, 1963.

GILLHAM, J. K., "A Semimicro Thermomechanical Technique for Characterizing Polymeric Materials: Torsional Braid Analysis," *A.I.Ch.E. Journal*, **20**, 1066 (1974).

GORDON, M., *High Polymers: Structure and Physical Properties*. Reading, Mass: Addison-Wesley, 1963.

HASSAN, A. M., "Application of Wide-Line NMR to Polymers," *CRC Critical Rev. Macromol. Sci.*, **1**, 399 (1972–73).

HILTNER, A., "Relaxation Processes of Cryogenic Temperatures," *CRC Critical Rev. Macromol. Sci.*, **1**, 215 (1972–73).

KARASZ, F., "The Glass Transition of Linear Polyethylene," *J. Macromol. Sci., Rev. Macromol. Chem.*, **C17** (1), 37 (1979).

KOVAKS, A. J., "Transition vitreuse dans les polymères amorphes. Étude phénoménologique," *Fortschr. Hochpolym.-Forsch*, **3**, 394 (1963).

MANDELKERN, L., *Crystallization of Polymers*. New York: McGraw–Hill, 1964.

MANDELKERN, L., "Thermodynamic and Physical Properties of Polymer Crystals Formed from Dilute Solution," *Progr. Polymer Sci.* (A. D. Jenkins, ed.), **2**, 163 (1970).

MARCHESSAULT, R. H., "Nascent Morphology of Polyolefins," *CRC Critical Rev. Macromol. Sci.*, **1**, 315 (1972–73).

MIRABELLA, F. M., AND J. F. JOHNSON, "Polymer Configuration and Compositional Variables as a Function of Molecular Weight," *J. Macromol. Sci.—Rev. Macromol. Chem.*, **C12**, 81 (1975).

PETRIE, S. E. B., "The Effect of Excess Thermodynamic Properties Versus Structure Formation on the Physical Properties of Glassy Polymers," *J. Macromol. Sci.-Phys.*, **B12**, 225 (1976).

SANCHEZ, I. C., "Modern Theories of Polymer Crystallization," *J. Macromol. Sci.—Rev. Macromol. Chem.*, **C10**, 113 (1974).

SAUER, J. A., G. C. RICHARDSON, AND D. R. MORROW, "Deformation and Relaxation Behavior of Polymer Single Crystals," *J. Macromol. Sci.—Rev. Macromol. Chem.*, **C9**, 149 (1973).

SCHUUR, G., *Some Aspects of the Crystallization of High Polymers.* Delft: Rubber–Stiching, 1955.

SHARPLESS, A., *Introduction to Polymer Crystallization.* London: Arnold (St. Martin's Press, New York), 1966.

WATTS, M. P. C., A. E. ZACHARIADES, and R. S. PORTER, "New Methods of Production of Highly Oriented Polymers by Solid State Extrusion," *Contemp. Top. Polymer Sci.* (M. Shen, Ed.), **3**, 297 (1979).

WILKES, G. L., "The Measurement of Molecular Orientation in Polymeric Solids," *Adv. Polymer Sci.,* **8**, 91 (1971).

WOODWARD, A. E., AND J. A. SAUER, "The Dynamic Mechanical Properties of High Polymers at Low Temperatures," *Fortschr. Hochpolym.-Forsch.,* **1**, 114 (1958).

YEH, G. S. Y., "Morphology of Amorphous Polymers," *CRC Critical Rev. Macromol. Sci.,* **1**, 173 (1972–73).

ZACHMANN, H. G., "Das Kristallisations und Schmelzverhalten hochpolymerer Stoffe," *Fortschr. Hochpolym.-Forsch.,* **3**, 581 (1964).

18

Conformational Analysis
of Polymers

THE ROLE OF CONFORMATIONAL ANALYSIS

As discussed elsewhere in this book, the primary motivation for most modern fundamental chemical research is the drive to relate the physical and chemical properties of materials to their molecular structure. This is true in spectroscopy, reaction mechanism studies, synthetic chemistry, and in many other areas. It is particularly true in polymer chemistry. One of the main purposes of science is to "make sense" of observable phenomena in terms of the behavior of microscopic, molecular, or atomic particles. In polymer science this need becomes manifest in attempts to relate properties, such as strength, toughness, solution viscosity, elasticity, crystallinity, or biological behavior to the composition, shape, and dynamic behavior of the component molecules.

As will be obvious from the earlier chapters, we now know a great deal about the ways in which changes in chemical composition affect the properties of polymers. However, we have only a fragmentary knowledge of the way in which the shape and flexural differences between molecules can be related to the physical properties. This, then, constitutes a considerable challenge.

Conformational analysis involves the study of the ways in which molecules can alter their geometry by torsional rotations (or "twisting" motions) of their covalent bonds. As such, it must be distinguished from *configurational* analysis which is concerned with the fixed geometric differences between closely related molecules as, for example, in the identification of *d*- or *l*-configurations in a molecule.

Conformational changes usually result in a change in the *shape* of the molecule without the cleavage of bonds having occurred. Because the torsional motions of many bonds take place readily at normal temperatures, conformational changes are responsible for many of the phenomena that we associate with changes in macroscopic physical properties. Thus, the glass and melting transitions of a polymer, the absence or presence of crystallinity, the extensibility or elasticity, and the ways in which polymers raise the viscosity of small-molecule solvents can all be related to changes in the conformation of the macromolecules. In fact, nearly all of the useful properties of polymers can be ascribed in some way to the conformational characteristics of the component molecules.

Thus, it is of vital importance that we attempt to understand the reasons why some polymers assume one conformation while others prefer another, or why a particular polymer will undergo a conformational change at a specific temperature or when the solvent or the pH is changed. Specifically, a number of good reasons exist for attempting to predict or rationalize polymer conformations. These are:

1. To explain why some polymers crystallize and others do not.
2. To explain why some polymers are glasses at room temperature while others are rubbery, thermoplastic, or even liquid.

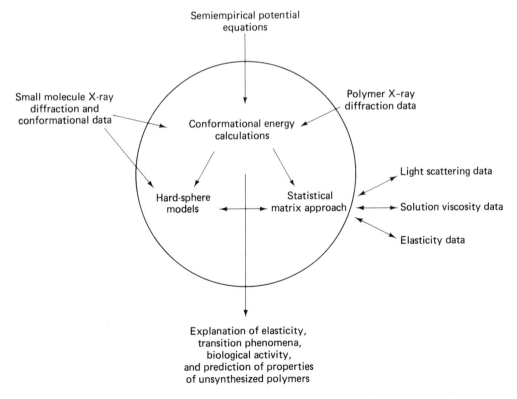

Scheme 1   Relationships among various aspects of conformational analysis.

3. To predict the properties of polymers not yet synthesized, in order to answer the question: Are they worth making?

4. To predict the solution properties of polymers, such as mean square end-to-end distance, solution viscosity, light scattering behavior, and so on.

5. To explain the biological functions of naturally occurring polymers, such as proteins and nucleic acids.

It would be misleading to imply that these objectives have been achieved on a broad scale. Nevertheless, significant progress has been made in the right direction, and the following sections summarize some of the approaches that have been taken.

Scheme 1 summarizes the relationships that exist between the various aspects of conformational analysis to be considered in this chapter. The three subjects enclosed by the circle represent the three main theoretical approaches to conformational analysis. The peripheral subjects to which they are connected by arrows are areas that provide input to and output from the theoretical treatments.

POLYMER CONFORMATIONS

A small molecule, such as ethane (1), can adopt relatively few different conformations, because the detailed molecular shape is affected by the torsion of one bond only (the C—C bond). (Of course, the shape will also be modified by vibrational motions, but these have relatively little effect on physical properties.) However,

propane (2), with two C—C bonds, is in theory capable of generating many more independent conformations in which the C—H bonds on different carbon atoms have different spatial relationships to each other. It is obvious that polyethylene (3) offers the possibility that an almost infinite number of different conformations can be generated from the same molecule.

The superimposition of *configurational* changes on to these conformational possibilities increases the complexity of the problem even further. Thus, poly(vinyl chloride) can exist in isotactic, syndiotactic, or atactic geometries, and each of these forms will offer the prospect of an entirely new set of conformational possibilities. Consequently, at first sight, an understanding of the conformational behavior of macromolecules may appear to be too complex a problem to warrant serious discussion.

However, a considerable simplification of the problem can be introduced if polymers are divided into three categories.

1. Macromolecules that show a strong preference to adopt one distinct molecular conformation in the crystalline solid state or in solution.
2. Macromolecules that assume no preferred conformation at all and adopt the overall shape of a random coil.
3. Those in which an individual chain assumes either random coil or a discrete conformational state in different sections along the chain.

Clearly, polymers that fall into category (3) can be treated as a composite of those in categories (1) and (2). Hence, in its simplest sense, the problem of polymer conformation can be reduced to the extreme states of the regular, repetitive conformation, on the one hand, and the random coil, on the other. As we shall see, both of these extremes can be treated satisfactorily.

SMALL-MOLECULE MODELS FOR POLYMERS

Given the complexity of macromolecular structures, it makes sense to consider first how conformational problems have been solved for small molecules and then to consider if such approaches can be extended to larger systems. The conformations of small molecules have been examined by means of two approaches, namely qualitative and quantitative methods, and these two methods will be considered in turn.

The Qualitative Approach

For many years organic and inorganic chemists have made predictions about the conformations assumed by simple, covalently bonded molecules. These predictions are based on the intuitively reasonable assumption that atoms forming part of the same molecule repel each other if they approach too closely. This is the idea behind the concept of *steric hindrance*. In fact, two guiding principles have become part of the folklore of chemistry. These are:

1. Large atoms or groups give rise to greater repulsions than small atoms for a given distance between the centers.
2. Staggered conformations are of lower energy than eclipsed conformations.

All this is responsible for the predictions that ethane should adopt a staggered conformation (**4**) (at least at low temperatures), while *n*-butane should occupy the conformation shown in (**5**).

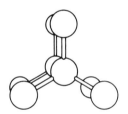

These arguments form the rationale behind the use of "hard-sphere" molecular models. The preferred conformation is predicted from these by the assumption that atoms and groups tend to avoid "collisions." Although these methods seem naive, they are remarkably effective for simple systems. Thus, the "hard-sphere" approach predicts that, in a molecule that can undergo conformational changes by torsion of the bond that connects two tetracoordinate atoms (such as carbon), the three "staggered" minima shown in Figure 18.1 will be accessible. As will be seen later, this leads immediately to the concept of the "threefold rotational isomeric model" that forms the basis of some of the more complex calculations.

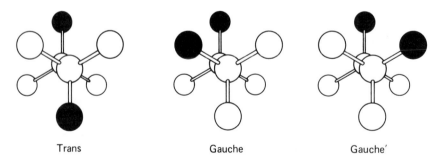

Fig. 18.1   The hard-sphere model predicts that eclipsed conformations can be discounted (because of intramolecular repulsions) in favor of the three staggered conformations (shown here as *trans*, *gauche*, and *gauche* primed).

However, the more complex the molecule, the more difficult it is to make valid predictions from the qualitative model. For this reason, more quantitative treatments have been developed.

Quantitative Approaches

Basic principles

It is assumed that a molecule will occupy that conformation which gives the lowest energy to the system. The main problem, then, is to calculate the energies of a wide variety of different conformations in order to decide which particular conformation yields the lowest energy. In practice, three factors must be taken into account:

1. Short-range nonbonding intramolecular forces between the atoms.
2. Long-range nonbonding intramolecular forces (this is a particularly important factor in large flexible molecules).
3. *Inter*molecular forces between atoms on adjacent molecules. (Intermolecular forces are very difficult to predict.)

For each of these three types, it is necessary to make decisions about:

(a) The distances between the atoms.
(b) The nature of the potential between the atoms.

For small molecules, information about the *distances* between atoms in the same molecule can usually be obtained from single-crystal X-ray data or from infrared-Raman or microwave experiments. However, there is very little agreement between different research workers on the best nonbonding *potential* to use, especially when atoms other than hydrogen are present. This aspect will be discussed later. Some investigators also believe that an "intrinsic" threefold rotational barrier is encountered when a bond undergoes torsion, and that this effect is connected more with the characteristics of the bond itself and the orbital arrangement than with the nonbonding interactions of the groups attached to the skeletal atoms.

The interatomic potential

If the assumption is made that short-range intermolecular forces dominate the energetics of conformational changes, then one of the critical problems is to identify the nature of the forces between the contributing atoms. Thus, for calculations on small molecules, use is made of quantum mechanical or semiempirical potentials such as the Lennard-Jones or Buckingham potentials (see later) which require the use of constants that have been derived from experimental data (usually from the behavior of gases).[1]

[1] J. O. Hirschfelder, C. F. Curtiss, and R. B. Bird, *Molecular Theory of Gases and Liquids* (New York: Wiley, 1954).

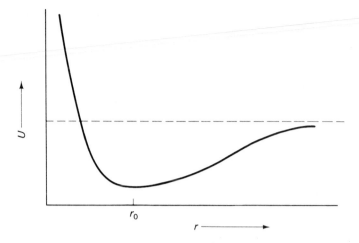

Fig. 18.2 Interatomic potential between two nonbonded atoms. The
energy (U) rises sharply as the nuclei approach each other at short
distances (r), whereas the interaction energy is zero at very large distances. The energy minimum occurs at the point r_0.

When individual nonpolar gas molecules approach each other, their energy
varies in the familiar way shown in Figure 18.2. As the two molecules approach, they
first experience a "London" attractive force, and eventually a repulsive force due
to nuclear–nuclear repulsion. The energy minimum at r_0 represents the position at
which these two opposing forces are in equilibrium. This distance often corresponds
closely to the "intermolecular contact distance" or van der Waals distance in solids.
It is generally assumed that the same type of potential curve describes the effects of
nonbonding *intra*molecular interactions in larger molecules. The interactions are
assumed to be attractive at large distances and repulsive at short distances. The
main problem is to predict the exact shape of the curve for different interactions, for
example, for a hydrogen–hydrogen, hydrogen–carbon, or chlorine–chlorine
interaction.

The Lennard-Jones potential is possibly the best known and most widely used
formula. It has the form shown in (1).

$$U_{ij} = \frac{B_{ij}}{r_{ij}^{12}} - \frac{A_{ij}}{r_{ij}^{6}} \tag{1}$$

where U_{ij} is the energy at any distance, r, between the atoms i and j, and A and B are
constants. The $-A/r^6$ term represents the "dispersion" attraction between the
atoms, and $+B/r^{12}$ term represents the repulsion. Note that the repulsive term is
steeper or "harder" than the attractive term. In fact, the "hardness" or "softness"
of both the attractive and repulsive terms will depend on the types of atoms being
considered. The equation given above is known as a "6–12" potential. However,
different investigators may use "7–11" or other variants of the formula. The
constants A and B determine the depth of the well and the distance of r_0. Thus, if the

appropriate constants can be found, it should be possible to calculate the energy of any nonbonded interaction for any distance, r. The sum of all the interactions in the molecule should then represent the energy of the molecule in a particular conformation.

Other equations can be used for the energy calculation. The modified Buckingham potential, the Stockmayer potential, and others are described in standard texts[1] and in the general literature.[2]

The choice of a suitable potential is very much a matter for individual preference. However, the use of constants taken from gas data may not always be valid in calculations for larger molecules. For example, the interaction of two H_2 molecules leads to significant repulsions at distances closer than ~ 2.5 Å with an r_0 minimum at distances greater than 3 Å. By contrast, it is known that when two C—H units interact, the $H \cdots H$ repulsions are significant only at distances less than 1.9 Å, with an r_0 minimum at 2.4 Å.

In addition to the use of potential functions that describe the dispersion attractions and van der Waals repulsions, it is often necessary to include a term to account for Coulombic forces. Point-charge equations can be used to simulate dipolar effects, but these are not entirely satisfactory.

Model calculations on small molecules

First, let us consider *ethane*. The barrier height for the torsion of the carbon–carbon bond is known from experimental work to be about 2.8 kcal/mol. The torsional potential in this molecule can be calculated by summing the potential energies of all the interactions shown in Figure 18.3 for all conformations generated

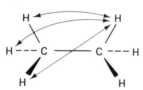

Fig. 18.3 Model for the ethane molecule, showing the intramolecular interactions that must be summed to calculate the overall energy for each conformation generated by torsion of the C—C bond.

by 360° torsion of the C—C bond in, say, 10° increments. The calculated potential curve appears as shown in Figure 18.4. It should be noted that a torsional barrier of about 2.8 kcal/mol in ethane means that at room temperature the molecule can switch from one minimum to another at a rate of about 10^{10} times per second.

As a preliminary exercise to aid in the visualization of conformational effects in polymers, it is instructive to consider also the torsional energy profile of 1,2-*dichloroethane*, as shown in Figure 18.5. Clearly, the replacement of a hydrogen

[1] J. O. Hirschfelder, C. F. Curtiss, and R. B. Bird, *Molecular Theory of Gases and Liquids* (New York: Wiley, 1954).

[2] See particularly publications by Birshtein and Ptitsyn; De Santis et al.; Flory (*Statistical Mechanics of Chain Molecules*); Hopfinger; Lowe; Scheraga; and Volkenstein in the Suggestions for Further Readings at the end of this chapter.

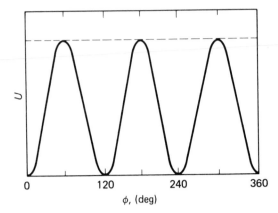

Fig. 18.4 Calculated dependence of the potential energy function, U, on the angle of internal rotation in ethane. [From M. V. Volkenstein, *Configurational Statistics of Polymeric Chains* (High Polymers, Vol. XVII) (New York: Wiley–Interscience, 1963).]

atom on each carbon by a chlorine atom markedly distorts the torsional profile in such a way as to favor the *trans*-conformation (Figure 18.1). The barrier designated U_1 is 3.05 kcal/mol, and U_2 is 5.58 kcal/mol.

Finally, as an additional model, we consider *n*-butane (Figure 18.6). This molecule possesses three skeletal bonds that can undergo independent torsion. Torsion of bond ψ, while bonds ϕ and χ are held in a fixed position, generates a regular, threefold profile, similar to the one shown in Figure 18.4, but with higher barriers. Torsion of bond ϕ, without torsion of ψ or χ generates a profile similar to Figure 18.5 with three stable conformers. However, if more than one bond is permitted to twist, the potential must be represented by a three-dimensional energy

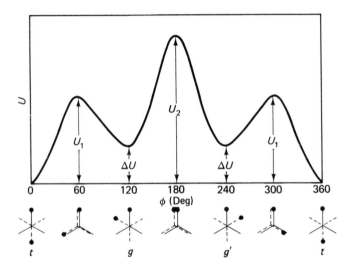

Fig. 18.5 Potential as a function of torsional angle for 1,2-dichloro-ethane. The chlorine atoms are represented by the solid circles. [From M. V. Volkenstein, *Configurational Statistics of Polymeric Chains* (High Polymers, Vol. XVII) (New York: Wiley–Interscience, 1963).]

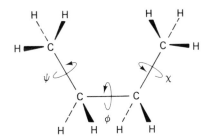

Fig. 18.6 Model for exploration of the effects on the potential energy of torsions of bonds
ψ, ϕ, and χ in n-butane.

surface (based on torsional angle ψ, angle ϕ, and energy) rather than by a profile. In the case of n-butane, the situation is more complicated because *three* bonds can presumably undergo independent torsional motions and a four-dimensional system would be needed to represent the overall conformational energy.

In practice, two bonds are allowed to undergo independent torsion and the third, fourth, fifth, and so on, bonds are kept fixed. Then after the energy has been minimized for the first two, successive additional bonds can be varied. Figure 18.7 shows a surface calculated for n-butane for independent torsion of bond ϕ, and synchronous torsion of bonds ψ and χ. Because the surface has a fourfold symmetry, only one-fourth need be shown. The calculated energies are indicated on the contours (in kcal/mol), and the *minimum energy path* or *energy pass* between two low energy sites is shown by the dashed line. When the real molecule undergoes torsional motions, it is likely to follow this path of least resistance. The preferred conformation of the molecule should in principle be represented by the ψ and ϕ values of a low-energy site on the surface.

An extension of these arguments to higher linear alkanes indicates that the number of possible stable conformers is given by $3^{(N-3)}$, where N is the number of carbon atoms in the chain. For polyethylene, the number of accessible conformational states reaches an astronomical number. Hence, simplifications are needed.

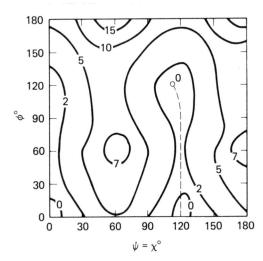

Fig. 18.7 One-fourth of the
energy surface calculated for the
n-butane. [From T. M. Birshtein and O. B. Ptitsyn, *Conformations of Macromolecules*
(New York: Wiley–Interscience,
1966).]

SHORT-CHAIN MODELS FOR POLYMERS

The Short-Chain Concept

One simplification is to assume that the conformational preference of the whole polymer chain can be represented by the preferred conformation of a short segment of that chain. This is a key concept on which a large amount of research work is based. It depends on the assumption that the conformation of the whole polymer is determined by a composite of all the *short-range* interactions in the molecule.

This allows an enormous simplification of the problem. The question is: How valid is this assumption? It is generally believed that if the chain assumes an extended type of conformation [such as (a) or (b) in Figure 18.8], the assumption is reasonably valid. However, if a tight helix is formed (Figure 18.8c), the long-range forces become important and cannot be neglected. It is almost certain that the *inter*molecular forces exert a significant effect in the crystalline state, especially when polar side groups are present. However, relatively few calculations take this into account. Moreover, it will be clear from the assumptions made at the beginning of this chapter that the concept cannot apply accurately to a polymer random coil,

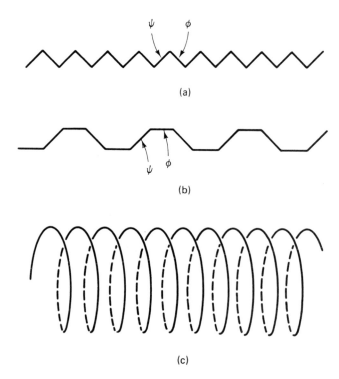

(a)

(b)

(c)

Fig. 18.8 Short-range intramolecular forces (and intermolecular effects) dominate the conformational energy preferences for extended conformations, such as (a) or (b). However, longer-range forces become important if the polymer occupies a tight helical conformation (c).

although it might apply in a statistical sense to suggest relative preferences of conformational states. As will be seen later, one of the purposes of short-chain conformational analysis is to identify accessible minima and estimate the probability that specific minima are occupied. Thus, the approach discussed in the sections that follow is of value *both* in relation to discrete, repeating conformational structures and to random coils.

Conformational Nomenclature

The conformation of the chain can be represented simply by quoting the torsional angles of the backbone bonds. Unfortunately, two quite different conventions are in use. In the first, all skeletal angles are quoted as deviations from the *trans–trans* (i.e., planar zigzag conformation of the chain). For example, the conformation shown in Figure 18.8a can be represented by the symbolism $\psi = 0°$, $\phi = 0°$. The *cis–trans*-planar conformation shown in Figure 18.8b is given the notation, $\psi = 180°$, $\phi = 0°$. Helices (Figure 18.8c) must be represented by ψ and ϕ values between $0°$ and $180°$, and these values may be positive or negative. However, some workers use the *cis–cis* conformation as the $\psi = 0$, $\phi = 0$ starting conformation, and this can lead to confusion.

Because torsional motions can be left- or right-handed, an additional convention is needed. One convention is as follows. If we look down a bond from atom 1 to atom 2 (Figure 18.9a), then ψ is *positive* if the bond (x) must be rotated *clockwise*

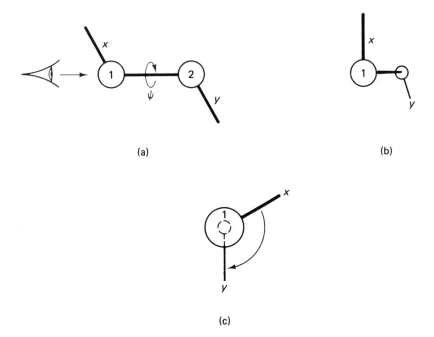

(a)   (b)

(c)

Fig. 18.9   Common sign convention for torsional rotation of bonds. If bond x must be rotated in a clockwise direction to superimpose it on bond y, the torsion is given a positive value.

by the angle needed to superimpose it on bond (y). Counterclockwise rotations of less than 180° are considered to be negative. Using this system, the *trans*-planar structure, shown in Figure 18.8a, would be described as $\psi = \phi = 180°$ rather than $\psi = \phi = 0°$. However, the *cis–trans* structure (Figure 18.8b) would continue to be described as $\psi = 0°$, $\phi = 180°$.

We prefer a system in which extended *trans*-planar structures are depicted as $\psi = \phi = 0°$, with the sign of the torsional angle determined as follows. A positive torsion angle is generated when the rear bond (y in Figure 18.9) has been moved less than 180° in a clockwise direction *from* the *trans–trans* ($\psi = \phi = 0°$) position to generate the observed conformer. Positive rotations generate right-handed (clockwise) helices—a scheme that is easy to remember.

Use of the Hard-Sphere Approach

The simple hard-sphere-model approach that was described earlier for small molecules can be used for short-chain segments of polymers also. For example, polyethylene is simply a logical extension of the *n*-butane molecule, and it requires only a brief examination of models or a few sketches with pencil and paper to see that the extended zigzag conformation (**6**) is generated by this scheme. The

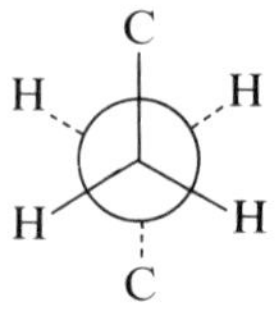

6

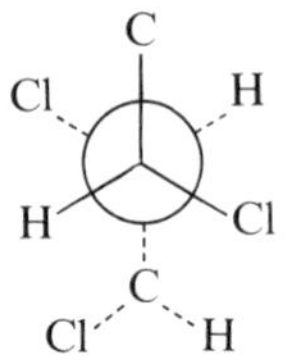

7

predicted conformational profile for torsion of one bond in polyethylene would resemble Figure 18.10.

If it is assumed that chlorine atoms are particularly repulsive to each other, the same argument can be used to predict that syndiotactic poly(vinyl chloride) (**7**) should also form a planar zigzag arrangement (**8**), since this conformation places the longest distance between the bulky chlorine atoms. This conformation is found in the solid state, although crystal packing forces may be partly responsible for this behavior.

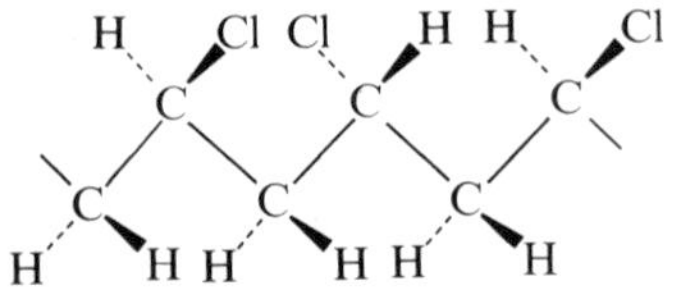

8

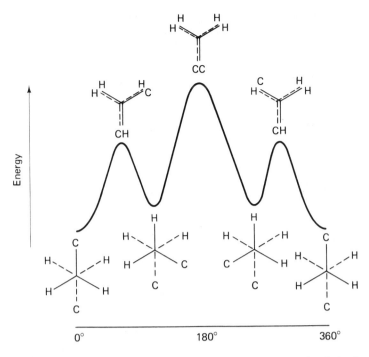

Fig. 18.10 Estimated torsional profile for 360° torsion of a skeletal bond in a short-chain segment of polyethylene, based on a hard-sphere rationalization.

Use of Intramolecular Potentials

It will be obvious that the approach discussed earlier for the calculation of torsional energy profiles and surfaces for small molecules can be extended in a relatively straightforward manner to short segments of a long chain. The assumption must be made once again that the conformational preference is dominated by short-range forces.

Consider the short segment of a polyethylene chain shown in **9**. The *intramolecular distances* between atoms can be guessed with some certainty from bond-

9

angle and bond-length data obtained from low-molecular-weight compounds (from single-crystal X-ray work) or from other polymers. For example, bond angles at tetrahedral carbon are often close to 109.5°, the C—H bond length in saturated organic compounds is 1.09 Å, and the C—C bond length is 1.54 Å. There are only 15 atoms in this segment, and the calculation of the distances and energies for all the nonbonding interactions (H · · · H, H · · · C, C · · · C, etc.) with the segment held in one chain conformation, is a trivial problem. Even if all the possible conformations which result from 360° incremental torsion of bonds A and B are considered, the calculations are quite manageable with the use of a modern computer.

Macromolecules that contain very polar side groups must be treated in a special way. This includes polymers, such as poly(methyl methacrylate), polyacrylonitrile, poly(vinyl alcohol), and so on. Polar or hydrogen-bonding forces are difficult to calculate.

The following sequence of operations would be followed in order to perform a preliminary calculation of the conformation of a polymer:

1. Consider a short segment of the chain and place it in a starting conformation. For example, a possible starting conformation for polyethylene would be the one shown in Figure 18.11.

2. Assign assumed bond angles and lengths. (A good starting point would be to use values obtained from X-ray single-crystal studies on smaller hydrocarbons.)

3. Set up a computer program to permit torsion of bonds C_2—C_3 and C_3—C_4 independently through increments of, say, 10°. At each conformational position, calculate the nonbonding distances (C_1 · · · C_5, H_1 · · · H_6, etc.).

4. Introduce a suitable nonbonding potential into the calculation and allow the computer to calculate the interaction energy for every calculated distance. Sum all energies for each conformation until C_2—C_3 and C_3—C_4 have both undergone 360° torsion.

5. Obtain a matrix printout from the computer giving the energies for all incremental values of ψ and ϕ, and plot energy contours at preselected intervals.

6. Select the preferred conformational minima from the surface, or use the selected minimum as a starting point for additional calculations to explore the effect of the torsional motions of the adjacent skeletal bonds. In practice, this latter procedure is often incorporated into a "hunting" program to find the overall conformational minimum if four, five, or more skeletal bonds are allowed to undergo independent torsion.

7. Find the minimum energy pass between the preferred minima, and calculate the barrier heights, Boltzmann distributions, and so on.

8. Calculate the number of monomer units per repeat. This information would be used in conjunction with X-ray crystallographic studies (see Chapter 19).

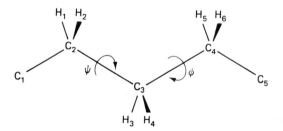

Fig. 18.11 Short-chain segment used as a model for a preliminary conformational energy calculation on polyethylene.

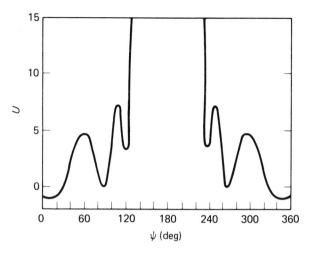

Fig. 18.12 Calculated conformational potential energy profile for a short segment of a polytetrafluoroethylene chain. [Modified from P. De Santis, E. Giglio, A. M. Liquori, and A. Ripamonti, *J. Polymer Sci. (A)*, **1**, 1383 (1963).]

Three examples of the results from this type of treatment will be mentioned.

1. A calculated potential-energy profile for a short segment of a poly(tetra-fluoroethylene) chain is shown in Figure 18.12. Note that the profile shows more detail than would have been predicted from hard-sphere models alone. Several pronounced minima are present and this suggests that several crystal–crystal transitions might occur in this polymer if the system can surmount the barriers. The principal minima lie just a few degrees on each side of the main $0°:0°$ minimum (planar zigzag). This implies that the chain should prefer a very loose helical structure. This result is compatible with X-ray diffraction data which indicate that, below 20°C in the crystalline state, the polymer assumes a helix with 13 monomer units in six turns.

2. Figure 18.13 shows an energy profile for polyisobutylene when a short segment of the molecule is allowed to find the *minimum energy path* by compensatory "avoidance" motions of adjacent units. Hard-sphere molecular models imply that this polymer is very highly hindered, an implication that is at variance with the elasticity of this material. However, the calculated torsional profile indicates that, in spite of the fact that the lowest minima correspond to a tight, helical arrangement, these minima form part of a broad, low-energy plateau that would permit torsional motions of $\sim 200°$ before a high barrier was encountered. Such synchronous bond movements can also be visualized as "crankshaft motions." Conversion of one conformer to another in polyisobutylene should, therefore, be a very facile process, and it is interesting to speculate that this feature may explain the elasticity and low glass transition temperature of this polymer.

3. Calculations of this type can be extended to polymers that have complex side groups. An example is poly[bis(trifluoroethoxy)phosphazene],

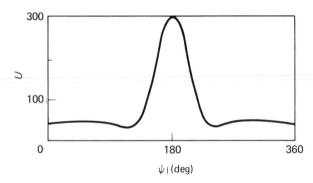

Fig. 18.13 Calculated energy profile for a short segment of poly-isobutylene following the minimum-energy pathway allowed by compensatory "avoidance" motions of the nearby chain units. [Modified from P. De Santis, E. Giglio, A. M. Liquori, and A. Ripamonti, *J. Polymer Sci. (A)*, **1**, 1383 (1963).]

$[NP(OCH_2CF_3)_2]_n$, in which the side groups can themselves occupy a variety of different conformational orientations for each conformation of the chain. An approach to this problem is to allow the side groups to undergo energy minimization for each tested conformation of the backbone. Figure 18.14 shows an energy surface calculated in this way for the segment shown in **10**. In spite of the bulkiness of the

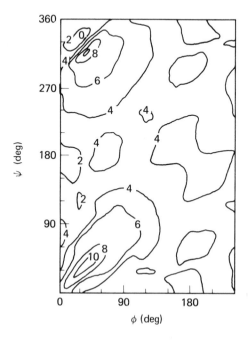

Fig. 18.14 Energy surface calculated for poly[bis(trifluoroethoxy)phosphazene]. [From R. W. Allen and H. R. Allcock, *Macromolecules*, **9**, 956 (1976).]

$$\text{OCH}_2\text{CF}_3 \qquad \text{OCH}_2\text{CF}_3$$
$$\text{OCH}_2\text{CF}_3 \qquad \text{OCH}_2\text{CF}_3$$

10

side groups, the surface shows broad areas of low potential that suggest considerable conformational mobility. This is compatible with the low T_g value of this polymer $(-66°C)$.

Longer-Chain Models

The simplest approach based on the foregoing treatment would involve a calculation of the potential surface resulting from the independent torsion of two adjacent skeletal bonds, as shown in 18.11 and **9**. Valuable data can indeed be obtained by the use of such techniques, especially when the polymer assumes an extended *trans*-planar or *cis–trans*-planar conformation. However, caution must be exercised during the choice of the actual energy minimum or minima occupied by the polymer chain. Some minima calculated for the short-chain segment may actually be eliminated by longer-range forces. For example, a model based on torsion of a short, two-bond segment of the chain might generate an energy minimum at a conformation that represents a very tight helix or even a *cis–cis* conformation. This conformation may be totally precluded by a 1:6 collision of the skeletal atoms. Thus, a more rigorous treatment would require that the torsion of a third adjacent skeletal bond should also be included in the treatment. Ideally, a segment of four or five skeletal bonds should be considered. However, the solution of a three- or four-bond problem seriously raises the complexity of the task. If the polymer possesses multiatomic side groups which can also undergo complex torsional motions, the calculations become almost prohibitively complicated.

Three approaches are possible:

1. The complete permutation of conformations for the three- or four-bond torsional segment could be calculated as an extension of the program outlined in the preceding section. In practice, this process would be exceedingly expensive in terms of computer time.

2. A "hunting" or a "searching" procedure may be instituted. In this, the minimum for a two-bond segment may be found first, and this minimum is then used as a basis for the search for a resultant minimum when the third or fourth bonds are considered. Multiatomic side group conformations can be energy minimized for each conformation of the skeleton. This process saves computer time. However, care must be taken to ensure that minima located early in the calculation remain the principal minima at the end. In other words, a danger exists that new minima (the "real" minima) may be present in conformational space not searched by the program. Thus, judgment, insight, and intuition play an important role in this approach. Moreover, it is essential that the calculations should be tied to

experimental facts at nearly every stage to prevent them from becoming unacceptably speculative. The results should be reviewed at each step in terms of actual polymer conformations, glass transition temperatures, barriers to rotation in small molecules, and so on, until the investigator is satisfied that he or she is on the right track. This back-checking procedure is essential if atoms other than carbon or hydrogen are involved since very little factual potential data are available for such atoms.

 3. The process can be reduced to a search for minima in a two-, three-, four-, or higher-fold rotational isomeric model. This procedure will be discussed in a later section.

RELATIONSHIP TO THE STATISTICAL COIL CONCEPT

As mentioned earlier, the regular, repetitive, extended, or helical chain conformation is only one extreme of the conformational spectrum. This is the type of conformation that is most readily treated by the techniques discussed in the previous sections. The other extreme is the random coil, in which a wide variety of different conformations is assumed to occur more or less randomly along the length of an individual chain.

 The random coil structure of a polymer has traditionally been treated by analogy with *random flight* calculations. In a random flight simulation, a particle is allowed to undergo a series of random displacements in three dimensions in a manner reminiscent of Brownian motion or the diffusion of gas molecules. The analogy with randomly coiled polymers comes from the supposition that the motion of such a particle mimics the path traced by a polymer chain, with the points of directional change corresponding to the bond angles at the atomic positions along the skeleton.

 Such a model for polymers seems to be rather fanciful until it is recognized that certain surprising similarities exist between the two ideas. Although polymers have fixed bond angles and bond lengths, the phenomenon of torsional mobility around the backbone generates a randomness of direction throughout the chain. The direction of, say, the first bond is rapidly "forgotten" at points farther and farther from it along the chain. Thus, the *statistical* properties of a polymer chain do, in fact, resemble those of a random flight or random coil.

 Perhaps the most important starting point for a comparison of the two systems is in relation to the idea of the *mean-square end-to-end distance*, $\overline{r_0^2}$, of a polymer. In a fully extended chain, the end-to-end distance, r, will be a measurable quantity, dependent on the number of repeating units and the type of conformation assumed by the polymer. Only a very low probability exists that a particle undergoing random flight would follow such a linear or regular extended helical path.

 In a random flight, such as the one in Figure 18.15, the mean-square end-to-end distance is given by (see also page 391)

$$\overline{r_0^2} = Nb^2 \tag{2}$$

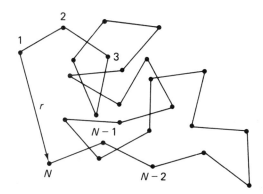

Fig. 18.15 End-to-end distance, r, for a freely jointed (i.e., no fixed valence-bond angles) chain made up of N atoms.

where N is the number of links in the chain, with a link length of b. The probability that a particular end-to-end distance will be found within the range of r to $r + dr$ is determined by the distribution function, $W(r)$, and is defined by

$$W(r)4\pi r^2 \, dr = (\tfrac{3}{2}\pi Nb^2)^{3/2} \exp\left(-\frac{3r^2}{2Nb^2}\right)4\pi r^2 \, dr \tag{3}$$

A familiar distribution plot is generated by this function, in which both very large and very small values of r have a low probability, and the maximum in the probability curve falls at an intermediate value (Figure 18.16). The actual shape of the curve and the position of the maximum for real molecules depends mainly on N and on several other factors, including the bond angles in the backbone and the torsional mobility of the skeletal bonds.

The fact that real polymers possess more-or-less fixed skeletal bond angles (θ) rather than freely jointed atoms requires a modification of equation (2). The modified equation is[1]

$$\overline{r_0^2} = \frac{1 - \cos\theta}{1 + \cos\theta} Nb^2 \tag{4}$$

Because bond angles are normally larger than $\pi/2$, the factor $(1 - \cos\theta)/(1 + \cos\theta)$ is greater than 1, and the bond-angle restriction will *increase* the value of $\overline{r_0^2}$. For an aliphatic carbon-atom chain, this increase will be by a factor of 2.

The value of $\overline{r_0^2}$ is also affected by restrictions to free torsion of the skeletal bonds. Any limitations to free 360° torsion of the skeletal bonds will serve to further increase the value of $\overline{r_0^2}$. Equations for an analysis of this influence will be given later. Hence, conformational analysis, which provides information about restrictions to free bond torsion, plays an important role in the study of random coil polymer systems as well as regularly extended chains. Thus, connections can be

[1] P. J. Flory, *Principles of Polymer Chemistry* (Ithaca, N.Y.: Cornell University Press, 1953), Chap. X.

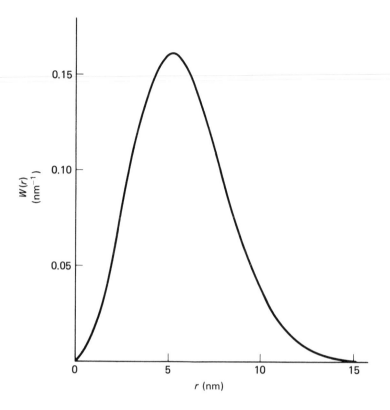

Fig. 18.16 Distribution function, $W(r)$, for the distance between chain ends of a statistical coil for the case $N = 10^4$ and $b = 2.5$ Å. [From P. J. Flory, *Principles of Polymer Chemistry* (Ithaca, N.Y.: Cornell University Press, 1953), p. 406.]

made between physical properties that depend on random coil arrangements (such as elasticity or dilute solution properties) and the results of conformational calculations. These will be discussed briefly later in this chapter.

INFLUENCE OF THE BOLTZMANN DISTRIBUTION

Calculated torsional potential profiles, such as Figures 18.12 or 18.13, or energy surfaces, such as Figure 18.14, are valuable because they suggest (1) the location of preferred (low-energy) conformations, (2) the heights of the barriers that must be surmounted by the molecule to get from one well to another, and (3) the minimum-energy pathway that the molecule probably follows in order to switch from one conformation to another.

At first sight, it might be imagined that a polymer molecule would always assume the one conformation that corresponds to the lowest-energy well. Yet the experimental facts tell us otherwise. The existence of random coil arrangements, the evidence that many extended polymer chains occupy supposed high-energy conformations, and the existence of conformational transitions all indicate that the lowest-energy well is not the only one occupied under all circumstances.

In fact, as is well known from elementary physical chemistry considerations, a molecule will "populate" different accessible energy levels to a degree that depends on the temperature. All other things being equal, the molecule will occupy the lowest energy well at low temperatures and will populate progressively higher levels as the temperature is raised. Thus, the probability that a particular segment of the chain will occupy a well that exists for a torsional angle, ψ, will be given by the Boltzmann equation,

$$P(\psi) \sim \exp\left[\frac{-U(\psi)}{RT}\right] \tag{5}$$

where U is the energy of that well. The higher the temperature, the greater is the probability that a chain will assume higher-energy conformations. The heights of the barriers between the wells should affect the *rates* at which bonds will switch back and forth from one conformation to another, but usually these barriers are low enough to permit many transitions per second at normal temperatures.

Thus, the important point to be noted is that conformational energy calculations can be used to estimate the *probability* that one or another conformational state will be occupied. Clearly, this has important ramifications for a consideration of the way $\overline{r_0^2}$ varies with temperature. It is also extremely valuable for the statistical matrix approach, discussed in the next section.

THE STATISTICAL MATRIX APPROACH

The Rotational Isomeric State Model[1]

By now it will be clear that any attempt to describe a polymer molecule in terms of one infinitely repeating chain conformation is probably not valid, except in the special case of polymer single crystals. Moreover, it will also be clear that the existence of different conformational states in the same chain adds enormously to the complexity of attempts to describe these systems, because changes in the conformation of one segment of the chain will affect the stability of the nearby conformational states. Thus, an *exact* description of the conformation of a macromolecule is beyond the capability of our present computational systems.

However, a *statistical* description is feasible if we can be satisfied with information about the *probabilities* that certain states will be occupied by the molecule.

[1] M. V. Volkenstein, *Configurational Statistics of Polymer Chains* (New York: Wiley–Interscience, 1963).

Moreover, an enormous simplification of the problem can be obtained if it is assumed that each skeletal bond has access to only a small and limited number of torsional states. For example, using the hard-sphere model mentioned earlier, these states could be the three staggered conformations (Figure 18.1) commonly used for organic compounds. Alternatively, on a more sophisticated level, they could represent the principal energy wells detected from conformational energy calculations. This approach is called the *rotational isomeric state model*. For polymers that contain carbon atoms in the backbone, three possible states only are considered to be accessible; *trans*, *gauche*(+), and *gauche*(−). For polymers that contain other atoms, the number of rotational isomeric states may be less than or more than three.

The Statistical Weight Matrix[1]

If a threefold rotational isomeric model is used, the conformation at each chain atom can be reduced to a choice between *trans*, *gauche*(+), and *gauche*(−). Thus, torsion of two adjacent bonds (say, *A* and *B*) generates nine possibilities which can be depicted by a 3 × 3 matrix. Examination of molecular models or a consideration of torsional energy maps allows the investigator to make a judgment about the relative probabilities of each of the nine possibilities. If a hard-sphere model is to be used, conformations that generate "collisions" (serious interpenetration of van der Waals radii) are assigned a probability of zero. For example, in saturated carbon-backbone polymers, successive *gauche*(+) and *gauche*(−) conformations cause serious steric hindrance. Those interactions that appear to involve nonbonding attractions are assigned a high probability, and so on. Alternatively, the probabilities may be estimated from the calculated energies of the wells and from the Boltzmann populations of these wells.

As an example, consider a segment of a chain in which two adjacent skeletal bonds are permitted to undergo threefold rotational isomeric torsion. Assume that the *trans* conformation of each bond represents the lowest-energy state. The statistical weight matrix is

$$
\mathbf{U}' = \begin{array}{c} \\ (t) \\ (g+) \\ (g-) \end{array}
\begin{array}{ccc} (t) & (g+) & (g-) \end{array}
\left| \begin{array}{ccc} 1 & u_{12} & u_{13} \\ u_{21} & u_{22} & u_{23} \\ u_{31} & u_{32} & u_{33} \end{array} \right| \tag{6}
$$

The *trans–trans* combination is assigned an arbitrary statistical weight of 1. Suppose now that an examination of molecular models or of a calculated energy surface reveals that combinations of the type g(+)g(−) and g(−)g(+) are prohibitively high in energy. These can then be assigned a statistical weight of 0. If the molecule is structurally symmetric, and the energy of the *gauche–trans* combination is assumed

---

[1] P. J. Flory, *Statistical Mechanics of Chain Molecules* (New York: Wiley–Interscience, 1969).

to be one-half of the energy of the *gauche–gauche* (all energies being compared to that of the *trans–trans* combination), the matrix can be simplified to

$$
\mathbf{U}' = \begin{vmatrix} 1 & \delta^{1/2} & \delta^{1/2} \\ \delta^{1/2} & \delta & 0 \\ \delta^{1/2} & 0 & \delta \end{vmatrix}
\tag{7}
$$

where $\delta = \exp\left[-\Delta U_{g(+)g(+)}/KT\right]$, and $\Delta U_{g(+)g(+)}$ is the energy of the *gauche*($+$) *gauche*($+$) well above that of the *trans–trans* well.

A second 3×3 matrix can then be calculated for bonds B and C. The product of the two matrices will reflect the probability of the various combined conformations of the two bonds. The process can be extended for four or more bonds for more complex repeating sequences. The results of such calculations have been used successfully to explain the mean square end-to-end distances of a number of polymers.

A calculation of $\overline{r_0^2}$ requires a statistical mechanical average projection of each bond on every other bond. These projections can be expressed as products of *transformation matrices*, with their averages determined with the use of statistical weight matrices.[1] This allows the ratio, $\overline{r_0^2}/nb^2$ to be given in closed form as a function of the transformation and statistical matrices.

The way in which the mean-square end-to-end distance, $\overline{r_0^2}$, can be related to the hindered rotation about the skeletal bonds is as follows. As was discussed earlier, equation (4) must be modified further to take into account the increases in $\overline{r_0^2}$ that results from the lack of totally uninhibited rotation about the backbone bonds. For a threefold rotational isomeric model of a vinyl polymer, $-CH_2CHR-$, if the chains are atactic, equation (8) holds.[2]

$$
\overline{r_0^2} = \left(\frac{1 - \cos\theta}{1 + \cos\theta}\right)\left(\frac{1 - \eta^2 - \varepsilon^2}{(1 - \eta)^2}\right)Nb^2
\tag{8}
$$

where θ is the bond angle, η the average value of the cosine of the torsional angle, and ε the average value of the sine of the torsional angle. For polymers of structure, $-CX_2-CY_2-$, in which the potential function is symmetrical, the term $\varepsilon = 0$, and equation (8) reduces to[3,4]

$$
\overline{r_0^2} = \left(\frac{1 - \cos\theta}{1 + \cos\theta}\right)\left(\frac{1 + \eta}{1 - \eta}\right)Nb^2
\tag{9}
$$

Polyethylene- and vinylidene-type polymers would fall into this category.

[1] P. J. Flory, *Statistical Mechanics of Chain Molecules* (New York: Wiley–Interscience, 1969).
[2] D. K. Carpenter, "Colloids," *Encyclopedia of Polymer Sci. and Technol.* (H. F. Mark, N. G. Gaylord and N. M. Bikales, eds.), **4**, 16 (1966).
[3] W. J. Taylor, *J. Chem. Phys.*, **15**, 412 (1947).
[4] P. J. Flory, *Principles of Polymer Chemistry* (Ithaca, N.Y.: Cornell University Press, 1953), Chap. 10.

Thus, with the use of equation (8) or (9), it is possible either to predict $\overline{r_0^2}$ from statistical matrix calculations, or to use experimental values of $\overline{r_0^2}$, derived from solution viscosity, light scattering, or elasticity measurements to estimate the average value of the torsional angle.

APPLICATIONS OF CONFORMATIONAL ANALYSIS

Our interest in conformational analysis is based on the theory that the physical properties of materials can be explained in terms of molecular shape or changes in molecular shape. Thus, it is worthwhile to examine briefly some of the areas in which conformational analysis has assisted in an understanding of physical phenomena.

Conformation and Crystallinity

Polymers crystallize because their chains can pack together in a regular manner. Such regular packing depends on two factors: the minimum-energy conformation of the chain and the presence or absence of tacticity.

A random coil will not crystallize. Only in extended chain conformations can the conditions be met for the parallel orientation of adjacent chains. Even so, some extended chain conformations are more prone to crystallize than others. Linear polyethylene crystallizes readily because its preferred conformation is an extended zigzag (3). Such a molecular arrangement can easily be packed in a regular manner in a crystallite. Poly(tetrafluoroethylene) also crystallizes, but the chain conformation is a twisted zigzag arrangement which repeats only once every 13 carbon atoms. Repulsions by the fluorine atoms are responsible for the slight twist of the backbone. The fluorine atoms also maintain the extended conformation and are responsible for the chain stiffness. All this is favorable for the generation of a close-packing arrangement between neighboring chains. Syndiotactic poly(vinyl chloride) crystallizes with the chains assuming a planar zigzag conformation. Again, the chain conformation is highly favorable for crystallization.

In these cases, and in many others, conformational analysis can be used to identify low-energy wells that might be occupied in the microcrystalline or single-crystal state. However, as discussed in the earlier sections, a preferred chain conformation alone is usually not sufficient to generate a long sequence of repeating units with one discrete repetitive conformation, because the higher-energy states also will be populated at normal temperatures. Thus, *inter*molecular forces appear to play a crucial role in stabilizing one conformation at the expense of equivalent or higher-energy states. Such crystal packing forces may even favor a higher-energy single-chain conformational state over a lower one.

The case of syndiotactic poly(vinyl chloride) provides a probable example of this effect. The *trans*-planar (planar zigzag) conformation of this polymer may not be the lowest-energy state for an isolated chain. However, this conformation packs so well into a crystalline matrix (see Figure 18.17) that the overall multichain arrangement undoubtedly has a lower energy than other, less efficient, packing modes for other chain conformational states.

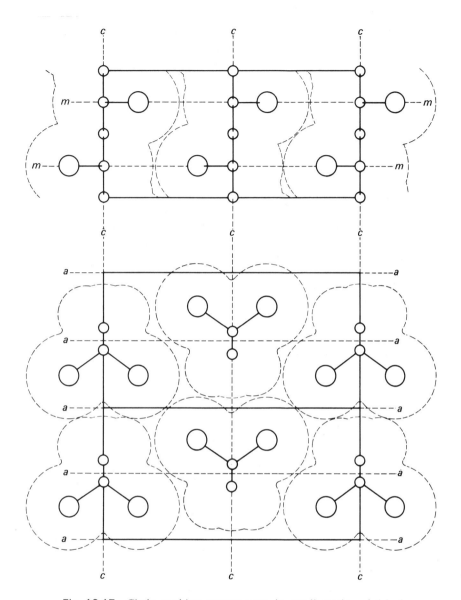

Fig. 18.17 Chain packing arrangement in syndiotactic poly(vinyl chloride). [From G. Natta and P. Corradini, *J. Polymer Sci.*, **20**, 251 (1956).]

Thus, conformational energy calculations cannot be used directly and indiscriminately to predict polymer conformations in the crystalline state, and conversely the conformational results from X-ray structure studies on polymers should not be applied without question to proving the validity of single-chain conformational calculations. Of course, such comparisons would be quite valid if the *inter*-chain forces could be incorporated into the conformational energy calculations. To do this would require not only an exploration of the possible single-chain conformational states, but also an investigation of a wide variety of packing arrangements for each individual chain conformation. Thus, the *prediction* of crystal structures from conformational calculations is a formidable problem. On the other hand, the *rationalization* of known crystal structures in terms of both single-chain calculations and intermolecular forces is more manageable.

The use of single-chain calculations to understand crystal conformations is probably least questionable if two conditions are met. First, if the calculated energy well occupied by a particular conformation is much deeper than competing wells, and the barriers to conformational changes are very high, the interchain packing forces may be insufficient to tip the balance in favor of another conformation. Second, if the cross-sectional profile of a chain in a low-energy conformation is close to circular, individual chains may be capable of undergoing independent rotations within the matrix without appreciable perturbation of the interchain packing forces (as if an individual cylindrically shaped pencil were to be turned in the middle of a stack of similar pencils). Some helices or extended conformations [e.g., the *cis–trans*-planar (0° : 180°) conformation of poly(dichlorophosphazene)] may fall into this category.

Chain Flexibility

It is generally assumed that a correlation exists between the bulk flexibility of a polymer and the flexibility of the molecular chains. The flexibility of a polymer chain can be defined as the ease with which a randomly coiled chain can be unraveled or stretched out. Because the elongation of a chain takes place by the "unwinding" or rotation of backbone bonds (see Figure 17.2), any factor that inhibits the torsion of the skeletal bonds should decrease the polymer flexibility. If this hypothesis is correct, it should be possible to predict the relative flexibilities of polymers from their torsional energy profiles.

Two approaches to this problem exist. First, it can be assumed that the torsional flexibility of a skeletal bond depends on the "span" of torsional oscillations *within* one particular energy well. Thus, the ease of torsion depends on the *steepness* of the walls of the energy well, and flexibility is only an indirect function of the barrier height. This viewpoint is illustrated in Figure 18.18.

The second approach assumes that the backbone bonds undergo torsion by *switching* from one well to another. In other words, the internal rotation of a bond requires a jump over the barrier. If this idea is correct, then flexibility should be

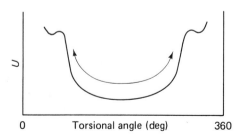

Fig. 18.18 Torsional energy profile, demonstrating the concept that a broad energy well will allow appreciable torsional freedom in the skeletal bonds. This may be responsible for polymer flexibility or elasticity on a macroscopic level.

0   Torsional angle (deg)   360

determined mainly by the ease with which a bond can jump from one well to another and this, in turn, will depend on the Boltzmann distribution. Furthermore, the presence of several closely spaced wells with comparable energy minima should generate a high degree of flexibility.

In practice, it can be shown that the stretching of natural rubber leads to a greater population of the *lowest*-energy well, and this provides an explanation for the fact that many polymers crystallize only when they are stretched. In the absence of definitive data favoring one viewpoint or another, it must be assumed that chain flexibility depends on *both* the energy difference between competing conformations and on the amplitude of torsional oscillations within the principal well.

Relationship to the Glass Transition Temperature

In Chapter 17 it was stressed that a rough relationship exists between the glass transition temperature and the flexibility of the polymer chain. The more flexible the polymer, the lower may be the T_g value. Hence, an important facet of polymer conformational analysis is the need to attempt to predict glass transition temperatures. At the present time, this part of the field is in its infancy and rough correlations only can be made. For example, the low glass transition temperature ($-76°C$) of polyisobutylene can perhaps be correlated with the broad well shown in Figure 18.13. The glass transition temperatures of poly(dihalophosphazenes) increase in the following order: $(NPF_2)_n$, $-90°C$; $(NPCl_2)_n$, $-66°C$; $(NPBr_2)_n$, $-8°C$. Calculations performed on these polymers show a striking increase in barrier height along this series and a narrowing of the principal well.[1]

Elasticity Measurements

The phenomenon of elasticity represents the tendency for a flexible polymer chain to reassume a random coil conformation after having been stretched. The driving force for this contraction is the increase in entropy that is associated with the formation of the random coil. The restoring force in elastic retraction is nearly always dependent on the temperature, and this temperature can be used as a measure of the mean-square end-to-end distance, $\overline{r_0^2}$. Moreover, by measurements of the elasticity as a function of temperature, it is possible to obtain information about the minimum-energy conformation of the polymer and the energy difference between the lowest well and the higher accessible minima.[2,3]

In practice, a crosslinked elastomer is used to prevent chain slippage during the experiment. The two ends of the stretched sample are clamped at a fixed distance from each other. The sample is then heated or cooled and the restoring force, f, between the clamped ends is measured as a function of temperature.

[1] H. R. Allcock, R. W. Allen, and J. J. Meister, *Macromolecules*, **9**, 950 (1976).
[2] A. Ciferri, *J. Polymer Sci.*, **54**, 149 (1961).
[3] A. Ciferri, C. A. J. Hoeve, and P. J. Flory, *J. Am. Chem. Soc.*, **83**, 1015 (1961).

Earlier theories of rubber elasticity were based on the supposition that the restoring force was due entirely to entropic influences. More recent treatments have postulated that the phenomenon is also affected by restrictions to free torsion about the backbone and, hence, an energetic component must also be taken into account.

Consider first an organic polymer that is in an extended all-*trans* conformation. Temperature increases should cause the higher-energy *gauche* states to be populated, and this conformational change must result in a decrease in $\overline{r_0^2}$, which becomes manifest in a contraction of the polymer or an increase in the restoring force if the sample is clamped at a fixed extension. Conversely, if the polymer normally occupies a low-energy *gauche* conformation, temperature increases should progressively populate the higher-energy *trans* conformations, and the $\overline{r_0^2}$ distance should increase. Only if all accessible conformational minima have the *same* energy will no change in $\overline{r_0^2}$ occur on heating and no change in the restoring force will be detected.

In these terms it is of great interest that polyethylene and polyisobutylene show an increase in restoring force (equivalent to a contraction) as the temperature is raised, whereas natural rubber and poly(dimethylsiloxane) rubber show the opposite effect. Thus, the first two polymers appear to possess a near-*trans* conformation as a minimum-energy state (compare with Figures 18.10 and 18.13), while the second two polymers possess their lowest-energy minima at *gauche* conformations.

Quantitatively, the factor measured in such experiments is f_e/f, where $f_e/f = -T[\partial(f/T)/\partial T]$ at constant length and volume. The ratio f_e/f is related to the mean-square end-to-end distance by

$$\frac{f_e}{f} = T \frac{d \ln \overline{r_0^2}}{dT} \tag{10}$$

One of the classical methods for the determination of $\overline{r_0^2}$ is by means of intrinsic viscosities (Chapter 15), because the elasticity data can be correlated with viscosities by means of (11),

$$\frac{f_e}{f} = \frac{2}{3} T \left[\frac{\partial \ln [\eta]}{\partial T} \right]_{T=\theta} \tag{11}$$

where θ is the temperature at which the solution becomes ideal.

The elasticity data can also be used to calculate the energy difference, ΔU, between the accessible torsional energy wells from the relationship

$$\frac{f_e}{Tf} = \frac{d \ln \overline{r_0^2}}{dT} = -\frac{2}{2+\delta} \left(\frac{\Delta U}{kT^2} \right) \tag{12}$$

where $\delta = \delta_0 \exp(-\Delta U/kT)$ and for the case where $\delta_0 = 1$. As an example, the use of this equation yields a value of $\Delta U = 500$ cal/mol for the *trans* to *gauche* energy difference in polyethylene.[1]

[1] A. Ciferri, *J. Polymer Sci.*, **54**, 149 (1961).

Dilute Solution Phenomena

As discussed in earlier chapters, both the scattering of light and the enhancement of solvent viscosity by dissolved polymer molecules are phenomena that can be related to the shape of the macromolecule in solution and to the mean-square end-to-end distance, $\overline{r^2}$. If the solvent is a theta solvent (see Chapter 15), the measured values of $\overline{r^2}$ are equal to the unperturbed value, $\overline{r_0^2}$. Even if the solvent is not a theta solvent, a value for $\overline{r_0^2}$ can be estimated by the use of corrections.

Now, it will be recalled from equation (2) that $\overline{r_0^2}/Nb^2 = 1$ for a freely jointed chain, and that $\overline{r_0^2}/Nb^2 = 2$ for a polymer that has fixed, tetrahedral bond angles. Thus, any deviation from the value of 2 for a carbon–chain polymer provides a measure of the restrictions to torsional motions of the skeletal bonds. Values for $\overline{r_0^2}/Nb^2$ of 6.55 for polyethylene,[1] 6.6 for polyisobutylene,[2] and 8.8 for isotactic poly(methyl methacrylate)[3] have been reported. Moreover, the change in this value as the temperature is varied provides an indication of the way in which higher torsional energy states are being populated. For example, for polyisobutylene[4], the term $(10^3 \, d \ln \overline{r_0^2})/dT = -1.1$. Thus, it is possible to relate actual experimental values, which are directly related to conformational changes, to calculated potential functions and, more particularly, to the results of statistical matrix calculations.

POLYPEPTIDES AND PROTEINS

The most sophisticated objective in current polymer conformational work is the drive to simulate the conformations of polypeptides and proteins by means of nonbonding energy calculations. This objective is important because the biochemical activity of many proteins is a function of chain conformation or conformational changes (see Chapter 8). In principle, it should be possible to calculate the conformation of a protein on the basis of nonbonding interactions. In practice, this is a very complex problem because:

1. A protein is a complex *copolymer*. Each different monomer unit in the chain will give rise to different nonbonded interactions.
2. Coulombic forces are important in these structures.
3. Hydrogen bonding exists.
4. Extensive coiling of the chains brings *long-range* forces into play.
5. The aqueous matrix inside a living system must affect the conformation.

In spite of these difficulties, steady progress has been evident in recent years toward the calculation of protein conformations. However, it must be stressed that the ultimate objective has not yet been achieved. Rather, the research in this area must be viewed as a steady progression in which successively more complicated biopolymers have been treated with increasing success by conformational analysis

[1] P. M. Henry, *J. Polymer Sci.*, **35**, 3 (1959).
[2] T. G. Fox and P. J. Flory, *J. Am. Chem. Soc.*, **73**, 1909 (1951).
[3] G. V. Schulz, W. Wunderlich, and R. Kirste, *Makromol. Chem.*, **75**, 22 (1964).
[4] T. G. Fox and P. J. Flory, *J. Am. Chem. Soc.*, **73**, 1909 (1951).

techniques. Because of the specialized nature of this research, the following summary is intended as an introductory outline only. The reader is encouraged to explore the subject in greater depth by reference to the sources listed at the end of this chapter.

The general strategy has involved first, conformational energy calculations on the simplest small-molecule model structures that might simulate the behavior of amino acid residues present as middle units in a protein chain.[1,2] Linear dipeptides or other simple, short-chain linear oligopeptides that contain glycine–glycine, glycine–proline, or proline–proline sequences, and related oligomers that possess end "blocked" units to eliminate end-group effects have been investigated. The purpose of these preliminary calculations is (1) to develop suitable potential functions that can be refined by comparison with the known experimental behavior of the linear oligomers, (2) to explore the preference of particular peptide linkage environments for *trans*-planar, *cis*-planar, or nonplanar conformations,[3] and (3) to identify sequences of residues that favor the formation of bends in the chain. For example, calculations on the tetrapeptide, Asp-Lys-Thr-Gly, provide an explanation for the existence of bending sites when this sequence appears in the protein α-chymotrypsin.[4]

Next, the calculations have been extended to synthetic linear, homopoly-peptides, such as poly(L-proline) or poly(L-valine). Such calculations, and a comparison of the results with experimental data, allow the role of an aqueous matrix or of *inter*chain packing forces to be evaluated. The interchain interaction potentials thus identified can then be employed later for the calculation of long-range *intra*molecular interactions in a folded chain. The helix-coil transition behavior of the homopolymers can also be assessed by conformational calculations.

The information derived from these types of calculations has then been used to predict the conformations of naturally occurring cyclic oligopeptide copolymers, such as, for example, the cyclodecapeptide, Gramicidin S.[5] The role played by coordinated water can also be assessed at this stage.

Finally, the calculations have been extended to long-chain copolymeric synthetic polypeptides or to proteins. Such studies take into account the interaction energies of the different amino acid residues, and the hydrophilic and hydrophobic character of the amino acid side groups. Known protein structures have been analyzed to ascertain the influence of different sequences in the generation of α-helix, extended chain, other coiled and bent states, and these data have been incorporated into the predictive framework.[6] The more sophisticated recent calculations have been directed toward understanding the reasons for cooperative helix-coil transitions in proteins, the speed and mechanism of protein folding in aqueous media, and the existence of coiled-coil structures in fibrous proteins. Eventually, such calculations will be extended to cover the role played by prosthetic groups, such as metalloporphyrins, in the structure of proteins.

[1] S. Tanaka and H. A. Scheraga, *Macromolecules*, **7**, 698 (1974).
[2] I. Simon, G. Nemethy, and H. A. Scheraga, *Macromolecules*, **11**, 797 (1978).
[3] S. S. Zimmerman and H. A. Scheraga, *Macromolecules*, **9**, 408 (1976).
[4] J. C. Howard, A. Ali, H. A. Scheraga, and F. A. Momany, *Macromolecules*, **8**, 607 (1975).
[5] M. Dygert, N. Go, and H. A. Scheraga, *Macromolecules*, **8**, 751 (1975).
[6] S. Tanaka and H. A. Scheraga, *Macromolecules*, **10**, 9 (1977).

STUDY QUESTIONS

1. If you have access to a computer, set up a simple program for calculation of the energy surfaces for *n*-propane, polyethylene, and polyoxymethylene. Choose your own potential function (variation of energy with interatomic distance) and examine how the energy surface changes with the type of potential used. (This problem could form the basis of a class project with different participants using different potential functions or structural parameters.)

2. What types of polymers might be treated more effectively by the use of a *two*fold rotational isomeric model rather than by the usual threefold rotational isomeric approach?

3. Most models for conformational energy calculations make the assumption of fixed bond angles in the skeleton. What effects on the physical properties of a polymer would you expect if the skeletal bond angles were capable of undergoing considerable valence-bond angle distortion (say over $\pm 20°$) without encountering an appreciable energy penalty?

4. Is it possible in principle to calculate a satisfactory intramolecular nonbonding potential from experimental physical data for polymers? If so, how would you do it, and what would be the weaknesses of the method?

SUGGESTIONS FOR FURTHER READING

BIRSHTEIN, T. M. AND O. B. PTITSYN, *Conformations of Macromolecules* (*High Polymers*, Vol. XXII). New York: Interscience, 1966.

BOVEY, F. A., "NMR Observations of Polypeptide Conformations," *J. Polymer Sci.* (*D*) (*Macromol. Rev.*), **9**, 1 (1974).

BOVEY, F. S., *Polymer Conformation and Configuration*. New York: Academic Press, 1969.

CARPENTER, D. K., "Colloids," *Encyclopedia of Polymer Sci. and Technol.* (H. F. Mark, N. G. Gaylord, and N. M. Bikales, eds.), **4**, 16 (1966).

CIFERRI, A., "Present Status of the Rubber Elasticity Theory," *J. Polymer Sci.*, **54**, 149 (1961).

CIFERRI, A., C. A. J. HOEVE, AND P. J. FLORY, "Stress–Temperature Coefficients of Polymer Networks and the Conformational Energy of Polymer Chains," *J. Am. Chem. Soc.*, **83**, 1015 (1961).

DE SANTIS, P., E. GIGLIO, A. M. LIQUORI, AND A. RIPAMONTI, "Stability of Helical Conformations of Simple Linear Chains," *J. Polymer Sci.*, **A1**, 1383 (1963).

FLORY, P. J., "Foundation of Rotational Isomeric State Theory and General Methods for Generating Configurational Averages," *Macromolecules*, **7**, 381 (1974).

FLORY, P. J., *Principles of Polymer Chemistry*, especially Chaps. X and XIV. Ithaca, N.Y: Cornell University Press, 1953.

FLORY, P. J., *Statistical Mechanics of Chain Molecules*. New York: Wiley–Interscience, 1969.

FLORY, P. J., A. CIFERRI, AND R. CHIANG, "Temperature Coefficient of the Polyethylene Chain Conformation from Intrinsic Viscosity Measurements," *J. Am. Chem. Soc.*, **83**, 1023 (1961).

FLORY, P. J., V. CRESCENZI, AND J. E. MARK, "Configuration of the Poly(dimethylsiloxane) Chain, III: Correlation of Theory and Experiment," *J. Am. Chem. Soc.*, **86**, 146 (1964).

FLORY, P. J., AND J. E. MARK, "The Conformation of the Polyoxymethylene Chain," *Makromol. Chem.*, **75**, 11 (1964).

HOPFINGER, A. J., *Conformational Properties of Macromolecules*. New York: Academic Press, 1973.

LOWE, J. P., "Barriers to Internal Rotation about Single Bonds," *Progr. Phys. Org. Chem.*, (A. Streitwieser and R. W. Taft, eds.), **6**, 1 (1967).

MORAWETZ, H., "Some Studies on the Rates of Conformational Transitions and of *Cis-Trans* Isomerizations in Flexible Polymer Chains," *Contemp. Top. Polymer Sci.*, **2**, 171 (1977).

ORVILL-THOMAS, W. J. (ed.), *Internal Rotations in Molecules*. New York: Wiley, 1974.

PETERLIN, A., "Conformation of Polymer Molecules," in *Polymer Science and Materials* (A. V. Tobolsky, and H. F. Mark, eds.), Chap. 3. New York: Wiley–Interscience, 1971.

POLAND, D. AND H. A. SCHERAGA, *Theory of Helix-Coil Transitions in Biopolymers*. New York: Academic Press, 1970.

SCHERAGA, H. A., and coworkers. A series of papers on the conformational analysis of polypeptide chains. *Macromolecules*, **3**, 178, 188, 628 (1970); **4**, 112 (1971); **5**, 455 (1972); **6**, 91, 447, 525, 535, 541 (1973); **7**, 137, 459, 468, 698, 797 (1974); **8**, 479, 491, 494, 504, 516, 607, 623, 750 (1975); **9**, 142, 159, 168, 395, 408, 812 (1976); **10**, 9, 291, 305, 1049 (1977); **11**, 9, 552, 797, 805, 812, 819, 1168 (1978).

TERAMOTO, A., AND H. FUJITA, "Conformation-dependent Properties of Synthetic Polypeptides in the Helix-Coil Transition Region," *Adv. Polymer Sci.*, **18**, 65 (1975).

TERAMOTO, A., AND H. FUJITA, "Statistical Thermodynamic Analysis of Helix-Coil Transitions in Polypeptides," *J. Macromol. Sci.—Rev. Macromol. Chem.*, **C15**, 165 (1976).

VOLKENSTEIN, M. V., *Configurational Statistics of Polymeric Chains (High Polymers*, Vol. XVII). New York: Wiley–Interscience, 1963.

19

X-Ray Diffraction by Polymers

INTRODUCTION

Many methods are available for the determination of molecular structures of polymers. These include infrared and ultraviolet spectroscopy, optical rotatory dispersion (ORD), nuclear magnetic resonance, light scattering, ultracentrifugation, gel permeation chromatography, conformational analysis, and X-ray crystallography. Ideally, as many techniques as possible should be applied to the structure solution of a given polymer. In practice, two techniques have proved to be especially powerful: nuclear magnetic resonance (nmr) and X-ray crystallography. For a discussion of the nmr approach, the reader is referred to more specialized texts (see pages 581–583). In this chapter we will discuss the use of X-ray diffraction.

The X-ray diffraction method is the most powerful technique available for the examination of polymers in the solid state. In general, useful information can be obtained only if the polymer forms oriented fibers, is microcrystalline, or yields single crystals (polyethylene, crystalline globular proteins, viruses, etc). The following types of information can be obtained from X-ray diffraction experiments:

1. Estimates of the degree of crystallinity of a polymer sample.
2. Determination of the extent of orientation of crystallites in a polymer.
3. Analysis of the "macrostructure" of the polymer—the way in which the crystallites or bundles of chains are packed together.

4. Determination of the *molecular structure*, including the chain conformation and the position of individual atoms. This aspect is the most involved but yields the greatest amount of useful fundamental information.

Sections of this chapter will be concerned with a brief discussion of items 1 to 3, but most of the discussion will deal with item 4.

EXPERIMENTAL ASPECTS

Sample Preparation

1. If the polymer is a naturally occurring fibrous protein (e.g., hair), a fiber or bundle of fibers can be clamped in a metal holder (Figure 19.1a). Tension is applied to the fiber, and the holder and fiber are then mounted in an X-ray camera.

2. If the sample is a synthetic polymer, a fiber is prepared by melt or solution extrusion, or a film is cast by solvent evaporation or melt techniques (see Chapter 20). The fiber or film is then stretched (oriented) by hand before being clamped in a holder (Figure 19.1b). Further stretching is effected by the slow actuation of the screw-clamp device. A 400% elongation of the sample is not unusual during this step. The sample may be alternately warmed and cooled during this process to encourage microcrystallite growth. It makes little difference whether a fiber or a film sample is used, because no overall crystalline order normally exists *between* the bundles of polymer chains. Thus, a cylindrical symmetry is found in both fiber and film samples. However, some differences do exist between the diffraction behavior of films and fibers. For example, a fiber or bundle of fibers may form a thicker sample than a thin film, and the intensity of X-ray diffraction may be greater with the fiber, thus permitting shorter exposures. On the other hand, very thick

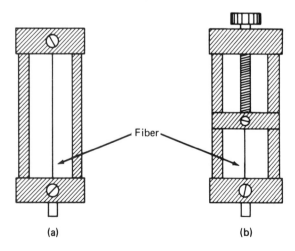

Fiber

Fig. 19.1 (a) Simple brass holder for orientation of a polymer fiber or film in the X-ray beam. (b) Device for stretching a fiber or film (by turning the knurled nut and screw arrangement) and for positioning of the stretched, torsioned sample in the X-ray beam.

(a)   (b)

fibers (~ 1 mm in diameter) sometimes yield doubled X-ray reflections because the X-rays passing through the center of the fiber are totally absorbed, but those passing through the edges are not. Biaxially oriented films may create special problems because the X-ray diffraction patterns may be dependent on the angle at which the beam enters the film. Fibers that have a cylindrical cross section yield data that can be corrected more easily for X-ray absorption effects than can thick films. Thus, unless special effects are to be studied, the sample should be a thin cylindrical fiber (~ 0.5 mm diameter) or a uniaxially oriented, very thin film.

3. When the polymer forms single crystals (globular proteins, nucleic acids, viruses, etc.), different techniques must be used. Biological polymers must be crystallized from aqueous media, and water molecules frequently form part of the lattice. Hence, it is usually necessary to mount the crystal in contact with its mother liquor. The crystal is wedged into a capillary tube together with the mother liquor, and the tube is sealed to prevent evaporation of water. The tube is then glued to a goniometer head, and movement of the head permits orientation of the crystal relative to the X-ray beam (Figure 19.2). If the crystal is stable to air, it can be cemented to the end of a glass fiber and the latter can be attached to a goniometer head (Figure 19.2). A suitable crystal size would be less than 0.5 mm in edge length.

Generation of X-Rays

X-rays are produced when a beam of high-energy electrons, generated in an evacuated tube, strikes a metal target. The wavelength of the X-rays depends on the target material used; copper or molybdenum targets are commonly employed. The beam of X-rays emerging from the X-ray tube passes through a safety shutter arrangement and a filter, and is then collimated to a narrow beam (~ 0.5 mm diameter) by passage through a brass tube or "collimator."

Because X-rays generated in this way are polychromatic, filtration of the beam is necessary to. remove unwanted wavelengths. A thin sheet of nickel is used as a filter for copper radiation. For some applications, filtration is insufficient to produce radiation of the necessary purity. A more monochromatic beam can be produced by diffraction of the main beam from the surface of a flat or curved crystal surface.

X-Ray Cameras and Diffractometers

A number of different camera or diffractometer designs are in use. The following are the most widely used instruments for polymer research:

Cylindrical cameras

This type of camera is also known as a rotation camera. It is used for the preliminary investigation of all polymers and for the total molecular structure investigation of oriented fibers or microcrystalline polymers. A cylindrical camera can also be employed to measure the degree of crystallinity and the degree of

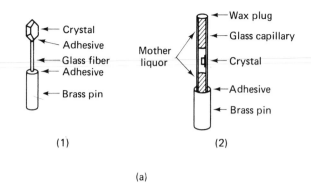

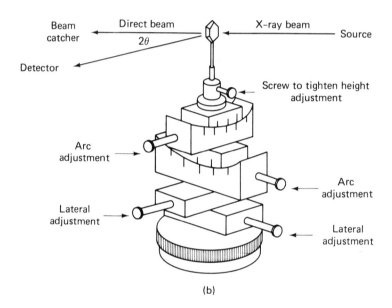

(b)

Fig. 19.2 (a1) Crystal of an air-stable compound can be cemented to a glass fiber in preparation for mounting on a goniometer head. (a2) Air-sensitive crystal of a protein can be wedged into a capillary tube together with the mother liquor solution. (b) Goniometer head for alignment of the crystal. [Reproduced with permission from J. Pickworth Glusker and K. N. Trueblood, *Crystal Structure Analysis* (New York: Oxford University Press, 1972).]

crystallite orientation. The camera consists of a cylindrical brass tube (often 57.3 mm in diameter) which acts as an outer support for a curved sheet of photographic film (Figure 19.3). X-rays enter the camera via the collimator, strike the polymer sample, and the undeflected beam is trapped by a "beam stop." Diffracted X-rays strike the photographic film, and subsequent development reveals the positions at which the diffracted rays passed through the film. The camera is usually capped

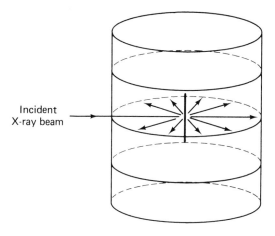

Fig. 19.3 Schematic representation of a polymer fiber mounted in a cylindrical or "rotation" camera. Only the diffracted rays falling on the zero level (equator) of the film are shown. However, upper and lower layer reflections will also be generated (e.g., along the upper and lower ellipses seen in the figure).

by a lid to exclude light, and the sample is mounted on a goniometer head for orientation purposes. If necessary, the sample can be rotated mechanically about its vertical axis, although in fiber crystallography this is usually unnecessary. It is sometimes advantageous to evacuate the camera or replace the air by a light gas, such as helium, to reduce the scattering of X-rays by air. A typical photograph obtained by the use of a cylindrical camera is shown in Figure 19.4.

A "powder" or Debye–Scherrer camera is simply a shallower version of a cylindrical camera. It is used mainly to obtain photographs of randomly oriented specimens. A Guinier–Wolff camera is a form of cylindrical camera in which the radiation is monochromatized and focused by a curved crystal. A Weissenberg

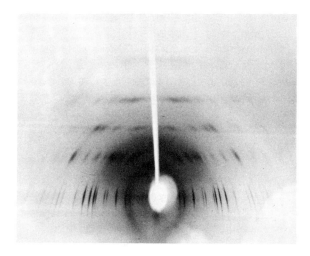

Fig. 19.4 X-ray fiber photograph of poly(dichlorophosphazene), $(NPCl_2)_n$. Note the arrangement of the arclike reflections along distinct layer lines. The layer line at the bottom of the photograph is the $l = 0$ layer. (Photograph obtained by R. A. Arcus.)

camera is a cylindrical camera in which the film is mechanically translated along the camera axis in synchronization with the rotational oscillation of the crystal.

The precession camera

The Buerger precession camera[1] is sometimes used as an alternative to a cylindrical camera for molecular structural work. It has the advantage that it reveals an undistorted picture of the reciprocal lattice. Precession cameras are used extensively in the crystallography of biological macromolecules such as enzymes or heme proteins. A flat sheet of photographic film is used and this, together with the fiber or crystal, is caused to undergo a precession motion.

A simple variant (and early precursor) of the precession camera is the Laué camera, in which a fixed specimen is allowed to diffract X-rays on to a fixed, flat piece of photographic film. The interpretation of Laué photographs is more complex than the analysis of precession photographs.

Diffractometers

An alternative to the photographic detection of diffraction is the use of a Geiger, proportional, or scintillation-type counter. In practice, the use of a counter-type instrument permits the diffraction angles to be determined quickly (if only a few reflections are present), and it allows greater accuracy in the measurement of diffraction angles and intensities. Two main types of diffractometer instruments are available: manual and automated. With manual instruments, the operator must move the detector to the appropriate position for detection and measurement of a diffracted beam. These instruments are generally of an older design and some safety problems are associated with their use. Automated or semiautomated diffractometers operate under the control of a small computer or, indirectly, under the control of an operator. In single-crystal work, the space group of the crystal is used by the computer to orient automatically the diffractometer components and measure reflection intensities at the correct angle. These instruments cannot normally be used for small-angle work. Only the so-called "Weissenberg geometry" diffractomers offer appreciable advantages over cameras for fiber diffraction work and relatively few of these instruments are in use.

Small-angle cameras

Wide-angle instruments, such as those described in the preceding sections, are useful for the detection of crystallinity and for molecular structure determination. However, information about the *macro*structure of a polymer (i.e., the packing of fibrils or crystalites) can only be obtained from narrow-angle diffraction, from reflections that fall very close to the beam stop. Special instruments are required to detect and measure reflections such as these. The Kratky small-angle camera is one such instrument.

[1] M. J. Buerger, *The Precession Method* (New York: Wiley, 1964).

Measurement of the Degree of Crystallinity

A crystalline or microcrystalline material will diffract X-rays at specific angles (Figure 19.5b and c). On the other hand, amorphous materials simply scatter X-rays in all directions (Figure 19.5a). Thus, in theory, if the sharp diffraction can be compared with the general scattering from a polymer, a measure of the degree of crystallinity can be obtained. Until recently, it was common practice to do just this. However, it is now recognized that the degree of coalescence of the diffraction halos into sharp arcs or spots depends, among other things, on the presence of chain folding, lamellar crystalline growths, lattice dislocations, and on the degree of ordering in an oriented specimen, and that some of the scattering from crystalline regions is, in fact, diffuse. Moreover, crystallinelike X-ray patterns can be generated by oriented helical polymer chains even when no intermolecular crystalline packing is present. Hence, a reliable measure of the degree of crystallinity must take into account the results from nuclear magnetic resonance, density determinations, and infrared spectroscopy, as well as X-ray measurements.

Degree of Crystallite Orientation

The orientation of the crystallites in a microcrystalline polymer has a profound influence on the polymer properties. Good orientation parallel to the direction of external stress can generate a high resistance to breaking and a correspondingly high tensile strength. The strength of a nylon fiber may in part be attributed to this effect. X-ray crystallography provides a convenient way for the estimation of the efficiency of orientation.

It was mentioned earlier that one of the main differences between single-crystal X-ray photographs and those derived from oriented polymers is the diffuseness and arclike qualities of the reflections often derived from polymers. This difference is illustrated by a comparison of a typical polymer photograph (Figure 19.4) with the diffraction pattern obtained from a small-molecule single crystal (Figure 19.6). Note also that in Figure 19.4 prominent, diffuse reflections are apparent above and below the beam stop position. These are known as meridianal reflections.

The diffuseness and arclike character of the reflections in polymer X-ray photographs is partly a consequence of nonideal orientation of the crystallites. Totally random crystallite orientation results in an extension of the arcs into complete circles, as shown in Figure 19.5b. Thus, some measure of the degree to which the crystallites are oriented parallel to each other can be derived from the shape and size of the individual reflections. A second clue to crystallite orientation is provided by the meridianal reflections. Perfect parallel orientation of the crystallites would result in a total disappearance of the meridianal reflections if the polymer specimen were exactly at right angles to the incident X-ray beam. However, vertical tilting of a perfectly oriented specimen, toward or away from the beam, would cause the meridianal reflections to appear on the photograph. This phenomenon provides a rather sensitive measure of the degree of orientation. In addition,

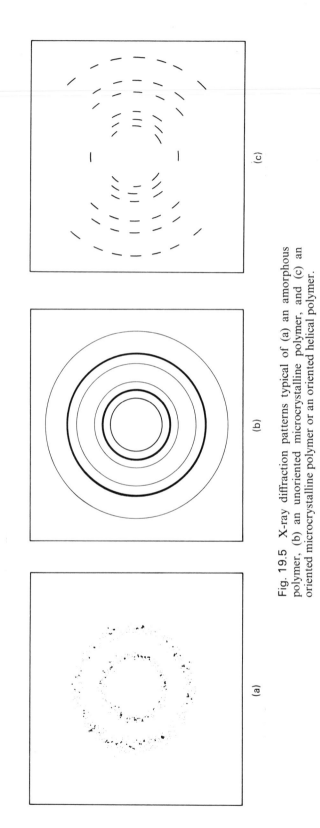

Fig. 19.5 X-ray diffraction patterns typical of (a) an amorphous polymer, (b) an unoriented microcrystalline polymer, and (c) an oriented microcrystalline polymer or an oriented helical polymer.

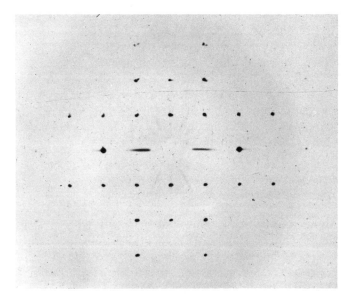

Fig. 19.6 X-ray diffraction pattern from a small-molecule single crystal. Note the sharpness of the "reflections" or diffraction spots and the variations in intensities.

the way in which the intensity varies *across* a meridianal reflection corresponds to the number of crystallites oriented along a particular azimuthal angle (see Figure 19.7).

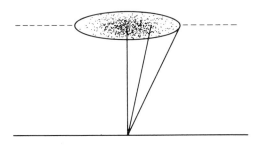

Fig. 19.7 Variation of intensity across a meridianal reflection provides information about the relative numbers of microcrystallites that are oriented along specific reflections, for example along the radial lines shown.

DETERMINATION OF MOLECULAR STRUCTURE AND CONFORMATION

Different Approaches

It is necessary for the reader to recognize that two closely related but quite distinct approaches are used for the structure solution of fibrous or microcrystalline polymers, on the one hand, and single-crystal materials, on the other. The single-crystal techniques used for the solution of relatively simple molecules can be expanded with some modification and an increased complexity to the structure solution of polymer single crystals. On the other hand, microcrystalline or fibrous polymers present special problems because of disorder in the packing of the chains. Preliminary structural information can be obtained quickly and easily, but a detailed analysis can be an exceedingly difficult task. The techniques used for single-crystal analysis are widely known and are described in numerous textbooks (see Suggestions for Further Reading). Here, we will touch briefly on these techniques but will concentrate mainly on the study of microcrystalline materials and noncrystalline fibrous polymers.

Interpretation of X-Ray Photographs from Microcrystalline Polymers

An oriented microcrystalline polymer can yield photographs that are superficially similar to those obtained from rotated single crystals, and for similar reasons. Figure 19.5c depicts the features of a photograph of a typical oriented microcrystalline material. The two most striking features are the existence of the meridianal reflections (discussed in an earlier section), and the presence of arcs arrayed along parallel horizontal lines. These horizontal lines are known as *layer lines* and they are numbered according to the notation, $l = 0, 1, 2, 3, \ldots, -1, -2, -3, \ldots$. The "zero layer" always includes the point where the direct beam passes through the film. This layer represents diffraction arcs or "reflections" that emerge from the sample at right angles to the fiber direction (see Figure 19.3). Arcs that lie on the higher and lower layer lines are formed from diffracted X-ray beams which emerge from the sample at an elevated or declined angle to the plane of the camera.

Individual reflections on the zero layer are formed by the following mechanism. Consider a section of a microcrystallite, as shown in Figure 19.8, with regularly packed chains oriented in a vertical direction (along the axis of orientation). Note that lines or planes can be drawn through individual atoms to generate a lattice.

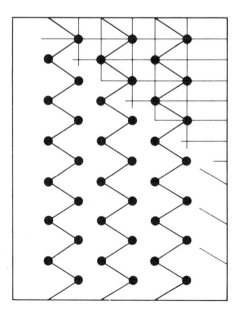

Fig. 19.8 Section through a microcrystalline domain, showing laterally packed polymer chains and the way in which various interplanar spacings can be drawn through individual atoms.

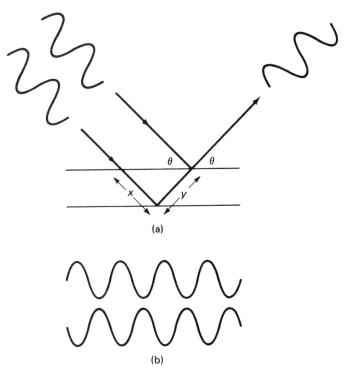

Fig. 19.9 (a) Diffraction of X-rays from parallel planes of atoms within a single crystal of polymer microcrystallite. The path-length difference $(x + y)$ must correspond to an integral number of wavelengths if destructive interference is to be avoided. (b) Two wave systems that are 180° out of phase. Total destructive interference occurs.

Consider now the path of an X-ray beam, incident to one set of planes in Figure 19.9a. According to the Bragg approach to diffraction, X-rays will be strongly "reflected" from the layers if the angle of incidence, θ, is such that an integral number of wavelengths exists in the "path difference" $x + y$. If this condition is met, the beam emerging from the crystal will be reinforced by superimposition of waves. At other angles the waves will interfere destructively with each other (Figure 19.9b). Thus, reflections emerge only at discrete angles, and these angles are determined by the path-length difference, and hence by the separation between the layers. The relationship between diffraction angle and layer spacing is given by the well-known Bragg equation

$$n\lambda = 2d \sin \theta \qquad (1)$$

where n is an integral number (1, 2, 3, 4, ...), λ the X-ray wavelength, and d the layer spacing.

It is obvious that the different reflections that appear on the zero layer of the photograph represent different separations between different vertical planes. Such planes can be visualized as formed in the way shown in Figure 19.10, which represents a view *down* the chain axes.

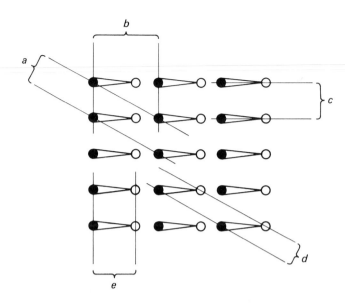

Fig. 19.10 View *down* the long axis of chains packed in a micro-crystallite. This view is at 90° to the one shown in Figure 19.8. The solid circles represent atoms that occupy a position closer to the viewer than those represented by the open circles. Note how sets of *vertical* planes can be drawn through individual atoms. These vertical planes are responsible for the reflections that appear on the zero-layer (equator) of the X-ray photograph. The larger lattice spacings (*b*, *c*) will yield reflections at a narrow angle (close to the beam stop). The smaller spacings, such as *d*, will yield diffraction spots at wider angles (nearer the edge of the photograph).

Those planes with the largest *d*-spacing will generate a reflection closest to the primary beam (small angle of 2θ). Those with the smallest spacings will yield reflections at wider angles.

The reflections which fall on the upper and lower layer lines ($l = 1, 2, 3, \ldots$, $-1, -2, -3, \ldots$) represent planes in a microcrystallite that are inclined to the fiber axis. The inclined planes in Figure 19.8 are an example. Finally, the meridianal reflections originate from planes that are at right angles to the fiber axis (see Figure 19.8). Because of the geometry of reflection (see Figure 19.9), it is clear that these planes can diffract only if the fiber is inclined at an angle to the beam. Hence, as mentioned previously, the presence or absence of a meridianal reflection can be used as a measure of crystallite orientation.

Preliminary Clues about the Structure

A simple analysis of the *d*-spacings can often provide valuable clues about the conformation of the chain or the packing of adjacent chains. For example, separation between the layer lines is inversely related to the repeating distance along the *c*-axis (fiber axis) of the sample. Hence, one measurement from the film can yield

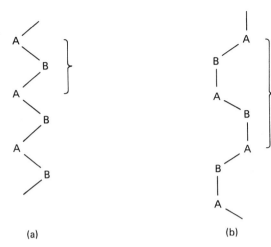

Fig. 19.11 Polymers containing repeating units A—B. (a) One monomer residue per repeat: meridianal reflection expected on the $l = 1$ layer line. (b) Two monomer residues per repeat: meridianal reflection expected on $l = 2$ layer line.

(a)   (b)

the polymer repeating distance. Second, meridianal reflections arise from planes that are perpendicular to the fiber axis. They represent reflections from planes of monomer residues. Hence, the number of the layer line on which a meridianal reflection falls is the number of monomer residues in the conformational repeat distance. This is illustrated in Figure 19.11.

Third, reflections on the zero layer of the photograph provide clues about the "width" of the zigzag, the helix, or other conformation of the chain. They may also indicate the distances between side-group atoms and the skeletal atoms, or the distances between atoms in neighboring chains. A careful analysis of these interplanar distances can suggest the dimensions and shape of the unit cell. Then, if the density of the crystalline regions can be estimated, the number of monomer residues per cell, Z, can be calculated from the formula

$$\text{Density} = \frac{(Z)(MW)}{V \times 6.023 \times 10^{23}} \tag{2}$$

where MW is the "molecular weight" (in g/mol) of a monomer residue and V the volume in cm^3 of the unit cell.

Some X-ray studies of microcrystalline materials are terminated at this point. The information obtained is sometimes considered sufficient to define the conformation of the chain if the bond angles and bond lengths of the skeleton are known. However, it is possible for an investigator to be misled by the preliminary data. To confirm a structure, a careful analysis of the *intensities* of the reflections is required.

Preliminary Analysis of Intensity Data

The most reliable clues about the details of the molecular structure are obtained by an analysis of the relative intensity of each reflection. By "intensity" we mean the density of silver deposited on the developed film, or the number of counts recorded in a given time by a counter device. Although the *positions* of the arcs on the film indicate the crude, overall orientation of the molecules in the unit cell, the intensities allow the positions of individual atoms to be identified accurately.

The preliminary intensity analysis can be carried out with the use of two approaches. First, a simple inspection of the photograph may reveal that some reflections are much more intense than others. Because each reflection represents a series of identical parallel planes and, because the intensity of scattering is related to the electron density in those planes, a strong reflection may indicate that many atoms lie on that plane.

Thus, crystallite planes that contain planar polymer chains may reflect X-rays with a particularly high intensity, especially if heavy elements, such as phosphorus or sulfur, form part of the backbone. Planes that contain aromatic rings may be revealed in the same way.

Second, certain expected reflections may be totally absent. These "systematic absences" indicate specific orientations of the contents of the unit cell, such as the presence of glide planes, screw axes, lattice centering, and so on. Such information can provide valuable evidence in favor of one structure or another.

Often, at the end of this phase of the investigation, the investigator will be cautiously optimistic that the correct conformation and the disposition of the chains in the unit cell have been found. The next step is to analyze the intensity data in a more systematic manner.

Structure Factor Analysis

Two main factors affect the intensities of the reflections from a crystal or a microcrystallite. First, as discussed above, a high intensity in a particular reflection may result from the presence of many atoms or heavy atoms in a particular set of planes. Second, the intensities may represent the effects of destructive or constructive interference between X-ray beams reflected from different parts of the same molecule.

Consider the repeating motif shown in Figure 19.12. Although each motif occupies a lattice point (heavy lines, A), parts of the motif lie above or below or to the left or right of the lattice point. Those other parts of the motif or molecule form additional planes (which may be depicted by the dashed lines, B). Reflections of X-rays will occur at the same angle, θ, from all layers that have the same interplanar spacings, but the reflection from one set of planes will interfere with that from the other. The interference can be completely destructive, in which case the intensity will be zero. Alternatively, it could be partially destructive due to a partially offset phase relationship. It will be clear that the intensity is a function of the distance between the different types of planes. Because each subsidiary plane represents a different set of *atoms*, the intensity of a given reflection will be a function of the distance between different atoms in the unit cell.

The intensities are readily measured, but the "phase difference" is not known. However, by postulating models for the structure, the investigator can predict what the intensities would be if the model were correct. If the calculated intensities match

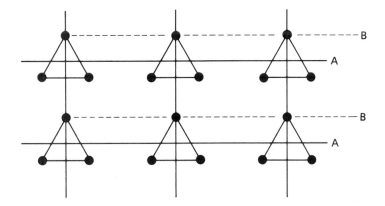

Fig. 19.12 Motif of atoms centered around lattice points (intersection of solid lines). All atoms in the motif cannot lie on the same
lattice points; hence the reflection of X-rays from the different resultant
planes will cause destructive wave interference unless the separation
permits exact phase matching.

the experimental ones, the model is presumed to be correct. If they do not match,
the model is altered until a closer correspondence is obtained. This process is known
as the *trial-and-error technique*, and the overall method is known as a *structure-
factor calculation*.

This method is the classical approach to the solution of single-crystal structures.
For biological macromolecules in single crystals, the structure analysis is usually
facilitated by the introduction of heavy atoms such as mercury into the structure.
These serve as reference points or "phasing atoms" for the structure analysis.

However, it should be recognized that this method cannot be applied satisfactorily to most oriented fibers or microcrystalline polymers. The lack of perfect
crystallite orientation, the resultant diffuseness of the arcs, the cylindrical symmetry
of the polymer structure, and the concentration of information in relatively few
reflections generate difficulties not normally found in single crystal analysis. Moreover, a tendency is often found for the available data to become superimposed and
concentrated in a few low angle reflections, on the $l = 0$ or $l = 1$ layer lines. These
problems virtually eliminate the possibility that a microcrystalline or noncrystalline
fiber structure can be solved with the high accuracy now common in single crystal
analyses. Nevertheless, structures of microcrystalline polymers have been, and are
now being, solved by this method.

Diffraction by Helices

Most synthetic polymers and some biological polymers (such as DNA, fibrous
proteins, and viruses) assume helical conformations in the oriented state. These
helices may or may not be packed together in microcrystallites. Even though the
helices may not form part of a crystalline array, they can generate a characteristic
X-ray pattern. The study of such patterns is known as *helical transform analysis*.

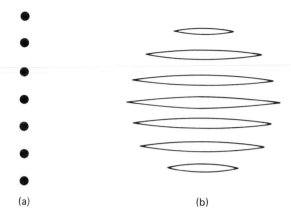

Fig. 19.13 (a) Single linear array of regularly spaced diffraction centers (atoms, motifs, or repeating units) that corresponds in a very crude way to a low-resolution representation of a polymer chain. (b) Diffraction pattern from a structure of type (a) consists of a series of parallel "envelopes."

(a)   (b)

The method was first introduced in 1952 by Cochran, Crick, and Vand[1,2] and by Stokes. The identification of the structure of poly(α-methyl-L-glutamate) and the double helix structure of DNA constitute two of the first successes of this approach.

The analysis of diffraction effects produced by isolated helical polymer molecules can be considered at three levels of sophistication: (1) in terms of diffraction by a one- or two-dimensional regular repeating arrangement along the axis of the chain; (2) in terms of the diffraction effects expected from a continuous helix; and (3) as diffraction by a discontinuous helix made up of individual atoms and side groups. These three approaches will be considered in turn.

Diffraction by one- or two-dimensional arrays

In many oriented polymer samples, the chains are extended in such a way that the macromolecular skeleton generates a regular repeating sequence along the fiber axis. This is true even if the side groups attached to the chain are disordered. A diffraction experiment will identify only what is ordered in the system and will ignore that which is disordered. If we cease to worry for a moment about the finer details of the structure, we can imagine that a polymer might be depicted in a low-resolution fashion by a one-dimensional sequence of "units" as shown in Figure 19.13a. Such a situation might arise if a marked periodicity existed *along* the chain axis, but with little or no periodicity in the directions at right angles to that axis. The Fourier transform (and hence the X-ray pattern) of such a one-dimensional array is a series of "envelopes," as shown in Figure 19.13b. These envelopes represent rudimentary layer lines that are related to those depicted earlier in Figure 19.5c. Such patterns are actually encountered in practice from some polymer systems.

Imagine now that we can view the polymer at a slightly higher resolution, in such a way that the individual monomer residues or atoms can be recognized at all points along the chain, and the helical conformation of the chain can be crudely discerned. The projection of the atoms or monomer residues along the helix onto

[1] W. Cochran and F. H. C. Crick, *Nature*, **169**, 234 (1952).
[2] W. Cochran, F. H. C. Crick, and V. Vand, *Acta Cryst.*, **5**, 581 (1952).

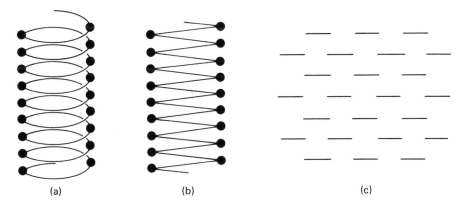

Fig. 19.14 Projection of the diffracting centers (atoms or monomer motifs) on a helix (a) onto a plane surface (b), and the diffraction pattern expected from such a projection (c).

a surface generates a zigzag pattern (Figure 19.14a and b). Thus, the diffracting sites on the polymer, when viewed from the side, are arrayed in a two-dimensional pattern (Figure 19.14b) and the diffraction pattern generated by this arrangement is shown in Figure 19.14c. Note that the layer-line envelopes have now separated into broad but discrete "reflections." This pattern has been generated by a "convolution" of the original one-dimensional atomic array with the ordered arrangement at right angles to it. Of course, the diffraction pattern depicted is only a very crude approximation of the expected Fourier transform because the three-dimensional character of the polymer structure has been ignored. The next section partially corrects this deficiency.

Diffraction by a continuous helix

Consider the diffraction pattern that would be generated by an isolated, continuous, three-dimensional helix (e.g., by an infinitely thin wire wound helically around the surface of a cylinder) (Figure 19.15a). A cross-sectional view down the helix is a circle. Therefore, the diffraction phenomenon can be considered in terms of diffraction at the surface of a cylinder, but a cylinder that possesses a regular periodicity along its axis. Diffraction by such systems can be analyzed most easily in terms of cylindrical waves rather than by the diffraction of plane waves as used in conventional X-ray analyses. Cylindrical waves can be visualized as resembling the concentric wave fronts that radiate out when a pebble is tossed into a pond, or the circular waves that form when ocean wave fronts strike an obstruction.

Diffraction by a continuous helix gives rise to a diffraction pattern similar to that shown in Figure 19.15b. Such patterns can be generated on a macroscopic scale by the use of an optical diffractometer (see a later section). The main features of this pattern are (a) the existence of layer lines (indicated by $n = 0, 1, 2, \ldots$), (b) the characteristic cross pattern formed by the most intense reflections (which results from [c] the fact that the strongest reflection on each layer line is found progressively farther and farther from the meridian for higher and higher layer lines).

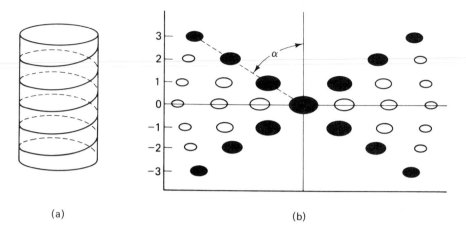

(a)

(b)

Fig. 19.15 Diffraction pattern of a continuous helix. The solid ellipses represent the largest peak in the Bessel function that contributes to each layer line. The open ellipses denote the subsidiary (attenuating) peaks of each Bessel function. A zero-order Bessel function is the principal contributor to the zero-layer line; a first-order Bessel function comprises the first layer line, and so on. The angle α on the diffraction diagram equals the pitch of the helix.

Such behavior can be rationalized and simulated by the use of Bessel functions. Bessel functions are special integrals that are used in physics to describe the behavior of wave motions generated by cylindrical objects. A cylindrical wave of order n is made up from a Bessel function of order n (written as J_n) multiplied by a cosine wave of period n

$$C_n(r, \phi) = J_n(2\pi Rr) \cos n(\phi - \theta + \pi/2) \tag{3}$$

where r and ϕ are the radial and azimuthal coordinates in real space, and R and θ are the appropriate coordinates in reciprocal space.

The Fourier transform of a continuous helix (i.e., one that does not possess discontinuities from the presence of atoms or repeating residues) for each layer line, n, is given by

$$G(R, \theta, z) = J_n(2\pi Rr)e^{in(\theta + \pi/2)} \tag{4}$$

The form of the $n = 0$, 1, and 2 Bessel functions is shown in Figure 19.16.

A Bessel function of order $n = 0$ generates an attenuated wave that has its maximum peak located at $X = 0$. Note that this behavior simulates the situation depicted along the zero layer line of the diffraction pattern in Figure 19.15b. Similarly, the Bessel function of order $n = 1$ has a maximum farther from the origin, and the position of this peak corresponds to the location of the strongest reflection on the $n = 1$ layer line. In fact, the construction of the whole diffraction pattern can be understood if the origin of each Bessel function occupies the meridian of the photograph, with higher layer lines corresponding to higher-order Bessel functions.

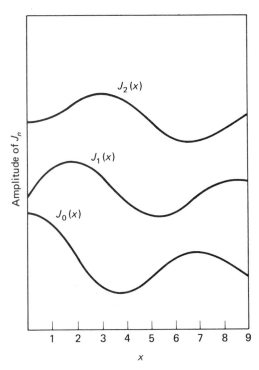

Fig. 19.16 Bessel functions J_0, J_1, and J_2, showing how the strongest peak moves farther from the origin for higher and higher values of n in J_n. A comparison of these profiles with Figure 19.15 indicates how the behavior of a Bessel function of order, n, corresponds to the diffraction density profile on the nth-layer line.

Diffraction by a discontinuous helix

A real polymer molecule differs from the two situations discussed above because it possesses both discrete atoms or monomer residues *and* a three-dimensional helical arrangement. Thus, the picture presented above for diffraction by a continuous helix must be modified to take into account the discontinuous structures. This is accomplished by combining or "convoluting" the two types of results. Descriptions of the mathematical procedures are beyond the scope of this book (for details see the list of books for suggested reading), but pictorially the result is shown in Figure 19.17.

Thus, the actual diffraction pattern will consist of multiple layer lines and multiple crosses. If the spacing between the origins of the crosses is small, the cross patterns will overlap to give an extremely complex pattern (Figure 19.18). The diffraction pattern may be even more complex if several turns of the helix are needed to accomplish one repeat. Straightforward techniques (such as the "$n - l$ plot," "the radial projection," and the "helical projection") have been developed to assist in the interpretation of such complex patterns.[1-4] The actual process of

[1] W. Cochran and F. H. C. Crick, *Nature*, **169**, 234 (1952).
[2] W. Cochran, F. H. C. Crick, and V. Vand, *Acta Cryst.*, **5**, 581 (1952).
[3] A. Klug, F. H. C. Crick, and H. W. Wyckoff, *Acta Cryst.*, **11**, 199 (1958).
[4] K. C. Holmes and D. M. Blow, *The Use of X-Ray Diffraction in the Study of Protein and Nucleic Acid Structure* (New York: Wiley–Interscience, 1966).

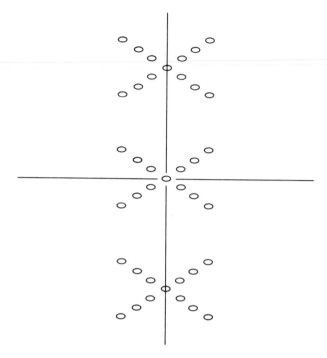

Fig. 19.17 Discontinuous helix (i.e., a helix made up of atoms or monomer repeat units) gives a diffraction pattern made up of multiple crosses arrayed along the meridian.

structure solution for oriented, noncrystalline helical polymers involves a trial-and-error matching of the geometric relationship on the diffraction photograph and the measured intensities with plausible trial structures.

Helices within microcrystallites

If individual helices are packed together in microcrystalline domains, additional reflections will appear from the *inter*molecular spacings. At the same time, the diffraction disks will shrink in size to resemble the spots found in single-crystal photographs. In such cases, a combination of space group information (i.e., the packing symmetry within the unit cell), standard structure factor, and helical transform calculations may be needed to solve the structure. This can be an exceedingly complex undertaking and, for this reason, the structures of many polymers have not yet been worked out in detail.

Optical Diffraction

We cannot see the structure of molecules simply by looking at them through a very powerful optical microscope because the distances between atoms are shorter than the wavelength of light. X-rays have wavelengths similar to the distances between atoms, but no lenses exist that are capable of focusing the diffracted X-ray

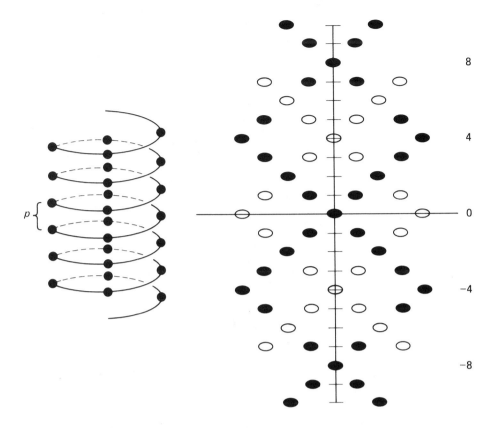

Fig. 19.18 Diffraction pattern of a discontinuous helix with four motifs per turn of the helix. The cross patterns are centered on the 0, 4, 8, −4, −8, . . . , layer lines. Hence, the number of monomer units per repeat (in this case four) can be determined by inspection of the photograph to ascertain the lowest layer lines above 0 that bear meridianal reflections (in this case, layer line 4). Open and closed ellipses simply indicate reflections from the different cross patterns and do not represent intensity differences.

beams. Hence, in X-ray crystallography, we are forced to analyze the diffraction pattern mathematically rather than to reconstruct the image directly.

However, devices exist that enable the optical diffraction patterns of scaled up "models" of the polymer to be studied. If the *optical* diffraction pattern from the model resembles the X-ray diffraction pattern from the polymer, the model is assumed to be a true representation of the polymer structure. If not, the model can be altered in a trial-and-error procedure until a good correspondence is obtained. The method is used more commonly with fibrous polymers than with single crystals because, although Fourier calculations are now so easy to perform for single-crystal materials, they are less applicable to fiber diagrams. The optical diffraction method is really a visualization procedure for helical transform calculations.

In practice, the "model" consists of holes punched in a piece of opaque photographic film, or a photographic transparency derived from a molecular model. The holes represent the atoms, with the size of the hole being roughly related to the size of the atom. Diffraction patterns generated by single chains can be studied, or the effects of chain packing can be simulated by a model which contains a number of motifs arranged side by side (Figure 19.19).

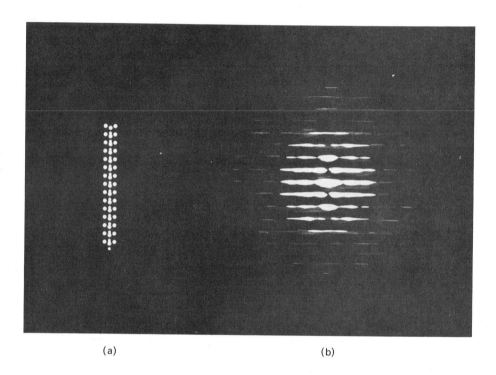

(a) (b)

Fig. 19.19 Optical diffraction pattern (b) obtained from the polymeric motif shown in (a). The motif represents one conformation of polydichlorophosphazene. (Motif and diffraction photograph prepared by R. A. Arcus.)

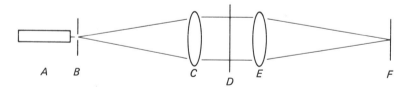

Fig. 19.20 Schematic view of an optical diffractometer. Coherent light from a laser (A) is converted into a point source by a pinhole (B) and a parallel beam is formed by lens C. After passage through the perforated mask (D), the beam is focused by lens E onto a piece of photographic film (F). For the best results, lenses C and E should be of long focal length. In a typical instrument the effective diameter of the lenses is about 10 cm.

An optical diffractometer of the Taylor and Lipson[1] design is illustrated in Figure 19.20. The light source is a low-power laser unit. The diffracted image can be examined through a small telescope or by means of a camera and photographic film. The advantage of the method is the speed with which a number of alternative model structures can be evaluated. Once a reasonable correspondence between the optical and X-ray diffraction patterns has been achieved, helical transform calculations can be performed to optimize the structure.

STUDY QUESTIONS

1. Discuss the reasons why the molecular-structure determination of a synthetic high polymer is nearly always a more involved process than the structure solution of a small molecule that forms part of a single crystal.

2. What supplementary (i.e., non-X-ray) information about a new polymer would you seek before you undertook an attempted X-ray structure analysis? Give reasons for your choices.

3. By an examination of molecular models, convince yourself that a three-dimensional repeating structure of two or more polymer chains can be interpreted in terms of *planes* of atoms. Attempt to correlate different sets of planes with the directions in which diffracted X-rays might be formed. Look for atomic planes that might generate reflections on the "upper" layer lines of an X-ray photograph.

4. Suppose that you know the repeating distance of a polymer chain, but you do not know the bond angles, bond lengths, or the chain conformation, what steps would you take in order to attempt to solve the molecular structure? What serious errors could you make?

5. After reading this chapter, consult several of the books and articles mentioned in the section Suggestions for Further Reading. Then write a research proposal on the subject of a proposed structure solution for any polymer of your choice. Stress experimental and structure analysis methods that you would plan to use.

6. Consider possible ways in which the projection of a polymer molecule might be depicted in two dimensions on a sheet of paper or a piece of photographic film. If this projection is used as a mask in an optical diffractometer, what effects on the optical transform would you expect from the following changes: 10 chains oriented side by side instead of one chain; increasing the separation between the chains; the use of one repeating unit instead of a long chain; use of a continuous line instead of individual atoms to represent the chain; elimination of the side groups from the structure; decreasing the wavelength of the incident light.

SUGGESTIONS FOR FURTHER READING

Polymer crystallography

ALEXANDER, L. E., *X-Ray Diffraction Methods in Polymer Science*. New York: Wiley–Interscience, 1969.

BRUMBERGER, H., (ed.), *Small-Angle X-Ray Scattering*. London: Gordon and Breach, 1967.

[1] C. A. Taylor and H. Lipson, *Optical Transforms* (Ithaca, N.Y.: Cornell University Press, 1964).

HARBURN, G., C. A. TAYLOR, AND T. R. WELBERRY, *Atlas of Optical Transforms.* Ithaca, N.Y.: Cornell University Press, 1975.

HOLMES, K. C., AND D. M. BLOW, *The Use of X-Ray Diffraction in the Study of Protein and Nuclei Acid Structure.* New York: Wiley–Interscience, 1966.

HOSEMANN, R., "The Paracrystalline State of Synthetic Polymers," *CRC Critical Rev. Macromol. Sci.,* **1**, 351 (1972–73).

KADUKO, M., *X-Ray Diffraction by Polymers.* New York: Elsevier, 1972.

LIPSON, H. (ed.), *Optical Transforms.* London: Academic Press, 1972.

SHERWOOD, D., *Crystals, X-Rays, and Proteins.* New York: Halsted Press (Wiley), 1976.

TADOKORO, H., "Structure of Crystalline Polyethers," *J. Polymer Sci. (D) (Macromol. Rev.),* **1**, 119 (1967).

TAYLOR, C. A., AND H. LIPSON, *Optical Transforms.* Ithaca, N.Y.: Cornell University Press, 1964.

VAINSHTEIN, B. K., *Diffraction of X-Rays by Chain Molecules.* New York: Elsevier, 1966.

WILSON, H. R., *Diffraction of X-Rays by Proteins, Nucleic Acids, and Viruses.* London: Arnold, 1966.

Single-crystal X-ray crystallography

BUNN, C. W., *Chemical Crystallography* (2nd ed.). London: Oxford University Press, 1961.

GLUSKER, J. AND K. N. TRUEBLOOD, *Crystal Structure Analysis—A Primer.* New York: Oxford University Press, 1972.

LIPSON, H., AND W. COCHRAN, *The Determination of Crystal Structures* (3rd ed.). Ithaca, N.Y.: Cornell University Press, 1966.

MILBURN, G. H. W., *X-Ray Crystallography.* London: Butterworths, 1973.

STOUT, G. H., AND L. H. JENSEN, *X-Ray Structure Determination.* New York: Macmillan, 1968.

FABRICATION, TESTING, AND USES OF POLYMERS

20

Fabrication of Polymers

Throughout this book an attempt has been made to emphasize the practical utility of polymers and the reasons for their importance to modern society. In this chapter, an emphasis is placed on the ways in which polymers are converted into useful products. To a very large extent, most polymer chemists and chemical engineers gain considerable pleasure from using polymers to make things—fibers, films, and molded objects of all kinds. These activities generate the kind of fundamental satisfaction that other scientists derive from growing crystals or observing color changes in a chemical reaction. Thus, the fabrication of polymers is as much a part of laboratory work as it is of the manufacturing process. In the following sections, when possible, both small-scale laboratory and large-scale industrial fabrication methods are mentioned.

Fabrication methods can be divided roughly into those which yield films, fibers, or bulk-molded objects. Areas of more specialized importance are elastomer technology, the formation of expanded polymers (foams), and surface coatings. Closely related to nearly all these topics are the problems of polymer compounding, blending, and curing, and the question of polymer reinforcement. Each of these topics will be considered briefly in turn.

PREPARATION OF FILMS

Polymer films can be made by two fundamentally different techniques—by solution-casting or by melt- or sinter-fabrication methods.

Solution Casting

The basic idea in solution casting is to dissolve the polymer in a suitable solvent to make a viscous solution. The solution is then poured onto a flat, nonadhesive surface, and the solvent is allowed to evaporate. The dry film can then be peeled ("stripped") from the flat surface.

One of the principal differences between high polymers and low-molecular-weight compounds is that polymers will form films, whereas small molecules are deposited as crystals or weak conglomerates. Thus, it is sometimes possible to confirm the formation of a high polymer by the fabrication of a strong, cohesive film.

On a laboratory scale, the solution casting of films is quite easy. Usually, the first problem is to find a suitable solvent for the polymer. An ideal solvent is one which is sufficiently volatile that it will evaporate at a reasonable rate at room temperature or slightly above, but not so volatile that it vaporizes rapidly and forms bubbles or semicrystalline precipitates. Rapid volatilization also causes cooling of the film, which could cause crazing or condensation of water from the atmosphere. A solvent that has a boiling point between about 60°C and 100°C will usually give good films. If the casting procedure is to be carried out on an open laboratory bench, some consideration should be given to the potential toxicity and flammability of the solvent.

High polymers dissolve only slowly in most solvents. Swelling of the polymer occurs first, and this is followed by dissolution from the edges of the polymer particles. Stirring—particularly high-speed stirring—accelerates the dissolution process. In the laboratory, a high-speed, shear disk stirrer is often used. This consists of a spindle to which is attached a metal disk, as shown in Figure 20.1. Stirred polymer solutions have a tendency to climb the stirrer spindle, and additional disks may be needed above the surface of the solution to prevent this action. The final polymer solution should be quite viscous—sufficiently fluid that slow liquid flow is possible, but not so fluid that the solution spreads out quickly on the casting surface. Solutions containing about 20 wt % of polymer often give a suitable viscosity.

If the polymer solution needs to be filtered before casting, filtration must be accomplished by *pressure* techniques rather than by gravity or vacuum filtration.

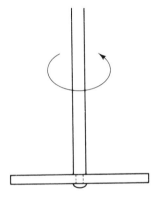

Fig. 20.1 High-speed metal disk stirrer to aid the solution of polymers in suitable solvents.

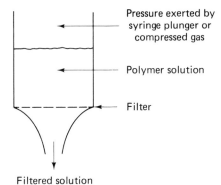

Pressure exerted by
syringe plunger or
compressed gas

Polymer solution

Filter

Fig. 20.2   Pressure filtration
unit for polymer solutions.

Filtered solution

Small quantities of polymer solutions can often be filtered in a hypodermic syringe-filtration unit. Larger quantities must use a filter unit operated by compressed air (Figure 20.2).

The simplest casting surface for laboratory work is a sheet of plate glass. The polymer solution is simply poured onto the glass and a uniform solution thickness is ensured by spreading of the film with a glass rod (Figure 20.3). Of course, the thickness of the solution film determines the thickness of the final film once shrinkage has occurred from the loss of solvent. Thus, for a 20% polymer solution, the initial film thickness should be about five times greater than that required for the final dry film.

A number of refinements are possible to this simple technique. First, the sheet of glass may be replaced by a film of poly(tetrafluoroethylene), which has a lower tendency to stick to polymer films, or a chromium-plated, heated "casting bench" may be employed. Second, the film thickness may be determined by the use of a "Gardner knife," a device with a micrometer adjustment of the blade height above the casting surface. Third, the casting surface may be covered by a removable lid to slow the rate of solvent evaporation and prevent dust particles from settling on the film. In the most sophisticated devices a transparent poly(methyl methacrylate) cover encloses the whole unit and only dry, filtered air is admitted to the system (Figure 20.4). Such devices often resemble large glove boxes. In the laboratory, a lid fabricated from aluminum foil, or an inverted cooking pan, may serve nearly as well.

On an industrial scale, the polymer solution is fed continuously through a slit die onto a large rotating drum (Figure 20.5), or onto a moving metal belt (Figure

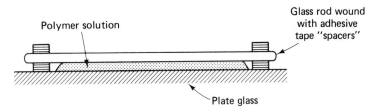

Polymer solution

Glass rod wound
with adhesive
tape "spacers"

Plate glass

Fig. 20.3   Laboratory spreading device for the solution casting of films.

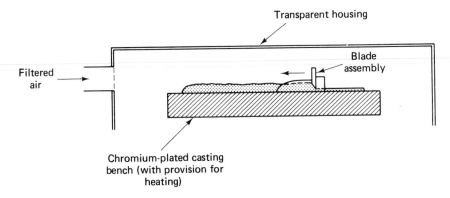

Fig. 20.4   Polymer casting bench.

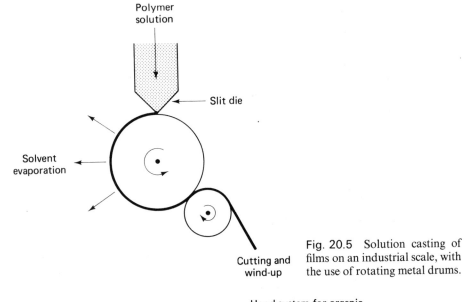

Fig. 20.5   Solution casting of films on an industrial scale, with the use of rotating metal drums.

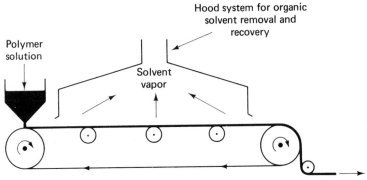

Fig. 20.6   Use of a moving-belt system for the continuous solution casting of polymer films.

20.6). A hood assembly may be used to remove organic solvents from the work area. Films of poly(vinyl alcohol), poly(vinyl chloride), or vinyl chloride copolymers are often manufactured by the use of these techniques.

Melt Pressing of Film

Polymers that are thermally stable above their melting or softening temperatures can be fabricated into films by a combination of heat and pressure. The melt-pressing technique is more often used in the laboratory than in the manufacturing plant because large films are difficult to prepare by this method, and the process is discontinuous rather than continuous. The apparatus shown in Figure 20.7 is employed. It consists of two electrically heated platens. One platen can be forced against the other by means of a hydraulic unit (usually hand-pumped). The powdered or subdivided polymer is placed between two sheets of aluminum or copper foil, and this sandwich is placed between the two heated platens. Pressure is then applied (~ 2000 to 5000 psi for about 30 s), whereupon the sandwich is removed and cooled, and the film is separated from the foil. In practice, the temperature and pressure must be determined by trial and error. If the temperature is too high, the polymer will simply flow out of the sandwich. If the temperature is too low, the film may be opaque or weak because of inadequate fusion. Metal shims or gaskets can be used in the sandwich to define the thickness of the film.

Sinter-fabrication of film

Some polymers, such as poly(tetrafluoroethylene) (Teflon), have melting points that are so high that melt-fabrication techniques at high pressures are not feasible. However, powders of such polymers can be "preformed" into weak films at high pressures. Subsequent heating above the melting point completes the sintering

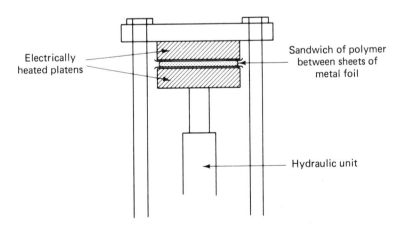

Electrically heated platens

Sandwich of polymer between sheets of metal foil

Hydraulic unit

Fig. 20.7   Hydraulic press for the melt pressing of polymer films.

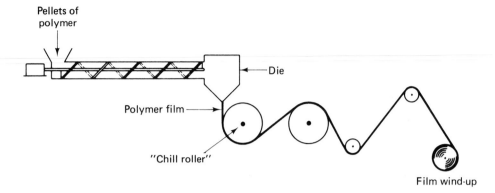

Fig. 20.8 Sequence of operations for the melt extrusion of polymer films.

process. This technique is reminiscent of those used for the fabrication of certain metals or ceramics. For poly(tetrafluoroethylene), the initial preforming is carried out at 1500 to 6000 psi, and the subsequent sintering takes place at 365 to 385°C during a brief exposure to the high-temperature conditions.

Melt-extrusion of films

Preferred manufacturing processes are those which are continuous. Melt-extrusion processes have this advantage. The overall sequence of operations is illustrated in Figure 20.8. Polymer pellets or powder are fed into a screw extruder. This is a device that heats the polymer and, by means of a rotating screw spindle, forces it under pressure into the die. The molten polymer is then extruded through the die slit. The flat sheet of molten polymer is collected by a rotating drum, which cools the film to below the melting point. Subsequent rollers complete the cooling and orientation processes. Typically, the sheet of molten polymer emerging from the die is 10 to 40 times thicker than the final film because the speed of the rotating drum exceeds the speed at which the polymer is extruded from the die. The final film may be 0.5 to 4 mils (~0.01 to 0.1 mm) thick.

Bubble-blown Films

An alternative method for the melt extrusion of films involves the extrusion of a *tube* of polymer, which is then expanded by compressed gas to form a tube of thin film. The process is shown schematically in Figure 20.9. Molten polymer from a screw extruder is forced through an annular die. Gas pressure inside the extruded tube blows the tube into a cylindrical bubble. The bubble is flattened by rollers, slit lengthways to form a continuous film, and then wound into a roll. Film made in this way has a high degree of biaxial orientation. Copolymers of vinyl chloride and vinylidine chloride (Saran) can be fabricated into films by this technique.

Fabrication, Testing, and Uses of Polymers / Part IV

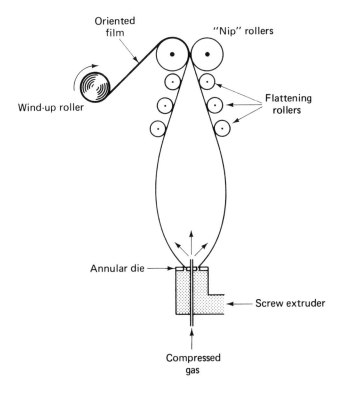

Fig. 20.9   "Bubble" blowing of films.

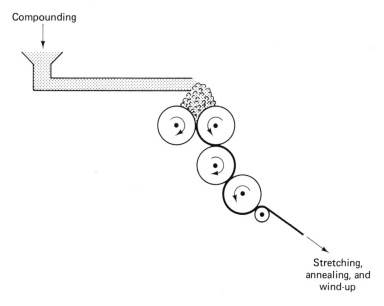

Fig. 20.10   Film manufacture by calendering.

Films by Calendering

The process of calendering consists of squeezing molten polymer into a thin sheet between heated rollers. This is a method normally used for the manufacture of thick films [2 mils (0.05 mm) to more than 10 mils (0.25 mm) thick]. The process is shown schematically in Figure 20.10. Calendering is a process much used for poly(vinyl chloride) and related copolymers.

FIBERS

The production of synthetic fibers for textile manufacture comprises one of the largest and most important branches of polymer technology. Even in the research laboratory, the preparation of fibers from new polymers is an important step in the physical evaluation of a polymer by mechanical or X-ray diffraction techniques.

Two fundamental techniques exist for the production of fibers: solution-spinning and melt-spinning techniques. In solution spinning, a solution of the polymer is extruded as a filament either into a nonsolvent (*wet* spinning), or solvent is removed from the filament by a stream of hot air or inert gas (*dry* spinning). In melt spinning, the molten polymer is extruded directly into filaments. After spinning, the fiber is usually *drawn* or *oriented* by stretching to improve its strength. Conditioners are usually applied also.

Filaments are manufactured in three main forms. *Filament yarn* consists of bundles of tens to hundreds of roughly parallel, continuous individual polymer filaments of great length. *Staple* is formed from a very large number (perhaps thousands) of shorter, randomly oriented fibers. *Monofilament*, as the name suggests, consists of individual fibers of great length. Monofilaments are much thicker (0.1 to 2 mm diameter) than the other types of fiber.

Solution Spinning

Wet spinning

Solution-spinning methods comprise the oldest processes used for the preparation of synthetic fibers. The wet-spinning modification requires the coagulation of a filament of the viscous polymer solution in a nonsolvent for the polymer.

On a laboratory scale, filaments can be wet-spun with the use of the apparatus shown in Figure 20.11. A viscous solution of the polymer is extruded into a continuous filament by means of a hypodermic syringe and needle. Coagulation takes place in a trough of nonsolvent, and the solid filament is wound continuously on to a spool. In practice, the hypodermic needle diameter should be larger than the diameter of the filament needed. Moreover, the choice of the nonsolvent and the temperature of the coagulation bath are critical. If coagulation is too abrupt, a weak or "granular" fiber will be formed. Hence, the coagulation bath often consists of a mixture of solvent and nonsolvent to effect a slower, more controlled

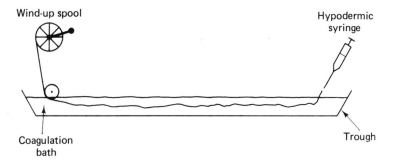

Fig. 20.11 Apparatus for the laboratory wet spinning of fibers.

precipitation. Refinements to the laboratory apparatus include the use of a motor-driven syringe (Figure 20.12) for a steady extrusion rate, and the addition of a motor-driven take-up spool for the filament. Fiber stretching spools may also be included in the sequence.

On an industrial scale, wet spinning is used to manufacture filament yarn from viscose rayon, proteins, poly(vinyl alcohol), polyacrylonitrile, poly(vinyl chloride), and other polymers. The industrial equipment is illustrated in Figure 20.13. The polymer solution is first pressure-filtered and then forced into filaments by passage through a spinneret. A spinneret is a metal plate with holes drilled in it. Tens to hundreds of holes may be present if the product is to be used for filament yarn; thousands of holes are present if staple is to be the product. The size of the holes does not determine the final thickness of the fiber; this factor is determined by the rate of fiber wind-up, the shrinkage during coagulation, degree of orientation, and so on.

As in the laboratory process, the viscosity of the solution, the nature of the nonsolvent, and the temperature all affect the properties of the fibers. Very viscous solutions are needed to prevent the filament from separating into droplets at the

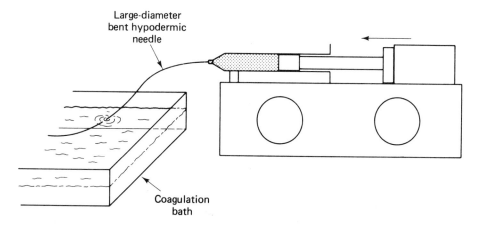

Fig. 20.12 Use of a motor-driven syringe pump for the laboratory preparation of wet-spun fibers.

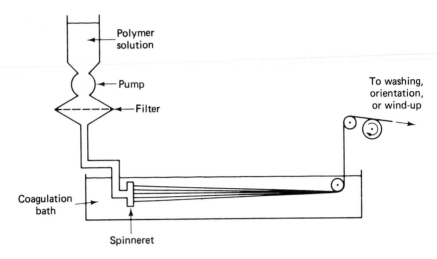

Fig. 20.13   Industrial wet spinning.

extrusion step. Some typical coagulation systems are as follows. Dilute aqueous sulfuric acid, sodium sulfate, and zinc sulfate form the coagulation bath for the spinning of viscose rayon xanthate solutions (see page 172). Cuprammonium rayon is spun into water. Polyacrylonitrile is spun from dimethylformamide into aqueous dimethylacetamide, or from aqueous 50% sodium thiocyanate into aqueous 10% sodium thiocyanate. Spandex (polyurethane–elastomer) fibers can be spun from dimethylformamide into water.

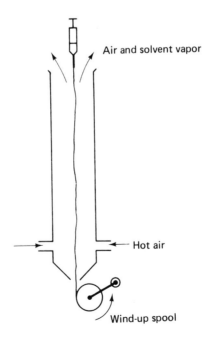

Fig. 20.14   Laboratory assembly for the dry spinning of fibers.

Although wet-spun fibers account for a large percentage of synthetic fiber production (mainly viscous rayon), they have certain disadvantages. Fibers with a uniform cross section are very difficult to produce by the wet-spinning process. The outer surface of each filament coagulates first to form a skin. When the core coagulates at a slower rate, the outer skin shrinks and becomes convoluted. Subsequent orientation by stretching may generate greater order in the sheath than in the core. A further disadvantage of the process is that it is slow. Low extrusion speeds are needed to permit precipitation in long coagulation baths. Some of these disadvantages can be overcome by the use of dry-spinning or melt-spinning techniques.

Dry spinning

The dry-spinning process involves the extrusion of a polymer solution through a spinneret into a hot air stream which volatilizes the solvent and leaves a dry polymer fiber. The technique can be carried out on a laboratory scale, but it is difficult. As shown in Figure 20.14, a polymer solution is extruded in the laboratory

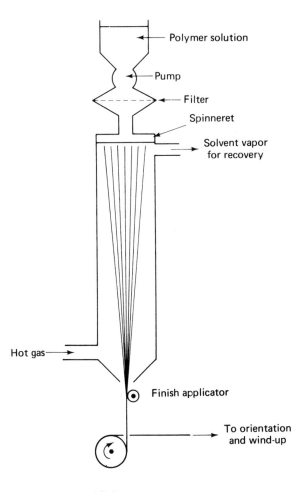

Fig. 20.15 Manufacturing equipment for the dry spinning of fibers.

from a hypodermic syringe into a glass or metal chimney containing a stream of hot air (possibly from a laboratory air blower). Problems encountered include the formation of droplets instead of fibers (if the solution is not viscous enough), and adhesion of the fiber to the wall of the chimney due to turbulence of the hot air stream. Moreover, inflammable or toxic solvent vapors must be removed effectively.

However, the process is carried out effectively on a large scale in industry. The sequence of operations is illustrated in Figure 20.15. A relatively concentrated polymer solution is filtered and pumped through a spinneret. The fibers pass down a vertical tube (which may be up to 25 ft long) countercurrent to a stream of hot gas. Often the polymer solution is heated before extrusion to lower the viscosity sufficient for passage through the spinneret holes. This enables smaller quantities of solvents to be used. Obviously, it is advantageous if the solvent used is quite volatile, such as acetone or carbon disulfide; but water has been used as a solvent for poly(vinyl alcohol), and dimethylformamide or dimethylacetamide are employed as solvents for polyacrylonitrile or Spandex.

Melt Spinning of Fibers

In this process, *molten* polymer is extruded through spinnerets. Immediate cooling causes solidification of the fibers, which can then be stretched or collected immediately on a bobbin. The advantages of the melt-spinning technique are that (1) the spinning process is extremely rapid, and (2) the fibers have a uniform, circular cross section. The disadvantage of melt spinning is that some polymers

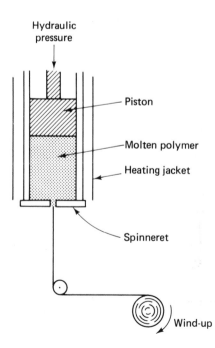

Fig. 20.16 Laboratory or pilot-plant equipment for the melt extrusion of fibers.

are not sufficiently stable above their melting temperature to survive the spinning process intact.

The simplest laboratory technique for the formation of fibers from a molten polymer is to insert a glass rod into the melt and pull out a long fiber. More sophisticated laboratory or pilot-plant methods make use of forced extrusion of the melt through a small orifice under pressure generated by nitrogen gas or a hydraulically operated piston (Figure 20.16).

On a large industrial scale, the equipment illustrated schematically in Figure 20.17 is employed. Solid polymer pellets or chips are melted by a heated grid, and air is removed. A nitrogen atmosphere is often maintained over the melt. After pressure filtration, the molten polymer is forced through spinnerets and the molten fibers solidify in a stream of cold air. The fibers are then collected on rollers and bobbins. Using these techniques, filament yarn can be produced at the rate of several thousand feet per minute. The melt-spinning process is used for the preparation of fibers from Nylons 6 and 66, poly(ethylene terephthalate), polyethylene,

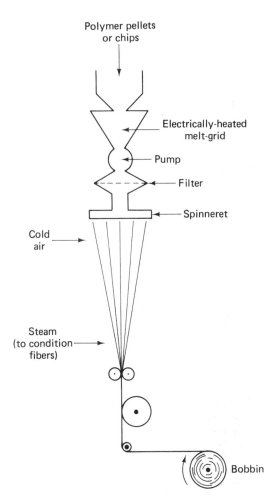

Fig. 20.17 Equipment for the melt spinning of fibers on an industrial scale.

vinylidine chloride copolymers, polyurethane, and polyacrylonitrile. The last polymer must be plasticized with 30 to 40% dimethylformamide before it can be spun.

Glass fibers are produced in the same way—both to form filament and staple. In this case the starting material is molten glass from a manufacturing furnace, or marbles that are melted before extrusion. The extrusion temperature for glass fibers is in the region of 1250 to 1450°C.

Monofilaments are often made by melt-spinning techniques. Nylon 610, Nylon 11, polyethylene, Saran, and polypropylene monofilaments are well known.

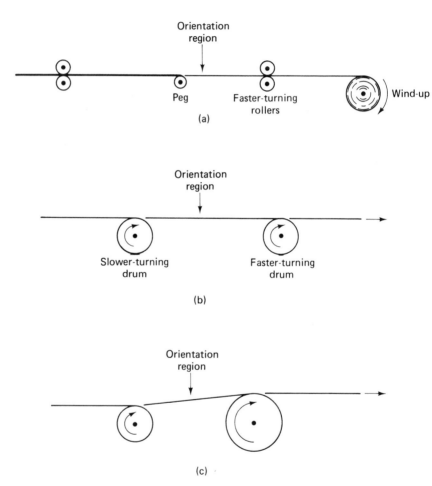

Fig. 20.18   Three methods for the continuous orientation of synthetic fibers. (a) The fiber passes round a "peg" which stabilizes the stretch orientation being induced by the faster-turning set of rollers. (b) The faster-turning drum stretches the polymer in the orientation zone. With this technique some difficulty may be experienced with stabilization of this zone. (c) The two drums turn at the same speed, but the fiber is stretched because of the greater circumference of the second drum.

Fiber Orientation and After-treatment

The maximum strength of a fiber is not realized until it has been drawn to orient the polymer molecules (see Chapter 17). On a large scale, orientation of the fibers is achieved by a continuous process in which the fiber passes round successive drums which either rotate at increasing speeds, or which rotate at the same speed but have increasing diameters (Figure 20.18). Polymers that have a high glass transition temperature (such as some polyesters) may need to be heated during the stretching process. In the orientation zone, the fiber diameter is markedly reduced as the polymer "necks down." Some polymers are oriented immediately after spinning. Others, such as Nylon 6, may be stored first to permit crystallization before orientation is carried out.

Many textile fibers are treated with lubricants and antistatic agents immediately after the spinning process. Some fibers are dyed after spinning, but it is usually much more convenient to incorporate the dyestuff into the polymer solution or melt before spinning takes place.

FABRICATION OF SHAPED OBJECTS

A wide variety of different techniques are available for the conversion of polymers or polymerization systems into shaped objects. With only a few exceptions these techniques require the use of complicated and expensive machinery. Hence, they are not suitable for laboratory fabrications. The types of shaped objects that can be made from synthetic polymers covers a vast range, from the plastic housings of ball-point pens or plastic caps for bottles to the interiors of refrigerators or the nose cones of missiles or supersonic aircraft. The following sections outline some of the techniques available.

Casting

Casting is a process in which a liquid monomer or prepolymer is polymerized inside a suitable mold. Initiation of polymerization is usually effected by the use of chemical reagents, although photochemical or high-energy radiation techniques have occasionally also been used. With chemical initiation, the "curing" or poly-merization step usually takes place when the mold is heated in an oven. The main advantage of the casting process is that intricately shaped objects can readily be made. Furthermore, the process is inexpensive and can easily be adapted to laboratory or small-scale production procedures.

The simplest and most straightforward application of casting is in the preparation of rigid polymer sheets. This type of procedure was mentioned earlier in Chapter 3 for the preparation of clear sheets of poly(methyl methacrylate). The mold consists of two sheets of plate glass, separated by a gasket (Tygon tubing), and held together by spring clips (Figure 20.19).

The polymer–catalyst mixture is poured into the mold (with careful removal of bubbles), the gasket is adjusted to seal the inlet, and the assembly is placed in an

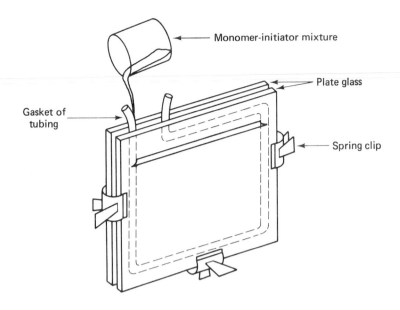

Fig. 20.19   Procedure for casting sheets of polymer.

oven to cure. Care must be taken to avoid overheating, to allow for shrinkage, and to permit annealing to occur. Large sheets of window-type material can be made in this way. Similar sheets can also be made from polystyrene, or epoxy resins. Polymer sheets of this kind are lighter than glass, but they are scratched more readily.

More intricately shaped objects can easily be made with the use of suitable molds. Plaster, clay, or wooden molds are often used if the surface of the mold is sealed with a nonadhesive coating. Such molds have a relatively short lifetime, but they are often used by hobbyists or artisans. Glass or metal molds are more durable, but molds made from Teflon or elastomers (such as silicone rubber) are preferred when a facile release of the polymer from the mold is needed. "Release agents," such as waxes, greases, or silicones, may be used to prevent adhesion of the polymer to any mold material. The casting process can also be used to encapsulate articles such as electronic components, or to produce lamp stands or paperweights which contain shells, pictures, and so on.

Ethyl or methyl methacrylate polymers are commonly fabricated by casting techniques. However, the method is also used to make objects from polystyrene, silicones, epoxy and phenol–formaldehyde resins, and from polyurethanes. Cast epoxy resins, in particular, yield very tough, durable castings.

Compression Molding

Compression-molding techniques are normally used for the fabrication of thermo-*setting* polymers, such as phenol–, urea–, or melamine–formaldehyde, alkyd, diallyl phthalate, or silicone resins. A charge of the molding powder (resin, fillers,

Fabrication, Testing, and Uses of Polymers   /   Part IV

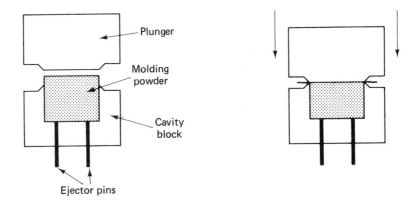

Fig. 20.20   Compression molding.

pigments, curing agents, and mold release agents) or a tablet of the mixture is placed in the lower half of a heated metal mold (Figure 20.20). The mold is closed, air and excess resin are forced out, and the mold is held shut until the resin has cured (30 s to several minutes). The mold is then opened and the object is released or ejected. Typically, the mold is made from chromium plated metal. Temperatures above 150°C, and pressures of 2000 psi or more, are usually employed. A modification of the compression molding technique is *transfer* molding. In this, the resin is melted outside the mold and is then rapidly injected into the mold by a plunger. This procedure causes less wear on the mold than does conventional compression molding. A disadvantage of the compression molding method is the time required for the curing step.

Injection Molding

Injection molding is a high-speed method that is used for the fabrication of both thermoplastic and thermosetting polymers. The equipment is illustrated in Figure 20.21. The powdered polymer or resin is heated above the melting or softening point, and the liquid is then forced by a plunger into a closed two-piece mold. The polymer cools or cures, solidifies, the mold opens, the product is ejected, and the cycle is repeated. The sequence may take only 10 to 30 s, which makes it particularly suitable for mass production. Often, the liquid is heated to temperatures above 250°C, and the injection pressure may be as high as 10,000 to 30,000 psi. Objects as large or larger than television cabinets can be made by injection molding.

Blow Molding

This technique is used to make bottles, toys, tanks, or other hollow objects from thermoplastic polymers. As shown in Figure 20.22, a thermally softened tube of polymer (known as "parison") is delivered from an extruder into an opened,

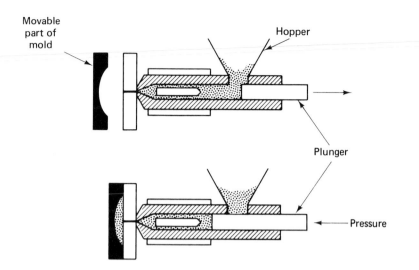

Fig. 20.21   Injection-molding cycle.

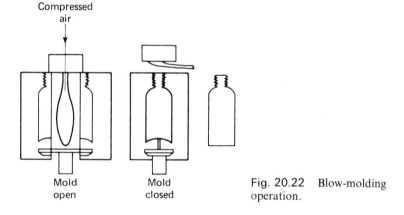

Mold
open

Mold
closed

Fig. 20.22   Blow-molding
operation.

two-piece mold. The mold closes around the parison, pinching it down at one or both ends. Compressed gas (25 to 100 psi) is then injected into the parison, which expands to line the inside of the cooled mold. The thermoplastic hardens, the mold opens, and the object is removed. The process is then recycled. In practice, all degrees of automation are possible from the manual transfer of individual parisons into the mold to a continuous but carefully synchronized extrusion of parison material into successive different molds. Blow molding is an inexpensive process. It allows facile changes to be made in the wall thickness of the product simply by making changes in the wall thickness of the parison. Moreover, the polymers used can have higher molecular weights than those used in injection molding, because high fluid viscosity is an advantage rather than a disadvantage. Polyethylene, poly(vinyl chloride), polycarbonates, methacrylate polymers, polyformaldehyde, and polystyrene are commonly blow-molded.

Thermofusion and Thermoforming

Thermofusion is a process used to make large objects, such as boats, barrels, and other large containers from a finely divided thermoplastic, usually polyethylene. The powder is placed in large sheet-metal molds, and the polymer is melted in an oven.

In thermoforming, a *sheet* of a thermoplastic is heated above the softening temperature, and is then pressed into a mold, often by the application of pressure or vacuum (Figure 20.23). The technique is used to make shallow trays, transparent skylight roof "blisters," or raised topographical relief maps. Polyethylene, poly(vinyl chloride), poly(methyl methacrylate), polystyrene, polycarbonate, or ABS polymers are suitable for this fabrication method.

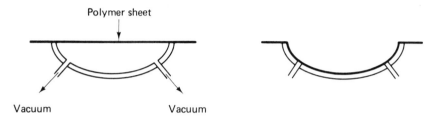

Fig. 20.23   Thermoforming of polymer sheets.

Rotational Molding

This is a technique for the fabrication of hollow objects from thermoplastic polymers. The solid polymer is melted inside a closed mold. The mold is then rotated simultaneously around its axial and equatorial axes so that the polymer coats the inside surface uniformly as it cools. The technique is used to make water tanks, fuel tanks, glove boxes, luggage, battery cases, and housings or ductwork.

Bag Molding

This technique is used for the preparation of large objects at a low production volume, especially when high-strength, high-performance structures are required. Prototype nose randomes for aircraft and missiles are made in this way. A thermosetting resin and a reinforcement material (glass cloth) are placed in an open mold and covered by a flexible diaphragm. The surrounding air chamber is then closed and air pressure is introduced. The pressure forces the diaphragm and resin into the mold. Curing then takes place as the temperature is raised.

Tube Fabrication

Tubes of polymer are normally prepared by extrusion from a heated screw extruder through an annular die. Thermoplastics that are suitable for tube extrusion include polyethylene, poly(vinyl chloride), polypropylene, polyformaldehyde, nylon, ABS terpolymers, and blends of rubber and polystyrene. Most of the tube-extruded thermoplastics are used for water pipes, drains, irrigation pipes, gas lines, and electrical or telephone conduits. Tygon tubing used in the laboratory is poly(vinyl chloride) plasticized by phthalate esters. Polymer rods and channels are also formed by extrusion techniques.

EXPANDED POLYMERS

Foam rubber and cellular insulation material are made by generating gas bubbles in a polymer and then stabilizing the expanded structure. Such processes are used to make polyurethane foams, polystyrene foam beverage cups and furniture, poly(vinyl chloride) fabric coatings, ordinary sponge rubber, and epoxy flotation devices. The cellular expansion process is illustrated in Figure 20.24. Three methods are available for the formation of gas bubbles:

1. Latex foam rubber is made by mechanically induced frothing of a latex or a liquid rubber, followed by crosslinking the polymer in the expanded state.

2. The liquid polymer or monomer is mixed with a chemical "blowing agent," which liberates a gas when heated. Azobis*iso*butyronitrile (AIBN) evolves nitrogen gas when heated. Sodium bicarbonate liberates carbon dioxide.

3. A low-boiling liquid or gas is dissolved in the liquid polymer under pressure. Heating of the polymer causes boiling of the liquid to generate bubbles. Pentanes, hexane, or halocarbons are commonly used expansion agents.

Once the polymer has been expanded or "blown," the cellular structure must be stabilized rapidly; otherwise it would collapse. Two stabilization methods are used. First, if the polymer is a thermoplastic, expansion is carried out above the softening or melting point, and the form is then immediately cooled to below the melting temperature. This is called *physical stabilization*. The second method—*chemical stabilization*—requires the polymer to be crosslinked immediately following the expansion step.

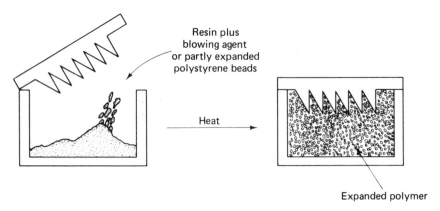

Fig. 20.24 Preparation of molded objects from expanded cellular polymers.

Polystyrene or poly(vinyl chloride) foams are usually stabilized simply by cooling. Such polymer-blowing agent mixtures are often extruded through a slit. Expansion and simultaneous cooling occur as the polymer is extruded. Polyurethane foams are expanded by carbon dioxide bubbles generated from the reaction of excess isocyanate with water or carboxylic acids in the system, and by the expansion of volatile organic expansion agents. Crosslinking occurs during foaming. Epoxy foams are stabilized by crosslinking, as are those from phenolic resins. Silicone foams are expanded by chemical blowing agents, and are stabilized by the reaction of a crosslinking catalyst (peroxide).

The individual cells within the foam may be separated from each other (closed-cell foams), or they may be interconnected (open-cell foams). The closed-cell variety are obviously better suited for thermal insulation purposes. Whether a closed- or open-cell structure is formed depends on a critical balance between the viscosity of the liquid polymer and the rate of decomposition of the chemical blowing agent.

A closed-cell system used in some flotation devices makes use of a cellular arrangement of glass or silica microballoons incorporated into the polymer.

REINFORCED POLYMERS

Many polymers must be reinforced before they can be used. For example, although most raw elastomers have an advantageous flexibility and impact strength, they are often too soft and delicate for use in rubber tires. The introduction of reinforcement introduces toughness and allows the basic shape of the object to be retained even under conditions of high stress. Reinforcement is used for solid, glassy polymers as well as for flexible or elastic polymers.

Two basic types of reinforcement are in common use: reinforcement by fabrics or cords, and reinforcement by finely divided fillers, such as carbon black, short glass fibers, carbon fibers, asbestos, mica, wood flour, calcium carbonate, and so on. For some types of application (e.g., rubber tires or Fiberglas boats), both types of reinforcement are used together.

Woven fabrics of glass fiber, nylon, polyester, or rayon are used as reinforcement in automobile tires. The fabric itself is coated with a rubber compound in a calendering operation. The tread and sidewalls are made by extrusion. A tire is fabricated first from a calendered rubber liner wrapped around a drum. The fabric reinforcement is then added and the tread and sidewalls are positioned outside the reinforcement. The unit is then heated to cure the elastomer.

Glass fabrics are also used to reinforce thermosetting polymers in laminates. A *laminate* is a sandwich of fabric layers bound together by the polymer. The laminate is built up layer by layer with each layer of fabric impregnated by and separated from the next one by a layer of resin. The sandwich is then compressed in a press and heated in an oven until the resin has cured. Laminates are characterized by toughness, dimensional stability, and a resistance to cracking or shattering when placed under stress.

Finely divided fillers in a polymer matrix serve the same purpose. Most fillers are bound to the polymer by van der Waals forces; hence the binding strength increases with surface area and with the degree of subdivision of the filler. However, carbon black is a particularly effective filler for rubber and other polymers, apparently because it participates in chemical grafting to the polymer. Chopped glass fibers are extensively used as filler materials for rigid thermosetting resins in the manufacture of boats, automobile bodies, building panels, crash helmets, and so on. A commonly used resin formulation is a styrene solution of the condensation product from ethylene glycol and maleic or phthalic anhydrides. Benzoyl peroxide is a suitable curing accelerator. Other resins include phenol-formaldehyde, epoxy, silicone, alkyd, and melamine–formaldehyde formulations. The polymer may constitute 10 to 60% by weight of the material. Often a mixture of the resin and the chopped glass fiber is sprayed onto an open mold, with further reinforcement sometimes introduced by the application of a woven fabric. Boats, large tanks and housings, and crash helmets are made in this way.

Thermoplastics, such as nylons, polyacrylates, polycarbonates, polypropylene, polyformaldehyde, or ABS terpolymers can also be strengthened by the incorporation of fillers. These materials are being used increasingly in the automobile industry.

Other forms of reinforcement include the use of metal honeycombs and filament-wound products. The filament-winding technique uses a continuous filament wound around a mandrel and impregnated with the resin. The cylindrical- or spherical-shaped products have a very high strength.

ELASTOMER TECHNOLOGY

Raw rubber, either natural or synthetic, is only rarely suitable for use in most applications. The pure elastomer is usually too soft and extensible, or too readily attacked by oxygen or ozone. For these reasons, a substantial number of additives are compounded into the polymer to improve its performance. These additives include:

1. *A vulcanizing agent*, such as sulfur, and related additives, such as mercaptothiazole (a vulcanizing accelerator), activators, or retarders. Metal oxides (ZnO) are used as activators. The vulcanization process is a thermally induced crosslinking step which imparts strength, resistance to viscous flow, and elasticity to the polymer. Natural rubber, SBR-butyl, or nitrile rubber can be vulcanized with sulfur. Silicone rubber is crosslinked by means of peroxides.

2. *Fillers*, such as carbon black, increase the tensile strength and elasticity (the rapidity of retraction). Apparently, carbon black functions by forming weak, covalent crosslinks between the chains. This effect stiffens the elastomer and generates toughness. Other fillers, such as clay, are sometimes added to rubber to improve its handling qualities before vulcanization. Silica may be used as a reinforcing filler for silicone elastomers.

3. *Pigments* to modify the color of the elastomer. Pigments cannot be used if carbon black is employed as a filler.

4. *Plasticizers*. These may be added to soften an elastomer.

5. *Antioxidants*. Most organic elastomers react slowly with oxygen or ozone. In this process they become either soft or hard and brittle. This effect is especially serious with elastomers made from dienes. An antioxidant is a compound added to protect the elastomers from this effect.

SURFACE COATINGS

Polymers are widely used as materials for coating metals, wood, paper, fabric, or even, in rare cases, for coating stone and masonry. Polymers are used as coatings in order to confer waterproofing, flameproofing, fungus-resistance, or corrosion-protection properties on to other substrates. The polymer coating may be an adhesive, insulating, decorative, or reflective coating. Such surface-coating properties are ultilized in the painting of metal, the coating of fabrics (to make rainwear or artificial leather), the production of magnetic recording tape, pressure-sensitive tape, book bindings, paper, wire enamels, floor protection, and photographic emulsions. Some of the methods employed are listed below.

Dipping

A mold is dipped first into a solution or emulsion of a polymer and then into a coagulation bath. The procedure is repeated until a thick-enough layer of polymer has been built up. In the manufacture of, for example, a rubber glove, a fabric glove placed over the mold actually serves as the reinforcing substrate on which the polymer is coated. Rubber and polyethylene coatings are often applied in this way.

Calender Coating

A film of the coating polymer is prepared by calendering. It is then applied in a continuous process to the film of substrate. The two films are then squeezed together by rollers.

Extrusion Coating

A thermoplastic is extruded through a slit to form a thin film. This film coats one side of the moving substrate film, and the two layers are forced together by passage through rollers.

Electrostatic Coating

This is essentially a spray gun method in which the coating material is given an electric charge before spraying. The object to be coated is either grounded or bears an opposite charge. The method facilitates a uniform coating procedure. It is used for coating automobile parts or even textiles or paper products.

Knife Coating

This method is a large-scale continuous process development of the solution film-casting technique discussed earlier. A film is cast continuously on a moving belt formed by the substrate film material. A schematic diagram of the process is shown in Figure 20.25. The same type of process can also be used for the melt casting.

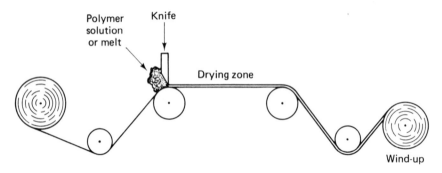

Fig. 20.25   Knife coating of a polymer onto a film.

Roll Coating

A continuous sheet or film of the material to be coated passes over a roller. The roller picks up a film of the molten or dissolved polymer and applies it to the film in a continuous process. The method closely resembles the techniques used in mass-production newspaper printing.

Fluidized-Bed Coating and Powder Molding

A preheated metal object is dipped into a fluidized bed of powdered polymer and a compressed gas. Polymer particles adhere to the hot object. The object is then passed through an oven to fuse the polymer and form a film. The coating may be left in place to protect the metal, or it may be removed from the metal to yield a hollow article.

Radiation-Cured Coatings

Although heat is the principal method used for the curing of thermosetting materials applied as coatings, experimental units have been tested that irradiate a coating with gamma rays or ultraviolet light in order to crosslink the polymer. Of course, an ideal process would bring about a chemical grafting of the coating to the substrate.

STUDY QUESTIONS

1. From what you know of the properties of the following polymers, suggest the types of fabrication methods that might be suitable for these materials on a large scale: poly(vinyl alcohol), polybenzimidazoles, polycarbonates, polymeric sulfur, borosilicate glass, polybutadiene, polystyrene.

2. Why does a volume contraction take place when methyl methacrylate polymerizes in a mold?

3. What precautions would be necessary if you were to carry out a solution-spinning experiment in an attempt to prepare fibers of polyethylene from hot xylene?

4. Suppose that you wished to prepare thin films of a polymer to test its ability to function as a barrier to gaseous diffusion. What fabrication problems and solutions to these problems do you foresee?

5. You have been given the task of planning the fabrication of the nose segment of a new rocket-propelled aircraft. Extreme thermal and dimensional stability of the unit will be needed. Prepare a written proposal for this work, justifying your decisions at each step.

6. Thermoplastics that are reinforced with glass fibers are being used increasingly for the fabrication of components for automobiles. What problems would you be likely to encounter if you wished to use conventional, high-volume fabrication techniques for the manufacture of such components?

7. Polymers are now used extensively for the manufacture of the hulls of motor boats, yachts, and so on. What reasons can you think of that might prevent an extension of this technology to the manufacture of ocean liners or freighters, or the manufacture of the outer shells of civil aircraft?

8. What fundamental physicochemical characteristics of a polymer are needed if it is to be used as (a) a textile fiber; (b) a film for use as a packaging material; (c) an artificial leather; (d) a surface coating for outdoor use; (e) a liner for a home refrigerator?

9. One fabrication technique that is not discussed in this chapter is the use of polymers as wire coatings. Suggest ways in which a metal wire might be coated with (a) a thermoplastic polymer, and (b) a thermosetting resin. What special properties of the polymer would you be looking for in this application?

10. Why do inorganic substances, such as glass fiber, asbestos, carbon black, boron fibers, carbon fibers, and so on, constitute materials that are added to polymers as reinforcement fillers? Why, for example, are nylon fibers not normally used to reinforce epoxy resins?

SUGGESTIONS FOR FURTHER READING

HATTORI, H., "Reinforced Plastics," in *Encyclopedia of Polymer Sci. and Technol.* (H. F. Mark, N. G. Gaylord, and N. M. Bikales, eds.), **12**, 1 (1967).

KRASSIG, H., "Film to Fiber Technology," *J. Polymer Sci.* (D) (*Macromol. Rev.*), **12**, 321 (1977).

Kraus, G., "Reinforcement of Elastomers by Carbon Black," *Adv. Polymer Sci.*, **8**, 155 (1971).

Manson, J. A., and L. H. Sperling, *Polymer Blends and Composites.* New York: Plenum Press, 1976.

Pearson, J. R. A., *Mechanical Principles of Polymer Melt Processing.* New York: Pergamon Press, 1966.

Penn, W. S., *PVC Technology.* New York: Wiley, 1971, 1972.

Rastogi, A. K., "Fiber-Reinforced Plastics Today," *Chemtech*, June, 349 (1975).

Sorenson, W. R. and T. W. Campbell, *Preparative Methods of Polymer Chemistry* (2nd ed.). New York: Wiley–Interscience, 1968.

Williams, H. L., *Polymer Engineering.* New York: Elsevier, 1975.

Wolinski, L. E., "Films and Sheetings," in *Encyclopedia of Polymer Sci., and Technol.* (H. F. Mark, N. G. Gaylord, and N. M. Bikales, eds.), **6**, 764 (1967).

21

The Testing of Polymers

Polymers are used in an enormous variety of different applications. However, each different use normally requires a polymer with very specific properties. Hence, an important aspect of polymer technology is the testing of new polymers to determine their advantages and disadvantages for different applications. The ultimate test of any polymer is for it to be fabricated into a suitable object and tested under the same operating conditions that it would encounter in normal use. This is an expensive and time-consuming procedure. Thus, considerable emphasis is placed on the generalized *laboratory* testing of new polymers in the hope that the results will provide guidelines to the types of applications for which the polymer is best suited.

Many of the tests that are conducted on a new polymer involve *engineering* evaluations. A detailed discussion of such procedures and the underlying theory is beyond the scope of this book. However, the following sections outline a few of the laboratory tests that are commonly applied to any new polymer. For a more detailed discussion of this topic, the reader is referred to the specialized references at the end of this chapter. Here we will review the role of fundamental physico-chemical tests, mechanical evaluations, thermal properties, electrical tests, and tests designed to predict the stability of a polymer to weathering, solar radiation, and other environmental influences.

FUNDAMENTAL PHYSICOCHEMICAL TESTS

Two characteristics of a polymer form the foundation of any use-oriented evaluation. These are the *glass transition temperature* (T_g) and the presence or absence of *crystalline melting transitions*. If a polymer has a high T_g (say, above 30°C), it will generally be unsuitable for use in applications that require flexibility and rubbery properties. If the material is contemplated for use as an elastomer in a low-temperature environment (e.g., in the arctic or in aircraft) it must have a very low T_g and have a low degree of crystallinity. For a polymer to be useful as a textile fiber, it should normally have a T_g that is below its normal operating temperature but a T_m that is above this temperature. A polymer that is to be used as a rigid structural material should have a high T_g (100°C or above).

Thus, the measurement of T_g and T_m transitions by the methods discussed earlier in Chapters 17 and 19 is the logical first step in any polymer evaluation program. This information must then be viewed in the light of the *molecular weight* of the polymer (Chapters 14 and 15), because physical properties usually depend on the average chain length for the lower-molecular-weight species.

MECHANICAL TESTS

Chemists normally seek to understand the physical properties of materials in terms of *molecular* features, for example, in terms of molecular shape, conformational mobility, crystal packing forces, and bond energies. On the other hand, the engineer or technologist, while being aware of such fundamental matters, is usually much more interested in the question: What are the special properties of this material that would favor its use in this particular application? Although marked advances have been made in recent years in the physics and chemistry of polymers, mechanical properties cannot yet be predicted from fundamental molecular structural principles, except in a few very favorable cases. For example, the actual strength of a polymer may be only $\frac{1}{10}$ to $\frac{1}{100}$ of the value calculated on the basis of bond strengths and intermolecular forces. Hence, for the present, mechanical tests provide the only method for obtaining engineering-type evaluations of new polymers. Such tests are performed routinely in materials laboratories and in industrial research laboratories. Because most of the tests result in destruction of the sample, and because such tests do not have a high degree of reproducibility, multiple tests on similar samples are needed before valid results can be obtained.

Four different groups of questions about the mechanical properties form the basis of an applications evaluation. These are:

1. What is the response of the material to stress and strain? Does the material distort or elongate easily when stretched? Does it remain elongated or does it snap back to its original form when the stretching force is released? How readily will it break if subjected to stretching (tensile) forces? Is the polymer rigid or does it flow under pressure?

2. What is the resistance of the material to impact? Does the polymer shatter like glass, or does it absorb the force and remain intact, for example, like rubber?

3. What is the strength of the material to flexural distortions? Does a rod or a plate of the material break when bent, does it remain bent, or does it spring back when the stress is removed? What is the response to continued long-term flexural distortion (fatigue)?

4. What is the hardness and abrasion resistance of the material? Is the surface of the polymer readily deformed when pressed into contact with sharp objects? Does the material abrade when used in bearings? Is the surface scratched easily when abraded by metals, other plastics, fabrics, grit, and so on?

It will be clear that answers to most of these questions must be obtained before a polymer can even be considered for use in any important or large-scale application, whether that application happens to be as a textile fiber or in the nose cone of a rocket.

Stress–Strain Curves

The tensile strength of a material provides a measure of its resistance to elongation or breaking when stretching forces are applied to it. The terminology used in this field relates the stress (the loading or stretching force applied to the sample) to the strain (the elongation of the sample under a given stress) (see Chapter 17). Because the stress–strain behavior of most materials is time-dependent, the speed at which the stress is applied must also be taken into account. For instance, a sudden, abrupt pull on a fiber may cause it to break. The same force applied slowly may result in slight elongation of the fiber and in a higher resistance to breakage.

In practice, stress–strain experiments are often carried out on a flat sample that has been shaped into the form shown in Figure 21.1. The ends of the sample are clamped into the jaws of a testing machine and the jaws are separated by the application of a known mechanical force. The test material usually elongates or breaks in the narrower, central region of the specimen. Both the stress and the strain can be read from the dials on the machine, or the data may be provided directly on chart paper.

The ratio of stress to strain is a measure of the *stiffness* (i.e., a stiff polymer will yield very little as the stress is applied). A soft polymer, such as an elastomer, will yield considerably under the same circumstances. The data are normally plotted in the form shown in Figure 21.2, which typifies the behavior of a thermoplastic material, such as polyethylene. Note that *toughness* is a measure of resistance to breaking. This property can be crudely estimated from stress–strain curves, as

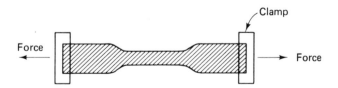

Fig. 21.1 Typical shape of a flat polymer sample used for stress–strain tests.

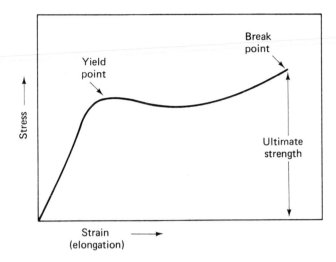

Fig. 21.2 Stress–strain curve for a thermoplastic material such as polyethylene. Note the initial section of the curve in which increased stress causes a moderate, but noncatastrophic, elongation. However, further applied stress causes appreciable elongation or "yield," without the application of comparable additional stress.

well as from impact tests. The stress–strain behavior of five different types of materials is illustrated in Figure 21.3. For example, an uncrosslinked elastomer might behave in the manner shown in curve (a) of Figure 21.3. A polymer for use as a tough, structural plastic (for use in housings, gear wheels, etc.) might conform to curve (d). Polystyrene is a hard, brittle polymer with stress–strain characteristics of type (c). One of the most important pieces of information derived from stress–strain

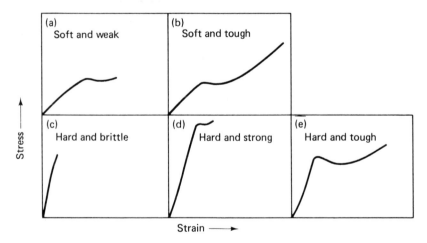

Fig. 21.3 Characteristic stress–strain curves for five different types of polymeric materials.

TABLE 21.1   Tensile Strengths of Various Polymers and Steels

POLYMER	TENSILE STRENGTH (psi)
Polyethylene (low to medium density)	1,000–2,4000
Poly(tetrafluoroethylene)	3,500
Polyethylene (high density)	4,400
Poly(dimethylsiloxane)	5,000
Polypropylene	5,000
Poly(vinylidene chloride)	8,000
Polystyrene	8,000
ABS terpolymer	8,500
Polyamides	9,000–12,500
Polycarbonate	9,500
Polyesters (cast)	$\sim 10,000$
Polysulfone	10,200–12,000
Poly(phenylene oxide)	10,500
Epoxy resin	
Cast	12,000
Molded	16,000
Glass-filled nylon	31,000
Fabric-reinforced epoxy resin	60,000–85,000
Carbon steel	80,000
Filament-wound epoxy resin	100,000–250,000
Type 420 stainless steel	250,000
Steel wire	500,000

curves is the value of the tensile strength at the breaking point. Some typical values for a number of different materials are given in Table 21.1.

Thermoplastic materials generally become less rigid as the temperature is raised, although the change in rigidity is not continuous if T_g or T_m transitions are encountered. Rigidity is clearly a favorable property for materials that will be used as structural plastics. Perhaps more important is the long-term ability of a rigid polymer to withstand *cold-flow* or *creep*. Tests of this property can be conducted by measuring the long-term elongation of a test sample, one end of which has been attached to a weight.

Impact Resistance

Impact resistance is normally associated with *toughness*. A polymer that is prone to shatter on impact cannot be used in many applications (although it may be a perfectly satisfactory structural material). Polymers in their glassy state (i.e., below their T_g) are particularly prone to shatter on impact. The standard impact test involves two types of experiments. Either a swinging pendulum is allowed to strike the sample from different displacements or a falling weight is dropped on to the sample from various heights. The impact force is increased until the sample breaks. Impact tests provide only rough guidelines at best to comparative toughness. Some polymers that are quite tough under actual operating conditions (nylon, for example) perform poorly in standard impact tests. Moreover, thinner samples are relatively tougher than are thicker specimens.

Flexural Strength

The ability of a material to undergo flexural distortions without weakening or shattering is a key property for the use of polymers in gear wheels, vibrating components, structural parts of automobiles, boats, or aircraft. A commonly used criterion of flexural strength is the force needed to cause a beam of the polymer to be deflected (i.e., bent) by a known amount. However, a more meaningful criterion is the ability of the material to withstand *fatigue* (i.e., multiple cyclic flexural motions). A graphical plot of the stress versus the number of flexural cycles needed to bring about failure of the sample provides an estimate of the fatigue characteristics of that particular polymer.

Hardness and Abrasion Resistance

Hardness, abrasion resistance, scratch resistance, and friction are related but not necessarily directly related properties. *Hardness* is measured by the distance of indentation and recovery that occurs when a steel ball (the indentor) is pressed into the surface under constant load, and is then released. In these terms, polyolefins are soft but polyimides and poly(ethylene terephthalate) are hard polymers.

Abrasion resistance and scratch resistance are subtly different properties. *Abrasion resistance* represents the ability of a polymer to retain a smooth surface while moving constantly in contact with another (smooth) surface, for example, when the polymer is used as a spindle bearing material, as sliding surfaces in reciprocating devices, in hinges, or in gears. A relationship appears to exist between abrasion resistance and the coefficient of friction. Those polymers that are both hard and have a low coefficient of friction (i.e., a slippery surface) in general have a high abrasion resistance. Abrasion tests may take the form of a measurement of the weight loss at the abrasion surface. Nylons, polyacetals, and poly(ethylene terephthalate) perform well in these tests.

The susceptibility of a smooth polymer surface being scratched may be critical in determining the use of a polymer in windows, lenses, or automobile windshields. *Scratch resistance* tests are highly subjective evaluations that are conducted differently in various laboratories. Some tests require the pressing and twisting of a piece of sandpaper on to the surface. Others require a scratching of the surface with pencils of differing hardness. Scratch resistance is frequently associated with rubbery, rather than hard or brittle surface properties, or with the ability of a surface to "heal" by cold-flow of the polymer.

THERMAL PROPERTIES

Perhaps the main reason why synthetic polymers have not yet replaced metals and ceramics in many applications is the inability of most polymers to maintain their advantageous physical properties at temperatures above 150 to 200°C. Other reasons include their high thermal expansion characteristics, their brittleness at low temperatures, their flammability, and poor chemical stability at high temperatures. One of the main thrusts in polymer research and technology is the drive to use synthetic polymers in ever more thermally hostile environments. Hence, thermal tests are vitally important in any polymer evaluation program.

Thermal Expansion

The thermal expansion coefficient of a polymer is measured by means of a dilatometer (see page 435) or by direct mechanical measurement of a length of a bar of polymer at different temperatures. The thermal expansion coefficients of most synthetic polymers may be as much as ten times greater than those of common metals. Hence, severe thermal distortions can arise when polymers are bonded to metals.

Mechanical Changes

The brittleness, rigidity, and strength of a polymer can be measured as a function of temperature change by the use of stress–strain or impact tests carried out on heated or cooled samples. Thermal softening can be examined with the use of penetration-type measuring devices.

Flammability

The flammability of a polymer is now a critical factor that determines its potential uses. Government flammability regulations increasingly control the types of materials that may be used in textiles, household furnishings, the interior components of civil aircraft, electrical insulation, and thermal insulation. Many organic polymers burn, and questions must be answered about the conditions under which burning can be initiated, the ability of the material to continue burning after ignition, and the generation of toxic fumes from a burning polymer.

Flammability tests in the laboratory sometimes bear little relationship to the possible burning behavior of the polymer in normal use. For example, a common flammability test for textile fabrics requires that a test sample of the fabric should be dried in an oven, cooled in a desiccator, and then ignited with a bunsen burner within a short time of exposure to the atmosphere. Another test of the polymer flammability involves an estimate of the minimum oxygen concentration needed for sample ignition. Toxicity tests on the products from polymer combustion require an involved laboratory procedure. Combustion products may be detected by vapor-phase chromatography or mass spectrometry. Animal toxicity tests are often needed to evaluate the possible physiological effects on humans following smoke inhalation from burning polymers.

Thermal Decomposition

The chemical decomposition of a polymer at elevated temperatures usually becomes evident in a practical sense by a deterioration of the physical properties. The chemical aspects of thermal degradation were discussed in Chapter 9. Such chemical information can often be used to predict the property changes expected when a polymer is used at high temperatures. For example, depolymerization of the polymer would be expected to result in a loss of strength, increasing brittleness, and perhaps even liquefaction. However, a more meaningful test of technological thermal stability is to examine the actual mechanical properties of the material after it has been heated ("aged") for various periods of time at elevated temperatures. Stress–strain experiments and impact tests may reveal more information about thermal stability than can be estimated from chemical facts alone.

ELECTRICAL TESTS

Polymers are used as electrical insulators, electric wire coatings, as dielectric materials, as electrets, and even as semiconducting or superconducting materials. Thus, an examination of the electrical properties of a polymer forms an important part of the evaluation procedure. The following properties are usually measured.

Resistivity

The resistance of a material to the flow of an electric current can be measured from the potential gradient developed between two electrodes applied to a polymer specimen. The volume resistivity is defined as the measured resistance times the distance between the electrodes, divided by the cross-sectional area. Clearly, a material that is a candidate for use as an electrical insulator should show a high volume and surface resistivity. The following factors are also important.

Dielectric Strength and Arc Resistance

As the voltage is increased across a polymer sample, a point is reached at which catastrophic electrical breakdown occurs. This point is determined by increases in the voltage applied to electrodes placed on opposite faces of a thin sheet of polymer film. Polymers frequently show electrical "fatigue" in which the repeated application of relatively low voltages eventually causes electrical breakdown. Organic polymers are prone to undergo surface arcing by the formation of carbonized spark pathways.

Dielectric Constant and Power Factor

A knowledge of the dielectric constant of a polymer is important if the material is to be used either as an insulator or as the dielectric material in an electrical condenser. In practice, the dielectric constant is measured from the capacitance of a condenser that contains the polymer as an insulating dielectric compared to the capacitance of the same condenser containing only air as the separation medium. A high dielectric constant is associated with the polarization and polarizability of the electrons that form individual bonds in the polymer matrix. This, in turn, depends on the *orientation* of polar groups in the matrix. The orientation motion of individual segments or component parts of each polymer chain depends on conformational changes and thermal motions. Hence, the ability of the polymer groups to switch orientations in phase with an alternating current may be limited, especially at high frequencies. This phase lag results in an absorption of energy by the polymer—the *loss factor*, which is a measure of the energy absorbed per cycle by the polymer from the field. The sine of the phase difference or loss angle is called the *power factor*, and this value multiplied by the dielectric constant yields a value for the loss factor.

A polymer that has a high power factor absorbs a considerable amount of heat from the alternating electrical field. Hence, such polymers may soften or melt and lose their insulation capability. Polyethylene has a low power factor and thus is suitable for use as an insulator. Another term, the *dissipation factor*, is the tangent of the loss angle. It measures the ratio of the in-phase to out-of-phase power.

ENVIRONMENTAL STABILITY

Most synthetic polymers are more stable than steel, copper, sandstone, or limestone in moist environments that contain dilute aqueous acid or inorganic salts. However, compared to most forms of stone and structural metals, synthetic polymers are quite unstable when exposed to solar radiation or to an ozone-containing urban atmosphere. For example, rubber automobile tires degrade quite rapidly in an atmosphere that contains photochemical smog. Polymeric surface coatings have only a limited outdoor life, especially in regions where exposure to intense sunlight is common. Even some of the most stable polymers, such as low-density polyethylene, crack and degrade after long exposure to the atmosphere, especially in sunlight. Many polymers are also affected adversely by contact with organic solvents, detergents, strong acids, or oxidizing agents. As polymers become more expensive and as the labor costs rise for the reinstallation of degraded polymeric materials, more and more emphasis will probably be placed on the monitoring of environmental stability and on the improvement of it.

Weathering Tests

The resistance of a polymer to weathering is often tested experimentally by long-term outdoor exposure of polymer samples to the atmosphere and sunlight followed by evaluation of the changes in mechanical or optical properties. Such tests may take years to complete. Preliminary evaluations can be carried out more conveniently in the laboratory with the use of an "accelerated" weathering unit. Such apparatus contains high-intensity lamps to simulate the effects of solar radiation. Polymers that are used for aircraft or space applications are often tested by irradiation with mercury vapor lamps that simulate the high ultraviolet-light content of unfiltered sunlight.

Many of the chemical reactions that take place in a polymer during exposure to sunlight are oxidation reactions. These reaction pathways were discussed earlier in Chapter 9.

Solvent Resistance

Some polymers are used throughout their working life in contact with organic fluids or hydrocarbon greases. O-rings in hydraulic systems must withstand the action of such fluids for long periods of time. The testing of polymers for their solvent resistance takes two different forms. First, it is necessary to establish if the polymer swells (or even dissolves) in a particular fluid. Second, it is essential to determine if the polymer cracks or crazes in contact with organic fluids. Many do, and such effects lower the strength and flexibility of these materials.

ADDITIONAL PROPERTIES

Optical Properties

Transparent polymers are used in lenses, prisms, bottles, or as a base for photographic film. The tests conducted on these materials are often designed to determine if yellowing of the polymer takes place over a long period of time, especially after exposure to sunlight. Other optical tests are designed to measure the gloss on the surface of a polymer.

Moisture and Gas Permeability

Films of hydrophilic polymers are often permeable to water vapor, whereas hydrophobic polymers form films that are impervious to water. Such considerations must be taken into account when choosing a polymer for packaging or building applications. Similarly, the use of polymer films to protect packaged items (food, oxidizable chemicals, etc.) must take into account the permeability of the polymer to gases such as oxygen. The water or oxygen permeability of specific polymers is also important when polymers are being considered for use in biomedical devices, for example, in dialysis membranes or heart-lung machines. This topic is considered further in Chapter 22.

STUDY QUESTIONS

1. What types of stress–strain behavior would you expect to be shown by materials that would be suitable for use as (a) automobile shock housings; (b) ball-point pen housings; (c) a plastic basin for the kitchen sink; (d) a cushioning material for delicate instruments; (e) the outer casing of a football; (f) a decorative paperweight; (g) an automobile bumper.

2. Why is the actual strength of a polymer nearly always far less than calculated on the basis of the skeletal bond strengths and intermolecular forces?

3. How would you explain the stress–strain behavior of a "soft and weak" polymer (Figure 21.3a) in terms of molecular phenomena? How might the same polymer be modified to change it to the "hard and tough" category?

4. In molecular terms, why should a polymer such as polyisobutylene or silicone rubber be more impact-resistant than, say, polystyrene?

5. The catastrophic failure of metals following multiple, apparently benign flexural motions has led to several engineering disasters in the past. Could polymers fail in the same way? If so, what might the mechanism be that gives rise to the ultimate failure? How would you set up a test procedure to evaluate polymers for possible uses (a) as materials for the construction of ships; (b) as structural materials in skyscrapers; (c) as the structural material in aircraft wings?

6. Suggest ways in which rigid polymers might be bonded to steel in such a way that the sandwich would be unaffected by large temperature fluctuations.

7. Suppose that you are engaged in a search for polymers that conduct electricity. What kind of apparatus would you construct to perform the tests? What problems might you encounter?

SUGGESTIONS FOR FURTHER READING

BAIRD, M. E., "Recent Developments in the Study of the Dielectric Behavior of Polymers," *Progr. Polymer Sci.* (A. D. Jenkins, ed.), **1**, 161 (1967).

BERRY, J. P., et al., *Fracture Processes in Polymeric Solids: Phenomenon and Theory*. New York: Wiley–Interscience, 1964.

BIKERMAN, J. J., "Sliding Friction of Polymers," *J. Macromol. Sci.—Rev. Macromol. Chem.*, **C11**, 1 (1974).

BLOCK, H., "The Nature and Application of Electrical Phenomena in Polymers," *Adv. Polymer Sci.*, **33**, 93 (1979).

DELMAN, A. D., "Recent Advances in the Development of Flame-retardant Polymers," *J. Macromol. Sci.—Rev. Macromol. Chem.*, **C3**, 281 (1969).

FRISCH, K. C., AND A. V. PATSIS, *Electrical Properties of Polymers*. Westport, Conn.: Technomic Publishing Co., 1972.

HANSEN, C. H., *The Three-Dimensional Solubility Parameter and Solvent Diffusion Coefficient. Their Importance in Surface Coating Formulation*. Copenhagen: Danish Technical Press, 1967.

HAYAKAWA, R., AND Y. WADA, "Piezoelectricity and Related Properties of Polymer Films," *Adv. Polymer Sci.*, **11**, 1 (1973).

HENISCH, H. K., "Amorphous-Semiconductor Switching," *Sci. Am.*, **221**, 30 (1969).

HERTZBERG, R. W., "Fatigue Failure in Polymers." *CRC Critical Rev. Macromol. Sci.*, **1**, 433 (1972–73).

KAELBLE, D. H., "Rheology of Adhesion," *J. Macromol. Sci.—Rev. Macromol. Chem.*, **C6**, 85 (1971).

KAMBOUR, R. P., "A Review of Crazing and Fracture in Thermoplastics," *J. Polymer Sci. (D) (Macromol. Rev.)*, **7**, 1 (1973).

KE, B., *Newer Methods of Polymer Characterization (Polymer Reviews*, Vol. 6). New York: Wiley–Interscience, 1964.

KRYSZEWSKI, M., AND A. SZYMANSKI, "Space Charge Limited Currents in Polymers," *J. Polymer Sci. (D) (Macromol. Rev.)*, **4**, 245 (1970).

LEE, L.-H., *ACS International Symposium on Advances in Polymer Friction and Wear, Los Angeles, 1974*. New York: Plenum Press, 1974.

LOCKETT, F. J., *Non-linear Viscoelastic Solids*. New York: Academic Press, 1972.

MANSON, J. A., "Fatigue Failure in Polymers," *CRC Critical Rev. Macromol. Sci.*, **1**, 433 (1972–73).

MATHES, K. N., "Electrical Properties," in *Encyclopedia of Polymer Sci. and Technol.* (H. F. Mark, N. G. Gaylord, and N. M. Bikales, eds.), **5**, 528 (1966).

MCCRUM, N. G., B. E. READ, AND G. WILLIAMS, *Anelastic and Dielectric Effects in Polymeric Solids*. New York: Wiley, 1967.

MEARES, P., *Polymers: Structure and Bulk Properties*. New York: Van Nostrand, 1965.

NATURMAN, L., (ed.), *Polymer-Plastics—Technol. and Eng.*, **2**(1), i (1973).

NIELSEN, L. E., *Mechanical Properties of Polymers*. New York: Rheinhold, 1962.
   Polymer-Plastics—Technol. and Eng., **1** (1972), and subsequent issues.

RABINOWITZ, "Craze Formation and Fracture in Glassy Polymers," *CRC Critical Rev. Macromol. Sci.*, **1**, 1 (1972–73).

REBENFELD, L., P. J. MAKAREWICZ, H.-D. WEIGMANN, AND G. L. WILKES, "Interactions between Solvents and Polymers in the Solid State," *J. Macromol. Sci—Rev. Macromol. Chem.*, **C15**, 279 (1976).

REICH, L., "Polymer Degradation by Differential Thermal Analysis Techniques," *J. Polymer. Sci. (D) (Macromol. Rev.)*, **3**, 49 (1968).

REICH, L., AND D. W. LEVI, "Dynamic Thermogravimetric Analysis in Polymer Degradation," *J. Polymer Sci. (D), (Macromol. Rev.)*, **1**, 173 (1967).

SEVERS, E. T., *Rheology of Polymers*. New York: Reinhold, 1962.

SEYMOUR, R. B., *Modern Plastics Technology*. Reston, Va.: Reston Publishing Co., 1975.

SLADE, E., AND L. T. JENKINS (eds.), *Techniques and Methods of Polymer Evaluation*. New York: Dekker, 1966.

VAN TURNHOUT, J., *Thermally Stimulated Discharge of Polymer Electrets*. New York: Elsevier, 1975.

VARGA, O. H., *Stress–Strain Behavior of Elastic Materials (Polymer Reviews*, Vol. 15). (New York: Wiley–Interscience, 1966).

VOYUTSKII, S. S., *Autohesion and Adhesion of High Polymers* (*Polymer Reviews*, Vol. 4). New York: Wiley–Interscience, 1963.

WILLIAMS, D. J., *Polymer Science and Engineering*. Englewood Cliffs, N.J.: Prentice–Hall, 1971.

WILLIAMS, J. G., *Stress Analysis of Polymers*. London: Longmans, 1973.

WRASIDLO, W., "Thermal Analysis of Polymers," *Adv. Polymer Sci.*, **13**, 1 (1974).

22

Biomedical Applications of Synthetic Polymers

USES FOR POLYMERS IN BIOMEDICINE

The widespread use of synthetic polymers in technology and in everyday life is an accepted feature of modern civilization. Polymers are now being used for almost every conceivable application, and there is every indication that these uses will continue to increase in future years. However, there exists one important area in which the use of synthetic polymers has generally been cautious and limited—the area of medicine. There are a number of scientific reasons for this, and these will be discussed below. The important point is that profound changes are expected in medical techniques as new synthetic polymers are developed. In fact, the application of polymers to medicine has become one of the principal challenges facing the polymer scientist.

The uses for synthetic polymers in medicine can be grouped roughly into four categories: (1) the use of polymers for the fabrication of artificial organs, (2) the use of polymer membranes for hemodialysis or oxygenation, (3) the development of polymeric substitutes for plasma or blood, and (4) the use of implanted or soluble polymers as substrates for the slow release of drugs or birth control agents. The following sections will deal briefly with these four areas in turn, although the development of artificial organs from thermoplastics is by far the most advanced area of this field.

ARTIFICIAL ORGANS

Uses of Polymers in Artificial Organs

Everyone is familiar with the idea that body organs can be transplanted from one person to another. Heart, kidney, and corneal transplants are now performed frequently. One of the problems with organ-transplant procedures is that there are never enough donor organs to meet the need. Another problem is that the antibodies of the recipient reject the donor tissues and attempt to destroy them. Although this effect can be suppressed by immunosuppressor drugs, the transplant may eventually be rejected. Moreover, immunosuppressor drugs reduce the body's ability to combat microorganisms or to destroy abnormal cells. Hence, a high risk of serious infection or even cancer is associated with their use.

For these reasons there has been an increasing drive to obviate the need for live organ transplants by the use of artificial organs made from synthetic elastomers and rigid polymers. Polymeric devices that can fulfill the functions of the heart, lungs, or kidneys have been under development for several years, but additional uses for polymers are now being found in heart valves, blood-vessel replacement tubes, temporary skin, bone replacement or sockets, replacement corneas, permanently implanted artificial teeth, and as synthetic suture materials.

In all these uses, synthetic polymers offer a broad range of advantages over metals, glass, or ceramics. Prominent among these advantages are their low density, chemical inertness, flexibility, elasticity, or rigidity according to need, and ease of fabrication into intricate shapes. Moreover, the texture, hardness, or softness of the original tissue can be mimicked by the choice of a suitable polymer. Almost all the major classes of polymers have been investigated for possible biomedical uses. However, a polymer must fulfill certain critical requirements if it is to be used in an artificial organ.

First, it must be physiologically inert. Nearly all synthetic polymers suffer from one common disadvantage—their ability to trigger off rejection mechanisms by the body. These rejection processes become manifest in the coagulation of blood in contact with polymers or the inflammation or even tumor formation which occurs when some polymers remain in contact with internal tissues for long periods of time. The overcoming of these deleterious interactions is one of the most urgent problems faced by the synthetic polymer chemist. Some of the approaches that have been tried will be outlined in the following sections.

Second, the polymer itself should be stable during many years of exposure to hydrolytic or oxidative conditions at body temperature. It must be resistant to enzyme attack, and it must not change dimensions, disintegrate, or dissolve in aqueous media or in contact with lipids or other fatty materials.

Third, if it is to be used as a structural material to replace bone, it must be strong and resistant to impact.

Fourth, the polymer must be sufficiently stable chemically or thermally that it can be sterilized by chemicals or by heat.

Stability of Polymers in Living Systems

It is important to recognize that the use of synthetic polymers in living systems revolves around one of two requirements: that the polymer should be totally inert, or that the polymer should be totally biodegradable. Unfortunately, most polymers fall between these two extremes.

Biological inertness is difficult to predict on the basis of intuitive chemical knowledge. Almost certainly, hydrolysis reactions form the first line of attack by the body on a polymer but, because most synthetic polymers are totally insoluble in aqueous media, the conventional reactivity relationships familiar to chemists are irrelevant. Nevertheless, it is a fact that some insoluble synthetic polymers more than others initiate inflammation of the surrounding tissues, blood clots, and so on, and even tumor formation.

The body has three basic responses to the implantation of a foreign body. First, it responds to the physical characteristics of the object (shape, roughness, presence of sharp edges, etc.). These responses may take the form of epithelial encapsulation of the foreign body, keratinization of the surrounding tissue, thickening of the connective tissue, or generation of giant cells. Second, the body reacts to the chemical toxicity (if any) of the polymer by the appearance of tissue inflammation, inhibition of epithelial growth, and other effects. Finally, there is a possibility of bacterial or viral infection originating at the surface of the implant or of the direct generation of an antigenic reaction by some chemical component of the polymer surface.

Thus, the comparison of different polymers for biomedical uses is not a straight-forward process. Research workers disagree about the relative significance of implant design and the chemical properties of the material. They also disagree on the question of whether demonstrated tumor formation in rodents means that some polymers will initiate tumor growth in human beings. Human metabolism is much slower than that of rodents, although human beings live much longer. Add to this the complication that many commercial polymers contain potentially toxic or carcinogenic monomers or additives that can be leached out easily in the body, and it will be seen that enormous difficulties face the researcher who wishes to answer the question: Which polymer is the best for a particular biomedical application?

With these uncertainties it will be recognized that the following observations are tentative indeed. Tissue culture experiments suggest that the following order represents an *increasing* degree of toxicity of various polymers:[1,2]

[1] H. Lee and K. Neville, *Handbook of Biomedical Plastics* (Pasadena, Calif.: Pasadena Technology Press, 1971), Ch. 14, p. 4.

[2] C. A. Homsy, K. D. Ansevin, W. O'Bannon, S. A. Thomson, R. Hodge, and M. E. Estrella, in *Biomedical Polymers*, A. Renbaum and M. Shen, eds. (New York: Dekker, 1971), p. 132.

Silicone rubber ≈ polyethylene < poly(tetrafluoroethylene)

≈ fluorinated poly(ethylene–propylene)

≈ poly(phenylene oxide)

≈ poly(methyl methacrylate) < poly(vinylidene fluoride)

≈ nylon ≈ polystyrene < polyurethane

≈ poly(vinyl chloride)

≈ ABS polymer.

Those toward the end of the list totally inhibited the growth of tissue culture cells.

Of course, the interaction between a polymer and the body may also lead to a weakening of the polymer itself. Polyurethanes disintegrate after only 16 months in the body. Nylon apparently loses 80% of its tensile strength after being implanted in the body for 3 years, polyacrylonitrile loses 24% of its strength in 2 years, and poly(tetrafluoroethylene) 6% in a year. Silicone rubber, on the other hand, is hardly affected at all in a year and a half in the body.

Heart Valves and Vascular Reinforcement

Two of the commonest circulatory disorders are associated with damaged heart valves and weakened arterial walls. The former disorder can be a result of rheumatic fever; the latter may result from arteriosclerosis (deposition of fats and minerals in the arterial wall), which raises the blood pressure, destroys the elasticity of the blood vessel wall, and can cause a balloonlike aneurysm in the main artery leaving the heart, in the brain, or elsewhere.

Artificial heart valves have been in use for a number of years. Two main designs have been employed. In one, a ball of silicone rubber is retained inside a stainless steel cage, as shown in Figure 22.1. Silicone rubber is used because of its inertness, elasticity, and low capacity to cause blood clotting. The failure of such valves is usually due to "wedging" of the ball (i.e., the formation of a trough at the points where the ball is forced against the cage) or more serious abrasion effects. The other main designs are butterfly or diaphragm valves made from reinforced silicone rubber of Kel F.

Aneurysms can be repaired by reinforcement of the artery with an external coating of a vinyl copolymer (vinyl chloride-vinylidine chloride), or epoxy cement, or by the use of woven Dacron or Teflon fabric. The latter materials are also used to replace sections of weakened blood vessels.

The Artificial Heart

Heart disease and circulatory disorders are responsible for more deaths in North America and western Europe than any other ailment. The most serious problems arise from arteriosclerosis and from the progressive narrowing of the cardiac arteries. A blockage of one of these arteries can precipitate a "heart attack." For patients with an irreversibly damaged heart, two prospects exist. First, a heart transplant may be possible, but this procedure is limited for the reasons discussed

Fig. 22.1 Starr–Edwards ball-type heart valves constructed from a silicone rubber ball, a chrome–cobalt cage, and a Teflon ring for suturing to the heart tissue. [Photograph by courtesy of Edwards Laboratories, American Hospital Supply Corporation.]

earlier. Second, the functions of the damaged heart may be taken over permanently or temporarily by an artificial pump. A considerable amount of research has been devoted to the design and testing of artificial heart pumps. Synthetic elastomers and rigid polymers have been used extensively for the construction of these devices. Unfortunately, most synthetic polymers accelerate the clotting of blood. This problem is so serious that animals on which the pumps are tested frequently die within hours from the massive, gelatinous blood clots that form in the pumps. Avoidance of the clotting process is a complex problem because it depends on the design of the pump and the presence or absence of turbulence as well as on the materials used for construction. In the following sections we will consider briefly first the design problem, and then the polymer problem.

Heart pump designs

Two types of pumps have been developed on an experimental basis: (1) auxiliary blood pumps to bypass or supplement the action of a damaged heart until it can repair itself, and (2) total artificial heart pumps that can completely replace the living organ. Many of the booster pumps have used a rigid housing, often made of reinforced epoxy resin, with an internal tube of silicone rubber (Figure 22.2). Compressed air applied inside the rigid casing compresses the silicone or polyurethane rubber inner tube and this forces blood from the pump. Valves may be used to prevent backflow, or the compression cycle may be synchronized with the pumping motion of the heart. A related device is the intraaortic balloon, a 25 cm × 2 cm polyurethane balloon inserted into the aorta which expands as compressed helium or carbon dioxide is pulsed in or out. Other devices use hemispheres of polycarbonate or poly(methyl methacrylate) containing a silicone rubber dia-

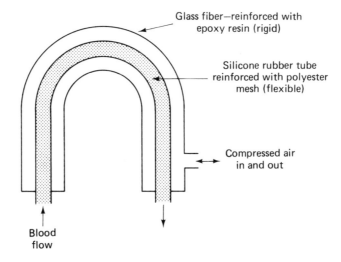

Glass fiber—reinforced with
epoxy resin (rigid)

Silicone rubber tube
reinforced with polyester
mesh (flexible)

Compressed air
in and out

Blood
flow

Fig. 22.2   Relatively simple "artificial heart" device designed for implantation in the body. Pulses of compressed air compress the silicone-
rubber inner tube, which is connected to the aorta. The phase of the
pumping cycle is synchronized with that of the patient's heart.

phragm.[1] Pulses of compressed air or carbon dioxide actuate the diaphragm and
cause the pumping of the blood (Figure 22.3). Total artificial hearts have been
constructed which resemble the general structure of a human heart but which
are actuated by compressed gas or oil.

Polymers for heart pumps

A wide variety of different polymers have been used for the fabrication of heart
pumps. These include silicone rubber, polyurethane rubber, Dacron polyester,
Teflon, polycarbonate, poly(methyl methacrylate), poly(vinyl chloride), and
pyrolytic carbon. Most of these materials cause blood clotting, destruction of red
cells, or alteration of the blood proteins, although some are markedly better than
others.

Silicone rubber is perhaps the most widely used material for this application.
It is soft, flexible, and unaffected by body fluids. However, it can promote blood
clotting if the blood is flowing slowly, and it can fail after continuous flexing. It has
been estimated that a silicone rubber diaphragm can be flexed about 90 million
times without breaking, but in artificial heart devices it would have to withstand at
least 400 million flexures over a 10-year period. Polyurethane and natural rubber
have better flexing strength, although not as much as is needed for a long-term
heart device. Another problem with silicone rubber is its tendency to absorb fats

[1] J. H. Kennedy, M. E. DeBakey, W. W. Akers, J. N. Ross, W. O'Bannon, L. E. Baker, S. D.
Greenberg, D. W. Wieting, C. W. Lewis, M. Adachi, C. P. Alfrey, W. J. Spargo, and J. M. Fuqua,
Biomat. Med. Dev., Art. Org., **1**(1), 3 (1973).

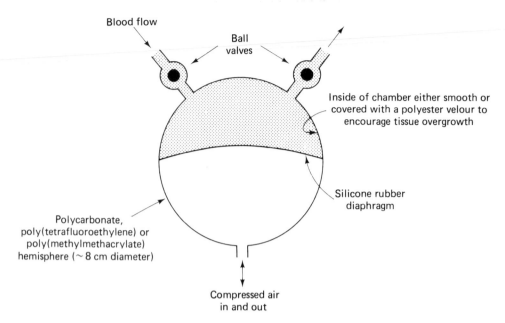

Blood flow

Ball
valves

Inside of chamber either smooth or
covered with a polyester velour to
encourage tissue overgrowth

Silicone rubber
diaphragm

Polycarbonate,
poly(tetrafluoroethylene) or
poly(methylmethacrylate)
hemisphere (~ 8 cm diameter)

Compressed air
in and out

Fig. 22.3 Design of a hemispherical "artificial heart" pump, designed to operate outside the patient's body. The silicone rubber diaphragm is actuated by compressed air. Turbulence of the blood as it
passes through the valves is a major cause of clotting.

from the blood, to swell, and eventually to weaken. Fluoroalkylsiloxane polymers
or polyphosphazenes may prove to be more suitable for artificial heart applications.

The ability of a synthetic polymer to initiate the clotting of blood depends on
the nature of the surface (smooth surfaces are better than rough) and on the
chemical and physical properties of the polymer. For example, highly water
repellent polymers appear to be among the best materials for contact with blood.
Because the inside lining of blood vessels is negatively charged, it has also been
speculated that polymers with a surface charge might be more effective than
neutral polymers.

Two additional approaches to the problem of blood incompatibility have
been examined. In the first, an anticoagulant, such as heparin, is bonded to the
surface or absorbed into the polymer. In the second, an attempt is made to use a
polymer that will facilitate tissue overgrowth to insulate the polymer from the
blood. A velvety velour of polyester fibers has been tested and has been found to
function in this way. Tissue overgrowth takes more than 16 weeks. "Seeding" of
the velour with living cells before implantation of the device speeds up this process.
Webs of fine polypropylene fibers function in the same way.

Tissue Adhesives and Artificial Skin

It has been recognized for many years that a need exists for synthetic polymers that
can be used to glue tissues together. The use of an adhesive would be much more
rapid and effective than the sewing of a wound with a suture. A group of polymers

based on the poly(α-cyanoacrylate) structure have proved to be effective for this purpose. α-Cyanoacrylates have the general formula shown in **1**, where the group,

$$CH_2{=}\underset{\underset{\underset{\underset{R}{|}}{O}}{\overset{\overset{\overset{C{\equiv}N}{|}}{\underset{|}{C}}}{\underset{|}{C{=}O}}}{C}$$

1

R, can be methyl, butyl, hexyl, octyl, and so on. These monomers polymerize by an anionic mechanism in the presence of water. Higher alkyl derivatives polymerize more rapidly on biological substrates and are less irritating to tissues than are the lower alkyl derivatives. However, their curing characteristics are somewhat unpredictable. In addition to their use as skin adhesives, they have been tested as adhesives in corneal and retinal surgery, and as an adjunct to suturing in internal surgery.

The search for polymeric materials that can be used as synthetic skin to cover large burns has led to the use of synthetic poly(amino acid) films for this purpose. Velours of nylon fiber have also been tested for this use, as have films of poly(α-cyanoacrylates).

Bones, Joints, and Teeth

Bone fractures are occasionally repaired with the use of polyurethanes, epoxy resins, and rapid-curing vinyl resins. Silicone rubber rods and closed-cell sponges have been used as replacement finger and wrist joints, and vinyl polymers and nylon have been investigated as replacement wrist bones or elbow joints. Furthermore, cellophane and, more recently, silicone rubber have been used in knee joints to prevent fusion of the bones. Dramatic advances have been made in hip-joint surgery with the use of stainless steel or polyethylene ball joints attached to the femur by means of a poly(methyl methacrylate) filler and binder. Teflon fabric and silicone rubber have been used to make synthetic ligaments and tendons.

Synthetic polymers have been utilized for many years in the fabrication of dentures. Poly(methyl methacrylate) is the principal polymer used both for acrylic teeth and for the base material. Acrylic resins are also used for dental crowns, and epoxy resins are sometimes employed to cement crowns to the tooth post. Considerable interest has been generated in the implantation of replacement teeth directly into the mandible or maxilla. A fused carbon base for the tooth is used to delay rejection.

More recent work and anticipated developments include the use of polymeric coatings or paint to prevent the decay of teeth and the development of thermosetting polymers to replace silver amalgam or gold as tooth-filling materials.

Some progress along these lines has been made with the use of inorganic powders bound together by means of a rapid-curing poly(acrylic acid) cement.

Synthetic Sutures

The use of sutures to close an internal or external wound is well known. Catgut was used for all sutures until recently. However, catgut is relatively inert, and post-operative procedures were usually necessary for the removal of the suture after the normal 15-day healing of the tissues. A recent replacement for catgut is synthetic poly(glycolic acid) (**2**). Poly(glycolic acid) has a high tensile strength and is

$$\left(CH_2-\overset{\overset{\displaystyle O}{\|}}{C}-O\right)_n$$

2

compatible with human tissue. However, it differs from catgut in being totally absorbable by many patients within 15 days, thus removing the need for a suture-removal operation. The polymer degrades by hydrolysis to nontoxic glycolic acid.

Contact Lenses

Polymeric materials, such as poly(methyl methacrylate) have traditionally been used for the manufacture of "hard" contact lenses. However, the increasing popularity of flexible or "soft" contact lenses has stimulated research into the development of some unusual polymer systems. The ideal material for this type of application is one that is hydrophylic and is swelled by but not dissolved by aqueous media. Lightly crosslinked water-soluble polymers have these properties.

MEMBRANES FOR DIALYSIS AND OXYGENATION

The Artificial Kidney

The function of a kidney is to remove low-molecular-weight waste products from the bloodstream. Artificial kidneys have been available on a limited scale for several years. They function by passage of the blood between the walls of a dialysis cell which is immersed in a circulating fluid. Because conventional hospital hemodialysis equipment is bulky and expensive, a continuing need exists to construct smaller and cheaper units. Synthetic polymers form the basis of these new developments.

Cellophane (regenerated cellulose) has been used for semipermeable dialysis membranes in conventional kidney machines. However, the need for miniaturization has been responsible for the use of bundles of hollow fibers as a dialysis cell. In one particular development, a bundle of 2000 to 11,000 hollow fibers of modified polyacrylonitrile (17 cm long and 300 μm diameter) are used. The polymer is "heparinized" to prevent blood clotting. Hollow rayon fibers or polycarbonate or cellulose acetate fibers have also been used for the same purpose.

Oxygen-transport Membranes

Surgical work on the heart frequently requires the use of a heart-lung machine to circulate and oxygenate the blood. A variety of devices have been developed, but many make use of a membrane through which oxygen and carbon dioxide must pass. Poly(dimethylsiloxane) membranes are highly efficient gas transporters. They are made by dip-coating a Dacron or Teflon screen in a xylene dispersion of silicone rubber. When dried, a film of 0.075 mm or more in thickness can be obtained, and this can be incorporated into the oxygenator. Silicone rubber membranes have also been tested in "artificial gills" for underwater breathing. It is of interest that silicone rubber has approximately six times the oxygen permeability of fluoro-silicones, nearly 80 times the value for polyethylene, and 150,000 times the permeability of Teflon.[1] This could be connected with the high torsional mobility of the siloxane chains.

POLYMERIC BLOOD SUBSTITUTES

Blood serves many functions, including the transport of oxygen and carbon dioxide, nutrients, minerals, and white cells. To fulfill these functions it must maintain a suitable viscosity to prevent turbulence. Synthetic polymers have been investigated for use in plasma substitutes and as volume expanders to reduce the amount of whole blood needed, for example, during the use of a heart-lung machine. Further-more, the transmission of hepatitis through the use of pooled plasma provides a continuing incentive for the development of a synthetic substitute for this fluid. Poly(vinyl pyrrolidone) (3) was used extensively by the Germans in World War II

3

as a colloidal plasma substitute for the treatment of casualties. Its disadvantages for this application are connected with its poor biodegradability. In fact, it is retained indefinitely in the spleen, lymph nodes, liver, and bone marrow, and it may initiate carcinogenic changes. Hence, there is a serious need for the development of a water-soluble or hydrophilic polymer that is nontoxic and biodegradable.

---

[1] H. Lee and K. Neville, *Handbook of Biomedical Plastics* (Pasadena, Calif.: Pasadena Technology Press, 1971), Ch. 2, p. 20.

On an even more ambitious level, certain polymers have been investigated as oxygen-transport compounds for use in blood. For example, an emulsion of poly(tetrafluoroethylene) particles (less than 0.001 mm in diameter) or liquid fluorocarbons in water, together with glucose, salts, and surfactants, has been used to replace the blood of rats. The rats remained alive and active for periods from 5 h to several days. The long-range possibility also exists that biodegradable, water-soluble macromolecules can be synthesized that possess oxygen-carrying side groups such as metalloporphyrins. Solutions of such polymers could be used as blood replacement fluids, but would be degraded and excreted as the body produced new blood cells. Prototypes of such polymers are under development.

SUBSTRATES FOR DRUG DELIVERY AND SLOW RELEASE

The treatment of many diseases requires the injection of drugs into the body. Frequently, much larger quantities of the drug are injected than are needed because a large fraction of the drug will be excreted before serving its function. As a result of this procedure, the patient is subjected to an alternating overdose and then a deficiency of the chemotherapeutic agent. It has been recognized for some years that a system for the slow, continuous release of drugs would be a decided advantage for the treatment of many ailments. For example, the slow release of insulin into the bloodstream would markedly improve the well-being of diabetics. Similarly, the continuous release of anticancer agents within the body might permit smaller doses of these highly toxic agents to be used. Advantages can also be foreseen for the slow, continuous release of birth control agents into the bloodstream over a period of months or years. On a more speculative level, the possibility exists that polymeric antioxidants introduced into the body could protect individuals against the effects of high-energy radiation.

Three approaches are being investigated in an attempt to use polymers to effect a slow release of drugs. These are (1) the use of polymer membranes as diffusion-controlling barriers, (2) the employment of solid but biodegradable polymers to effect the controlled release of an encapsulated drug, and (3) the use of drugs that are chemically bound to water-soluble polymers.

Diffusion-Controlling Membranes

Many chemotherapeutic drugs are relatively small molecules that can diffuse slowly through polymer membranes. Thus, if an aqueous solution of a drug is enclosed by a polymer membrane, the drug will escape through the membrane at a rate that can be controlled by membrane thickness and composition. A device that employs this principle is in use for the slow, controlled release of the antiglaucoma drug, pilocarpine, from a polymer capsule placed beneath the eyelid (Figure 22.4).

The same principle applies if a film, rod, or bead of a polymer is impregnated with a drug and is then implanted in the body at a site where the drug can have the

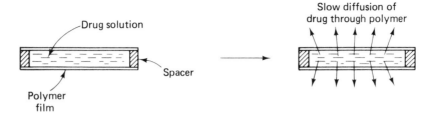

Fig. 22.4 Slow, controlled release of a drug by diffusion through a polymer membrane.

maximum beneficial effect. Diffusion of the drug from the polymer matrix permits a continuous controlled release to be achieved over a period of weeks or months. Regular injection or oral ingestion of the drug is no longer needed. This technique has been used for the slow release of birth control drugs and it has been suggested for possible use in cancer chemotherapy.

One important medical advantage in controlled-release devices of this kind is that the drug delivery system can be removed at any time when the therapy is no longer needed.

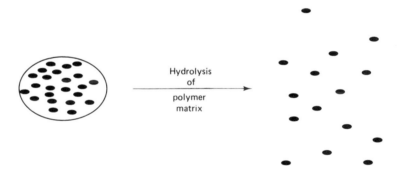

Fig. 22.5 Controlled release of a chemotherapeutic drug by dissolution of an encapsulating polymer matrix.

Biodegradable Solid Polymers

An alternative way to achieve the slow, controlled release of a drug from a solid matrix is to use a biodegradable polymer as the matrix. As the polymer degrades slowly (usually by hydrolysis), the chemotherapeutic molecules are released (Figure 22.5). An important requirement is that the hydrolysis products from the polymer should be nontoxic and readily excreted. One experimental polymer of this type (based on a polyphosphazene skeleton) is designed to hydrolyze to phosphate, ethanol, and an amino acid, all of which can be utilized by the body, and ammonia which can be excreted.

Polymer-Bound Drugs

As discussed earlier, two major reasons for the inefficiency of conventional chemotherapy are the rapid dilution of the drug as it diffuses from the required target site, and the ease of excretion of small-molecule drugs through the kidneys. Thus, much larger quantities of the drug must be introduced into the body than should be needed to correct the medical problem.

Water-soluble polymers diffuse only slowly through the tissues and, moreover, will not be excreted as rapidly as small molecules because macromolecules cannot normally pass through semipermeable membranes. Thus, a polymer-bound drug should offer considerable advantages over a small molecule drug.

Two possibilities exist. First, the drug could be linked to a relatively stable molecule, in which case the activity of the drug and its entry into the cell may be modified by the presence of the polymer. Alternatively, if the polymer degrades in the body and concurrently releases the drug, the chemotherapeutic activity of the drug will be unchanged. It will be clear that the design and synthesis of polymers that have the correct water solubility, lack of toxicity, and an appropriate rate of hydrolytic decomposition at body temperature is one of the most demanding challenges faced by the polymer chemist during the coming years.

STUDY QUESTIONS

1. Of all the polymers discussed in this book, which ones appear to you to be the most suitable for the construction of (a) artificial heart pumps; (b) surgical tapes for covering superficial wounds; (c) contact lenses for the eye; (d) replacement tubes for varicose veins; (e) a tympanic membrane for the ear; (f) replacement aqueous humor of the eye; (g) a replacement for nose cartilage. In each case give the reasons for your choice.

2. How would you undertake the task of purifying the following commercial polymers to prepare "surgical-grade" material for use in implanted devices: poly(vinyl chloride), poly(dimethylsiloxane), a fluoroalkoxyphosphazene rubber, polystyrene, a polyester? What effect might the purification have on the properties in each case?

3. Which synthetic polymers might be suitable for use in replacement nerve fibers, and why?

4. Design a miniature artificial kidney using synthetic polymers for all the components.

5. What water-soluble polymers are known that might be used as carrier molecules for chemotherapeutic drugs? Suggest ways in which drugs such as steroids, antibiotics, or anticancer agents might be linked to the polymer.

SUGGESTIONS FOR FURTHER READING

Batz, H. G., "Polymeric Drugs," *Adv. Polym. Sci.*, **23**, 25 (1977).

Biomaterials, Medical Devices, and Artificial Organs, **1** (1973), and subsequent issues.

Braley, S. A., "Acceptable Plastic Implants," in *Modern Trends in Biomechanics* (D. C. Simpson (ed.), pp. 25–51. London: Butterworths, 1970.

Braley, S. A., "The Chemistry and Properties of the Medical-Grade Silicones," *J. Macromol. Sci. Chem.*, **A4**, 529 (1970).

Breslow, D. S., "Biologically Active Synthetic Polymers," *Pure Appl. Chem.*, **46**, 103 (1976).

Bruck, S. D., "Polymeric Materials in the Physiological Environment," *Pure Appl. Chem.*, **46**, 221 (1976).

Chang, T. M. S., *Artificial Cells*. Springfield, Ill.: Charles C Thomas, 1972.

Donaruma, L. G., "Synthetic Biologically Active Polymers," *Progr. Polymer Sci.*, (A. D. Jenkins, ed.), **4**, 1 (1974).

Donaruma, L. G., and O. Vogl (eds.), *Polymeric Drugs*. New York: Academic Press, 1978.

Hopfinger, H. B. (ed.), *Permeability of Plastic Films and Coatings*, Part III, pp. 409–469. New York: Plenum Press, 1974.

Kammermeyer, K., "Biomaterials—Developments and Applications," *Chem. Tech.*, **1**, 719 (1971).

Lee, H. and K. Neville, *Handbook of Biomedical Plastics*. Pasadena: Pasadena Technology Press, 1971.

Lindsey, A. S., "Polymeric Enzymes and Enzyme Analogs," *J. Macromol. Sci.—Rev. Macromol. Chem.*, **C3**, 1 (1969).

Lyman, D. J., "Biomedical Polymers," *Rev. Macromol. Chem.*, **1**, 355 (1966).

Paul, D. R., and F. W. Harris (eds.), *Controlled Release Polymeric Formulations* (ACS Symposium Series No. 33), 1976.

Ringsdorf, H., "Structure and Properties of Pharmacologically Active Polymers," *J. Polymer Sci., Symp.*, **51**, 135 (1975).

Sanders, H. J., "Artificial Organs," *Chem. Eng. News*, p. 68, April 12 (1968).

Shalaby, S. W., and E. M. Pearce, "The Role of Polymers in Medicine and Surgery," *Chemistry*, **51**, 20 (1978).

APPENDICES

Polymer Nomenclature

Polymer nomenclature is not yet completely systematic and some aspects of the subject are in a state of flux. Different naming systems are often used in research and in technology, and occasionally some of the well-established names for polymers are confusing or even ambiguous. Some aspects of nomenclature were dealt with in Chapters 1 and 4. This appendix provides a summary of the main principles that are involved.[1]

Structural and Source Names

Two fundamentally different systems of polymer nomenclature are in widespread use—the "structural" and "source" (or "derivative") methods. A structural name emphasizes the *actual* structure of the polymer without reference to the monomer from which it was derived. The source name is based on the name of the original monomer and is obtained simply by addition of the prefix, *poly*. For example, structures **1** and **2** represent the same polymer. The name "polymethylene"

$$+CH_2\frac{}{}_n \qquad +CH_2-CH_2\frac{}{}_n$$

1	**2**
Polymethylene	Polyethylene

[1] Further details will be found in *Macromolecules*, **1**, 193 (1968); *Macromolecules*, **6**, 149 (1973); *J. Polymer Sci., Polymer Lett.*, **11**, 389 (1973).

is a fundamental structural description.[1] The term "polyethylene" tells how the polymer was made. In general, source names are more common, but structural names are preferred in the newer nomenclature systems. However, both types of names are used interchangeably at the present time. For example, the names "poly(oxymethylene)" and "polyformaldehyde" are both in widespread use.

Parentheses

Parentheses and square brackets are used only to prevent confusion. The name, "polyethyl acrylate" is ambiguous. "Poly(ethyl acrylate)" is not. Neither is "polyacrylonitrile." In general, parentheses or brackets are needed if the monomer name consists of two or more words or if the root name itself contains numbers. An extreme example is the name poly[bis(2,2,2-trifluoroethoxy)phosphazene], which, because of the placing of the brackets and parentheses, is unambiguous.

End Groups

End groups are not specified for high polymers. In low polymers (telomers) the end groups may be depicted by the symbols α and ω as in α-chloro-ω-(trichloromethyl)-polymethylene, $Cl + CH_2 \rightarrow_n CCl_3$.

Polymers Prepared from Unsaturated Monomers

Addition polymers are usually described by the source name. Examples are poly(vinyl chloride) (*not* polyvinyl chloride), poly(methyl methacrylate), polystyrene, polypropylene, and so on.

Two complications may be encountered. First, if the monomer is a diene, the polymerization may involve different addition pathways (1,2 or 1,4, for example). The name, poly(1,2-butadiene) is ambiguous as a source name, since the polymer was not derived from 1,2-butadiene. A more correct name is 1,2-polybutadiene. The polymer of formula, $+CH=CHCH_2CH_2\rightarrow_n$ is named poly(1-butenylene) to avoid ambiguity.

The second problem is concerned with the use of an unambiguous description of side groups attached to the main chain. Consider the polymer repeating structures shown in **3** and **4**. The name poly(methylstyrene) would apply to both structures.

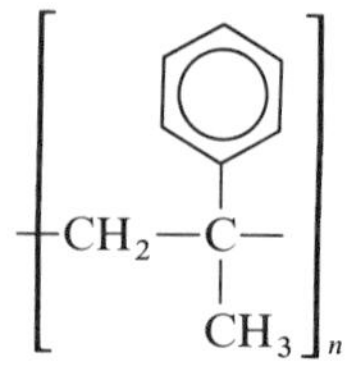

3 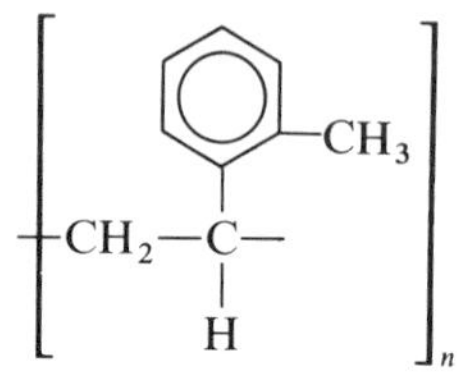   4

---

[1] Of course, the name "polymethylene" would be both a structural *and* a source name if the polymer were to be synthesized from carbene, CH_2.

The name, poly(2-methylstyrene) is also ambiguous. Structure **3** is best described by the name poly(α-methylstyrene), and structure **4** by poly(o-methylstyrene). Note that conventional organic nomenclature describes compound **5** as 1,1-dichloroethylene (also known as vinylidine chloride), whereas compound **6** is 4-chloro-1-butene. These designations should be included in the source name of the polymer.

$$CH_2{=}C\overset{\displaystyle Cl}{\underset{\displaystyle Cl}{|}} \qquad CH_2{=}CH{-}CH_2{-}CH_2{-}Cl$$

$$\textbf{5} \qquad\qquad\qquad \textbf{6}$$

Polymers Prepared by Step-Type Processes

The nomenclature used for these systems can be confusing. Polyamides and polyesters are named by both structural and derivative methods. The polymer depicted in **7** could be called poly[imino(1-oxohexamethylene)], poly(6-hexanoamide),

$$\left[\overset{\displaystyle H}{\underset{}{N}}{-}(CH_2)_5{-}\overset{\displaystyle O}{\overset{\displaystyle \|}{C}}\right]_n \qquad \left[O{-}CH_2{-}CH_2{-}O{-}\overset{\displaystyle O}{\overset{\displaystyle \|}{C}}{-}\!\bigcirc\!{-}\overset{\displaystyle O}{\overset{\displaystyle \|}{C}}\right]_n$$

$$\textbf{7} \qquad\qquad\qquad\qquad \textbf{8}$$

poly(6-aminocaproic acid), or poly(ε-caprolactam). In practice, the source name polycaprolactam is in more common use. The polyester (**8**) is usually called poly(ethylene terephthalate)—a name that is based mainly on structural considerations. Particular ambiguity is evident with the polymer of structure **9**, which is

$$+CH_2{-}CH_2{-}O\!+_n \qquad \left(\!\!\bigcirc\!\!\right)_n \qquad \left(O{-}\!\bigcirc\!\right)_n$$

$$\textbf{9} \qquad\qquad \textbf{10} \qquad\quad \textbf{11}$$

called poly(ethylene oxide), poly(oxyethylene), or poly(ethylene glycol). Compounds of structure **10** are called poly(p-phenylenes), and those of structure **11** are known as poly(phenylene oxide) or poly(oxy-1,4-phenylene).

Inorganic Polymers

The field of inorganic polymers is still in its infancy and the nomenclature is in a "trial-and-error" phase. The problem arises because developments in the inorganic macromolecular field have involved contributions by researchers trained in mineralogical, small-molecule inorganic, and organic disciplines, each making use of a different nomenclature system. However, an attempt is being made to establish a uniform system,[1] and some of the suggestions are mentioned below.

[1] B. P. Block and G. Donaruma, Committee on Nomenclature, American Chemical Society Division of Polymer Chemistry.

First, a few one-dimensional inorganic macromolecules are known that are constructed from covalent bonds and have well-established structural names derived from the conventions of organic polymer nomenclature. Examples are polysiloxanes [e.g., poly(dimethylsiloxane)]; polyphosphazenes [e.g., poly(diphenoxyphosphazene)]; polythiazene or polythiazyl [poly(sulfur nitride)]; and polysulfur.

Second are those inorganic polymers that possess coordinative binding either alone or together with covalent binding. The multiplicity of coordination numbers and structures that is theoretically possible within this class of compounds is almost too large for comprehension. Hence, the nomenclature problems are staggering.

The names of one-dimensional polymers are preceded by the prefix *catena-*. Two-dimensional polymers are designated by the prefix *phyllo-*, and three-dimensional macromolecules are indicated by the prefix *tecto-*. The fundamental nomenclature is based on the name of the *central atom* and all groups attached to the central atom are named as ligands. Bridging groups are indicated by the symbol μ. Points of attachment of a bridging group are indicated by the addition of the italicized symbols for the atom or atoms through which attachment to each center occurs. The symbols for atoms attached to different centers are separated by colons. Two examples are:

catena-Poly(difluorosilicon)   catena-Poly[bis[μ-diphenylphosphinato(1-)O:O']beryllium(II)]

Abbreviations, Acronyms, and Trade Names

Polymers are often named by abbreviations. For example, poly(methyl methacrylate) is known as PMMA, polyacrylonitrile as PAN, polyoxymethylene as POM, and so on. If these represented a universal "shorthand" notation, they would be useful as a nomenclature system. Unfortunately, no agreement exists on the preferred acronym for every polymer, and the same abbreviations are often used to represent different polymers. Communication in acronyms across a language barrier (e.g., in scientific journals) can lead to confusion. Hence, our advice is to avoid such abbreviations except in casual conversations between close colleagues.

Some trade names for polymers are now part of the language (nylon and rayon, for example). The problem with trade names is that different names are often coined for the same polymer. For this reason, trade names are generally not used in the fundamental scientific journals, although they are employed in the technological literature.

Properties and Uses
of Selected Polymers

A problem encountered by many newcomers to polymer chemistry is the need to remember the structures, properties, and technological uses of a broad range of synthetic polymers. The data in this Appendix have been compiled as a reference source to which the reader can refer as he or she encounters new polymer names or structures in the main text or in the study questions. It is hoped that this compilation will also stimulate the reader to think about the relationship between polymer structure and uses in order to propose new polymer structures that might have more favorable properties for specific applications.

Name	Repeating Unit	T_g (°C)	T_m (°C)	Properties and Uses
Polyamides				
Poly(decamethylene carboxamide) or poly(11-aminoundecanoic acid) (nylon 11, Rislan)	$-N\!\!+\!\!CH_2\!\!\overset{}{)}_{10}\!\!\overset{O}{\overset{\parallel}{C}}-$ with H on N	46	198	Manufacture of fishing lines, bristles, gunstocks, and gasoline lines.
Poly(hexamethylene adipamide) (nylon 66, Bri-Nylon)	$-N\!\!+\!\!CH_2\!\!\overset{}{)}_6\!\!-N-\overset{O}{\overset{\parallel}{C}}\!\!-(CH_2)_4\!\!-\overset{O}{\overset{\parallel}{C}}-$	45	267	Fibers used in textiles, tire cords, rope, thread, belting, and fiber cloth. Polymer also used in molded objects such as high-impact gear wheels and electrical insulators.
Poly(hexamethylene sebacamide) (nylon 610)	$-N\!\!+\!\!CH_2\!\!\overset{}{)}_6\!\!-N-\overset{O}{\overset{\parallel}{C}}\!\!+\!\!CH_2\!\!\overset{}{)}_8\!\!-\overset{O}{\overset{\parallel}{C}}-$	50	165, 226	Used in sports equipment and bristles for brushes.
Poly(nonamethylene urea) (Urylon)	$-N\!\!+\!\!CH_2\!\!\overset{}{)}_9\!\!-N-\overset{O}{\overset{\parallel}{C}}-$		236	Fibers.
Polycaprolactam, poly(pentamethylene carboxamide), or poly(6-aminohexanoic acid) (nylon 6, Perlon, Caprolan)	$-N\!\!+\!\!CH_2\!\!\overset{}{)}_5\!\!-\overset{O}{\overset{\parallel}{C}}-$		223	Fibers used in textiles and tire cords. Polymer also molded into gears, cams, and shoe heels.
Poly(*m*-phenylene isophthalamide) (Nomex)	$-N\!\!-\!\!\langle\text{aryl}\rangle\!\!-N-\overset{O}{\overset{\parallel}{C}}\!\!-\langle\text{aryl}\rangle\!\!-\overset{O}{\overset{\parallel}{C}}-$		390	Heat-resistant polymer: retains dimensional stability and mechanical properties up to or above 250°C; decomposes above 370°C. Used as fiber in manufacture of heat-resistant textiles for use in space suits, filter fabrics for high-temperature filtration, conveyer belts, parachute cables, and aircraft tire cords. Also used in electrical insulation and aircraft panels.
Polyesters and Polycarbonates				
Poly(cyclohexane-1,4-dimethylene terephthalate) (Kodel)	$-O\!\!-\!\!CH_2\!\!-\langle\text{cyclohexane}\rangle\!\!-CH_2\!\!-O\!\!-\overset{O}{\overset{\parallel}{C}}\!\!-\langle\text{aryl}\rangle\!\!-\overset{O}{\overset{\parallel}{C}}-$	92 (*cis*)	318 (*cis*), 256 (*trans*)	Fabricated into textile fibers. As isophthalate copolymer, used in "blister" packaging material.

Polymer	Repeating unit			Uses
Poly(ethylene terephthalate) (Dacron, Terylene, Fortrel, Mylar)	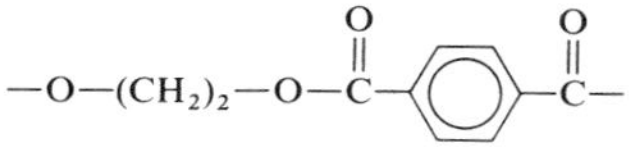 $-O-(CH_2)_2-O-C(=O)-\phi-C(=O)-$	69	270	Strong, tough, thermoplastic with surface lubricity and resistance to wear. As textile fibers, used in tire cords, yacht sails, and electrical insulation. Films used as base for photographic film, magnetic tape, or typewriter ribbon. Bulk polymer used to make gear wheels and structural objects.
Poly(butylene terephthalate)	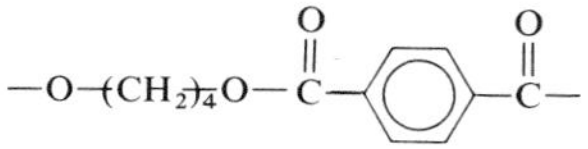 $-O-(CH_2)_4O-C(=O)-\phi-C(=O)-$	150	267	Tough, solvent resistant, thermoplastic with good fatigue resistance, and low moisture absorption. Used in automobile ignition systems.
Poly(4,4'-isopropylidine-diphenyl carbonate) or poly(4,4'-carbonato-2,2-diphenylpropane) (Lexan)	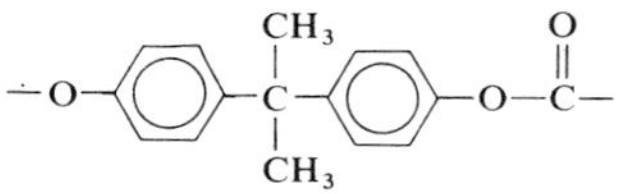 $-O-\phi-C(CH_3)_2-\phi-O-C(=O)-$		267	Tough, transparent polymer with high impact strength and tensile strength near 5500 psi; fire resistant. Used as safety glass, bullet-proof windows, skylights, bathroom fixtures, plumbing, automobile components, lighting fixtures, food containers, and expanded foams. Also used in automobile doors and as expanded foam in automobile roofs.

Polyethers

Polymer	Repeating unit			Uses
Poly(butylene glycol) (Polyglycol B)	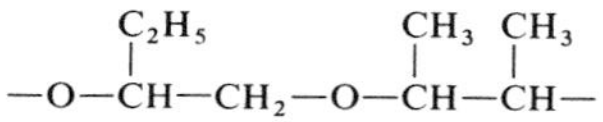 $-O-CH(C_2H_5)-CH_2-O-CH(CH_3)-CH(CH_3)-$			Used as an additive to gasoline, oils, greases, antifoaming agents, and detergents.
Poly(epichlorohydrin) (Polyglycol 166)	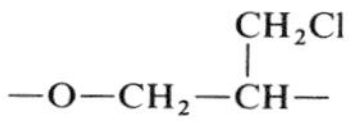 $-O-CH_2-CH(CH_2Cl)-$	121 (Isotactic)		Used in the manufacture of urethanes, coatings, resins, and surfactants, specialty elastomers, fuel-pump diaphragms, oil and fuel hoses, and oil-well equipment.
Poly(epichlorohydrin-ethylene oxide) copolymers (ECO, Hydrin)	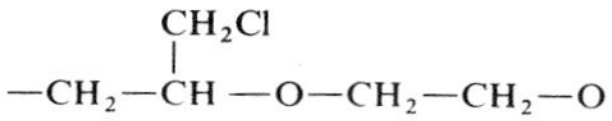 $-CH_2-CH(CH_2Cl)-O-O-CH_2-CH_2-O-$			Used in seals, gaskets, and hoses for automobiles.
Poly(ethylene oxide) (Carbowax)	$-O-CH_2-CH_2-$	-67	66.2	Used as a thickening agent; textile sizing agent, polymer intermediate; water-soluble films, and pharmaceutical binder.

Name	Repeating Unit	T_g (°C)	T_m (°C)	Properties and Uses
Polyformaldehyde (Delrin, Celcon)	—O—CH₂—	(−30)	(182.5)	Tough plastic used for fabrication of gears, brushes, pipes, molded articles, pens, and carburetor components.
(Nitroso rubber)	—O—CF₂—CF₂—N— (CF₃ on N)	(−82)	(60)	Solvent-resistant, low-temperature elastomer used for the manufacture of nonburning molded objects.
Poly(tetramethylene oxide), poly(tetrahydrofuran)	—(CH₂)₄—O—			Used as a plasticizer for cellulose or chlorinated rubber. Used in artificial leather.
Poly(2,6-xylenol) or poly(2,6-dimethyl-1,4-phenylene oxide) (Parlene)	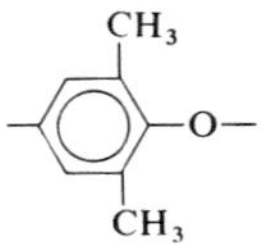 CH₃, O, CH₃ substituted ring		338	Structural plastic used in appliances, business machines, automobile components, water distribution equipment, and high-temperature applications.
Poly(phenylene sulfide)	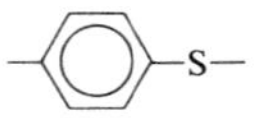 —S— ring	85	288	Solvent-resistant polymer below 200°C, with high thermal stability. Used in protective coatings for valves, pumps, pipes, and tanks, and in the manufacture of injection-molded articles.

Polyimides

Name	Repeating Unit	T_g (°C)	T_m (°C)	Properties and Uses
Poly(pyromellitimide) (Kaptan)	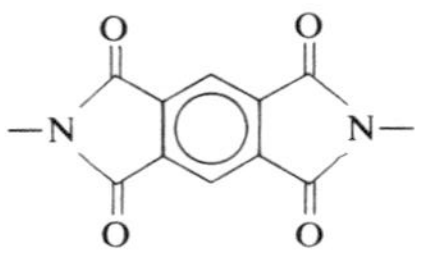			Used in high-temperature applications.
Polyimide	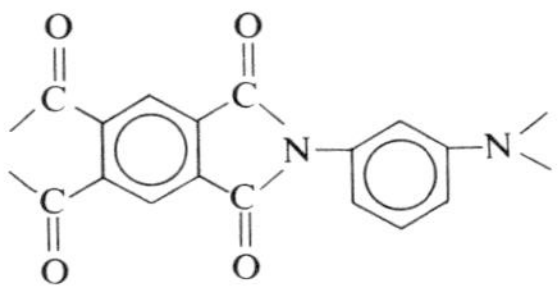			Used in wire enamels, laminates, and high-temperature applications.

Polyimines

Poly(ethylene imine)	$-CH_2-CH_2-\overset{H}{\underset{}{N}}-$			Wet-strength improver for paper.

Inorganic

Poly[bis(aryloxy)phosphazenes]	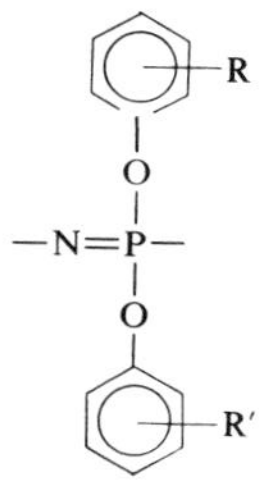			Nonburning materials.
Poly[bis(methylamino)phosphazene]	$-N=P-$ with NHCH$_3$, NHCH$_3$	14	—	Water-soluble polymer. Experimental carrier polymer for chemotherapeutic agents.
Poly[bis(trifluoroethoxy)-phosphazene]	$-N=P-$ with OCH$_2$CF$_3$, OCH$_2$CF$_3$	−66	242	Highly water-repellent, ultraviolet stable, nonburning, film- and fiber-forming polymer. Projected uses in biomedical implantation devices.
Poly[bis(fluoroalkoxy)phosphazene] mixed-substituent polymers (PNF rubber)	$-N=P-$ with OR$_F$, OR$'_F$	~ −80	—	Low-temperature, solvent and oil-resistant, nonburning elastomers, used in O-rings, gaskets, seals, fuel lines, pipes, carburetor components and other automotive and aircraft applications. Possible biomedical applications.
Poly(carborane-siloxanes) (Dexsil)	$-m\text{-carborane}-\left[\overset{CH_3}{\underset{CH_3}{Si}}-O\right]_x$ (Others contain phenylsiloxane units)			Thermal stability up to 300–500°C. Used in heat-stable elastomers, O-rings, etc., and as a stationary phase in chromatography.

571

Name	Repeating Unit	T_g (°C)	T_m (°C)	Properties and Uses
Poly(dimethylsiloxane) (silicone rubber)	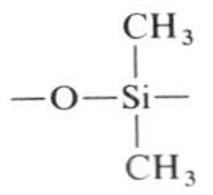 $-O-\overset{\overset{\displaystyle CH_3}{\mid}}{\underset{\underset{\displaystyle CH_3}{\mid}}{Si}}-$	−123	−29	Oxidation-resistant elastomer, used in seals, hoses, biomedical devices, mold releases, and waterproofing agents.
Carbon fibers				Strengths of up to $100 \times 10^6 \, lb/in^2$ in extension. Used as reinforcing agent in high-strength, heat-resistant composites. Some use in heat-resistant fabrics.
Poly(sulfur nitride), polythiazyl	$-S=N-$			Purple-gold fibrous polymer that is an electrical conductor at room temperature and a superconductor at very low temperatures.

Phenol– and amine–formaldehyde

Name	Repeating Unit	T_g (°C)	T_m (°C)	Properties and Uses
Poly(phenol–formaldehyde) resins (Bakelite)	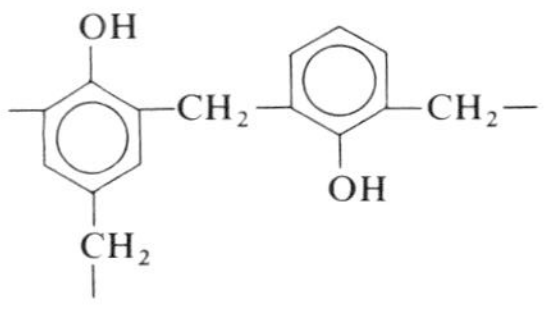 (Three-dimensional network)			Hard thermosetting polymer with high resistance to deformation under load. Used for the manufacture of cast and molded articles such as telephones, electrical insulators, buttons, laminates, and heat-resistant objects.
Poly(melamine–formaldehyde) resins	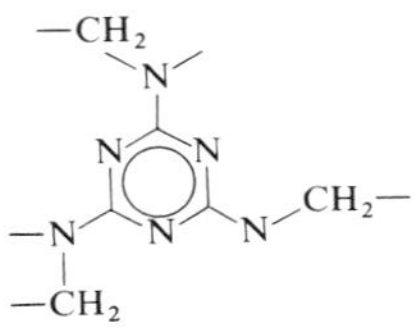			Thermosetting polymer fabricated into molded objects (e.g., dinnerware), laminates, surface coatings, table and countertops; used in textile and paper-treatment agents, adhesives, and wall paneling.
Poly(urea–formaldehyde) resins	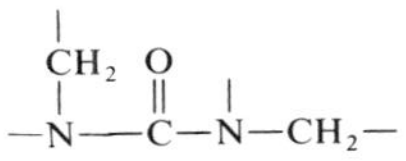 (Three-dimensional network)			Used in the manufacture of molded objects, thermal insulation, adhesives, lighting fixtures, and plywood.

Polymer	Structure	Softening/melting temperature (°C)	Uses
Cellulose (R = H)	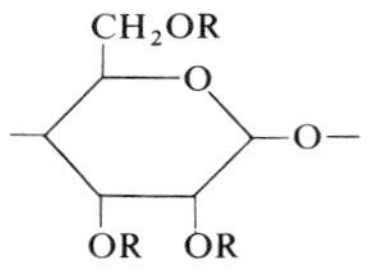 CH_2OR ring, with OR, OR substituents (pyranose)	>270	Used in paper, textiles, and wood products; employed as starting material for rayon manufacture; as rayon, used in textile fiber and tire cord; as Cellophane, used as packaging film.
Carboxymethyl cellulose	R = H, CH_2COOH		Used as adhesive and emulsifying agent; utilized in pharmaceuticals
Ethylcellulose	R = H, C_2H_5		Transparent film used for packaging, molded articles, lacquers, and as a printing ink stabilizer.
Cellulose acetate	R = H and $COCH_3$	306	Clear, transparent material used as fibers, or as injection-molded, extruded, and sheet plastics.
Cellulose nitrate	R = H, NO_2	157	Used in lacquers, adhesives, and molded objects.
Ethylcellulose / Methylcellulose	R = H, C_2H_5 ; R = H, CH_3	43 / 165	Employed as textile finish; as an adhesive or sizing agent, and as a thickening agent.

Polymer	Structure	Softening/melting temperature (°C)	Uses
Poly(diphenylether sulfone) (polyether sulfone)	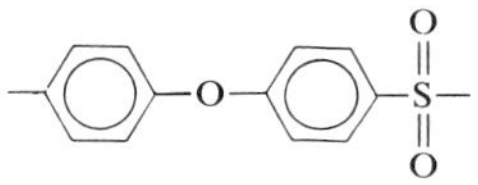 aromatic ether–sulfone repeat unit	230	Tough polymer with good electrical insulation and self-extinguishing flame behavior. Used for the manufacture of heat-stable, injection-molded articles.
Poly(diphenyl sulfone–diphenylene oxide sulfone) copolymer (Astrel 360) (polyether sulfone)	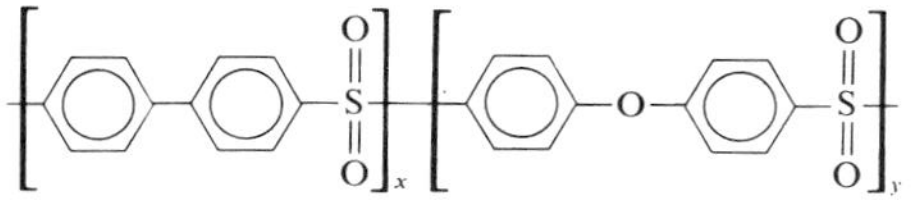 aromatic sulfone–ether copolymer repeat unit	250–285, (depending on copolymer composition)	Tough polymer with good electrical insulation and self-extinguishing flame behavior. Used for the fabrication of heat-stable, injection-molded articles.
(Udel polysulfone)	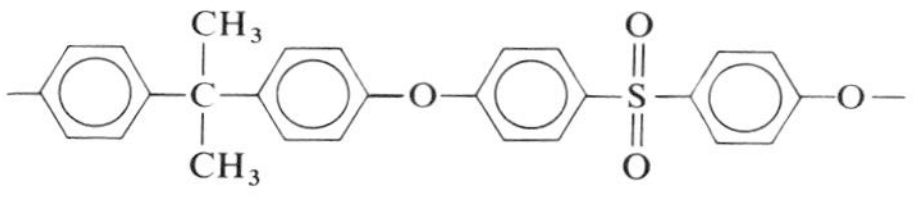 bisphenol-A–ether–sulfone repeat unit (CH_3, CH_3 on central carbon)	190	Tough, impact-resistant polymer with good electrical insulation and fire-resistant properties. Manufactured into injection-molded articles, housings for power tools, computer parts, circuit breakers, meter housings, battery cases, etc.

NAME	REPEATING UNIT	T_g (°C)	T_m (°C)	PROPERTIES AND USES
Polyurethanes				
Polyurethane	$-(CH_2)_3N-C-O-$ (with H and O)		148	Foam rubber, synthetic leather. Used as segmented copolymers with poly(tetramethylene oxide) in Spandex elastomers.
Polyurethane (Lycra)	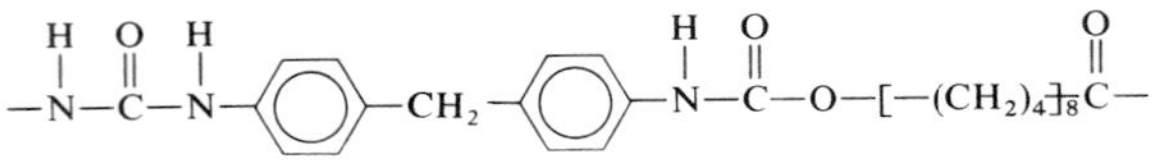 $-N-C-N-\!\!\!\bigcirc\!\!\!-CH_2-\!\!\!\bigcirc\!\!\!-N-C-O-[-(CH_2)_4\!-]_{18}C-$			Expanded foam rubber; textile laminates; carpet underlays; thermal insulation.
Poly olefin compounds				
Polyacrylamide	$-CH_2-CH-$; $C=O$; NH_2	165		Water-soluble polymer used in paper treatment or as a thickening agent; used as photopolymerized layers on lithographic plates.
Poly(acrylic acid)	$-CH_2-CH-$; $COOH$	106		Water-soluble polymer used as an adhesive or thickening agent.
Polyacrylonitrile (Orlon, Acrilan, Creslan)	$-CH_2-CH-$; $C\equiv N$	85	317	Often copolymerized with small amounts of acrylamide. A fiber-forming polymer with extensive use in textiles, netting, and as a precursor for the pyrolytic preparation of carbon fibers.
Poly(acrylonitrile–butadiene) copolymers (nitrile rubber)	$-CH_2-CH-$; $C\equiv N$; $-CH_2-C=C-CH_2-$ (with H, H) ; $-CH_2-CH-$; $CH=CH_2$			Solvent-resistant elastomer used in gaskets, oil hoses, oil seals, fan belts, oil-well components, adhesives, and in tank linings.

Polymer	Structure	T_m (°C)	T_g (°C)	Uses
Poly(acrylonitrile–butadiene–styrene) copolymers (ABS polymers)	$-CH_2-CH-$ (C≡N); $-CH_2-C=C-CH_2-$ (H, H); $CH=CH_2$; $-CH_2-CH-$ (phenyl)			Tough structural plastic or rubber used in the manufacture of telephones, pipes, and a wide variety of molded articles.
Poly(acrylonitrile-vinyl chloride) copolymer (Dynel)	$-CH_2-CH-$ (C≡N); $-CH_2-CH-$ (Cl)			Used as a textile fiber material.
Polybutadiene (butadiene rubber)	$-CH_2-C=C-CH_2-$ (H, H); $-CH_2-$ with $CH=CH_2$; $-CH_2-CH-$ etc.	125(1,2-isotactic); 154(1,2-syndiotactic); 6.3(1,4-cis); 148,109(1,4-$trans$)	−58(1,3); −65(1,2-isotactic); −102(1,4-cis); −10, −48 (1,4-$trans$); −85(20°, 1,2)	Rubbery polymer used as an alternative to natural rubber or SBR in the manufacture of footwear, belting, hoses, pneumatic tires, and toys.
Butadiene–acrylonitrile copolymers			−56(80 − 20 copolymer); −41(70 − 30 copolymer)	Used as adhesives.
Poly(1-butene)	$-CH_2-CH-$ (C_2H_5)	142, 126, 106, 65(isotactic)	−45, −24(isotactic)	Rubbery polymer used in heavy-duty plastic sheet, as a base for pressure-sensitive tapes, and in pipes and tubes.
Poly(butyl-α-cyanoacrylate)	$-CH_2-C-$ (OC_4H_9, C=O; CN)		85	Used as an adhesive.
Polychloroprene (neoprene)	$-CH_2-C=C-CH_2-$ (Cl, H)	43; 70 (cis-1,4)	−45 (85% $trans$-1,4); −20 (cis-1,4)	Solvent-resistant elastomer used in adhesives and cable jackets, seals, and golf-ball covers.
Poly(chlorotrifluoroethylene–vinylidene fluoride) copolymers (Kel-F)	$-C-C-$ (F, F; F, Cl); $-CH_2-C-$ (F, F)			Solvent-resistant, high-temperature elastomer used in rocket motors, molded articles, O-rings, and pipe lining.

Name	Repeating Unit	T_g (°C)	T_m (°C)	Properties and Uses
Poly(ethyl acrylate)	$-CH_2-CH-$, with $C=O$ and OC_2H_5	-22		Used in varnishes and printing inks.
Poly(ethyl vinyl ether)	$-CH_2-CH-$, with OC_2H_5	-42		Elastomeric polymer used in adhesives and as a nonmigratory plasticizer.
Polyethylene or polymethylene	$-CH_2-CH_2-$	$-125, -20$	$\sim 140, 95$	Tough plastic with extensive uses in monofilament fibers, films, extrusion-molded objects, electrical insulation, bottles, and toys.
Poly(ethylene–vinyl acetate) copolymers	$-CH_2-CH_2-$ $-CH_2-CH-$, with $O-C=O-CH_3$			Used in medical tubing and syringes, toys, and cable insulation.
Poly(ethylene–propylene) copolymers (Noedel) (EPR)	$-CH_2-CH_2-$ $-CH_2-CH-$, with CH_3	-60 (50–50 copolymer)		Used in high-pressure steam hoses, automobile parts, appliances, and seals.
Fluorinated ethylene–propylene copolymers (Teflon FEP)				Used for electrical cable insulation.
Polyisobutylene (butyl rubber)	$-CH_2-C-$, with CH_3 and CH_3 (often copolymerized with small amounts of isoprene)	-70	1.5	Rubbery elastomer used in adhesives, tire inner tubes, caulking compounds, dairy hoses, raincoats, and seals.
Poly(cis-1,4-isoprene) (natural rubber)	$-CH_2-C=C-CH_2-$, with CH_3 and H	-70	36	Extensive uses in automobile tires and in a wide range of other industrial products.
Poly(trans-1,4-isoprene) (gutta percha)	$-CH_2-C=C-CH_2-$, with CH_3 and H	-68	74	Used in toys, balloon, golf-ball covers, and automobile equipment.

Polymer	Structure	T_g (°C)	T_m (°C)	Uses
Poly(methacrylic acid)	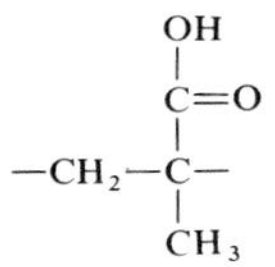 $-CH_2-C(CH_3)(C{=}O)(OH)-$			Adhesive and thickening agent.
Poly(methyl acrylate)	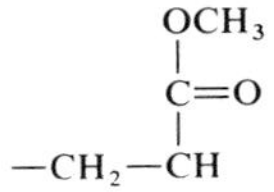 $-CH_2-CH(C{=}O)(OCH_3)-$	5–9 (atactic)		Used in surface coatings.
Poly(methyl-2-cyanoacrylate)	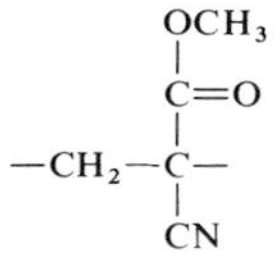 $-CH_2-C(CN)(C{=}O)(OCH_3)-$			Adhesive, especially for metals.
Poly(methyl methacrylate) (Plexiglas, Lucite, Perspex, PMMA)	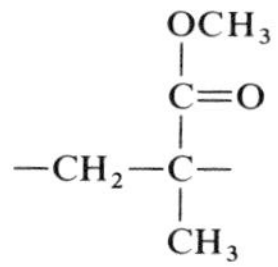 $-CH_2-C(CH_3)(C{=}O)(OCH_3)-$	105 (114); (60) (−7); 48 (isotact); 128 (syndio)	160 (isotact); 200 (syndio)	Clear, transparent, glassy polymer used extensively in castings, lenses, roof "bubbles," windows, dentures, fiber optics, and illuminated signs.
Poly(styrene–butadiene) copolymers (SBR polymers)	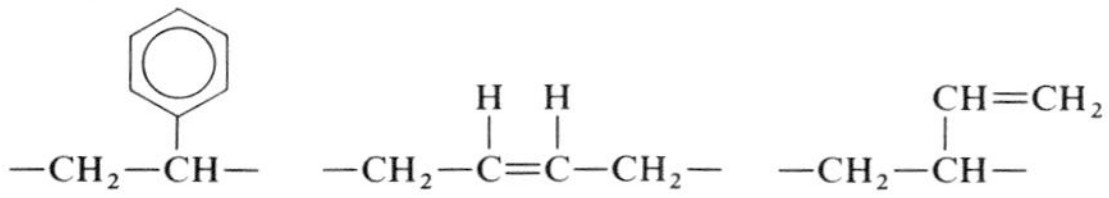 $-CH_2-CH(C_6H_5)-$ and $-CH_2-C(H){=}C(H)-CH_2-$	−56, (23–77 copolymers); −41 (30–70 copolymer)		Elastomeric polymers used as a replacement for natural rubber: for example in footwear, latex paints, and tire treads.
Poly(styrene-α-methylstyrene) copolymer	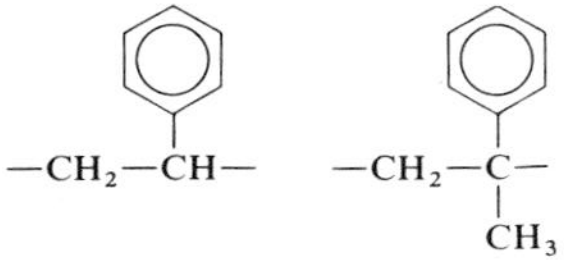 $-CH_2-CH(C_6H_5)-$ and $-CH_2-C(C_6H_5)(CH_3)-$			Used in electrical appliances and refrigerator linings.
Poly(tetrafluoroethylene) (Teflon)	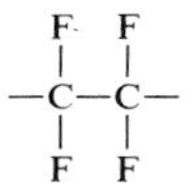 $-CF_2-CF_2-$	130; −113	327; 30	Highly water repellent polymer with a surface lubricity. Extensively utilized in bearings and other sliding surfaces, nonsticking cooking utensils, seals, machined parts, and protective liners.

577

Name	Repeating Unit	T_g (°C)	T_m (°C)	Properties and Uses
Poly(tetrafluoroethylene–hexafluoropropylene) copolymers (Teflon FEP)	$-CF_2-CF_2-$ $-CF_2-CF-$ with CF_3			Used in capacitors, printed circuits, mold liners, textile finishes, wire insulation, tubing, and in fluid power transmission.
Poly(vinyl acetate)	$-CH_2-CH-$ with $O-C(=O)-CH_3$	30		Used in emulsion paints and as a precursor for poly(vinyl alcohol) manufacture: a component of chewing gum, drinking straws, and adhesives.
Poly(vinyl alcohol) (Vinylon)	$-CH_2-CH-$ with OH	99	258	Water-soluble or hydrophilic polymer used in fibers, aqueous adhesives, sizing agents for textile fibers, as a binder for the fluorescent layer in TV tubes, as a thickening agent, as wet-strength adhesives. and films.
Poly(vinyl butyral)*	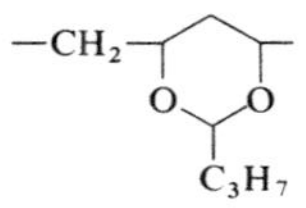 $-CH_2-$ ring with C_3H_7			Adhesive used in laminated safety glass.
Poly(N-vinylcarbazole) (Luvican, Polectron)	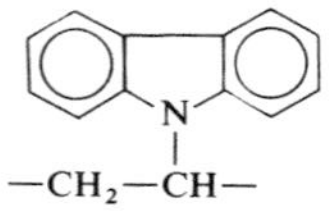 $-CH_2-CH-$ with carbazole N	200		Used for high-temperature electrical insulation, injection-molded products, and as an asbestos substitute.
Poly(vinyl chloride) (PVC)	$-CH_2-CH-$ with Cl	+78–81 (atactic)	285 (extrapolated)	Hard, inflexible polymer in the unplasticized state. When plasticized (usually with phthalate esters). used in Tygon tubing and films. automobile seat covering, electrical insulation, floor tiles, molded and extruded objects, coated fabrics, and plumbing pipes.

Polymer	Structure	T_g (°C)	T_m (°C)	Uses
Poly(vinyl chloride–vinyl acetate) (Vinylite)	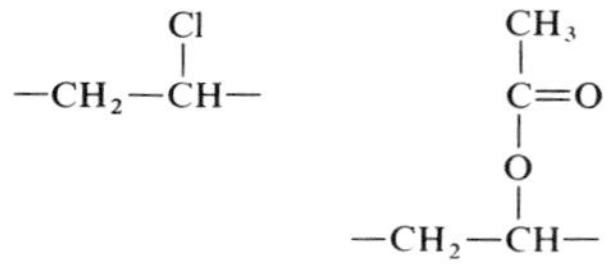 $-CH_2-CH-$ Cl ; $-CH_2-CH-$ O–C=O–CH$_3$			Used for the manufacture of phonograph records, and in coatings for cans and other metal containers.
Poly(vinyl cinnamate)	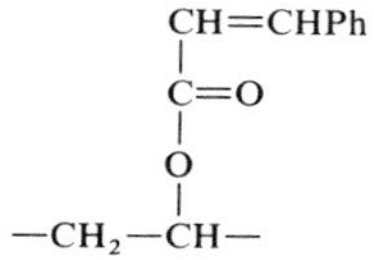 $-CH_2-CH-$ O–C=O–CH=CHPh			Photocrosslinking polymer for use in the preparation of photoresist printing plates and printed circuit boards.
Poly(vinyl fluoride)	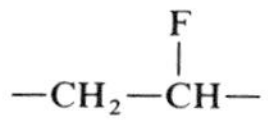 $-CH_2-CH-$ F		200	Used in films, window glazing, and in coatings for aluminum and wood.
Poly(vinyl pyrrolidone) (Kollidon, Periston)	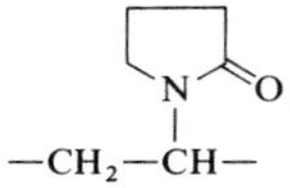 $-CH_2-CH-$ N (pyrrolidone ring, C=O)			Used formerly as a blood plasma extender. Used as a protective colloid, thickening agent, emulsion stabilizer in cosmetics, surfactant in dyeing, clearing agent in beer and other beverages, and in hair sprays.
Poly(vinylidine chloride)	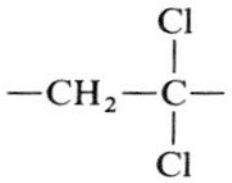 $-CH_2-C-$ Cl / Cl	-18	210	Polymer with low-gas permeability. A component of Saran copolymer films [with poly(vinyl chloride)]: used in blow-molded bottles, pipes, and tape.
Poly(vinylidine fluoride)	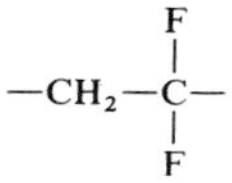 $-CH_2-C-$ F / F	-39 (13)	171	Piezoelectric polymer used in microphone diaphragms, molded and extruded objects.
Poly(vinylidine fluoride–hexafluoropropylene) copolymer (Viton)	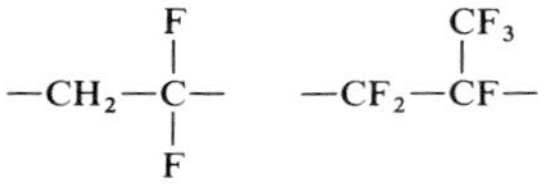 $-CH_2-C-$ F / F ; $-CF_2-CF-$ CF$_3$	-55		Solvent-resistant elastomer used for the manufacture of O-rings, seals, hose, tubing, and fuel-resistant diaphragms.
Poly(methyl vinyl ether)	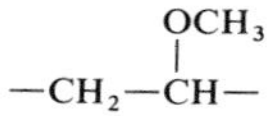 $-CH_2-CH-$ OCH$_3$	-13 (-31) (atactic); -21 (isotactic)	150 (isotactic)	Water-soluble polymer used in adhesives and nonmigratory plasticizers.

* Prepared from poly(vinyl alcohol) and butyraldehyde.

Name	Repeating Unit	T_g (°C)	T_m (°C)	Properties and Uses
Polypropylene (Herculon)	—CH₂—CH— 　　　| 　　　CH₃	26, −35 (isotactic)	183, 130; 150 (isotactic)	Tough plastic widely used as fibers for ropes, seat covers and carpets; as films for packaging, and in injection-molded articles, especially for automotive applications.
Polystyrene	—CH₂—CH— 　　　| 　　　⬡	100 (atactic and isotactic)	240 (isotactic)	Clear, transparent, glasslike polymer used widely for the fabrication of molded objects and foamed insulation. Also used as a substrate (when crosslinked) for polymer-bound transition metal catalysts or as a gel permeation chromatography substrate.

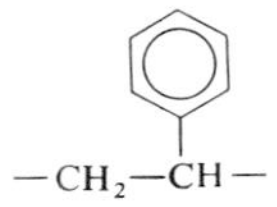

Polyalkynes

Name	Repeating Unit	T_g (°C)	T_m (°C)	Properties and Uses
Polyacetylene	H H | | ═C══C═			Silver- or gold-colored, metallic-type polymer that conducts electricity.

References to Topics
Not Discussed
in This Book

This appendix contains a listing of sources for further reading on topics not covered in detail in this book because of space limitations. The reader is encouraged to use this list as a starting point for broader reading once the material in this text has been assimilated.

Spectroscopy of polymers

BOVEY, F. A., *High Resolution Nuclear Magnetic Resonance of Macromolecules*. New York: Academic Press, 1972.

BOVEY, F. A., "The High Resolution of Nuclear Magnetic Resonance Spectroscopy of Polymers," *Prog. Polymer Sci.* (A. D. Jenkins, ed.), **3**, 1 (1971).

BOVEY, F. A., AND G. V. D. TIERS, "The High Resolution Nuclear Magnetic Resonance Spectroscopy of Polymers," *Fortschr. Hochpolym.-Forsch.*, **3**, 139 (1963).

CAMPBELL, D., "Electron Spin Resonance of Polymers," *J. Polymer Sci.* (*D*) (*Macromol. Rev.*), **4**, 91 (1970).

CLARK, D. T., "ESCA Applied to Polymers," *Adv. Polymer Sci.*, **24**, 125 (1977).

CLARK, D. T., AND W. J. FEAST, "Application of Electron Spin Spectroscopy for Chemical Applications (ESCA) to Studies of Structure and Bonding in Polymer Systems," *J. Macromol. Sci.—Rev. Macromol. Chem.*, **C12**, 191 (1975).

ELLIOTT, A., *Infrared Spectra and Structure of Organic Long-Chain Polymers*. New York: St. Martin's Press, 1969.

FRUSHOUR, B. G., P. C. PAINTER, AND J. L. KOENIG, "Vibrational Spectra of Polypeptides," *J. Macromol. Sci.—Rev. Macromol. Chem.*, **C15**, 29 (1976).

HENDRA, P. J., "Laser-Raman Spectra of Polymers," *Adv. Polymer Sci.*, **6**, 151 (1969).

HENNIKER, J. C., *Infrared Spectrometry of Industrial Polymers*. New York: Academic Press, 1967.

HUMMEL, D. O., *Infrared Analysis of Polymers, Resins, and Additives—An Atlas*. New York: Wiley–Interscience, 1969.

HUMMEL, D. O., *Infrared Spectra of Polymers* (*in the Medium and Long Wavelength Regions*), (*Polymer Reviews*, Vol. 14). New York: Wiley–Interscience, 1966.

INGHAM, J. D., "Free Radical Spin Labels for Macromolecules," *J. Macromol. Sci.—Rev. Macromol. Chem.*, **C2**, 279 (1968).

IVIN, K. J. (ed.), *Structural Studies of Macromolecules by Spectroscopic Methods*. New York: Wiley, 1976.

JENNINGS, B. R., "Electro-Optic Methods for Characterizing Macromolecules in Dilute Solution," *Adv. Polymer Sci.*, **22**, 61 (1977).

JIRGENSONS, B., *Optical Rotatory Dispersion of Proteins and Other Macromolecules*. New York: Springer, 1969.

KAUSCH-BLECKEN VON SCHMELING, H. H., "Application of Electron Spin Resonance Techniques to High Polymer Fracture," *J. Macromol. Sci—Rev. Macromol. Chem.*, **C4**, 243 (1970).

KITAGAWA, T., AND T. MIYAZAWA, "Neutron Scattering and Normal Vibrations of Polymers," *Adv. Polymer Sci.*, **9**, 335 (1972).

KOENIG, J. L., "Raman Spectroscopy of Biological Molecules: A Review," *J. Polymer Sci. (D) (Macromol. Rev.)*, **6**, 59 (1972).

KRIMM, S., "Infrared Spectra of High Polymers," *Fortschr. Hochpolym.-Forsch.*, **2**, 52 (1960).

LIU, K.-J., AND J. E. ANDERSON, "Proton Magnetic Resonance Studies of Molecular Interactions in Polymer Solutions," *J. Macromol. Sci.—Rev. Macromol. Chem.*, **C5**, 1 (1970).

O'KONSKI, C. T. (ed.), *Molecular Electro-Optics*. New York: Dekker, 1976–77.

OSTER, G., AND Y. NISHIJAMA, "Fluorescence Methods in Polymer Science," *Fortschr. Hochpolym.-Forsch.*, **3**, 313 (1964).

RAMEY, K. C., AND W. S. BREY, "Application of High Resolution Nuclear Magnetic Resonance to Polymer Structure Determination," *J. Macromol. Sci.—Rev. Macromol. Chem.*, **C1**, 263 (1967).

RANBY, B., AND J. F. RABEK, *Electron Spin Resonance Spectroscopy in Polymer Research*. New York: Springer–Verlag, 1977.

SAFFORD, G. J., AND A. W. NAUMANN, "Low Frequency Motions in Polymers as Measured by Neutron Inelastic Scattering," *Adv. Polymer Sci.*, **5**, 1 (1967).

Appendices

SEANOR, D. A., "Charge Transfer in Polymers," *Adv. Polymer Sci.*, **4**, 317 (1965).

SLICHTER, W. P., "The Study of High Polymers by Nuclear Magnetic Resonance," *Fortschr. Hochpolym.-Forsch.*, **1**, 35 (1958).

TOSI, C., AND F. CIAMPELLI, "Applications of Infrared Spectroscopy to Ethylene–Propylene Copolymers," *Adv. Polymer Sci.*, **12**, 87 (1973).

WILLIAMS, J. L. R., AND R. C. DALY, "Photochemical Probes in Polymers," *Progr. Polymer Sci.* (A. D. Jenkins, ed.), **5**, 2, 1 (1977).

WOODY, R. W., "Optical Rotary Properties of Biopolymers," *J. Polymer Sci.* (*D*) (*Macromol. Rev.*), **12**, 181 (1977).

YA SLONIM, I., AND A. N. LYUBIMOV, *The Nuclear Magnetic Resonance of Polymers.* New York: Plenum Press, 1970.

ZBINDEN, R., *Infrared Spectroscopy of High Polymers.* New York: Academic Press, 1964.

Rheology

BARAMBOIM, N. K., *Mechanochemistry of Polymers* (W. F. Watson, ed.), London: Rubber and Plastic Research of Great Britain, 1964.

BIRD, R. B., H. R. WARNER, AND D. C. EVANS, "Kinetic Theory and Rheology of Dumbell Suspensions with Brownian Motion," *Adv. Polymer Sci.*, **8**, 1 (1971).

CANNON, S. L., G. B. MCKENNA, AND W. O. STATTON, "Hard-Elastic Fibers (A Review of a Novel State for Crystalline Polymers)," *J. Polymer Sci.* (*D*) (*Macromol. Rev.*), **11**, 209 (1976).

CASALE, A., R. S. PORTER, AND J. F. JOHNSON, "Dependence of Flow Properties of Polystyrene on Molecular Weight, Temperature, and Shear," *J. Macromol. Sci.—Rev. Macromol. Chem.*, **C5**, 387 (1971).

ESTES, G. M., S. L. COOPER, AND A. V. TOBOLSKY, "Block Polymers and Related Heterophase Elastomers," *J. Macromol. Sci.—Rev. Macromol. Chem.*, **C4**, 313 (1970).

FAN, L. T., AND J. S. SHASTRY, "Polymerization Systems Engineering," *J. Polymer Sci.* (*D*) (*Macromol. Rev.*), 155 (1973).

GOLDING, B., *Polymers and Resins: Their Chemistry and Chemical Engineering.* Princeton, N.J.: Van Nostrand, 1959.

GRAESSLEY, W. W., "The Entanglement Concept in Polymer Rheology," *Adv. Polymer Sci.*, **16**, 1 (1974).

JANACEK, J., "Mechanical Behavior of Hydroxyalkyl Methacrylate Polymers and Copolymers, *J. Macromol. Sci.—Rev. Macromol. Chem.*, **C9**, 1 (1973).

KAWABATA, S., AND H. KAWAI, "Strain Energy Density Functions of Rubber Vulcanizates from Biaxial Extension," *Adv. Polymer Sci.*, **24**, 89 (1977).

KUHN, W., A. RAMEL, D. H. WALTERS, G. EBNER, AND H. J. KUHN, "The Production of Mechanical Energy from Different Forms of Chemical Energy with Homogeneous and Cross-striated High Polymer Systems, *Fortschr. Hochpolym.-Forsch.*, **1**, 540 (1960).

LIPATOV, Y. S., "Relaxation and Viscoelastic Properties of Heterogeneous Polymeric Compositions," *Adv. Polymer Sci.*, **22**, 1 (1977).

MANSON, J. A., AND L. H. SPERLING, *Polymer Blends and Composites*. New York: Plenum Press, 1976.

MARK, J. E., "Thermoelastic Results on Rubberlike Networks and Their Bearing on the Foundations of Elastic Theory," *J. Polymer Sci. (D) (Macromol. Rev.)*, **11**, 135 (1976).

NAUMAN, E. B., "Mixing in Polymer Reactors," *J. Macromol. Sci.—Rev. Macromol. Chem.*, **C10**, 75 (1974).

RIO, A., AND E. M. CERNIA, "Polyblends of Cement Concrete and Organic Polymers," *J. Polymer Sci. (D) (Macromol. Rev.)*, **9**, 127 (1974).

SCHILDKNECHT, C. E., *Polymer Processes, Chemical Technology of Plastics, Resins, Rubber, Adhesives, and Fibers*. New York: Interscience, 1956.

SEMJONOW, V., "Schmelzviscositäten Hochpolymerer Stoffe," *Adv. Polymer Sci.*, **5**, 387 (1968).

SHEN, M., AND M. CROUCHER, "Contribution of Internal Energy to the Elasticity of Rubberlike Materials," *J. Macromol. Sci.—Rev. Macromol. Chem.*, **C12**, 287 (1975).

SHEN, M., W. F. HALL, AND R. E. DEWAMES, "Molecular Theories of Rubber-like Elasticity and Polymer Viscoelasticity," *J. Macomol. Sci.—Rev. Macromol. Chem.*, **C2**, 183 (1968).

SPERLING, L. H., "Interpenetrating Polymer Networks and Related Materials," *J. Polymer Sci. (D) (Macromol. Rev.)*, **12**, 141 (1977).

SPERLING, L. H. (ed.), *Recent Advances in Polymer Blends, Grafts, and Blocks*. New York: Plenum Press, 1974.

TOBOLSKY, A. V., AND D. B. DUPRÉ, "Macromolecular Relaxation in the Damped Torsional Oscillator and Statistical Segment Models," *Adv. Polymer Sci.*, **6**, 103 (1969).

VASKO, M., T. BLEHA, AND A. ROMANOV, "Thermoelasticity in Open Systems," *J. Macromol. Sci.—Rev. Macromol. Chem.*, **C15**, 1 (1976).

WILKES, G. L., "Rheo-Optical Methods and Their Application to Polymeric Solids," *J. Macromol. Sci.—Rev. Macromol. Chem.*, **C10**, 149 (1974).

YANNAS, I. V., "Nonlinear Viscoelasticity of Solid Polymers (in Uniaxial Tensile Loading), *J. Polymer Sci. (D) (Macromol. Rev.)*, **9**, 163 (1974).

Polymer separations

BRAUN, J.-M., AND J. E. GUILLET, "Study of Polymers by Inverse Gas Chromatography," *Adv. Polymer. Sci.*, **21**, 107 (1976).

CANTOW, M. J. R. (ed.), *Polymer Fractionation*. New York: Academic Press, 1967.

INAGAKI, H., "Polymer Separation and Characterization by Thin-Layer Chromatography," *Adv. Polymer Sci.*, **24**, 189 (1977).

STEVENS, M. P., *Characterization and Analysis of Polymers by Gas Chromatography*. New York: Dekker, 1969.

TUNG, L. H., "Recent Advances in Polymers Fractionation," *J. Macromol. Sci.—Rev. Macromol. Chem.*, **C6**, 51 (1971).

Polyelectrolytes

BERGSMA, F., AND CH. A. KRUISSINK, "Ion Exchange Membranes," *Fortschr. Hochpolym.-Forsch.*, **2**, 307 (1961).

CONWAY, B. E., "Solvation of Synthetic and Natural Polyelectrolytes," *J. Macromol. Sci.—Rev. Macromol. Chem.*, **C6**, 113 (1972).

CRESCENZI, V., "Some Recent Studies of Polyelectrolyte Solutions," *Adv. Polymer Sci.*, **5**, 358 (1968).

EISENBERG, A., AND M. KING, *Ion Containing Polymers: Physical Properties and Structure.* New York: Academic Press, 1977.

ISE, N., "The Mean Activity Coefficient of Polyelectrolytes in Aqueous Solutions and Its Related Properties," *Adv. Polymer Sci.*, **7**, 536 (1971).

MORAWETZ, H., "Specific Ion Binding by Polyelectrolytes," *Fortschr. Hochpolym.-Forsch.*, **1**, 1 (1958).

OOSAWA, F., *Polyelectrolytes.* New York: Dekker, 1971.

OTOCKA, E. P., "Physical Properties of Ionic Polymers," *J. Macromol. Sci.—Rev. Macromol. Chem.*, **C5**, 275 (1971).

REMBAUM, A., AND E. SELEGNY, *Polyelectrolytes and Their Applications.* Boston: D. Reidel, 1975.

SOUCHAY, P., *Polyanions et Polycations.* Paris: Gauthier–Villars, 1963.

ZANA, R., "Studies of Aqueous Solutions of Polyelectrolytes by Means of Ultrasonic Methods," *J. Macromol. Sci.—Rev. Macromol. Chem.*, **C12**, 165 (1975).

Other topics

ALLEGRA, G., AND I. W. BASSI, "Isomorphism in Synthetic Macromolecular Systems," *Adv. Polymer Sci.*, **6**, 549 (1969).

BARRETT, K. E. J. (ed.), *Dispersion Polymerization in Organic Media.* New York: Wiley, 1974.

BERGER, M. N., "Addition Polymers of Monofunctional Isocyanates," *J. Macromol. Sci.—Rev. Macromol. Chem.*, **C9**, 269 (1973).

BERLIN, A. A., AND N. G. MATVEYEVA, "The Progress in the Chemistry of Polyreactive Oligomers and Some Trends of Its Development. Synthesis and Physico-chemical Properties," *J. Polymer Sci. (D) (Macromol. Rev.)*, **12**, 1 (1977).

BERRY, G. C., "Properties of Rigid-Chain Polymers in Dilute and Concentrated Solutions." *Contemp. Top. Polymer Sci.*, **2**, 55 (1977).

BOGUSLAVSKII, L. I., AND A. V. VANNIKOV, *Organic Semi-conductors and Biopolymers.* New York: Plenum Press, 1970.

Bradford, E. B., and L. D. McKeever, "Block Copolymers," *Progr. Polymer Sci.* (A. D. Jenkins, ed.), **3**, 109 (1971).

Butler, G. B., G. C. Cornfield, and C. Aso, "Cyclopolymerization," *Progr. Polymer Sci.*, (A. D. Jenkins, ed.), **4**, 71 (1975).

Cann, J. R., *Interacting Macromolecules*. New York: Academic Press, 1970.

Casale, A., and R. S. Porter, "Mechanical Synthesis of Block and Graft Copolymers," *Adv. Polymer Sci.*, **17**, 1 (1975).

Cassidy, H. G., and K. A. Kun, *Oxidation–Reduction Polymers* (*Redox Polymers*) (*Polymer Reviews*, Vol. 11). New York: Wiley–Interscience, 1965.

Coleman, L. E., and N. A. Meinhardt, "Polymerization Reactions of Vinyl Ketones," *Fortschr. Hochpolym.-Forsch.*, **1**, 159 (1959).

Cooper, W., and G. Vaughan, "Recent Developments in the Polymerization of Conjugated Dienes," *Progress in Polymer Science* (A. D. Jenkins, ed.), **1**, 91 (1967).

Cotler, R. J., and M. Matzner, *Ring-forming Polymerizations*. New York: Academic Press, 1969.

Crank, J., and G. S. Park (ed.), *Diffusion in Polymers*. New York: Academic Press, 1968.

Elias, H. G. (ed.), *Polymerization of Organized Systems* (Midland Macromol. Meeting, 1974). New York: Gordon and Breach, 1977.

Fendler, J. H., and E. J. Fendler, *Catalysis in Micellar and Macromolecules Systems*. New York: Academic Press, 1975.

Fujita, H., "Diffusion in Polymer-Diluent Systems," *Fortschr. Hochpolym.-Forsch.*, **3**, 1 (1961).

Goodman, M., A. Abe, and Y.-L. Fan, "Optically Active Polymers," *J. Polymer Sci.* (*D*) (*Macromol. Rev.*), **1**, 1 (1967).

Heitz, W., "Polymeric Reagents. Polymer Design, Scope, and Limitations," *Adv. Polymer Sci.*, **23**, 1 (1977).

Lee, C.-D. S., and W. H. Daly, "Mercaptan-containing Polymers," *Adv. Polymer Sci.*, **15**, 61 (1974).

Lumley, J. L., "Drag Reduction in Turbulent Flow by Polymer Additives," *J. Polymer Sci.* (*D*) (*Macromol. Rev.*), **7**, 263 (1973).

Millich, F., "Rigid Rods and the Characterization of Polyisocyanides," *Adv. Polymer Sci.*, **19**, 117 (1975).

Morawetz, H., "Comparative Studies of the Reactivity of Polymers and their Low Molecular Weight Analogs," *J. Polymer. Sci., Polymer. Symp.*, **62**, 271 (1978).

Overberger, C. G., A. C. Guterl, Y. Kawakami, L. J. Mathias, A. Meenakshi, and T. Tomono, "Recent Developments in the Use of Polymers as Reactants in Organic Reactions," *Pure Appl. Chem.*, **50**, 309 (1978).

Pino, P., "Optically Active Addition Polymers," *Adv. Polymer Sci.*, **4**, 393 (1965).

Plate, N. A., and V. P. Shibaev, "Comb-like Polymers. Structure and Properties," *J. Polymer Sci.* (*D*) (*Macromol. Rev.*), **8**, 117 (1974).

POSTELNEK, W., L. E. COLEMAN, AND A. M. LOVELACE, "Fluorine Containing Polymers. Fluorinated Vinyl Polymers with Functional Groups, Condensation Polymers, and Styrene Polymers," *Fortschr. Hochpolym.-Forsch.*, **1**, 75 (1958).

ROHA, M., "Ionic Factors in Steric Control," *Adv. Polymer Sci.*, **4**, 353 (1965).

SAEGUSA, T., "Spontaneous Alternating Copolymerization by Zwitterion Intermediates," *Angew. Chem., Int. Ed. Engl.*, **16**, 826 (1977).

SCHULZ, R. C., AND E. KAISER, "Synthese und Eigenschaften von Optisch aktiven Polymeren," *Adv. Polymer Sci.*, **4**, 236 (1965).

SMALL, P. A., "Long-Chain Branching in Polymers," *Adv. Polymer Sci.*, **18**, 1 (1975).

SMETS, G., AND R. HART, "Block and Graft Copolymers," *Fortschr. Hochpolym.-Forsch.*, **2**, 173 (1960).

SOLOMON, D. H., AND D. G. HAWTHORNE, "Copolymerization of Diallylamines," *J. Macromol. Sci.—Rev. Macromol. Chem.*, **C15**, 143 (1976).

TANI, H., "Stereoscopic Polymerization of Aldehydes and Epoxides," *Adv. Polymer Sci.*, **11**, 57 (1973).

TOSI, C., "Sequence Distribution in Copolymers: Numerical Tables," *Adv. Polymer Sci.*, **5**, 451 (1968).

TSUCHIDA, E., AND H. NISHIDE, "Polymer-Metal Complexes and Their Catalytic Activity," *Adv. Polymer Sci.*, **24**, 1 (1977).

WOHRLE, D., "Polymere aus Nitrilen," *Adv. Polymer Sci.*, **10**, 35 (1972).

WUNDERLICH, B., "Crystallization During Polymerization," *Adv. Polymer Sci.*, **5**, 568 (1968).

Author Index

Subject Index

Lennard–Jones potential, 454
Lewis acids, 88
Light-scattering cells, 357
Light-scattering dissymmetry, 351
Light-scattering measurements, 345–361
Light-scattering photometer, 356, 357
Linear polymers, 6
Liquefaction temperature, 12
Liquid state, 428
Living polymers, 72
   distribution of DP, 316
   kinetics, 311
Lucite, 6
Lysozyme, 178

Macromolecular hypothesis, 18
Mark–Houwink equation, 386, 388, 401
Materials science, 21
Mean square end-to-end distance,
   353, 355, 390–391, 466–468, 477
Mechanical shearing, 13
Mechanical testing of polymers, 534–538
Mechanism:
   of cationic polymerization, 88
   of condensation polymerization, 27
   of free-radical polymerization, 53
   of photopolymerization, 99
   of radiation-induced polymerization,
     107–110
   of ring-opening polymerization,
     125–130
   of siloxane polymerization, 141
   of Ziegler–Natta catalysis, 79–81
Mechanisms in polymerization-
   depolymerization processes, 241
Mechanistic pathway for step reactions, 23
Melamine–formaldehyde polymers, 41
   general structure of, 24
Melting temperatures, *table,* 429, 568–579
Melt polymerization, 33
Membranes, 554–556
Meridianal reflections, 489, 493
Merrifield synthesis, 181
Messenger RNA, 190
Metal coordination polymers, 160, 161
Metallic properties of covalent polymers,
   155–158
Metallocene polymers, 161
Methacrylonitrile, polymerization of,
   282, 298
p-Methoxystyrene, polymerization of,
   304, 317, 320
Methyl acrylate, polymerization of,
   282, 298, 304
Methylcellulose, 173
Methyl methacrylate:
   bulk polymerization of, 51
   chain transfer, 294
   initial polymerization rates, 282
   polymerization, 50, 51, 64, 69, 75,
     97, 100, 104, 116, 225, 280, 282,
     298, 302, 304, 320, 521, 522
   rate data for polymerization, 280
Methylolation, 15, 41
Methylolphenols, 39
α-Methylstyrene, 50
   cationic polymerization of, 87
   polymerization of, 225, 317, 325,
     326
p-Methylstyrene, polymerization of, 317,
   320
Methylvinyl ether, polymerization of,
   325
Methylvinyl ketone, polymerization of,
   304
Microbial attack on polymers, 206
Microcrystallinity (*see* Crystallinity)

Mineralogical polymers, 134–138
Modulus, 425
Molding of polymers, 522–525
Molecular conformation, 16
Molecular structure determination, 489–502
Molecular weight (*see also* Degree of
   polymerization)
   absolute, 331–377
   accuracy of, 343, 359–361, 372, 392
     393
   changes of due to heating, 28
   in condensation polymerizations,
     27, 28
   distribution for condensation
     polymers, 257–261
   distribution for step and chain
     reactions, 25
   in free-radical polymerizations, 285
   in step and chain polymerizations,
     25
Monofilaments, laboratory preparation
   of, 36
Monomer–polymer equilibria, 217–244
Monomers, 4
   for anionic polymerization, 69, 71
   for cationic polymerization, 86
   in condensation reactions, 245
   for free-radical polymerization, 50
   for photopolymerization, 97
   for radiation polymerization, 104,
     106
   reaction with free radicals, 60
   reactivity in copolymerization, 302
   for solid-state polymerization, 111
Morphology of polymers, 423–447
Mutations, 194
Mylar, synthesis of, 14
Myoglobin, 178, 179

Neoprene rubber, synthesis of, 17
Network polymer, 7, 264
Newtonian flow, 380
Nitration of polymers, 201
Nitriles, polymerization of, 220
Nomenclature, 563–566
   in conformational analysis, 459
Novolac resins, 40
Nuclear magnetic resonance and glass
   transition temperature, 434
Nucleic acids (*see* Polynucleotides)
Number average molecular weight, 257,
   260, 288–290, 339, 341, 342
Nylon 6:
   from caprolactam, 124
   synthesis of, 17
Nylon 66, 6
   crystallinity, 439
   synthesis of, 14, 27, 33
Nylon 610, synthesis of, 35
Nylons:
   melt spinning, 519
   synthesis, 14, 33

Octamethylcyclotetrasiloxane,
   polymerization of, 18
Olefin:
   addition reactions, 48–90
   polymerization of, 16
Oligomers, 4
   by depolymerization, 217–244
Optical birefringence, 442
   and crystallinity, 12
Optical diffraction analysis, 500–502
Optical diffractometer, 501, 502
Optical properties, 542
Organometallic initiators, 49, 67–83

Orientation:
   of fibers, 520, 521
   of crystallites, 481, 487
   of polymers, 437, 441
Orlon, 6
Osmometer, 336, 340
Osmometry, 340–345
Osmotic pressure, 335–345, 418
Ostwald viscometer, 382
Oxepanes, polymerization of, 122
Oxetanes, polymerization of, 122
Oxidative coupling, 42
Oxidative degradation, 207

Parentheses in polymer nomencalture,
   6
Penetrometer, 430
Permeability, 542
Peroxide initiators, 49, 55
Perspex, 6
Phase difference in X-ray
   diffraction, 494
Phenol–formaldehyde polymers, general
   structure of, 24
Phenol–formaldehyde resins, 15, 39
Phosgene, 31
Phosphate in polynucleotides, 188
Phosphates, 15
Phosphazene elastomers, 151, 153
Phosphazene polymers, 146–154 (*see
   also* Poly(organophosphazenes))
Phosphazenes:
   polymerization of, 146, 147
   pseudoaromaticity, 237
Phosphine ligands, 201, 202
Photographic processes based on
   polymerization, 95–97
Photolytic degradation, 207
Photolytic intitiation, 59
Photolytic polymerization, 95–103
Photosensitizers, 98, 101–103
Phthalic anhydride, 15
Plasticizer, 13
Plasticizers, 428
Plexiglas, 6
Poiseuille's law, 380
Polarizing microscope, 442
Polyacetals, 41
   general structure, 24
Polyacetylene, 5, 157
Polyacrylonitrile, 6
   fiber spinning, 516
   pyrolysis, 39, 158, 204
   by solid-state polymerization, 111
   synthesis of, 16
Polyaldehydes, depolymerization of,
   218
Polyamides:
   general structure of, 24
   rates of formation, 253
   synthesis of, 26, 33, 257, 258
   thermal decomposition, 210
Poly(aminophosphazenes), 148–150
Poly(aminostyrene), 201
Polyanhydrides:
   general structure of, 24
   synthesis of, 32
Polyarylenes, general structure of, 25
Polybenzimidazoles, 37, 38
   general structure, 24
   synthesis, 27
Poly[bis(trifluoroethoxy)phosphazene]:
   conformational energy surface, 464
   synthesis, 149
Polybutadiene:
   chlorination, 205
   synthesis of, 17